Online Credibility and Digital Ethos:

Evaluating Computer–Mediated Communication

Moe Folk
Kutztown University of Pennsylvania, USA

Shawn Apostel
Eastern Kentucky University, USA

Managing Director:	Lindsay Johnston
Editorial Director:	Joel Gamon
Book Production Manager:	Jennifer Romanchak
Publishing Systems Analyst:	Adrienne Freeland
Development Editor:	Austin DeMarco
Assistant Acquisitions Editor:	Kayla Wolfe
Typesetter:	Lisandro Gonzalez
Cover Design:	Nick Newcomer

Published in the United States of America by
Information Science Reference (an imprint of IGI Global)
701 E. Chocolate Avenue
Hershey PA 17033
Tel: 717-533-8845
Fax: 717-533-8661
E-mail: cust@igi-global.com
Web site: http://www.igi-global.com

Library of Congress Cataloging-in-Publication Data

Online credibility and digital ethos: evaluating computer-mediated communication / Moe Folk and Shawn Apostel, editors.
    p. cm.
  Includes bibliographical references and index.
  ISBN 978-1-4666-2663-8 (hbk.) -- ISBN 978-1-4666-2694-2 (ebook) -- ISBN 978-1-4666-2725-3 (print & perpetual access) 1. Digital media--Evaluation. 2. Electronic information resources--Evaluation. 3. Telematics--Social aspects. 4. Information technology--Moral and ethical aspects. 5. Mass media--Technological innovations. 6. Mass media--Moral and ethical aspects. 7. Trust. I. Folk, Moe, 1972- II. Apostel, Shawn, 1971-
  ZA4045.O55 2013
  302.23'1--dc23
                                       2012031816

British Cataloguing in Publication Data
A Cataloguing in Publication record for this book is available from the British Library.

All work contributed to this book is new, previously-unpublished material. The views expressed in this book are those of the authors, but not necessarily of the publisher.

Table of Contents

Section 1
Design and Arrangement

Chapter 1
Todd S. Frobish, Fayetteville State University, USA

Chapter 2
Natasha Dwyer, Victoria University, Australia

Chapter 3
Nathan Johnson, Purdue University, USA

Chapter 4
Kevin Brock, North Carolina State University, USA

Section 2
Perceptions of Online Information

Chapter 5
Joe Erickson, Angelo State University, USA
Kristine Blair, Bowling Green State University, USA

Section 3
News/Primary Research

Section 4
User-Generated Content

Section 5
Games

Detailed Table of Contents

Section 1
Design and Arrangement

Chapter 1

    Todd S. Frobish, Fayetteville State University, USA

This chapter situates classical notions of ethos and describes how they can be used to understand how ethos functions in computer-mediated communication (CMC). A four-part model of online ethos based on classical rhetoric is put forth; the model hinges on assessing the attempts of content producers to develop online identities based on: (1) Community Identification and Goodwill, (2) Moral Character and Virtue, (3) Intelligence and Knowledge, and (4) Verbal and Design Competence.

Chapter 2

    Natasha Dwyer, Victoria University, Australia

Interactivity has posed all sorts of problems for establishing and understanding trust in digital spaces. This chapter examines the importance of trust in digital contexts and questions the traditional reliance that designers have placed on trust as a static display of evidence. In looking at the effects of aesthetics and dynamic interaction, this chapter advocates for approaches that focus on the evolving enablement of trust rather than simply presenting a static façade of trust.

Chapter 3

    Nathan Johnson, Purdue University, USA

Most people have sustained connections with browser technology almost non-stop throughout their days, but few people question how those browsers are configured and why, particularly with regard to Web standards. This chapter examines how information infrastructure influences ethos in information labor and focuses on how different browsers approach the ACID3 test for infrastructural standards while also analyzing conference presentations, job announcements, and Web pages to theorize ACID3 in a broader way to shed light on how to approach the rhetorics of infrastructural standardization.

Chapter 4

Open source software (OSS) continues to gain prominence while disrupting the conventional approaches to software development and marketing that grew up around proprietary software. While much scholarship has focused on the differences between the development of OSS and proprietary software, this chapter fills a gap by scrutinizing the rhetorical appeals made in promoting both types of software with particular attention paid to competitors' efforts in operating systems, Web browsers, and image manipulation programs.

Section 2
Perceptions of Online Information

Chapter 5

In many areas of academia, print publications are often regarded as more rigorous and prestigious than digital publications. This chapter argues for the importance of online academic journals because of their intellectual legitimacy, sustainability, collaborative nature, and ability to provide important professional development opportunities for graduate students as future faculty members; taken together, these elements of online academic journals prove crucial in maintaining—and positively transforming—a discipline's ethos. The chapter recommends three benchmarks to foster the professional development of both graduate students and advanced faculty by working on online academic journals.

In the field of Information Retrieval, one of the main research areas focuses on how relevance factors into the decision-making activity of users on a search engine results page (SERP). However, much research has indicated that users have limited knowledge about how search engines work, thereby limiting the effectiveness of determining credibility in this preliminary information-gathering stage. As a result, many studies rely on user reactions to the credibility of the actual websites they select; however, this chapter argues that the primary credibility judgments take place on the SERP, not on the websites they lead users to. By conducting a study that observed the search behaviors of undergraduate participants ranging from 18-25 years old, the chapter illuminates how users infer contextual ethos from snippets and titles (including the absence of certain words and phrases). Their findings help to consider the importance of previous familiarity and tacit knowledge in making credibility judgments on SERP, as well as when and where the idea of credibility originates in the information-gathering process.

Some colleges have banned its use, and many people view it as a vast credibility wasteland where any-body can publish anything they want: It's hard to imagine anything used more widely but more widely loathed than Wikipedia. This chapter examines the credibility of Wikipedia by using Aristotelian ethos and Ian Bogost's concept of procedural rhetoric. In approaching Wikipedia from these lenses, the chapter makes the case that looking beyond surface content to Wikipedia's infrastructure and its easily acces-sible sets of rules and user data can help people make more informed decisions when evaluating other constantly evolving pseudonymous documents as well.

This chapter discusses the ways that Web search engines apply a variety of ranking signals to garner the trust of users. While the hallmark of most search engines is trust built on implicit markers of credibility such as page rankings, this chapter considers explicit markers of credibility such as expert mediation and increasing the prominence of quality-related cues.

Section 3
News/Primary Research

By presenting the findings of a large-scale survey on the credibility beliefs of U.S. children ranging from 11-18 years old, this chapter makes a significant contribution to the understanding of how an age group that has grown up immersed in digital content deals with credibility in online spaces. The sur-vey and its findings are important in light of the fact that youth lack the same cognitive and emotional development, personal experience, and familiarity with historical media that adults bring to credibility evaluation. Given the ever-increasing reliance on computers in children's education, the findings from the study can be used to inform theoretical, practical, and policy approaches to digital literacy skills that rely on nuanced notions of credibility.

The widespread use of social media and user-generated content presents opportunities and pitfalls galore for journalists. By surveying Portuguese journalists about their habits in assessing and using Web 2.0 sources such as YouTube, Facebook, Twitter, and Flickr, this chapter sheds light on how credibility and information are perceived in these contexts. Among the discoveries in the data is that the Portuguese journalists did not find Web 2.0 sources very credible as a genre unto themselves, but the journalists used such sources heavily nonetheless. The chapter addresses what such discrepancies mean relative to the training and working conditions of modern journalists, as well as the potential advantages and disadvantages Web 2.0 sources provide for researchers in terms of content-gathering and their own ethos in using such information.

Chapter 11

In a world of constant upheaval brought about by natural causes such as tsunamis but more often by economic and political factors, large communities of people have left their homelands but maintain intense interest in the social and political happenings back "home." To stay connected, these diaspora populations often need to rely heavily on online news sources, and this chapter proposes a framework that can be used to analyze the credibility of online news about their home countries. Using specific examples related to the Zimbabwean diaspora, this chapter demonstrates the importance of credibility factors such as accuracy, authority, design, comparison, and corroboration in assessing the complex factors that shape online communication used and produced by diaspora populations.

Chapter 12

Fake grassroots communication about a public interest issue, astroturfing is particularly problematic in this day and age because of the ease of producing it in digital spaces where anonymity and pseudonyms are commonplace, not to mention the large amount of grave issues so many countries now find themselves negotiating means a vast amount of opportunities to skew discussions toward particular viewpoints present themselves. By applying Michel Foucault's articulation of the ancient concept of parrhesia to digital contexts, this chapter theorizes digital parrhesia and develops a techno-semiotic methodological approach that researchers and the public in general can use to better consider online advocacy discourse.

Section 4
User-Generated Content

Chapter 13

Facebook, Twitter, and blogs have long been shunned as sources of information because of the difficulty in establishing the credibility of the author. However, the CRAAP test is flexible enough to evaluate the ethos of many posts while promoting critical thinking skills with novice researchers. This chapter provides a detailed description of CRAAP and also shows how viral flogs can be exposed by using this test.

With more than 161 million members and growing daily in the United States and abroad, LinkedIn is the most successful online professional networking site and one of the most-visited sites on the Web. This chapter examines the drawbacks of treating LinkedIn as an extension of professional print practices that revolve around the ethos of static documents. Instead, the author draws on personal experiences within pet industry circles on LinkedIn to demonstrate how to build a positive ethos that capitalizes on the dynamic qualities of the site. These strategies are presented in a way that can be adapted for a multitude of goals and purposes.

The Chinese blogosphere is the largest blogging space in the world, and by examining some of the most popular Chinese blogs, this chapter illuminates their various rhetorical strategies. Given the intricate nature of blogging in a country with a state-controlled media, this chapter sheds light on a number of persuasive features that can connect with diverse audiences in a variety of contexts.

As scholars continue to carve out online identities by sharing their work and interacting with peers in social networking spaces, the benefits and drawbacks of putting forth a searchable, archivable public persona become readily apparent. This chapter addresses the ways in which scholars can present a credible online identity and construct an ethos that is sustainable and beneficial.

Because traditional mainstream media continue to struggle with the challenges to their former domination posed by new means of accessing and providing news, they have gradually encouraged newer models of content-generation in order to compete for existing users' loyalty and to attract new users as well. To develop their audiences, mainstream media are increasingly reaching out to formerly "passive" readers and making them true participants in media. Adopting the audience as content producers, however, creates many new ethical and legal challenges for media companies. This chapter examines the boundaries and contracts that some companies have tried to use to accomplish their goals relative to user-generated content and points out the practices—in a practical, theoretical, and legal sense—that contribute to ethos in this area.

Chapter 18

By analyzing examples of social media that factored heavily into the Egyptian Revolution, this chapter explores the complexity of digital ethos construction in activist digital discourses. While much scholarship about online spaces laments the lack of credibility of pseudonymous communication (which is also manifested in the academic and popular disdain for using and trusting Wikipedia, for example), this chapter argues that the power of anonymity—as deployed in certain Facebook pages relating to the Egyptian revolution—was a key element in forming a positive, powerful ethos that helped unite a wide array of factions and ultimately topple a despot. This forces us to re-consider narrow conceptions about the ethos of anonymity in online spaces and the ways a strong communal ethos can be built around shared identity when the rhetor's identity is purposefully effaced. The chapter also suggests ways these ideas could be used to refashion approaches to research and writing in the classroom.

Section 5
Games

Through the lens of the popular massive multiplayer online game *World of Warcraft*, this chapter examines the importance of ethos developed and enacted within local contexts, in particular those of virtual environments. By distinguishing such game ethos from that of websites, which are more static, and other social media, where the environment is not as much of a defining factor, the chapter demonstrates how players—as inhabitants of the game ecology—establish ethos. The authors collected in-game chat and near-game forum posts in addition to conducting surveys with players; their results point to the importance of specificity, demonstrated expertise, and experience as the keys to ethos in these realms. The discussion surrounding these elements is important not only because of the widespread appeal of gaming to legions of current and developing gamers but because of the increasing use of gaming strategies and worlds in various contexts due to the amount of scholarship that has demonstrated the efficacy of video games as learning tools.

While much video game scholarship concerns itself with massively multiplayer online-role playing games (MMORPGs) that hinge on fantastical elements and virtual worlds, this chapter focuses on docu-games that attempt to depict and reflect on aspects of reality such as military conflicts, historical periods, or contemporary political and cultural issues. In going beyond examining such games as simply simulated conveyors of statements about reality, the chapter applies documentary film theory and game studies to relate how credibility and reality interact in such contexts. By examining how credible references to reality and credible game play are established in the eyes of the player, we gain a clearer picture of how persuasion functions in games in general and docu-games in particular.

Video games have become ubiquitous cultural engines, their profits even eclipsing those of movies as their popularity increases. One element of video games that has resonated in multiple contexts is video game music. Though many people view game music as unworthy of attention when viewed through the lens of classicism, this chapter argues for the artistic seriousness of game music and for its own powerful ethos in particular contexts as well. The chapter draws on musica poetica and digital musicality to set the stage for the importance of game music, and it uses concrete examples of game music to suggest the rhetorical power of game music and to connect music and rhetoric in terms of ethos in order to transcend their traditional connection of pathos.

Foreword

One important advantage of scholarly edited collections is that although the chapters are linked by a common topical thread (or threads), a reader can nevertheless experience a wide range of sub-topics, perspectives, research strategies and conclusions from the work of the varied contributors. Often, this allows for a wider view of the topic as a whole, while at the same time offering interesting specificity as each author examines his/her particular area of research. This big-and-small picture view is perhaps even more important for rhetorical studies of digital technologies, because of the myriad combinations of tools, spaces, users and texts that can come into play in different electronic settings. And yet, digital spaces do make use of specific technologies that enable certain similar kinds of interactions, which means that our use of these digital spaces binds us together in ways that are interesting for their commonalities as well as their differences. Thankfully, the plethora of different digital settings available in *Online Credibility and Digital Ethos: Evaluating Computer-Mediated Communication* offers both *specificity*, as authors offer details about their different research settings, and *commonalities*, as the editors have worked to create an overall look at the different ways that human beings (individually and in groups and institutions) work to shape their ethos in these different types of electronic settings.

In terms of its topic(s), *Online Credibility and Digital Ethos* is a collection that centers on the production of ethos and credibility in online spaces – possibly one of the most critical topics for scholars, teachers, and citizens as we continue to move into ever-more complicated digital spaces for our work and leisure. These issues move well beyond simple notions of individual rhetors establishing ethos with their audiences through spoken or written prose (even digital prose). Instead, the chapters in this collection tackle extremely complicated issues related to how credibility is built through interactions between authorial agencies (examples include individuals, collaborative non-professional contributors, combinations of commercial and non-commercial contributors and various types of other corporate entities) and the affordances of the technologies and spaces they use.

The breadth of the research in this collection, and the different sites the authors present, leads me to a brief discussion of a common (I believe) disadvantage to collections on digital tools and technologies, which is that sites, software, and tools move in and out of use so quickly that some collections risk becoming outdated before they are even published. However, I believe that *Online Credibility and Digital Ethos* avoids that pitfall, because the focus of the collection is not on specific technologies or tools, or even (necessarily) on particular digital spaces. Instead, the text as a whole surrounds and intersects with issues of meaning-making that are inherent to digital spaces generally. Indeed, various chapters in the collection consider these issues from multiple angles: how readers decide what to believe, how authors do (and can learn to) present themselves as believable/ethical, and how different authors or collaborations of authors work to build or contribute to ethos in larger and more complicated productions such as

Wikipedia or major news sites. As consumers and producers of digital content, we are all embedded in these issues – we make these kinds of choices on a daily basis, and we are affected by the productions of ethos and credibility that we encounter online.

As scholars interested in these issues, we must also work to track changes that are taking place in our human understanding of ethos and credibility precisely because these issues haven't remained static – digital productions present complications in such critical areas as communal/private resources, personal/public personas, expert/non-expert contributors and social/political interactions. The distinctions that used to be more visible (at least in certain socially constructed ways) between these dichotomous areas are at present extremely unstable – interconnected in ways that create moral and legal complications that impact both our access to information and the ways we are expected (culturally) to use it. This collection provides useful discussions of locations where these distinctions are actively in flux.

For teachers, this collection offers perhaps the most valuable resource. The varied articles and sub-topics presented, along with the different research approaches represented throughout the collection, make the text a useful resource for approaching the important topics related to online identity, ethos, and credibility. Additionally, the articles deal with a range of online spaces that can be connected directly to different digital teaching scenarios, which means the text could also be useful for discussions of digital pedagogy – in particular the ways that teachers can promote and support more nuanced understandings of digital spaces and the decisions about credibility that readers/users must make as they work and live in these spaces.

Joyce Walker
Illinois State University, USA

Joyce Walker *is an Associate Professor at Illinois State University in Normal, IL. She is the Director of the Center for Writing Research and Pedagogy and the Director of the Illinois State University Writing Program. She teaches both graduate and undergraduate courses in Rhetoric & Writing Studies. Her research is primarily concerned with the ideas, productions, and communities that can be developed and facilitated through interactions between humans and their composing tools. Her recent article with James Purdy (Duquesne University), "Valuing Digital Scholarship: Exploring the Changing Realities of Intellectual Work," which appeared in MLA Profession 2010, won the Ellen Nold Award for best article from the Computers & Composition Conference in 2011. She is currently at work on a book-length project with Kevin Roozen (Auburn University), Tracing Trajectories of Practice and Person: Sociohistoric Perspectives of Literate Activity.*

Preface

With the near ubiquity of smartphones, tablets, and laptops, acquiring and publishing online information has never been easier; however, increased access to consuming and producing digital information raises new challenges when establishing and evaluating online credibility. These challenges are important because they affect a broad range of meaning-making, both inside and outside of academia. We stand at an important moment in human history, a time when we have markedly increased access to an almost infinite amount of information—and not just access to consuming information that was molded by editorial elites and other gatekeepers, but access to producing and sharing information that suits any needs we might have, no matter how banal, impish, beneficial, or revolutionary those impulses are. Because the great hope of interconnected computers resides in openness, a technological utopianist vision where people can give voice to their own neglected viewpoints or subjugated stances, most societies err on the side of openness and freedom when it comes to producing and sharing information on the Web. Today, in an era where many phones are, in effect, more powerful computers than most people had access to a mere 15 years ago, the major challenge facing us now and in the conceivable future is not how to continue producing and sharing information, and not simply sifting through it, but discerning the qualities that make information trustworthy, usable, and deployable amongst the vast amounts of competing information available. This problem is not simply an academic one (though it is a problem across all disciplines) so much as it is a human one: As access to interconnected information proliferates and the ways people hope to use that ever-growing information evolve, it affects all people. For example, the events of the Arab Spring show that in the absence of what were traditionally seen as relatively reliable information sources, "unofficial" online sources deemed credible by a wide range of actors played a key role in successful uprisings; these developments also rippled out to affect those not directly involved—both people hoping to use such models for further democratic pressuring in a wide range of contexts and those seeking to prevent it.

Though this conundrum of too much information and what exactly to trust can be partly addressed through technical means such as tagging and, in certain contexts, traditional methods such as strict gatekeeping, the constantly evolving nature of digital information and the digitally-infused societies producing and assessing that information means that no fixed solution will be a panacea. However, this problem of trusting information is not simply an element of ubiquitous technology so much as it is an ongoing aspect of human nature, one that has been evident since the origins of debate and open societies. For that reason, we find ourselves returning to the ancient concept of ethos because of its flexibility and connection to persuasion in diverse contexts. Given the long tradition of rhetoric as a useful academic discipline, as well as the powerful concepts of ethos related to oratory that were developed by ancient scholars such as Aristotle and Isocrates and further developed over the centuries, ethos has grown to include many different conceptions relating to ethical communication and trust. Michael Hyde (2004), for example, drawing on the "primordial" etymological origins of ethos, argued that we should see ethos more as a dwelling place that "define[s] the grounds, the abodes or habitats, where a person's ethics and moral character take form and develop," a view that helps us approach the importance of digital contexts for ethos. However, on the whole, many people tend to associate ethos with credibility based on establishing a trustworthy personal character, thus often using ethos as shorthand for credibility. Though current views of ethos might deviate from classical notions, both classical and emergent notions of ethos provide a powerful, flexible tool to consider nuances of credibility in digital contexts.

In editing this book, we encouraged authors to develop their own connections to ethos instead of us as editors force-feeding them particular definitions and approaches, in part because of the plasticity of ethos as a concept and—given the importance of ethos and credibility to digital information— in part because we wanted flexible approaches based on specific contexts. The goal of this book is to offer chapters written by scholars from across the disciplines and from across the world that provide approaches to evaluating the credibility of digital sources, specific advice for negotiating popular websites, and useful techniques for a wide variety of digital genres and contexts. We wanted diverse approaches to the problem of online credibility from different disciplines and from different countries so that unique local strategies with which we as readers may not be familiar could be adapted to shape, or to inspire, new approaches in our comfort zones.

This book provides a much-needed resource for anybody who conducts online research. In particular, the book serves as a handy reference for a variety of academic disciplines, since both faculty and students continue to utilize online sources in their research and the reliance on those types of sources appears to be waxing, not waning. Information literacy specialists would find the chapters which focus on particular types of popular Web spaces like LinkedIn, Wikipedia, World of Warcraft, and Facebook useful. Journalists and educators in the field of Mass Communication and Library Sciences would find the book useful in establishing protocols for approaching a wide variety of sources because of chapters that cover blogs, microblogs, diasporic news, astroturfing, and documentary games. Web designers and writers could use this book to establish a more credible online presence. Any instructors of courses that involve research, particularly composition courses, could profit from this book as well, and graduate students and academics of all stripes could utilize certain chapters to help establish their own methods for determining the credibility of a source they hope to use for research purposes.

The power of our age is interconnection, and we hope this collection pulls together different and distant voices that make a significant contribution to moving forward together as we meet the many challenges of digital information proliferation.

Moe Folk
Kutztown University of Pennsylvania, USA

Shawn Apostel
Eastern Kentucky University, USA

REFERENCES

Hyde, M. J. (Ed.). (2004). *The ethos of rhetoric*. University of South Carolina Press.

Acknowledgment

While this collection began about 18 months before publication, our interest in digital ethos and online credibility began in 2004 with a conversation about how ill-equipped the "accepted rules" for evaluating book-based criteria were to handle online sources. We would like to thank Cindy Selfe for encouraging us to explore this topic and Dennis Lynch for conducting an independent study on ethos with us; their patience and expertise have been instrumental to our development in this topic. In addition, we would like to thank all of the people who gave so generously of their time during this process, especially the members of the editorial review board, reviewers, authors, and the people at the Noel Studio for Academic Creativity who provided feedback on this collection. We would also like to thank our families for their support during this period.

Section 1
Design and Arrangement

Chapter 1
On Pixels, Perceptions, and Personae:
Toward a Model of Online Ethos

Todd S. Frobish
Fayetteville State University, USA

ABSTRACT

This chapter works toward a four-part model of online ethos connecting classical rhetorical theory to the new age of computer-mediated technology with particular attention paid to the challenges, complications, and possibilities of this evolving rhetorical environment. This model demonstrates the usefulness of updating our understanding of ethos and its place in computer-mediated communication (CMC). The proposed model allows for the assessment of digital ethos by examining others' attempts to develop an online identity based upon: (1) Community Identification and Goodwill, (2) Moral Character and Virtue, (3) Intelligence and Knowledge, and (4) Verbal and Design Competence.

INTRODUCTION

The World Wide Web (WWW) and other online media are no longer infant technologies. They have matured to become an instrumental part of our social, economic, and political reality. While there are many wonders left to be borne out of them, and they certainly have no shortage of cheerleaders, they are not perfect. Viruses and worms infect them. Pornographic material contaminates them. Confidence men corrupt them. Radical doublespeak inflames them. And stupidity

DOI: 10.4018/978-1-4666-2663-8.ch001

seems to multiply around them. Yet, one problem seems to serve as a herald—a lack of institutional credibility. Media before the online environment—newspaper, radio, television—all reached a rather high level of credibility in their journeys as purveyors of information in the eyes of the public. Yet, the new online media, especially the Web, have not been elevated to such an esteemed position, perhaps due to the above reasons, but this general lack of credibility problematizes them as persuasive media.

Traditionally connected to our notion of credibility is the very ancient, but still potent, concept of ethos. This very versatile concept has its roots in ancient Greek philosophy and rhetoric, and it has evolved from meaning a habit of animals to gather in familiar places to the habits of men to act in familiar ways to more simply habits of human character. When these habits of character are perceived favorable by the community, one is said to have positive ethos, and, thus, some level of credibility from which to speak on matters of importance. Without ethos, no degree of logic or passion would be enough to persuade. Therefore, ethos had a special place in the philosophies of persuasion in the ancient Greek world and beyond. This chapter theorizes on the nature of ethos and how the online medium has complicated—or perhaps freed us from—our traditional understanding of it. Following this discussion, we will explore a basic model of online ethos that can be used to assess online ethos-based appeals.

GREEK AND ROMAN PERSPECTIVES ON CHARACTER

The Greek and Roman rhetorical traditions serve as important touchstones for our discussion of credibility and character. Aristotle is probably one of the most systematic thinkers in this area, but even he has been shadowed by others. Homer, in the *Iliad*, was the first to offer a basic account of character. While Homer did not develop such an explicit theory of ethos, we can speculate on what traits of character were esteemed in pre-Aristotelian times by carefully reading the stories of his protagonists and antagonists. Homer (trans. 1939) stated in the *Iliad*, for example, that King Nestor was "that grand old man whose counsel was always thought the best. He spoke with honesty and good courage setting out his thoughts neat and clear, like a weaver weaving a pattern upon his loom" (p. 104). Homer suggested in the *Iliad*, for instance, that character is highly valued and fully realized in public contest. We know Nestor best when he is giving advice, in moments of *agon*, to kings and warriors about the war and to Achilles

when arguing with Agamemnon or slaughtering soldiers on the battlefield. The Homeric character exuded wisdom, courage, style or eloquence, patience, foresight, bravery, skill, circumspection, honesty, and graciousness, always during times of conflict.

Homer's focus on character as an element in persuasion was an anchor for more advanced conceptualizations of ethos found centuries later. His version of character, according to Rowe (1983), centered "on the demands of the individual rather than on those of society in the broad sense" (p. 269) and assumed that the man of good character is good insofar as he is useful. As Yamagata (1994) has put it: "a man useful in battle is a good man" (p. 222). Yet, a positive character is an ends, not a means or a rhetorical tool that is explicitly used to gain power or prestige. Homer's notion of character assumed as important that which Aristotle explicitly excluded—the idea of prior action and reputation. According to Homer's proto-theory of character, there is an impact on a man's ability to be listened to, to be taken seriously, to be regarded as worth attending to, by his reputation for wisdom, or perhaps solely of his age precisely because it is perceived as a sign of wisdom, sagacity, "fair-mindedness," and good counsel.

While Homer wrote about practical skill, Plato spoke of ideals to characterize men of credibility. Plato illustrated his theory of character in his dialogues about dialectic, linking the good rhetor with philosophical rigor and a love for the audience. DuBois (1991) argued that Plato's notion of dialectic, Plato's tool for truth, is not unlike torture and can be used to test the character of men: "Philosophy becomes a method of arrest and discipline; philosophical argument is a dividing, a splitting, a fracturing of the logical body, a process that resembles torture" (p. 113). Plato (trans. 1925) wrote in his *Sophist*, "so we have the great man's testimony, and the best way to obtain a confession of the truth may be to put the statement itself to a mild degree of torture" (237b). Plato then wrote, "let us examine the opinion-imitator as if he were a piece of iron, and see whether he is sound or there is still some seam in him" (268a). "What would happen," asked Plato in his allegory of the cave, "if one of these prisoners were released from his chains, were forced to stand up, turn around, and walk with eyes lifted up toward the light of the fire? All of his movements would be exceedingly painful" (Stumpf, 1975, p. 53).

Pain, however, can be physical or psychological. As DuBois (1991) poignantly stated, "only relations of force and labor, the coercion through questioning to arrive at truth, the pushing of the young philosopher to the realm of the meta-physical, the power of the master, can enable the achievement of the truth, of the philosophical life" (p. 122). Yet even if we find truth, it is only the rhetorician, the effective user of language, who can bring that truth to the public. Plato (trans. 1925), therefore, simultaneously acknowledged and constrained the value of rhetoric in asserting that "any man who does not know the truth, but has only gone about chasing after opinions, will produce an art of speech that will seem not only ridiculous, but not art at all" (262c). Plato's notion of character, then, was necessarily tied to a good man's ability to withstand rigorous testing of his ideas and his ability to deliver artfully the truth to his community.

Isocrates's beliefs on the matter of character extended throughout his treatises on rhetorical education. Isocrates insisted that good conduct should be the end of education (see Isocrates, trans. 1980, 1.4.n2). In his speech to Demonicus, he wrote "I have not invented a hortatory exercise, but have written a moral treatise. . . . For only those who have traveled this road in life have been able in the true sense to attain to virtue—that possession which is the grandest and the most enduring in the world" (1.5). His moral beliefs are also evident in "Against the Sophists," where he discussed good versus bad leaders; good leaders were not those who possessed profound knowledge, but those with sound judgment—those who could best anticipate the contingencies of life (Norlin, 1980).

These earlier writers laid much conceptual ground for Aristotle's philosophical writings on ethos, and Aristotle's treatise on rhetoric remains one of the best accounts of character. For him, rhetoric was grounded upon an understanding of the enthymeme, deliberative, legislative, and forensic forms of oration, and three proofs of persuasion (*logos*, *pathos*, and *êthos*). Aristotle (trans. 1991) identified the proof based on ethos as a matter of being persuasive because "the speech is spoken in such a way as to make the speaker worthy of belief. For we believe fair-minded people to a greater extent and more quickly [than we do others]" (p. 38). In fact, Aristotle declared ethos "the controlling factor in persuasion" (p. 38). Kennedy (1991) argued that Aristotle dismissed that "the authority a speaker possesses is due to his position in government or society, previous actions, reputation for wisdom, or anything except what is actually contained in the speech and the character it reveals" (p. 38, n. 43). Character, then, was believed to be a perception of virtue, intelligence, and goodwill evinced from the speaker at the time of oration. As Kennedy (1991) has suggested, this may have been Aristotle's attempt to give laymen, who did not have the advantage of being well known, an equal chance to defend themselves against more reputable individuals (p. 38, no. 43).

Common to the Greek classical age, and every age thereafter, is this linkage between character and the public arena. Though more concerned with the practical over the philosophical, the later Roman rhetoricians asserted in even stronger terms that character is the life-blood of public life. In first century Rome, for example, citizens saw considerable struggle over control and power, mostly by generals like Julius Ceasar who commanded the loyalty of great armies. Rome was often threatened not by foreign forces, but by intrastate politics. The demand for men of good character was paramount, and individuals like Cicero and Quintilian strove to demonstrate to others how to become good orators and statesmen.

Cicero's *Of Oratory* stressed that the man of good character would possess broad knowledge—comprising history, law, and literature. Cicero (trans. 1959) also argued "that the wise control of the complete orator is that which chiefly upholds not only his own dignity, but the safety of countless individuals and of the entire State" (1.8). Cicero's conception of character was based upon the belief that an audience gains its morality from the orator and thus it is the orator's obligation to study a wide range of topics so that he may move his audience to goodness.

Quintilian's work, like Cicero's, was borne from this same political milieu. Quintilian, who has been called "the last great rhetorician of the classical period" (Bizzell and Herzberg, 1990, p. 35), despised where Rome was headed. His *Institutes of Oratory* (trans. 1958) was a reaction to the Second Sophistic and was intended to help mold those who would embrace classical virtues and a love for community welfare. Quintilian's ambition to write on good leadership is unparalleled in the classic period:

This man is not a plodder in the forum, or a mercenary pleader, or, to use no stronger term, a not unprofitable advocate that I desire to form, but a man who, being possessed of the highest natural genius, stores his mind thoroughly with the most valuable kinds of knowledge; a man sent by the gods to do honor to the world, and such as no preceding age has known; a man in every way eminent and excellent, a thinker of the best thoughts and a speaker of the best language. (12.1.25)

His attributed statement of the orator being a "good man speaking well" highlights the essence of this moral philosophy.

While Greek and Roman philosophers existed in differing cultural and political times, each offered a glimpse into the ancient ideal of good character. While concentrating efforts on moral identity, these theories teach us about life in the public sphere and the importance of community in construction of ethos. Conceptualizations of character have evolved over time, though, suggesting that we might yet find room for interpretation, especially in light of our newest online media.

POSTMODERN PERSPECTIVES ON CHARACTER

Internet technologies, like the Web, promise much in terms of what they could do for us as a society of individuals. And though much early conjecturing positioned hypertext and other forms of computer-mediated communication as revolutionary, other research tempered such utopian claims: "New technologies," like hypertext, "are commonly integrated into cultures in conservative ways, strengthening rather than defying relations of social and political force" (Johnson-Eilola and Selber, 1996, p. 117). In fact, Johnson-Eilola and Selber argued that "the contemporary state of hypertext contrasts sharply with the revolutionary potential prophesied by some of its originators" (p. 117). Schumpeter (2010) wrote "What sounds wonderful for the digital elite could be a nightmare for less-skilled workers." For example, "The internet of everything will render millions of people who currently look after buildings or perform low-level medical services redundant." "People who praise the role of information technologies and social networks in fostering democratic movements," wrote Swayne (2012), "often ignore how the technology has been used to suppress those movements" (n.p.). Swayne added that "Hate groups and extremist religious factions are also using information technologies to spread both information and disinformation" (n.p.). Instead of breaking social norms, destroying social-economic class structures, and leading to a more participatory democracy, computer-mediated communication has simply reinforced society's structures.

Our new communication technologies, however, are not at all devoid of rhetorical power. On the contrary, hypertext and other new computer media limit our discursive choices in new ways, changing how we project ourselves in the online public sphere. The key to understanding the online environment as a rhetorical arena is that hypertext is not a broadcast medium like television or radio, but more like a narrowcast or even manycast medium. Narrowcasting as a persuasive strategy enables rhetors to target a very specific audience and craft discourse that is very personal. A manycast or multicast medium is one that targets multiple audiences via multiple mediated strategies (e.g., convergence). Instead of employing general tactics, rhetors can adopt messages to address a specific audience's needs, values, and expectations. So while political rivals may shake hands on television, which is viewed by people of all parties, they are able to vilify each other online in a way that is not possible in broadcast media. "It is not just program content that affects identity," wrote Grodin and Lindloff (1996), "but also the use and presence of various technologies" (p. 4, 7). Online media can, therefore, have very powerful influences on public discourse and the construction of character.

Hypertextual documents may complicate, then, our classical understanding of character. Postmodern theory calls our attention to and challenges these traditional ways of knowing and seeing: How can one attribute intentionality to an author whose identity is constructed online and is masked from view? Can a critic get a sense of the author's convictions based upon a reading of the text? Can one talk of techniques if an author's intentions to employ a particular tactic are concealed? Since hypertext is defined by the interconnectedness of its information, and rhetors can construct their identities by linking their messages to other websites, does this devalue the subject-position of the author or does it just reaffirm the social-construction approach to identity? How can a critic talk of a hypertext's effect upon an audience when it is unclear to whom an online rhetor speaks? In other words, since, for most people, it is impossible to answer at this point

exactly who, from where, when, how often, and how long a person accesses a given website, are there ways of talking about effectiveness without a fixed audience to measure? Can the choices a rhetor makes still be rhetorically significant despite an ineffective website?

Postmodern theorists have grappled with many of these issues. McNamee (1996) asserted that the web enables users to create "a multitude of identities, all constructed in the ever expanding relational possibilities we engage" that challenge the classical notions of identity. (p. 142). The web has expanded the possibilities of self-construction. Online, individuals can (re)create who they are, create anyone new they wish to be, and make that new identity appear credible. Psychologist and CMC researcher Sherry Turkle (1996) wrote that "authorship not only is displaced from a single solitary voice, it is exploded . . . the self is not only decentered but multiplied without limit" (p. 158). For some people, their online identities are not simply temporary exercises in gender-bending or role-playing but become a part of who they are offline as well: "you are what you pretend to be" (p. 158). Turkle concluded that "working with computers has led me to underscore the power of this technology as a medium not only for getting things done but for thinking through and working through personal concerns" (p. 164). "I have found," she argued, "that individuals use computers to work through identity issues that center on control and mastery . . . to think through questions about the nature of self, including questions about definitions of life, intentionality, and intelligence" (p. 164).

Web technologies are, as Levi-Strauss (1960) might have referred to them, "objects to think with." If we agree with Turkle's postmodern perspective, no other medium has displayed the concept of the postmodern self so obviously as the WWW has. On the web, posited Grodin and Lindlof (1996), "self becomes multivocal as we carry a number of voices with us. Individuals, then, may find that they no longer have a central core with which to evaluate and act, but instead find themselves 'decentered'" (p. 4). Barthes (1977) has argued that "The removal of the Author is not merely a historical fact or an act of writing; it utterly transforms the modern text" (p. 145). Not only can an online author or designer construct various types of identities online, but those who browse the web often will acknowledge that they are pulled and pushed in multiple directions. These users often have several windows or pages open at the same time—a concept called multi-tasking—and may experience, as they browse the web, the uncomfortable sensation of being forced to see advertisements and unwanted pages by being trapped in a continual loop of pop-up screens. These phenomena destabilize a sense of being and obfuscate the concept of self. Michel Foucault (1997) argued the author's importance has disappeared in modern society: "A name," he wrote, "makes reading too easy" (p. 321). Knowing the author of a text, according to Foucault, will only guide the reader toward a limited view of the discourse, closing the mind of the reader and thus hindering the opportunities for a more encompassing interpretation. The important marker of a text, therefore, is not its author, since concentrating on the author's identity clouds the real issue: How does the technology of the medium alter the identity of the reader? In "Technologies of the Self," Foucault (1997) asserted this claim by announcing that he is more interested in "the interaction between self and others, and in the technologies of individual domination, in the mode of action that an individual exercises upon himself by means of the technologies of the self" (p. 225). By technologies of the self, he means those abilities "which permit individuals to effect by their own means, or with the help of others, a certain number of operations on their own bodies and soul, thoughts, conduct, and way of being, so as to transform themselves in order to attain a certain state of happiness, purity, wisdom, perfection, or immortality" (p. 225). "From the eighteenth century to the present," he wrote, "the

techniques of verbalization have been reinserted in a different text by the so-called human sciences in order to use them without renunciation of the self but to constitute, positively, a new self" (p. 249).

To say that readers of online sources have limited options for action is simplistic; they have a variety of methods by which to express themselves online, freed by the medium. Verbalization is not the chief method of expressing the self online, though users can usually write electronic mail, join listservs, post on blogs, and the like. Users have innovated a new technology of the self online: They have the option of easily leaving the conversation that exists between them and the webpage designer by moving to a new page or retracing one's steps to a previously visited site. There are no social obligations or cultural forces insisting on a particular manner of online behavior. This method of expression is, according to Foucault, a form of confession in which the user admits his contempt or boredom. The user can reify his or her online identity by entering or leaving these webpages and thereby constructing for him or herself the type of online user he or she wishes to be. One's history file on the computer becomes an identity of sorts. Foucault's perspective, therefore, can complicate our understanding of the audience in ways that webpage designers have taken note. These rhetors do appeal to our impatience by constructing pages that invite us inside their self-created worlds. Their worlds are real because audience members answer this invitation—rhetor and audience blur in these moments because the audience creates the rhetor and rhetor, thus inventing the audience. New media have faded the clear line between author and reader. In the hypertextual world, the "audience" does not exist in the classical sense. While knowing the audience is impossible, this does not mean that audience is irrelevant. Online rhetors make very real choices based on their perceived audience, which provide insight into how various web designers think. The once popular hit counter, for instance, might suggest a few ideas about a web designer's intentions. The web designer's reasons may be pure, which is to say the designer may wish to assess the popularity of his or her website over time, and may make changes to the site depending upon the number of such hits. On the other hand, a designer may wish to project the image of popularity instead, revealing doubts of self-confidence or suggesting what the designer believes his or her audience thinks of his or her character. Regardless of our assessment, such technologies offer us a glimpse into the mind of the rhetor and what he or she thinks about the perceived audience and self. Whether or not the WWW is Baudrillard's (1988) fated simulacrum—an illusory world in which signifiers have more realism than the signified—the decisions one makes online are rhetorically important (pp. 170-2). What is worth noting is that readers influence the online text so significantly that they simultaneously become part reader and part author. This suggests that no webpage is the result of a single author, but instead co-created by designer and users, or those we may now call participants. Any assessment of an online text, then, should not ignore the online participants that helped shape the text. A judgment of the author is, in part, also a judgment of the reader.

DIGITAL ETHOS AND ITS ASSESSMENT

Although rhetoricians are interested in how technologies such as the web become appropriated as discursive arenas, little attention has been paid to how the medium complicates our understanding and assessment of ethos. In fact, according to Chung et al (2012), "Although many studies have described the characteristics of online media, few have considered how the unique technological elements afforded by the Internet may affect their perceived credibility" (p. 174). Online ethos is no longer a face-to-face phenomenon and stable identities are rarer as we venture onto websites, chatrooms, and email. Indeed, following their

analysis of forty-two online learning community evaluation studies, Ke and Hoadley (2009) asserted that "the idea of creating a monolithic, one-size-fits-all . . . evaluation model may not work" (p. 505). Furthermore, "Given the infinite nature of CMC," argued Apostel and Folk (2008), "an easily defended checklist, cannot . . . be developed nor should it be" (p. 19). So, whatever the final design, a useful model needs to be flexible and draw from multiple approaches. We might start, for instance, by re-assessing our research on identity, persona, and credibility in light of the Internet. As a start, traditional argumentation theory, for example, shows us that credibility can be enhanced or limited by our assessment of the sender and the quality of the message. The online environment complicates these two elements, posing problems for online groups that wish to create a credible identity.

Online, it is often extremely difficult to identify and thus evaluate the credibility of "authors" because the identities of online authors are almost always obscured by the online environment. On some pages, the source of the information might be named yet not further identified. On other pages, there may not be any attribution at all. One website about Martin Luther King, Jr. (2012) boasted that they deliver the "Truth about King," information about his speeches and writings, and even include suggested readings. Though it will not take most people long to realize that the site is highly critical of King, denouncing his work and influence, many people could still be fooled. There is no author listed, but a link to a group called "Storm Front" is present at the bottom of the page in fine print, linking to a page with a large logo with the words "White Pride World Wide" on it. Sites that make the average reader hunt for authorship, such as this, confuse and complicate our ability to assess credibility, which is obviously non-existent in this case. Plus, most of the time, there is no phone number, mailing address, or link that might allow the user to follow-up on the information. In any case, names and institutions can be easily falsified. Even on Facebook, which has a minimum age limit of thirteen for members, manipulation is high. Falsifying one's age is very easy, and there is no real way to know if the person you are talking to is a minor or adult, which makes communication through this very popular medium very tricky. As Wright (2010) has stated, "Evaluating information—especially information available online—is truly an art, and it takes practice" (p. 65). Since identifying sources is difficult, readers are sometimes forced to make judgments about the quality of the information alone. Yet, even this too is difficult with this environment. According to Wu et al. (2010), two major trust antecedents can aid in establishing online credibility. Specifically, they found that "perceived interactivity has a positive impact on consumers' initial online trust" in a lesser-known online source, and "perceived Web assurance is a robust institution-based antecedent to consumers' initial online trust" (p. 16). As a result, creating a secure and engaging environment seems key.

Basing our judgment of an online source purely on the quality of the information itself is suspect. Web-based information, like the identity of the author, may be authentic or totally contrived. Users should, therefore, be skeptical of all web material. One of the first steps in assessing the credibility of information should be determining whether or not the information can be verified. Freeley (1990) asserted that verification is critical for assessing the credibility of evidence (pp. 109-12) and Newman and Newman (1969) highlighted the importance of a corresponding concept—authenticity (p. 66-71, 77). "If a statement has no potential for verification," wrote Ziegelmueller et al (1990), "it is not factual" (p. 58).

How does one verify online information? "The key to successful Internet use," said Carlson (2009), "is knowledge of what to look for as you evaluate the quality of the information offered" (p. 200). The problem online is that everything, from music to images to text, is comprised of digital bits that can be falsified and made to look as authentic as the real thing. Images on one online source can be made to look exactly like the images on another or faked entirely. The "Dihydrogen Monooxide" (2012) website claims to be raising awareness of a very dangerous chemical compound that is apparently everywhere in our environment. Too much of it and we could die. The authors state that groups such as the Environmental Protection Agency have ignored this problem. The site, which may appear professionally designed to some, even has a logo affixed to the main page by a non-existent United States Environment Assessment Center, which makes it look even more credible. What is noteworthy is that, despite the fact that Dihydrogen Monooxide is scientific jargon for water, visitors have been routinely donating money to the authors of the manipulative site for more than a decade. Beyond this form of "creative artistry," web designers can easily "publish" articles that have been plagiarized from other sources or they can cite research and statistics without attributing the correct source. In fact, much of the information available online is without citation. How can one verify information that has no link to its source? There is so much information online, and it changes so frequently, that close scrutiny is excessively complicated. December (1996) stated that "the Web is characteristically, notoriously changeable, with new technologies (servers, browsers, network communication) as well as new content being introduced continuously." CMC researcher George Landow (1992) once wrote that online materials "by definition are open-ended, expandable, and incomplete" (p. 59). Some webpages change daily and often radically. Though argumentation theories note that the concept of recency is critical in assessing the credibility of information (see Ziegelmueller et al., 1990, p. 96-99; Freeley, 1990, p. 118; Newman & Newman, 1969, p. 84), the unpredictable, quick-changing nature of online information creates special credibility problems for assessing web-based and perhaps other online materials.

We need to come to grips with changes wrought by the online technology for the creation of online ethical appeals. Heim (1988) asserted that face-to-face theories of character were appropriate for the "classical age when direct human contact predominated in the public process," but now our understanding of how online identity is created should be revised as "more than a theory of direct verbal persuasion among human beings" (p. 57-8). Online users may employ visual images, animation, sound, video, hypertextual links, games, chatrooms, real-time interaction, and many other technologies in order to identify with their interlocutors and establish ethos. In her questioning of the classical concept of ethos as it relates to the problem of authorship in digital poetics, Fleckenstein (2007) suggested that Aristotle's classical concept actually works quite well: "Lodged within a context that blurs seeing and saying, Aristotle constructs a concept of *ethos* that emphasizes the liquid movement among speaker, audience, scene, and context, offering a powerful lens for re-seeing author positions." Therefore, we should keep those parts of traditional concepts and assessment paradigms that are relevant, but be ready to adopt new measurements as appropriate for the new medium. In his analysis of the Vatican's website, for instance, Frobish (2006) demonstrated "how a group's special circumstances could call for a different set of rules regarding ethos and identity" (p. 67). In fact, the Vatican's site "did not need to personalize the site for its users, link itself with outside groups, or maintain an attractive background to draw and sustain the interest of its users" (p. 67), and yet it was still able to build substantial ethos for its efforts.

So what may be considered an online ethical appeal? Nothing in the online environment should be overlooked as an attempt by the rhetor to persuade its users of his or her credibility. The "visual cues and hypertextual structure used to construct a Web site," for example, "create a site that conveys a specific ethos, or character, for the organization" (Hunt, 1996, p. 378). A political site, for example, may use a combination of words, background and foreground images, online games, e-mail links, hypertextual links, and so forth, in order to convince us that it is, first, an official political site and, second, that its policies are honorable, community-oriented, beneficial, worth voting for, and so on. These appeals may even occur simultaneously. Employing an image of the U.S. flag on a political website, for instance, may speak to all at the same time. Further, one could look at a basic e-mail link and interpret that as an attempt by the online organization to demonstrate concern and accessibility—that it cares about the ideas of and feedback from the community. It could also be read as an attempt to show basic computer literacy: An online group that does not include a feedback mechanism on its website may look rather antiquated or unprofessional. A third reading may suggest no deliberate ethical appeal at all, but merely the existence of standard, routine technologies or website design. It is difficult to judge intent in these cases. Instead, only the possibilities of how such a thing might be read should be evaluated. We could just look to the textual content of the page to assess strategies of credibility, yet, as Warnick (1998) argued, "mere attention to the words on the web page will not suffice, since the images are so important to textual meaning. Even in texts without image, the way that the text is displayed on the screen has rhetorical impact" (p. 77).

When assessing whether or not a given word, image, or technology is an ethical appeal, it may be more helpful to distinguish between what is artistic and inartistic. In Aristotelian terminology, this suggests a difference of design. For example, the fact that a web designer creates or employs a hypertext link is not as rhetorically interesting as to how that particular link associates the site with the credibility of others. In traditional rhetorical criticism, scholars would look to one's words. When someone says "I have worked for charities for 20 years," for example, it seems apparent that those words constitute an ethical appeal, specifically virtue. Is it any less apparent when an organization places a photograph on its website showing its employees supporting Big Brothers/Big Sisters, or perhaps building houses for Habitat for Humanity? What if a commercial organization creates a link on its Facebook page to the Big Brothers/Big Sisters' site? Is that any less of an ethical appeal seemingly designed to build an image of virtue, communal welfare or goodwill, and so forth? Web technologies, like words, can be strategically employed by online groups to build ethos. Any word, phrase, image, or web technology may constitute an ethical appeal, and is worthy of critique. A web assurance seal, for instance, may help establish a secure look and create additional trust with users, but only in certain circumstances. As Wu et al. (2010) have shown, "it is not the displaying of a Web assurance seal itself but the perception of it that matters in trust formation." If users "pay no attention to a Web assurance seal or if they fail to understand its purpose, the seal may not achieve its intended effect even if it is displayed prominently on an e-vendor's Web site" (p. 17).

Rather than attempt to classify thousands of web technologies and speculate as to how they might be used as appeals to ethos, it seems more reasonable to reorganize the types of ethos appeals that might be possible within the online environment. Instead of relying upon Aristotle's tripartite theory of ethos, which may be too limiting, we may wish to investigate other systems of thought. Credibility expert Sharron Kenton (1989) developed a four-part typology of ethical appeals that appears more suited for the analysis of web discourse. Her four-part system consists, first, of goodwill and fairness, which includes a rhetor's focus on the receiver, displays of concern for the receiver's welfare, and an unselfish attitude. Second, she includes expertise, which incorporates appeals based upon the training, experience, qualifications, intelligence, competency, and achievements of the rhetor. The third factor is prestige, suggesting those appeals relating to the rank, power, position, or

status of the rhetor. The final factor in her typology is self-presentation, or the verbal abilities, platform skills, dynamism, energy, charisma, and confidence of the rhetor. Beaston (1991), who adapted Kenton's typology for his analysis of ethos in business speeches, removed goodwill and fairness, prestige, and self-presentation. He adds deference or respect for the audience, self-criticism or one's humility, similitude or attempts to create interpersonal relationships with audience members or to build a sense of community, and the inclination to succeed, or one's confidence and drive. Kapoun's (1998) evaluative system for online credibility has become very popular as a teaching tool at universities, particular within library sites. His assessment model includes the concepts of accuracy, authority, objectivity, currency, and coverage. Tillman's (2003) model asked the evaluator to assess the quality of the online source by questioning the appropriateness of the source for the user's needs, ease of identifying the currency and authority of the source, stability of the information, and the ease of navigation and speed of the connection. There is no shortage of assessment models. In fact, Apostel and Folk (2008) argued that "as CMC technology and content continues to evolve, different means of negotiation need to be developed and continually re-evaluated that address the concern of finding reputable, reliable information and re-presenting it in a world where the nature of information is radically different" (p. 19). Assessing previous models and discovering and combining the best of each may be one reasonable way to move forward.

Each of these previous methods has strengths and weaknesses that must be considered in the design of this chapter's model of digital ethos. Kenton's typology better addresses the technological aspects that classical theories of ethos cannot presume to consider, but her categories seem to overlap in some crucial areas. How do competency, a component of her "expertise" category, and platform abilities, a component of her "self-presentation" category, differ exactly?

It seems possible that, online, one's ability to deliver a dynamic website might equate to a degree of technical competency. So, competency and self-presentation could be connected. Beaston's typology is helpful since it allows for the criticism of community-based appeals, but fails to consider the moral component or virtue of the rhetor.

"Given the constantly changing state of digital culture and the ramifications of increased multimodality, access, diversity, viewpoints, and information styles on the web," Apostel and Folk (2005) advocated against "replacing one set of rigid criteria with another." A more general scheme is necessary, then—one that can avoid the problems of these earlier methods and allow for some flexibility when evaluating the new online media. As the previous sections have demonstrated, a rhetor's ability to show online

participants that its members constitute a common community and that he or she is willing to adapt and respond to that community's needs, is the first step toward building trust and gaining the assent of the audience. Also important is the morality or virtue of the rhetor. Showing that one celebrates the values of the community and that one is fairminded, honest, and not self-interested can help establish the kind of virtuous ethos discussed by Aristotle, Quintilian, and others. A rhetor's ethos, furthermore, is tied to his or her intelligence and knowledgeability—proof that one is qualified and able to deliver on promises. Finally, the technological means of building ethos and identity are especially salient. As both Warnick (1998) and Hunt (1998) have been cited as suggesting, the power and visual impact of the text and structure of the site can suggest important things about the

Figure 1. Model of online ethos

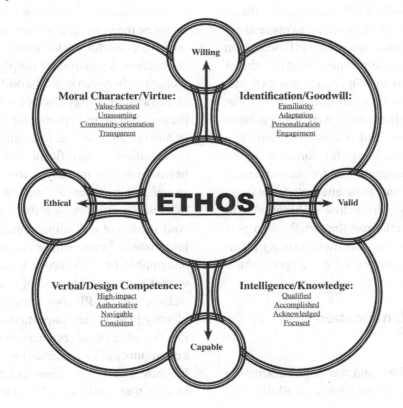

character of an online user or group. Therefore, assessing one's verbal and design competency is necessary if we are to accurately evaluate a website's appeals to ethos and identity.

Based upon a review of these prior issues, I propose a four-part system for the assessment of online ethos and identity, consisting of appeals to (1) Community Identification and Goodwill, (2) Moral Character and Virtue, (3) Intelligence and Knowledge, and (4) Verbal and Design Competence. This typology considers both classical speaker-oriented factors and the more visual and technological components of ethos and identity. The above model of online ethos is demonstrative. By developing an identity based upon Identification and Goodwill, for example, an online rhetor shows a willingness to engage the community. Be developing an identity based upon Moral Character and Virtue, an online rhetor shows the proper ethical awareness and moral compass to guide others. By developing an identity based upon Intelligence and Knowledge, an online rhetor shows a valid position from which to make arguments. By developing an identity based upon Verbal and Design Competence, an online rhetor shows the capability to deliver on promises made. The creation of a *willing*, *ethical*, *valid*, and *capable* persona can lead to a powerful and persuasive ethos that can lead others to action. The model can be especially useful, furthermore, by working as an assessment tool for the evaluation of online sources and their attempts to develop ethos and identity. The below elaboration will further demonstrate the usefulness of this four-part process, taking into consideration the unique possibilities and challenges of developing these rhetorical traits.

Community Identification and Goodwill

Though having reliable and verifiable information is important to an online user's credibility, the preceding review of ethos shows us that establishing a community-oriented identity is essential. In classical times, orators were expected to show to audience members, by means of face-to-face communication, that they spoke either for their benefit or for that of the larger community. This is ever more important today as people often feel less connected. The WWW was prophesied to bring people together, but many users feel more distanced in online settings. In fact, Cheseboro and Bonsall (1989) have said that CMC seems to "promote efficiency at the expense of special contact" (p. 221). It makes sense then to more fully examine how building that community-based identity is possible via CMC and what obstacles they might pose for groups wishing to establish this form of identity. This is important since, as CMC scholar Steve Jones (1998) has remarked, "conspicuously absent is an understanding of how computers are used as tools for connection and community" (p. 5).

Establishing a connection with one's audience might be the most effective means of establishing positive ethos for online users. To establish this connection, a group may simply point out commonalities between it and the online users to create identification. It may also occur by either making the user feel a part of particular larger community or making the user feel important, as if his or her contribution is significant and needed for the health of the community. Certainly, showing the audience that you are willing to adapt and respond to its needs can significantly help establish trust. And while not all online sources are equal, the importance of establishing a communal identity is undeniable. Frobish (2004), for example, showed us that even with the case of a sexually oriented website such as Playboy's creating a communal identity plays a very important role. In this case, Playboy offers its users what it believes they want, a community of like-minded people who share the Playboy 'lifestyle' . . . and focuses on the sort of identity that would attract a variety of users to the site—one designed to assuage their concerns over privacy and security so that they feel comfortable buying the Playboy merchandise.

Groups can appeal to community verbally. Often, a rhetor might say things such as "I understand where you are coming from," or "My parents, like yours, were not rich people," or "Your vote is needed to sustain the type of society that you and I, and now our children, have come to love and appreciate." Furthermore, "whenever the speakers use the term 'we' to refer to both themselves and their audience," says Beaston (1991), "they are claiming group membership with that audience," which "can be a subtle type of affiliation" (p. 331). Linguistic appeals that vilify a common enemy may also help build community. Rhetorical scholar Roderick Hart (1998) has argued that "feelings of antipathy directed toward some person or group . . . has long been the handmaiden of community" (p. xxv-xxvi). These verbal appeals work to build a sense of identification and trust between the rhetor and audience and are no less powerful in CMC. As Wu et al. (2010) have shown with their research, "In a computer-mediated communication environment that often tends to be impersonal, visitors desire to be treated as a human being or a guest" and as "message personalization increases, the perceptions of interactivity and Web site effectiveness are enhanced" (p. 18). Yet, digital designers do not have to rely upon these verbal appeals as the only way to build this connection.

Perhaps the clearest way to build communal identity online is for groups to embrace technologies that link people together. This can be done in many ways. Websites may provide e-mail links that can connect users to site administrators or directly to those in the group's hierarchy. It may allow users to register to receive periodical community newsletters, register on a community e-mail list, join with others in online chat rooms, allow users to sign a guestbook or blog, and so on. Facebook allows users to create multiple friend networks consisting of thousands of other users. Your success in many of Facebook's games, such as Farmville and SuperHeroCity, is even dependent upon the number of friends in your network who are linked to the same game. In their analysis of political websites, Frobish and Thomas (2012) revealed how "social networking capabilities, specifically, have certainly allowed website developers to incorporate more advanced interactive options, increasing the possibilities for user engagement, personalization, and eventual contribution." As Chung et al. (2012) demonstrated, "just 'being online' does not add much credibility if . . . users fail to choose multimedia features, control the flow of additional information, and actively respond to content" (p. 182). The good news is that the possibilities for this sort of community interaction are continually growing as the online environment develops. Online sources that can connect users to some larger community of people may make those same users more prone to trust the group that designed it.

Another way in which an online source might establish a communal identity may be to offer users free gifts. An online source that offers it users free services (e.g., books, videos, memorabilia, free online virtual pets, discounts), especially those that increase brand recognition for the group (e.g., books about the group, buttons with company logos), may help to establish a sense of goodwill. Additionally, blogs, Facebook pages, or Twitter feeds that act as informational news resources might appear altruistic and could establish the same communal goodwill identity.

Online sources can create a communal identity by incorporating visually appealing items such as pictures or videos with which users can identity. Images of flags on a political website, for instance, might suggest that the group is patriotic and cares about the larger American community. Pictures of children on a school's Facebook page might promote a child-centered teaching philosophy, showing that it cares about the community of the future. Certainly, a commercial shopping site that allows its users to view products or provides its users coupons for these products might represent a goodwill effort on the behalf of the group to the larger community. Even professors who build websites and post their academic resumes or syllabi might be showing that they belong to, and work for, the academic community.

It may seem that creating a community-based online identity is an easy process, but there are obstacles to overcome. First, the previously examined online security and anonymity problems may create obstacles for groups that wish to build a communal identity. Second, for groups or users who have credibility problems offline, it is unlikely that users would easily assent to community-based appeals, especially if that group has a self-serving reputation or if other parts of the group's website suggest an anti-community persona. Third, sites that rely upon web-based interaction or connectivity as their sole rhetorical strategy might not sufficiently develop sufficient ethos for persuasion. CMC scholar Steve Jones (1998) has written that "interaction ought not be substituted for community, or, for that matter, for communication" (p. 31). Though establishing a communal identity is important and has withstood the test of time as a vital factor in building ethos, it is not the only available ethical strategy online. There are other CMC-based appeals that are especially relevant and deserve notice.

Moral Character and Virtue

A second factor in building *ethos* and identity online is that of moral character and virtue. One's moral character is typically evaluated by whether that person has followed standards of good or just behavior, or if that person appears to have good or just intentions. Moral character or moral purpose is linked to virtue such in that one must embrace certain virtues to be considered moral. A person who has murdered another is not moral because he or she has violated specific virtues such as the love for life, respect for another's self-autonomy, and self-restraint—virtues that form the crux of civil society. Many general virtues such as honesty, candor, humility, and sincerity might play a role in the perception of a rhetor's moral character. In fact, Aristotle (trans. 1992) wrote that justice, courage, temperance, magnanimity, and magnificence, were the "virtues of the soul" (p. 64).

Online communicators, regardless of their purpose online, must communicate in a way that proves to their interlocutors that they are not above the needs of the larger community or its interests. They must, furthermore, appeal to what that specific community believes is right, just, or the correct course of action. In case of a political website that is promoting a candidate for office, all efforts should attempt to show that the candidate is honest, sincere, fair-minded, dedicated to the party, and courageous enough to take action. This may take the place of video testimonials from others that discuss qualities of character, textual proof of voting records, pictorial and narrative biographical information, and so on. For a religious blog seeking to create identification with followers to maintain their fellowship, the moral dimension of ethos is crucial, which may call for linguistic and technological appeals that show a leadership that is kind, honest, humble, and community-focused. In this case, the blogger may wish to focus its narrative on the church's charity work, links to important church information, and pictures or video of church events. Even sexually oriented online groups might wish to establish a common ground of anti-conventional morality with users by posting pictures and video that celebrate sexual behavior as perhaps indicative of personal freedom, free expression, art, self-empowerment, and so on.

Scholars have long argued that demonstrating one's moral character or virtue has an important role in a rhetor's ethos and identity. In his *Nichomachean Ethics*, Aristotle (trans. 1934) wrote that "Mock-modest people, who understand things, seem more attractive in character" (p. 102). "For the man who loves truth," he wrote, "inclines rather to understate the truth; for this seems in better taste because exaggerations are wearisome" (p. 101). Modern communication research has continued to demonstrate the value of self-critical honesty. Norton (1983) has explored the importance of an "open style," one in which the person communicates personal and sometimes risky information about him or herself. Interpersonal scholars have recognized that increased self-disclosure is typically connected to an increase in trust. Beason (1991) has argued that rhetors can develop ethos "by openly admitting the shortcomings of their claims, abilities, or affiliated organizations" (p. 336). Too much self-criticism, however, can be counterproductive, especially for groups that also need to show some expertise. Therefore, humorous or light-hearted self-criticism may be necessary for some groups who wish to develop moral character, and maintain their expert status.

Moral character or virtue may be established verbally through several means. First, online rhetors may directly talk about the importance of communal values, such as "liberty" in the case of political groups, "kindness" in the case of the religious groups, or "economic empowerment" in the case of commercial groups. Language that seeks to remind audience members about or reinforce the importance of these virtues might help build trust and credibility. Second, downplaying one's successes linguistically may serve to show audience members that the group or rhetor is neither presumptuous nor self-promoting. A political candidate could say, for instance, "while I have worked hard to eliminate poverty in the community, there is still much more work yet to be done." A religious leader who has received an award could show humility by stating the names of others who should share in the recognition. Commercial organizations may show their history of charity in support the appearance of moral character and virtue.

There is no simple list of online technologies that necessarily suggest moral character or virtue online. This does not mean, of course, that online communicators cannot use new media to demonstrate high morality, but high morality may be more tied to how they use those technologies. Awards or proof of achievements do not have to be emphasized, for instance, or presented at all. Perhaps, a better way to think of moral character and virtue in a technological sense is to imagine modest online discourse—one that, according to Hogel (2000), did not "get wrapped up in all the latest high tech bells and whistles." So, for groups that are already perceived by users as credible, hit counters, guestbooks, awards, flash animations, and so forth, have the potential to make them seem arrogant and dated. Online moral character and virtue, therefore, is probably evaluated best by the absence or moderate use of such web technologies.

Intelligence and Knowledge

Appeals designed to demonstrate one's intelligence, knowledge, or general qualifications may help to build a communicator's ethos and identity. Someone who discusses his or her prestige, experience, competence, qualifications, legitimacy, or authority is likely seeking to establish trust with its users. One's prestige (e.g., rank, seniority, reputation, and status), for example, suggests something about the quality of work done by the rhetor. One's prestige is, by definition, based upon one's past efforts and is, therefore, important for a group whose future success is dependent upon past success. Indeed, many people see past performance as the sole basis for ethical judgment.

Listeners and readers seem more likely to believe and trust those who can verbalize that they are intelligent and knowledgeable about issues important to them. Politicians often use their family history, education, voting record, and so on, to build the perception that they have qualities worthy of the office and are, therefore, worthy of their supporters' votes. Religious groups might speak of their divine knowledge or inspiration, which, in turn, gives them authority. They may also talk about how many supporters they currently have, as evidence that established authority. Commercial organizations may discuss their financial successes, stock performances, or how many years they have been in business as indicative of their corporate intelligence and business knowledge. One obvious way to demonstrate one's qualifications is to show a resume. Although placing a resume on a personal website can highlight one's effort to the larger community, it also highlights one's qualifications. And although talking about your intelligence on a Facebook posting may provide a textual means of establishing ethos, and probably no small amount of arrogance, there is also a whole visual and technical realm of possibilities.

Online communicators have many new possibilities to build the perception of intelligence and knowledgeability through visual and technological means. In fact, there are times when one might wish to eschew verbal appeals all together for the visual or technological. Beason (1991) has written that verbal statements of one's reputation or past achievements can be dangerous, especially since they may be perceived by audience members as conceited. Alternatively then, websites may display their technological accomplishments to suggest, perhaps more delicately, their qualifications or expertise. Some of these are content-related, emphasizing, perhaps, the intellectual expertise of an individual's efforts or that the individual has contributed significantly in some way to society and has been acknowledged for that work. Others may speak to technical competence or design ability. There are quantitative endorsements such as those provided by counting awards, hit counter clicks, guestbook, blog comments, and Facebook "likes," or YouTube views, which might be employed by those wishing to demonstrate the support of thousands, or in some cases millions, of users. Such appeals might suggest, in some way, a certain level of legitimacy or authority. These web technologies, however, are easily manipulated and can easily be made to exaggerate one's claim to legitimacy and authority. There are qualitative endorsements as well, which include such things as image and video testimonials. Certainly, im-

ages showing politicians or religious leaders surrounded by bands of supporters might work better to show popularity and therefore a higher degree of legitimacy. But, even here, online images can be distorted or falsified. Because of these complicated issues, users may look to other types of appeals for evidence of character and credibility. Hypertextual links, as yet another option, may also be employed to create the appearance of intelligence and knowledgeability by borrowing the ethos of other individuals or groups. Not only do links provide users a path to additional information and resources, which may prove some amount of knowledgeability on the part of the originating source, but they may also give unofficial support to these linked-to individuals and organizations. Much like naming powerful friends in an election speech, a new local political campaign website that aims to build credibility can link to national party sites, elected officials, appropriate think tanks, and so on, attempting to create the appearance of legitimacy and authority. In her review of the traditional rhetorical cannons and how podcasting may have complicated concepts like ethos, Bowie (2012) suggested that

In previous media, especially print texts and speeches, the testimony would only be quotes, which are helpful but not as persuasive as the actual voice or image of the speaker giving the testimony. Likewise, including the spoken words of witnesses or authorities will make the argument stronger and more reliable and trustworthy.

The possibilities of building intelligence and knowledgeability through the use of visual and technological means may, therefore, surpass those possible through verbal means alone.

Verbal and Design Competence

A fourth factor in building ethos online is the concept of verbal and design competence. This section is critically important to an understanding of how groups are attempting to build ethos and identity online. Certainly, a student that builds a personal website with his or her professional portfolio on it might seem more technologically proficient to prospective college recruiters than a student who does not, even if the website was not professionally developed. On the other hand, universities and corporations are expected to have a professionally designed online presence on multiple platforms, and then compete with each other to attract customers. So what is considered "competence" may not be whether one has a website, Facebook page, or YouTube account, but, instead, whether one's online presence projects a sense of quality about the hosting individual or organization. The shopping site that looks more attractive than its competitors is certainly more likely to gain the trust of users than a shopping site that is poorly designed and difficult to navigate. What is most important is that, online, there is definitely "a relationship between format and source credibility" (Armstrong and Adams, 2009). The higher the attractiveness or quality of the format, the stronger the perceived credibility of the online source.

We may consider many elements when assessing appeals to verbal and design competence, all of which can aid in building trust between online communicators and users. We may assess, for example, a group's verbal skills, including appropriate word choice, use of active voice, clarity, boldness, brevity, use of effective metaphors, and so on. These are traditionally considered stylistic choices, and the masculine form, characterized by brief, declaratory statements, active prose voice, a loud or projected delivery, and fewer pauses is perceived more credible (Kenton, 1989). The same may be assumed of online discourse. Communicators who employ high-impact language, active voice, and rich metaphors might create a more textually attractive, and thus more credible-sounding, site. In fact, in one analysis of blog posts, a masculine writing style was found to increase the perceived credibility of the postings (Armstrong

and McAdams, 2009). Political groups whose goal may be to provide information about policies or futures political objectives, for example, may not gain the trust, and, therefore, the vote of a user if the group is not able to forcefully articulate its message, or if they present an online message so terribly designed that users are not willing to read the message at all. In fact, the same could be said of any Facebook page, blog, or gaming platform that wishes to gain the trust and assent of their audiences—an online presence that appears incoherent and weak linguistically might, in many ways, be perceived as untrustworthy, as if the written word is a projection of potential ability.

The substance or quality of a group's discourse relate to the group's verbal competence. Building a verbally competent identity by means of quality content means providing audience members with useful information that is interesting and potent. This is nothing new to traditional rhetorical scholars who have always noted the importance of audience-centered information in persuasive discourse. Online, however, communicators may provide a much broader range of information and in a variety of styles. The types of information made possible by the online environment seem endless, including product information, corporate publications, resource links, engaging news, chat sessions, advice and support, entertainment and games, audio and video clips, planners and calculators, customer reviews, blogs, and so on (Hogle, 2000). Providing user-centered information can engage their users and establish trust. In fact, Hogel suggested that when online communicators provide informational and not promotional items, users are more willing to perceive them as competent.

Using well-formed, strategic discourse is only one part of demonstrating competence. Design competence can be established by an appropriate and productive use of web technologies, but technologies that cloud the substance or make it difficult for users to accomplish their goals will result in lost credibility. One way to assure users that the online resource is useful might be to display quality content that has an appropriate structure and arrangement. A website, for example, should be easily navigable. All links should work (i.e., should connect to another working site), the site should have at least one accessible internal search engine, and the site should be constructed logically and with purpose. Users, then, should never feel lost when looking for information. Offering the possibility for technical support can also help, but, in general, a site's architecture is very important for establishing design competency. Those that do display a clear and organized arrangement of the site's content, while adding well-designed backgrounds and images will make the hosting group look more capable and more worthy of trust. Critics should examine not just whether a group uses images on its website, but the way in which those images interact with the text. Regardless of the platform, groups that effectively use images to strengthen their textual appeals or effectively employ images as the message itself may appear more competent. This pertains to websites, but also online advertisements, blogs, Facebook pages, and even gaming media. Even one's download speed can be important, which may show an awareness of the medium and a concern for users. Still another issue is that of consistency. Hunt (1996) contended that "many of the sites on the Web today are inconsistent in look and feel, and are illogical in information structure. As a result, the information these sites convey loses credibility" (p. 379). Ultimately, users might put more trust in an online organization that shows thought, rather than those that show carelessness. It seems reasonable to assume that users will return to those online platforms that have provided them with attractive, useful information and the means to find it.

Developing a website, Facebook page, Youtube account, Twitter feed, or blog is relatively easy, but designing a highly impactful and potent resource takes talent and hard work. Those who take their sites to this next level will be perceived as more competent. Verbal and design competency is certainly a valid component of ethos since online communicators must "look" like they can deliver on their promises to the community. Without it, credibility will wane: Political groups will lose supporters, religious groups will lose followers, university sites will not recruit new students, and commercial groups will lose customers.

CONCLUSION

Despite new ways to establish ethos and identity, and new ways to assess such attempts, the basic nature of ethos may not have changed much at all. Online media that represent the beneficial interests of some larger community still have a stronger ethos than those that do not. We also believe that, though the web has radically altered the ways in which we think of "author" and "message," we still look for things such as moral character/ virtue and intelligence/ knowledgeability when trying to assess another's credibility. So while early conjecture sought to persuade us that the online environment will change our entire way of thinking on the issue, it has not. The environment may have changed the ways in which users create ethos and identity, but what readers look for in a credible source has remained essentially the same since ancient times.

Some effort has been spent exploring the basic tenets of classical ethos. It seems safe to say that the classical concept of ethos as an audience's perception of a speaker's character is not outdated but perhaps limited given new media possibilities. The digital environment offers us new ways of building ethos, and now we have a new way to assess those capabilities. The exception to traditional theories of credibility is our addition of the concept of verbal and design competence, which must be taken into consideration if we are ever able to answer some reasonable questions about ethos and computer-mediated communication. When online rhetors design websites with several hundred or even several hundred thousand pages, how are users to form judgments about their ethos? How does one get the "complete" picture when the majority of a group's discourse is hidden from view or at least so massive in size that one cannot possibly attend to all of the information? Furthermore, whose ethos is really being judged when a group could have people external to the organization designing its identity or a number of outsourced writers who are under contract? Does it even matter who designs a group's online presence? Will the future of new online media bring even more challenges to our concepts of character, trust, or credibility? During these complex transitions in the digital environment, the proposed model of assessing online ethos makes sense. Evaluating one's attempts to create an identity based upon Community Identification and Goodwill, Moral Character and Virtue, Intelligence and Knowledge, and Verbal and Design Competence works well for CMC scholars who are trying to keep up with the constantly changing technological environment. CMC as we know it today will not exist in even five years. But no matter how the environment evolves, identifying how ethos works online will always be of consequence. Continued study of this new universe of possibilities is vital to realize those factors that are unchanging in the nature of persuasion, but also what factors are worth investigating as having altered our understanding of rhetoric.

REFERENCES

Apostel, S., & Folk, M. (2005). First phase information literacy on a fourth generation website: An argument for a new approach to website evaluation criteria. *Computers and Composition Online*. Retrieved from http://www.bgsu.edu/cconline/apostelfolk/c_and_c_online_apostel_folk/apostel_folk.htm

Apostel, S., & Folk, M. (2008). Shifting trends in evaluating the credibility of CMC. In Kelsey, S., & St. Amant, K. (Eds.), *Handbook of research on computer mediated communication* (pp. 185–195). Hershey, PA: Idea Group Reference. doi:10.4018/978-1-59904-863-5.ch014

Aristotle. (1934). *Nichomachean ethics* (Rackham, H., Trans.). Cambridge, MA: Harvard University Press.

Aristotle. (1991). *On rhetoric: A theory of civic discourse* (Kennedy, G., Trans.). New York, NY: Oxford University Press.

Armstrong, C., & McAdams, M. (2009, April 16). Blogs of information: How gender cues and individual motivations influence perceptions of credibility. *Journal of Computer-Mediated Communication*, *14*, 435–456. doi:10.1111/j.1083-6101.2009.01448.x

Barthes, R. (1977). *Image, music, text*. New York, NY: Hill and Wang.

Baudrillard, J. (1988). Simulacra and simulation. In Poster, M. (Ed.), *Jean Baudrillard: Selected writings* (pp. 166–184). Cambridge, MA: Polity Press.

Beason, L. (1991). Strategies for establishing an effective persona: An analysis of appeals to ethos in business speeches. *Journal of Business Communication*, *28*, 326–347. doi:10.1177/002194369102800403

Bizzell, P., & Herzberg, B. (Eds.). (1990). *The rhetorical tradition: Readings from classical times to the present*. Boston, MA: Bedford Books.

Bowie, J. (2012, Spring). Rhetorical roots and media future: how podcasting fits into the computers and writing classroom. *Kairos: A Journal of Rhetoric, Technology, and Pedagogy*, *16*. Retrieved from http://kairos.technorhetoric.net/16.2/topoi/bowie/index.html

Carlson, E. (2009, July/August). What to look for when evaluating web sites. *Orthopedic Nursing, 28*, 199–202. doi:10.1097/NOR.0b013e3181ada7a0

Cheseboro, J., & Bonsall, D. (1989). *Computer-mediated communication*. Tuscaloosa, AL: University of Alabama Press.

Chung, C., Nam, Y., & Stefanone, M. (2012, January 13). Traditional and technological factors. *Journal of Computer-Mediated Communication, 17*, 171–186. doi:10.1111/j.1083-6101.2011.01565.x

Cicero. (1959). *De oratore* (Sutton, E., & Rackham, H., Trans.). Cambridge, MA: Harvard University Press.

December, J. (1996, August). Living in hypertext. *Ejournal, 6*. Retrieved from http://www.ucalgary.ca/ejournal/archive/v6n3/december/december.html

Dihydrogen Monoxide. (n.d.). *DHMO homepage*. Retrieved from http://www.dhmo.org

DuBois, P. (1991). *Torture and truth*. New York, NY: Routledge.

Fleckenstein, K. (2007). Who's writing? Aristotelian ethos and the author position in digital poetics. *Kairos: A Journal of Rhetoric, Technology, and Pedagogy, 11*. Retrieved May 19, 2012, from http://kairos.technorhetoric.net/11.3/binder.html?topoi/fleckenstein/index.html

Foucault, M. (1997). The masked philosopher. In Rabinow, P. (Ed.), *Ethics: Subjectivity and truth* (pp. 321–328). New York, NY: The New Press.

Freeley, A. (1990). *Argumentation and debate: Critical thinking for reasoned decision making* (7th ed.). Belmont, CA: Wadsworth.

Frobish, T. (2004, April). Sexual profiteering and rhetorical assuagement: Examining ethos and identity at Playboy.com. *Journal of Computer-Mediated Communication, 9*(3). Retrieved from http://jcmc.indiana.edu/vol9/issue3/frobish.html

Frobish, T. (2006, March). The virtual Vatican: A case study regarding online ethos. *Journal of Communication and Religion, 29*, 38–69.

Frobish, T., & Thomas, W. (2012). Crafting an online political ethos: Resurrecting direct mail tactics on the Web. *Proceedings of the 2012 Hawaii University International Conference on Arts and Humanities, USA*, [CD].

Grodin, D., & Lindlof, T. (Eds.). (1996). *Constructing the self in a mediated world*. Thousand Oaks, CA: Sage.

Hart, R. (1998). Introduction: Community by negation—An agenda for rhetorical inquiry. In Hogan, J. M. (Ed.), *Rhetoric and community: Studies in unity and fragmentation* (pp. xxv–xxxviii). Columbia, SC: University of South Carolina Press.

Heim, M. (1988). The technological crisis of rhetoric. *Philosophy and Rhetoric, 21*, 57–58.

Hogle, C. (2000). *Seeking superethos on the web: A guide for technical writers*. Retrieved from http://www.cas.ucf.edu/english/publications/enc4932/connie.html

Homer. (1939). *Iliad* (W. Rouse, Trans.). In W. Rouse (Ed.), *Homer*. New York, NY: T. Nelson and sons.

Hunt, K. (1996). Establishing a presence on the World Wide Web: A rhetorical approach. *Technical Communication, 43*, 376–386.

Isocrates. (1980). Speeches and letters. In Norlin, G. (Ed.), *Isocrates* (Norlin, G., Trans.). Cambridge, MA: Harvard University Press.

Johnson-Eilola, J., & Selber, S. (1996). After automation: Hypertext and corporate structures. In Sullivan, P., & Dautermann, J. (Eds.), *Electronic literacies in the workplace: Technologies of writing* (pp. 115–141). Urbana, IL: National Council of Teachers of English.

Jones, S. (1998). *Cybersociety 2.0: Revisiting computer-mediated communication and community*. Thousand Oaks, CA: Sage.

Kapoun, J. (1998, July/August). Teaching undergrads web evaluation: A guide for library instruction. *C&RL News, 59*, 522–523.

Ke, F., & Hoadley, C. (2009, August). Evaluating online learning communities. *Educational Technology Research and Development, 57*, 487–511. doi:10.1007/s11423-009-9120-2

Kennedy, G. (1991). *Aristotle on rhetoric: A theory of civic discourse*. New York, NY: Oxford University Press.

Kenton, S. (1989). Speaker credibility in persuasive business communication: A model which explains gender differences. *Journal of Business Communication, 26*, 143–157. doi:10.1177/002194368902600204

Landow, G. (1992). *Hypertext: The convergence of contemporary critical theory and technology*. Baltimore, MD: John Hopkins University Press.

Levi-Strauss, C. (1960). *The savage mind*. Chicago, IL: University of Chicago Press.

MartinLutherKing.org. (2012, February 14). *Martin Luther King, Jr.: A true historical examination*. Retrieved from http://www.martin-lutherking.org

McNamee, S. (1996). Therapy and identity construction in a postmodern world. In Grodin, D., & Lindlof, T. (Eds.), *Constructing the self in a mediated world* (pp. 141–155). Thousand Oaks, CA: Sage.

Newman, R., & Newman, D. (1969). *Evidence*. Boston, MA: Houghton Mifflin.

Norlin, G. (1980). *Isocrates*. Cambridge, MA: Harvard University Press.

Norton, R. (1983). *Communicator style: Theory, application, and measures*. Beverly Hills, CA: Sage.

Plato. (1925). *Sophist* (H. Fowler, Trans.). *Theaetetus*, Vol. VII. Loeb Classical Library. Cambridge, MA: Harvard University Press.

Quintilian,. (1958). Institutes of oratory. In Butler, H. (Ed.), *The institutio oratoria of Quintilian* (Butler, H., Trans.). Cambridge, MA: Harvard University Press.

Rowe, C. J. (1983). The nature of Homeric morality. In Rubino, C., & Shelmerdine, C. (Eds.), *Approaches to Homer* (pp. 248–275). Austin, TX: University of Texas.

Schumpeter, J. (2010, December 9). The "Internet of things": The Internet of hype. *The Economist*. Retrieved from http://www.economist.com/blogs/schumpeter /2010/12/internet_things

Stumpf, S. (1975). *Socrates to Sartre: A history of philosophy*. New York, NY: McGraw-Hill.

Swayne, M. (2012, April 20). Internet hype may blur fiction-fact line. *Futurity*. Retrieved from http://www.futurity.org/society-culture/internet-hype-may-blur-fiction-fact-line/

Tillman, H. (2003). *Evaluating quality on the Net*. Retrieved from http://www.hopetillman.com/findqual.html

Turkle, S. (1996). Therapy and identity construction in a postmodern world. In Grodin, D., & Lindlof, T. (Eds.), *Constructing the self in a mediated world* (pp. 141–153). Thousand Oaks, CA: Sage.

Warnick, B. (1998). Rhetorical criticism of public discourse on the Internet: Theoretical implications. *Rhetoric Society Quarterly*, *28*, 73–84. doi:10.1080/02773949809391131

Wright, A. (2010). A new literacy for the digital age. *Journal of Special Education Technology*, *25*, 62–68.

Wu, G., Hu, X., & Wu, Y. (2010, October 29). Effects of perceived interactivity, perceived web assurance and disposition to trust on initial online trust. *Journal of Computer-Mediated Communication*, *16*, 1–26. doi:10.1111/j.1083-6101.2010.01528.x

Yamagata, N. (1994). *Homeric morality*. New York, NY: E.J. Brill.

Ziegelmueller, G., Jay, J., & Dause, C. (1990). *Argumentation inquiry and advocacy*. Englewood, NJ: Prentice Hall.

ADDITIONAL READING

Baumlin, J., & Baumlin, T. (Eds.). (1994). *Ethos: New essays in rhetorical theory and critical theory*. Dallas, TX: Southern Methodist University Press. doi:10.2307/358718

Bolter, J. (1991). *Writing space: The computer, hypertext, and the history of writing*. Hillsdale, NJ: L. Erlbaum Associates.

Burgess, J., & Green, J. (Eds.). (2009). *YouTube: Online video and participatory culture*. Malden, MA: Polity Press.

Cheseboro, J., & Bonsall, D. (1989). *Computer-mediated communication*. Tuscaloosa, AL: University of Alabama Press.

Ess, C. (Ed.). (1996). *Philosophical perspectives on computer-mediated communication*. New York, NY: State University of New York Press.

Foley, J. (2012). *Oral tradition and the Internet: Pathways of the mind*. Champaign, IL: University of Illinois Press.

Hills, K., & Hughes, J. (1998). *Cyberpolitics: Citizen Activism in the Age of the Internet*. Lanham, MD: Rowman & Littlefield.

Holeton, R. (1998). *Composing cyberspace: Identity, community, and knowledge in the electronic age*. Boston, MA: McGraw Hill.

Howard, P., & Jones, S. (2004). *Society online: The Internet in context*. Thousand Oaks, CA: Sage.

Jones, S. (Ed.). (1995). *CyberSociety: Computer-mediated communication and community*. Thousand Oaks, CA: Sage.

Jones, S. (Ed.). (1998). *Cybersociety 2.0: Revisiting computer-mediated communication and community*. Thousand Oaks, CA: Sage.

Landow, G. (1992). *Hypertext: The convergence of contemporary critical theory and technology*. Baltimore, MD: John Hopkins University Press.

Landow, G. (Ed.). (1994). *Hyper/text/theory*. Baltimore, MD: John Hopkins University Press.

Lovink, G. (2007). *Zero comments: Blogging and critical internet culture*. New York, NY: Routledge.

Mazzarella, A. (2005). *Girls wide web: Girls, the Internet, and the negotiation of identity*. NY: Peter Lang.

Miller, V. (2011). *Understanding digital culture*. Thousand Oaks, CA: Sage.

Morozov, E. (2012). *The net delusion: The dark side of Internet freedom*. New York, NY: Public Affairs. doi:10.1017/S1537592711004026

Nakamura, L. (2007). *Digitizing race: Visual cultures of the Internet*. Minneapolis, MN: University of Minnesota Press.

Rheingold, H. (1993). *The virtual community: Homesteading on the electronic frontier*. Reading, MA: Addison-Wesley Publishing.

Shyles, L. (2003). *Deciphering cyberspace: Making the most of digital communication technology*. Thousand Oaks, CA: Sage.

Thurlow, C., Lengel, L., & Tomic, A. (2008). *Computer mediated communication: Social interaction and the Internet*. Los Angeles, CA: Sage.

Tredinnick, L. (2008). *Digital information culture: The individual and society in the digital age*. New York, NY: Neal-Schuman.

Vitanza, V. (Ed.). (1996). *Cyberreader*. Needham Heights, MA: Allyn & Bacon.

Wittkower, D. (2010). *Facebook and philosophy: What's on your mind*. Chicago, IL: Open Court.

Zaleski, J. (1997). *The soul of cyberspace: How new technology is changing our spiritual lives*. New York, NY: HarperEdge.

KEY WORDS AND DEFINITIONS

Assessment Model: A visual representation of an evaluative instrument meant to be applied to real-world events or activities.

Community Identification and Goodwill: An ethos-based appeal focused on building trust with a targeted audience by highlighting the similarities and common interests between rhetor and audience, and elaborating on the ways in which audience members may benefit through participation or interaction with the rhetor.

Computer-Mediated Communication: Either synchronous or asynchronous communication among individuals mediated through networked computer technology.

Ethos: Classical rhetorical concept formally defined by Aristotle, and related to an audience's perception of a speaker's character, derived through appeals to practical intelligence, goodwill, and virtue. Modern instances of the term are often connected to the ideas of identification, trust, dynamism, authority, habits of thought, attitude, and behavior, and a rhetor's projection of self.

Intelligence and Knowledge: An ethos-based appeal focused on building trust with a targeted audience by highlighting the rhetor's subject-based expertise, qualifications, and specialized knowledge.

Moral Character and Virtue: An ethos-based appeal focused on building trust with a targeted audience by highlighting the virtues that rhetor and audience have in common or those that appeal to an audience's sensibilities.

Verbal and Design Competence: An ethos-based appeal focused on building trust with a targeted audience by highlighting the verbal and nonverbal skills of the rhetor for the purpose of establishing a professional and competent identity.

Chapter 2
Online Trust:
A Moving Target

Natasha Dwyer
Victoria University, Australia

ABSTRACT

The design of trust in digital environments shapes how users relate. By reducing complexity, trust expedites transactions, and thus, some developers of online spaces seek to resolve their users' trust issues quickly. Some designers approach the problem of trust as a display of evidence. This chapter questions this approach, because it is easy to present a façade of trust on the Internet, and users are increasingly aware of this fact. Alternative design directions are reviewed, such as the enablement of trust, which may feature prominently in the future of trust design.

INTRODUCTION

A problem that online developers need to solve is how users explore trust in interactive spaces. Before online spaces, those in an interaction had methods to determine trust. For instance, Schneidermann (2000) described how the modern day gesture of clinking glasses derived from an ancient tradition whereby traders would swap each other's glasses to disprove an attempt to poison. Some on-line designers approach the problem of trust as a display of evidence and signs. This chapter problematizes the strategy of designing

signs to trust and concludes that signs of trust are a moving target. It argues that once a certain design feature is understood by users to be a sign of trust, untrustworthy parties adopt it. The feature then loses its value and users are no longer convinced.

New digital environments allow users to enter into a diverse range of transactions with people from all over the globe, sometimes in a one-off transaction with a fast result. However, the information that helps an individual create a context that includes rules and customs is limited. There may be little or no build-up to the situation or opportunity to gather more information (Cheshire & Cook, 2004). There may be a limited number of exchanges for parties to get to know each other and

DOI: 10.4018/978-1-4666-2663-8.ch002

understand each other's perspectives (Nooteboom, 2005). Within the area of information technology and trust there are also a range of perspectives. There is, for instance, the element of computer-to-computer trust, where the trust issue is whether technology used for a particular communication is secure and reliable (see for instance, Lenzini et al., 2008; Anderson et al., 2008). Another aspect is why and how users trust the technology they use, which was explored by researchers such as Clarke, Hardstone, Hartswood, Procter and Rouncefield (2006) and Lacohée, Crane, and Phippen (2006).

This chapter explores from an interactive design perspective—a discipline that investigates how users experience digital environments. Other fields have examined how people communicate via an exchange of indicators. The field of semiotics revolves around the issue. Post-structuralist theorists have examined how people anticipate, hold expectations, and base decisions upon what they think others might do. Bourdieu (2005), for instance, proposed the idea of the field, in which participants hold a position that they represent to others. Interactive design is a bowerbird discipline; knowledge from a variety of fields is drawn together to justify and explore a range of situations. A focus on the perspectives contributing to interactive design research is beyond the scope of this chapter.

BACKGROUND: WHAT IS TRUST AND WHY IS IT AN IMPORTANT PROBLEM TO SOLVE?

Although it can and has been argued that "trust" is central to the functioning of society (Watson 2009), it is difficult to apply a static and complete definition of the term. Trust is a concept that is defined and understood differently. Those defining trust emphasized different elements of the complex concept. Cofta (2007, p. 14) has argued that trust can be loosely and informally described as a relationship within which a trustor is confident

that another party (the trustee), to whom a trustor is in a position of vulnerability, will respond in the trustor's interests.

Luhman (1979) raised the difficulty of defining trust. He pointed to society as a place where trust interactions are grounded. Building on this notion, Clarke et al., (2006) described attempts to deal with trust in the abstract as a "pitfall" and instead call for investigations into trust "in action" that are embedded in a social world. Exploring trust in an interaction context calls for a definition that focuses on trust as a relationship rather than a "mental disposition."

There is an element of trust that is always unaccountable and mystical (Möllering 2006), otherwise the concept under consideration is not trust and could more aptly be defined in terms such as "calculation." Relying solely on "rationalism" will always "explain trust away or explain everything but trust" (James, 2002). Trust is at the very least a combination of "rational" thinking and what is known as "feeling" or intuition (Möllering, 2001). Although it may be argued that it is impossible to separate these different ways of thinking, it is clear that trust draws on a collection of sensibilities. Möllering (2006, p. 371) argued that an unusual function of trust is that a trustor can often justify and account for a decision, and that this explanation is to "uphold" self-respect. This adds a level of difficulty when undertaking first-hand research into how trust is understood and negotiated.

Some researchers, including Fukiyama (1995), tend to assume that trust is positive. As mentioned above, Cofta (2007, p. 14) side-stepped a judgment on what is good for the user by arguing that trust can be loosely and informally described as a relationship within which a trustor is confident that another party (the trustee), to whom a trustor is in a position of vulnerability, will respond in the trustor's interests.

Trust research literature has often raised the link between understanding and trust, which equates to familiarity as a key ingredient of trust.

Nooteboom (2006) maintained that trustors and trustees exchange information in order to establish insights into each other and the grounds for familiarity. Luhmann (1979) argued that the familiarity process in itself begets trust. Interactors communicate behavioral cues with each other, either consciously or not, regarding their intentions and how they perceive others' intentions (Six & Nooteboom, 2005). Thus a communication cycle is generated in which interactors respond to each other. In this view, people relate to each via a range of "modes," for instance, self-interest, or solidarity. Because trust can need maintenance and may dissipate if not tended to, a regular exchange of cues between those involved in the interaction can support the trust relationship (de Laat, 2005). Additionally, as trust is linked to an acceptance of vulnerability and is a reciprocal relationship, a participant can communicate both a willingness to trust and trustworthiness by exchanging communications that indicate vulnerability (Six & Nooteboom, 2005).

Trust is closely associated with a series of other notions that are difficult to conceptualize such as credibility, authenticity and risk (Marsh & Dibben, 2003). Credibility can be defined as present when a trustor believes information is reliable and correct, perhaps because the information resource is considered reputable or the claim is backed by evidence. Trust and credibility are differentiated by McKnight et al., (2002) who argue that trust is within the initial interaction between trustors but credibility occurs when some level of familiarity exists from the trustor's previous knowledge. The link between trust and authenticity is central to several researchers, including Bacharach and Gambetta (2000), and Riegelsberger, Sasse, and McCarthy (2005). Trustors automatically sort and synthesize evidence, deciphering whether it is genuine or not. Authenticity is an embedded dimension of trust, associated with the credibility of information, which is also affected by judgments about the expertise, believability, and depth of a message. It is difficult to pinpoint what is at risk

or lost when one is duped, as the event moves beyond rational explanations (Möllering, 2008). Trustors often understand trust in terms of risk, according to Lacohée et al., (2006).

Some of the risks or losses at stake for an individual engaging in a trust interaction in an online environment can be identified, while others cannot be foreseen (Lacohée et al., 2006). In the area of health information digital exchange, for example, a user might be given incorrect information that may have adverse effects. Within the context of on-line dating, somebody may pretend to have certain qualities (for instance, being single rather than married) that may result in disappointment for another. Online dating might result in a meeting that could result in physical danger. Other risks are more difficult to identify. These risks include violations of privacy (control over personal information) and security (safety to self and devices). For instance, a user with a mobile device enters a shopping center, which is wirelessly connected to the Internet. Unknown to the user, her contact details are taken from the device and listed in a database that can be accessed by a range of people. Past purchases made by the user are automatically calculated and the user is given advertisements tailored specifically to her desires. She makes a new purchase of an item flagged as an object that can be utilized by terrorists, thus an alert is activated that her behavior should be documented. Her photo and purchase time are uploaded to a database. Other risks include "information injustice," when information provided by an individual is taken out of context and used to draw interpretations that may be unfair (van den Hoven, 1999). For instance, a comment reflecting religious beliefs on a social networking site could be used in a job selection process. There may be an imbalance in the amount and access different parties in an interaction hold. Also, information provided by individuals may be used against their interests. Data on individual preferences is used by advertisers to target campaigns.

Trust and distrust are core issues for the designers of digital environments because they affect how users form and maintain relationships. Trust is an attractive problem to solve for developers (particularly those working in the area of e-commerce) because one of the functions of trust (and distrust) in society is the lessening of uncertainty, which in turn expedites transactions and greases the wheels of e-commerce (Fukuyama, 1995). Without the reduction of complexity, an individual might be "paralysed," unable to proceed or decide, weighed down by exploring options and possible outcomes (Cofta, 2007, p. 11). Trust results in the foreclosure of some future possibilities (Goffman, 1974, p. 499). This is why trust is viewed as a type of confidence (regardless of whether the confidence is well placed or not). Although it is often assumed that trust is a positive quality for a user, distrust can work in a similar fashion to trust, closing down options for the user to consider (Cofta, 2006).

ISSUES, CONTROVERSIES, AND PROBLEMS: NEGOTIATING TRUST IN A DIGITAL ENVIRONMENT

In order to understand and manage trust in online-environments, some developers and researchers focus on signs of trust and their exchange (Donath, 2006). The underlying premise is that users engage with each other, either consciously or unconsciously, in a constant process of signal display and interpretation. For instance, a man on a dating website may make mention of an expensive car to signal that he is moneyed.

Cofta (2006) provided a distillation of the work undertaken in the trust and digital research area and argued that trust evidence that users value fall into three dimensions: continuity, competence, and motivation. Continuity evidence is connected with time and is a reflection of how long someone has been a member of communities relevant to the trust interaction. The dimension of continuity

is shared interest beyond the current encounter: What is the shape of the "shadow of the future" (Axelrod, 1984)? A judgment using continuity uses the length of time the trustee has existed in a certain community (Cofta, 2006). Social embeddedness allows for the exchange of information about a trustee's performance. Trustees, knowing that trustors could exchange information about their behaviour, have an incentive to fulfil an arrangement, even if they do not expect future interaction with a given trustor (Riegelsberger et al., 2005). Second-hand trust, reliance on the reputation and advice information from other parties, plays an important role within this category of trust information (Lacohée et al. 2006, p. 19). This advice is informed by the previous experience (both positive and negative) of the contact.

Competence evidence indicates whether someone has the skills to deliver on a trust interaction. The dimension of competence refers to whether the trustee has the ability and skill to fulfill the requirements of the interaction (Cofta, 2007, p. 111). Competence includes the ability to competently negotiate a relationship. This involves the trustee as providing indications of understanding the trustor's perspective. The competence dimension emphasises that trust is a two-way relationship: Each participant provides an offering that is built on by the other. This process is fragile. The French theorist Bourdieu (1986) described competence as "cultural capital" and explored how this form of capital is heavily mediated by culture. He pointed out that the amount of cultural capital a person may hold is a function of what class they are from because it is an acquisition that can be passed through family and associates (which links this concept with the notion of community and social capital). For instance, access to education can determine the type of qualifications and knowledge one holds. However, this example also demonstrates how complicated and problematic the link is between competence and social capital. On one hand it may determine an individual's ability to access resources, but in actuality, many more factors shape an individual's place in society. Across different contexts, individuals may not agree on what constitutes an ideal social capital profile. In a particular community, an individual with a high level of education may be regarded as having wasted his time and not to hold the relevant life experience that others have.

The dimension of motivation has to do with shared interest (Cofta, 2007, p. 111): Does the trustee have an interest in working towards the welfare of the trustor? When the trustee's intentions are assessed as positive, this results in the "belief of unharmfulness." Motivation between two parties is explored by the sharing of a connection based on what they have in common and what is important to them. The process allows the trustor to ask if the trustee shares a world-view, feels a bond or connection, and thus is motivated to work in their interests. The familiarity process can beget trust. According to Luhmann (1979), when people start to cooperate, they get the chance to adopt each other's perspectives. Arguably, according to research undertaken by Dwyer (2011), continuity is the most important dimension, and the different dimensions interrelate and affect each other.

Although it is impossible to draw a stark distinction between different elements of a communication exchange, some researchers focus on what the delivery of a message reveals about its sender (how a message has gotten to the receiver) while other researchers focus on the content (the story held in the exchange). Pentland (2008) is in the group of researchers who are concerned with how messages are delivered, and he argued that the elements of a message that are revealing include how the message was sent (for instance, what technology was used and how quickly a message was delivered), how much influence a message is having over others (for instance, whether a message is widely distributed and whether a message is being referenced by others). Gambetta and Hamill (2005) focused more on the content of the communication and the intention that can

be read into a communication. In particular, they looked at how people judge the authenticity of messages. Trust evidence is, then, a fragment pointing to a larger story with which the trustor builds a picture of the trustee.

According to the human computer interaction researchers (Riegelsberger et al., 2005, p.75), the micro-evidence of trust divides into two categories, "symbols" (direct) and "symptoms" (indirect). Symbols are deliberately intended to communicate trust and are thus "learned signifiers." Examples are warranties, seals, and ratings. On the other hand, there are indirect indications of trust. "Symptoms" are by-products of trustworthiness. For instance, Gambetta and Hamill (2005, p. 13) gave the example of how a black man in New York City has worked out a way to successfully flag cabs in his city. He flags the cab with a copy of the *Wall Street Journal*, a newspaper with an educated and wealthy readership. By using a newspaper in such an integrated fashion, the cab driver understands the man to be educated and moneyed, and in a cab driver's view, less likely to be trouble. Gambetta and Hamill (2005) discussed how different taxi drivers, trustees, try to de-emphasize what might be at risk for them in an interaction or what might be their points of vulnerability. For instance, taxi drivers in areas divided by religious strife remove the symbols of their religious allegiance. Other taxi drivers say that they mention to new rides that they have just dropped off their takings for the night, to give the impression of not being worth robbing.

The Aesthetics of Trust

Working with the notion that there are some visual elements and other design features that indicate trust, some writers and designers attempt to develop guidelines that help impart the notion of a site as worthy of trust. As claimed by Luhmann (1979, p. 26), "trust always extrapolates from the available evidence." The aim of the design approach is to provide the user with enough cues to feel confident enough to trust. The "guideline" approach doesn't focus on whether the site is actually worthy of trust but instead aims to present a site as trustworthy. The intention is that a user will come to a website, read the signs and symptoms of trust favorably, and then feel comfortable enough to trust. Shankar, Urban, and Sultan (2002) claimed that users look beyond the website presence to establish trust and distrust. Users have a realm of offline cues they look for and they seek evidence for the existence of these cues in the content and presentation of the site. A potential e-commerce customer may believe that larger firms are more worthy of trust than smaller companies. The user may look for evidence of the size of the company to establish trust. This may be in the form of multiple addresses on the contact page. Some brands may be more trustworthy than others, and this sense of trust will transfer to their website presence.

As Dwyer (2011) pointed out, these guidelines are often in the abstract, thus containing very little detail. A popular recommendation is to develop a "professionally designed" site, one that a designer holding traditional graphic skills has created in order to provide an aura of authority (e.g., Swaak, de Jong, & de Vries, 2009; Djamasbi, Siegel, Tullis, & Dai 2010). Egger (2003) is an influential writer in the field who provided principles for designers to employ in order to provide the façade of trust. He suggested that the designer should provide "attractive" designs that are easy, consistent, and predictable to use. He argued that text should be checked for grammar and spelling mistakes and advertising should be separated from content wherever possible. Lynch (2010) contributed that users form an impression of trust via a quick overall assessment of a site and first impressions weigh heavily in the trustor's mind (Lindgaard, Dudek, Devjani, Sumegi, & Noonan, 2011). Color plays a role in the formation of a professional site, and Alberts and van der Geest (2011) recommended the use of blue to promote

trust and to avoid black. However, what is considered a "professional appearing" site changes across contexts, and the recommendation needs to be adapted across situations. For example, a site that looks too corporate deters users seeking online health advice (Sillence, Briggs, Fishwick, & Harris, 2004). The users in this study understood the "professional" style to be associated with sites sponsored by a vested interest that may provide biased information.

Inclusion of photographs of company people on a website, including their names, can also build a trustworthy picture. Users require human warmth in order to trust, and Hassanein and Head (2004) have explored how to engender a "human touch" into a site. They suggested the use of photos that represent people in "everyday situations" can improve trust relations. However, Riegelsberger, Sasse, and McCarthy (2003) pointed out that the use of photographs of people on a website to generate trust is fraught with uncertainty—within sites that do use photos of people as design elements, users find it hard to distinguish between website owners who are worthy of trust and those who are not.

Researchers agree that it is not just the presentation of the content but the information that is available at the site that affects trust. Where relevant, designers of online environments should include peer reviews and statements describing guarantee arrangements. Communication channels between users and the site owners remain open. Schneiderman (2000), another influential author, provided guidelines on how to present content to give the appearance of trust. He suggested certifications from third parties that can verify the legitimacy of the site. Disclosure of past performances of the site owner can help engender trust between user and site owner (Schneiderman, 2000). The provision of privacy statements can also instill confidence in users (Lin & Wu, 2008). However, all these recommendations to engender trust can be adopted by parties who are not trustworthy but who wish to offer the appearance of trustworthiness. Research such as Schneidermann's does not address whether the owner of the website, for which the guidelines will be applied, is trustworthy or not.

SOLUTIONS AND RECOMMENDATIONS

Trust as a Moving Target

Research outlining how trust can be designed into a site, such as the material described in the above section, demonstrates how trustworthiness can be faked. In this section, I argue that rather than regarding trust as a static entity, we should view trust as a moving target. Users are savvy and realize that trust can be faked and are continually seeking new means to ascertain whether a trustor deserves trust.

Via news reports and word of mouth, users hear of cases when a user's trust was misplaced because of the user's understanding of trust. Users are aware of risks that can happen via the Internet and understand that there are external parties on the Internet who aim to be trusted in order to deceive (Ivanitskaya et al., 2010). As users become aware of cases of deceit, the trust evidence used to deceive lose their legitimacy for all who demonstrate that sign of trust. An example is the case of websites promoting accommodations for the 2012 Olympics. Following the advice that trustworthy sites demonstrate official symbols to indicate trustworthiness, some illegitimate sites used the logo of the Olympic rings to appear as a trustworthy official site to lure in customers. Hopkins (2011) advised users to be careful when dealing with sites demonstrating this symbol. In this instance, the value of the Olympic logo is devalued, and the display of other official logos is put in question.

Verification by third parties has also been challenged recently, as demonstrated by a case of adopted identity on Twitter. A bored man on holiday set up an account under the name of Wendi Deng, wife to the media tycoon Rupert Murdoch. Twitter incorrectly verified this account. There was a ripple effect from this acceptance. The account holder proceeded to send out Tweets that gave followers the impression that they should trust that it was really Wendy Deng sending out communiqués. Adhering to the type of behavior expected on Twitter, "Wendi" sent friendly and flirtatious messages to acquaintances of her husband. The press and even Wendi's husband were initially deceived. As a result, it is possible that in future, users will question whether the identity behind any verified Twitter account is genuine (O'Carroll and Halliday, 2012).

As a result of awareness about Internet fraud and risk, users look further to find symbols of trust. For instance, users of medical sites look within the "fine print" of a site to detect whether the research has been sponsored and thus perceived as clearly biased. Within the information that is difficult to access on a site, hidden under several layers of interface, disclaimer information can be found that explains who has sponsored the research. Users seek out this information in order to discover the bias of the content provider and from then make assessments regarding the trustworthiness of the information (Sillence et al., 2004). Some users may only find information presented in a corporate style trustworthy (Swaak et al., 2009). Other users may distrust corporate communication at first glance because they may think this style reflects an adherence to business values at the expense of relationships. Different nationalities have different design styles and values. Something deemed as "accessible" from one culture may appear incomprehensible and alienating to someone from another culture (Bourges-Waldegg & Scrivener, 1988). As Cofta (2011) pointed out, researchers cannot agree on a definition of trust. How can disparate groups of users agree?

Once trust is understood as a moving target, a solution to negotiating trust in digital environments becomes clear. Digital environments that are dynamic, rather than static displays of trust signs, and that can switch in response to the needs of the user and work in the user's interest need to be designed.

What Can a Designer do?

Assuming it is correct that it is futile to contrive trust evidence in a digital environment, what can a designer do? Design can problem set: It can offer approaches that problematize a design situation, as well as alternatives for how to navigate through a design problem. So the aim of design, I argue, is not to close down a decision-making space and to de-limit options but rather to open up possibilities. Thus design is about problem setting rather than solving (Schön, 1983). Rather than providing recommendations, design can uncover motivations for action, needs and deeper human rationale, and explore the implications of these drivers. The approach of problem setting suits the influx and configurable nature of contemporary digital environments. This perspective acknowledges that the design of new spaces is never complete, is shaped by those who use them, and is always contested.

Rather than presenting a façade of trust, what approaches can designers adopt to allow users to negotiate trust? Some developers work with the notion of reputation and harness what others have found about a particular user and make that information widely available. Drawing on past behavior by a user, a profile is generated automatically by the software that can help other users predict how the user in question might behave in the future. The publishing of reputation has been cited as a reason behind the success of e-commerce and auction sites such as eBay (Cofta, 2007, p. 226). Buyers on eBay can access a value rating that gives a sense of the honesty and reliability of different sellers. Ahn and Esarey (2008) explained that systems that provide some sort of information about what is expected as trustworthy behavior, for example,

Wikipedia, actually foster trustworthy behavior because users can learn from the system. However, there are several limitations to the strategy of reputation profiling. Firstly, within an ever-expanding Internet, a user is likely to encounter an interaction that has not been recorded previously, and a trustor needs to make a trust decision without relying on any previous opinion. Secondly, a user needs to rely on the judgment and criteria of others. Marsh (1994) argued that if reputation is used as a means to explore trust, then a balance of other forces should be brought into play to allow depth to the judgment. In particular, he believes that notions of regret and forgiveness should be allowed consideration in a trust interpretation.

An alternative approach draws on the history of activity left by a participant to draw inferences about how that participant might construct trust (or distrust). The data is not interpreted by the system on the user's behalf. Tummolini and Castelfranchi (2007) examined how a history of past interactions and encounters of a certain user can be read as traces by those interacting with that user. In this model, there is a minimum amount of context and little opportunity for individuals to query or explain evidence. Social networks are an example of interfaces that allow users to review the history of other users' interactions, and judgments can be made that have serious repercussions, including, for example, whether or not to employ someone.

The inclusion of "catalysts" in a design, events in an interface within which the users have the opportunity to exchange evidence spontaneously, is another approach posited by Karahalios (2004). The catalyst approach has the potential to garner evidence that is more authentic because there is little opportunity for the user to contrive evidence through thought and planning. It also allows participants to explain or provide reasoning around the information they communicate. However, catalysts require a level of resources and commitments from those involved and are thus expensive to undertake.

Online Trust

FUTURE RESEARCH DIRECTIONS

Allowing a user to negotiate a trust interaction on his/her own terms is a recent notion posited in the literature by Cofta (2007) and may represent the future of trust design in digital environments. Trust-enablement is when a user is assisted to make a decision on his/her own terms. This could involve enabling users to locate their most preferred trust evidence or allowing users to query in order to clarify a trust interaction. Digital technology is now at a point where the creation of trust-enabling environments is possible. Current digital environments allow the user to configure and re-purpose (to varying extents) technology to meet their own needs and preferences. The open source and free software communities, groups who work collaboratively to create free and accessible source codes, are a model of how users with technical knowledge can together shape digital environments (Schweik 2006). The configuration movement affects more than software. The realm of computing hardware was once the province of highly specialized experts. Now that hardware creation is becoming more accessible, developers are making products that allow others to create their own systems (Kobayashi 2009). The configuration trend is happening across user groups with various levels of expertise in technology. In the near future, it is possible that "average" users will be able to configure their digital environments in fashions not imaginable now. The design of trust within digital environments will be in the control of users.

An interface that enables trust, rather than presenting the façade of trust, allows users to negotiate trust on their own terms and to configure an environment. The elements in an interface relevant to trust that a user may wish to configure include the following:

- *The allowance for monitoring and intervention. Instead of the interface simplify-*

ing issues, the user should be able to monitor and intervene on the workings of the interface.
- *The allowance of diversity. Rather than assuming what the user wants in regards to trust, an interface should ask.*
- *The allowance of querying. A system should be able to provide more information on request, so the user can clarify and explore what a piece of trust evidence means.*
- *The allowance of incompleteness. Rather than proposing a final solution, the interface should enable trust by keeping options open as long as possible, letting users re-examine opinions and "back track".*

(Basu, Marsh & Dwyer, 2012)

More research into the processes users adopt to engage trust and also distrust is needed to inform the design of interfaces that can handle the intricate and idiosyncratic nature of trust.

CONCLUSION

This chapter has reviewed how designers and developers of digital environments are responding to the problem of trust. Trust, a notion interconnected with other concepts such as risk and credibility, is an important problem for developers to solve because trust can reduce complexity and expedite transactions. Some approach the problem by providing and applying guidelines that can provide the façade of trust. For instance, a professionally appearing site can engender the confidence of trust. However, untrustworthy as well as trustworthy parties can adopt the guidelines and signs of trust. When working in digital environments, it is cheap to adopt the signs of trust. Users are savvy about the problems of trust when interacting via digitally mediated environments. What is considered a sign of trust inevitably shifts. What approaches are appropriate for a complicated concept such as trust? Some developers draw on the digitizing of reputation. Others try to provide traces of previous interactions as a means for a user to establish trust. Trust-enablement, when users are enabled to make a trust decision on their own terms, could represent the future of trust design in digital environments. Technologies are being developed that allow users to configure their environments to the extent that a nuanced treatment of trust is possible.

Thus, any attempt to reverse engineer the grounds to trust into a design are limited. Approaching the design of trust as a dynamic rather than static interaction can allow users to negotiate trust on their own terms. Designers can create digital environments that allow users to configure elements to meet their needs. Some features relevant to trust that users may wish to configure include how the interface is monitored, the diversity of preferences available, and how questions are managed. Rather than seeking to resolve a trust issue quickly, an interface could keep options open as long as possible, with guidance from the user.

REFERENCES

Ahn, T., & Esarey, J. (2008). A dynamic model of generalized social trust. *Journal of Theoretical Politics*, *20*(2), 151–180. doi:10.1177/0951629807085816

Alberts, W., & van der Geest, T. (2011). Color matters: Color as trustworthiness cue in web sites. *Technical Communication*, *58*(2), 149–160.

Anderson, R., Chan, H., & Perrig, A. (2004). *Key infection: Smart trust for smart dust*. Paper presented at the 12th IEEE International Conference on Network Protocols, Santa Barbara, CA.

Axelrod, R. (1984). *The evolution of cooperation*. New York, NY: Basic Books.

Bacharach, M., & Gambetta, D. (2000). Trust in signs . In Cook, K. (Ed.), *Trust in society* (pp. 148–184). New York, NY: Russell Sage.

Basu, A., Marsh, S., & Dwyer, N. (2012). *Rendering unto Caesar the things that are Caesar's: Complex trust models and human understanding*. Paper presented at the 6th IFIP WG 11.11 International Conference on Trust Management, Surat, India.

Bourdieu, P. (1986). The forms of capital . In Richardson, J. (Ed.), *Handbook of theory and research for the sociology of education* (pp. 241–258). New York, NY: Greenwood.

Bourdieu, P. (2005). The political field, the social science field, and the journalistic field . In Benson, R., & Neveu, E. (Eds.), *Bourdieu and the journalistic field* (pp. 29–47). Cambridge, UK: Polity Press.

Bourges-Waldegg, P., & Scrivener, S. (1998). Meaning, the central issue in cross-cultural HCI design. *Interacting with Computers*, *9*(3), 287–309. doi:10.1016/S0953-5438(97)00032-5

Clarke, K., Hardstone, G., Hartswood, M., Procter, R., & Rouncefield, M. (2006). Trust and organisational work . In Clarke, K., Hardstone, G., Rouncefield, M., & Sommerville, I. (Eds.), *Trust in technology: A socio-technical perspective* (pp. 1–20). Dordrecht, The Netherlands: Springer. doi:10.1007/1-4020-4258-2_1

Cofta, P. (2006). *Distrust*. Paper presented at Eighth International Conference on Electronic Commerce, Fredericton, Canada.

Cofta, P. (2007). *Trust, complexity and control: Confidence in a convergent world*. Hoboken, NJ: John Wiley and Sons.

Cofta, P. (2011). The trustworthy and trusted Web. *Foundations and Trends in Web Science*, *2*(4), 243–381. doi:10.1561/1800000016

De Laat, P. (2005). Trusting virtual trust. *Ethics and Information Technology*, *7*(3), 167–180. doi:10.1007/s10676-006-0002-6

Djamasbi, S., Siegel, M., Tullis, T., & Dai, R. (2010). *Efficiency, trust, and visual appeal: Usability testing through eye tracking.* Paper presented at the 43rd Hawaii International Conference on System Sciences (HICSS), Kauai, Hawaii.

Donath, J. (2006). *Signals, truth and design* [Video Recording]. Retrieved from http://www.ischool.berkeley.edu/about/events/dls09272006

Dwyer, N. (2011). *Traces of digital trust: An interactive design perspective.* Unpublished doctoral dissertation, Victoria University, Australia.

Egger, F. (2003). *From interactions to transactions: designing the trust experience for business-to-consumer electronic commerce.* Unpublished doctoral dissertation, Eindhoven University of Technology, The Netherlands.

Fukuyama, F. (1995). *Trust: social virtues and the creation of prosperity.* New York, NY: Free Press.

Gambetta, D., & Hamill, H. (2005). *Streetwise: How taxi drivers establish their customers' trustworthiness.* New York, NY: Russell Sage Foundation.

Goffman, E. (1974). *Frame analysis.* Cambridge, U.K: Harvard University Press.

Hassanein, K., & Head, M. (2004). *Building online trust through socially rich Web interfaces.* Paper presented at the 2nd Annual Conference on Privacy, Security and Trust, Fredericton, Canada.

Hopkins, N. (2011, December 26). Websites targeting Olympics visitors closed down by police. *The Guardian.* Retrieved from http://www.guardian.co.uk

Ivanitskaya, L., Brookins-Fisher, J., O'Boyle, I., Vibbert, D., Erofeev, D., & Fulton, L. (2010). Dirt cheap and without prescription: How susceptible are young US consumers to purchasing drugs from rogue internet pharmacies? *Journal of Medical Internet Research, 12*(2), e(11).

Karahalios, K. (2004). *Social catalysts: Enhancing communication in mediated spaces.* Unpublished doctoral dissertation, Massachusetts Institute of Technology, U.S.

Kobayashi, S. (2009). *DIY hardware: Reinventing hardware for the digital do-it-yourself revolution.* Paper presented at the ACM SIGGRAPH ASIA 2009 Art Gallery and Emerging Technologies: Adaptation, Yokohama, Japan.

Lacohée, H., Crane, S., & Phippen, A. (2006). *Trustguide.* Hewlett-Packard Laboratories. Retrieved March 2, 2012, from http://www.trustguide.org.uk/Trustguide%20-%20Final%20Report.pdf

Lenzini, G., Martinelli, F., Matteucci, I., & Gnesi, S. (2008). *A uniform approach to security and fault-tolerance specification and analysis.* Paper presented at the Workshop on Software Architectures for Dependable Systems, Anchorage, U.S.

Lin, Y., & Wu, H. (2008). Information privacy concerns, government involvement and corporate policies in the customer relationship management context. *Journal of Global Business and Technology, 4*(1), 79–91.

Lindgaard, G., Dudek, C., Devjani, S., Sumegi, L., & Noonan, P. (2011). An exploration of relations between visual appeal, trustworthiness and perceived usability of homepages. *ACM Transactions on Computer-Human Interaction, 18*(1), 1–30. doi:10.1145/1959022.1959023

Luhmann, N. (1979). *Trust and power.* Chichester, UK: Wiley.

Lynch, P. (2010). Aesthetics and trust: Visual decisions about Web pages. *Proceedings of the International Conference on Advanced Visual Interfaces*, New York, NY.

Marsh, S. (1994). *Formalising trust as a computational concept.* Unpublished doctoral dissertation, University of Stirling, Scotland.

Marsh, S., & Dibben, M. (2003). The role of trust in information science and technology. *Annual Review of Information Science & Technology, 37*(1), 465–498. doi:10.1002/aris.1440370111

McKnight, D., Choudhury, V., & Kacmar, C. (2002). Developing and validating trust measures for e-commerce: An integrative typology. *Information Systems Research, 13*(3), 334–359. doi:10.1287/isre.13.3.334.81

Möllering, G. (2001). The nature of trust: From Georg Simmel to a theory of expectation, interpretation and suspension. *Sociology, 35*(2), 403–420.

Möllering, G. (2006). Trust, institutions, agency: Towards a neoinstitutional theory of trust . In Bachmann, R., & Zaheer, A. (Eds.), *Handbook of trust research* (pp. 355–376). Cheltenham, UK: Edward Elgar.

Möllering, G. (2008). *Inviting or avoiding deception through trust? Conceptual exploration of an ambivalent relationship* (MPIfG Working Paper 08/1). Max Planck Institute for the Study of Societies. Retrieved from http://www.mpifg.de/pu/workpap/wp08-1.pdf

Nooteboom, B. (2005). *Framing, attribution and scripts in the development of trust*. Discussion Paper, University of Toronto. Retrieved March 20, 2012, from http://www.bartnooteboom.nl/

Nooteboom, B. (2006). Forms, sources, and processes of trust . In Bachmann, R., & Zaheer, A. (Eds.), *Handbook of trust research* (pp. 247–263). Northhampton, UK: Edward Elgar Publishing.

O'Carroll, L., & Halliday, J. (2012, January 03). Wendi Deng Twitter account is a fake. *The Guardian*. Retrieved from http://www.guardian.co.uk

Pentland, A. (2008). *Honest signals: How they shape our world*. Cambridge, UK: MIT Press.

Riegelsberger, J., Sasse, A., & McCarthy, J. (2003). *Shiny happy people building trust? Photos on e-commerce websites and consumer trust*. Paper presented at CHI '03: ACM Conference on Human Factors in Computing, Fort Lauderdale, USA.

Riegelsberger, J., Sasse, M., & McCarthy, J. (2005). The mechanics of trust: A framework for research and design. *International Journal of Human-Computer Studies, 62*(3), 381–422. doi:10.1016/j.ijhcs.2005.01.001

Schneidermann, B. (2000). Designing trust into online experiences. *Communications of the ACM, 43*(12), 57–59. doi:10.1145/355112.355124

Schön, D. (1983). *The reflective practitioner*. New York, NY: Basic Books.

Schweik, C. (2006). Free/open-source software as a framework for establishing commons in science . In Hess, C., & Ostrom, E. (Eds.), *Understanding knowledge as a commons from theory to practice* (pp. 277–309). Cambridge, MA: MIT Press.

Shankar, V., Urban, G., & Sultan, F. (2002). Online trust: a stakeholder perspective, concepts, implications, and future directions. *The Journal of Strategic Information Systems, 11*(3), 325–344. doi:10.1016/S0963-8687(02)00022-7

Sillence, E., Briggs, P., Fishwick, L., & Harris, P. (2004). *Trust and mistrust of online health sites*. Paper presented at the SIGCHI Conference on Human Factors in Computing Systems, New York, NY.

Six, F., & Nooteboom, B. (2005). *Trust building actions: A relational signalling approach*. Retrieved March 20, 2012, from http://www.bartnooteboom.nl/FrederiqueOS.PDF

Swaak, M., de Jong, M., & de Vries, P. (2009). *Effects of information usefulness, visual attractiveness, and usability on web visitors" trust and behavioral intentions*, Paper presented at the IEEE International Professional Communication Conference, Waikiki, Hawaii.

Tummolini, L., & Castelfranchi, C. (2007). Trace signals: The meanings of stigmergy. In D. Weyns, H. Parunak, & F. Michel (Eds.), *E4MAS'06 Proceedings of the 3rd International Conference on Environments for Multi-Agent Systems* (pp. 141-156). Berlin, Germany: Springer.

Van den Hoven, M. (1999). Privacy or informational injustice? In Pourciau, L. (Ed.), *Ethics and electronic information in the 21st century* (pp. 140–150). West Lafayette, IN: Purdue University Press.

Watson, R. (2009). Constitutive practices and Garfinkel's notion of trust: Revisited. *Journal of Classical Sociology*, 9(4), 475–499. doi:10.1177/1468795X09344453

ADDITIONAL READING

Ashraf, N., Bohnet, I., & Piankov, N. (2006). Decomposing trust and trustworthiness. *Experimental Economics*, 9(3), 193–208. doi:10.1007/s10683-006-9122-4

Barber, B. (1983). *The logic and limits of trust*. New Brunswick, NJ: Rutgers University Press.

Binder, T. De Michelis, G. Ehn, P., Jacucci, G., Linde, P., & Wagner, I. (2012). *Design things*. Cambridge, MA: MIT Press.

Castelfranchi, C., & Falcone, R. (2010). *Trust theory: A socio-cognitive and computational model*. Hoboken, NJ: John Wiley and Son.

Fry, T. (2011). *Design as politics*. New York, NY: Berg.

Harper, R. (2010). *Texture: Human expression in the age of communications overload*. Cambridge, MA: MIT Press.

Luhmann, N. (1979). *Trust and power*. Chichester, UK: Wiley.

Lyon, F., Möllering, G., & Saunders, M. (Eds.). (2011). *Handbook of research methods on trust*. Cheltenham, UK: Edward Elgar.

Marsh, S., & Dibben, M. (2003). The role of trust in information science and technology. *Annual Review of Information Science & Technology*, 37(1), 465–498. doi:10.1002/aris.1440370111

Murray, J. (2012). *Inventing the medium: Principles of interaction design as a cultural practice*. Cambridge, MA: MIT Press.

Pentland, A. (2008). *Honest signals: How they shape our world*. Cambridge, MA: MIT Press.

Rothstein, B. (2005). *Social traps and the problem of trust*. Cambridge, UK: Cambridge University Press. doi:10.1017/CBO9780511490323

Thaler, R., & Sunstein, C. (2008). *Nudge: Improving decisions about health, wealth, and happiness*. New Haven, CT: Yale University Press.

Chapter 3
Online Credibility and Information Labor:
Infrastructure Reverberating through Ethos

Nathan Johnson
Purdue University, USA

ABSTRACT

This chapter examines how information infrastructure influences ethos in information labor. The primary text is discourse about ACID3, a web page created by members of the Web Standards Project. ACID3 tests the compliance of infrastructural standards for web browsers. In addition to analyzing ACID3 code, several other related conference presentations, job announcements, and web pages are analyzed to theorize ACID3 as a rhetorical text. This chapter argues that three rhetorical commonplaces (mastery, purity, infallibility) are central for the credibility of ACID3 as a text of legitimacy. This study provides a better understanding of rhetoric and infrastructure. To understand rhetorics of infrastructural standardization is to understand the power structures embedded within the modern world. ACID3 is a significant case because of its criticality for standards that enable publics to publish Web content. This chapter contributes to literature in information infrastructural studies, science and technology studies, and the rhetoric of science.

INTRODUCTION

A library's worth of material has been devoted to understanding information infrastructure. The National Science Foundation, for example, has an office devoted to cyberinfrastructure, a term that can be used synonymously with information infrastructure (Edwards, Jackson, Bowker, & Knobel, 2007). The NSF's Office of Cyberinfrastructure

has provided ample funding for research about infrastructure, particularly infrastructure related to science, technology, engineering, and medicine (STEM) that has provided the impetus for much research (Edwards, Jackson, et al., 2007). In parallel, the American Council of Learned Societies (ACLS) has an office devoted to infrastructure for the humanities and social sciences that aids and encourages publishing research on infrastructure (American Council of Learned Societies Commission on Cyberinfrastructure for the Humanities

DOI: 10.4018/978-1-4666-2663-8.ch003

and Social Sciences, 2006). It seems that many see the study of information infrastructure as central to better understanding the conditions of modern societies (Edwards, 2003), and this has generated the large corpus of research on information infrastructure.

This research has been diverse, but one thing writers tend to agree on is that standards and classifications are defining aspects of infrastructure (Star & Ruhleder, 1996). Standards and classifications allow resource sharing by providing stability across diversities in time, space, and populations (Feng, 2003). The existing research on infrastructural standards has tended to be of pluralistic methods: action-based, critical, design, ethnographic, and historical have been popular approaches. Within this standardization research, critical theorists have been particularly good at demonstrating the dark side of standardization, showing how standardization creates disparities by trying to obfuscate social differences along with time, space, and populations (Bowker & Star, 1999; Timmermans & Berg, 2003). One good example is work that shows UK census classifications playing a key role in sustaining racism in genetic research (Smart, Tutton, Martin, Ellison, & Ashcroft, 2008).

Critical scholarship of standardization has a strong foundation, but one place that should be further explored is rhetorics that accompany how infrastructural components like standards are built into popular public information infrastructures in the first place. This study provides a better understanding of the rhetoric of standardization through a case study analysis of ACID3, which is particular significant because of the way ACID3 works to standardize writing for the World Wide Web.

BACKGROUND

The evolution of the web browser as a technology has been tumultuous. Different vendors have historically clashed over what capabilities to include in their browser technologies. Some have used differences in browser standards as a type of competitive advantage (Miller, 2005). Developing standards that can only be used with browser-specific code writing practices has been a technique to increase market share. Microsoft has often been criticized for this practice. When designers create for one browser, it benefits the single vendor, but not necessarily the larger community of web developers (or consumers, for that matter). For instance, when one vendor becomes more powerful, like Microsoft was in the mid 1990s, that vendor is better able to dictate how the web should be rendered.

Web developers are among the first to notice this issue, because as some vendors become more powerful, the developers, distributed across a multitude of organizations, lose voice in conversations that shape the web and browser capabilities. To assuage the issue, gatekeepers within the community have often participated in online protests. These have included "CSS Naked Day," when developers stop using the CSS standard for their web pages to shows how important the standard is. ACID tests show how well the browsers adhere to standards and also show weaknesses in each standard by providing a benchmark for visualization.

ACID3 is a web page that tests many of the standardized technical components of web browsers ("The Acid3 Test," n.d.). It primarily tests compliance to the World Wide Web Consortium's (W3C) Document Object Model (DOM) and Ecma's ECMAscript (popularly known as JavaScript) standards, although ACID3 also tests other standards written by the W3C: HTML, CSS, JavaScript, SVG (a vector image format), and XML, for example. The primary standard being tested, the DOM, is a conceptual map of several data formats. Another way to say this is that the DOM charts the different types of objects that may appear within an HTML page. In order to make sure that all browsers uniformly work the same way, ACID3 asks and tests browser technologies with a simple query: Do you meet the standards set out by the institutional law-making body of the Web, the World Wide Web Consortium (W3C) and Ecma?

The design work for ACID3 was led by Ian Hickson, a technical writer with a work history that includes time with Netscape, Opera, and Google, all of which are companies that develop browser technologies. To say that Hickson knows web browser technology would be a bit of an understatement. His professional life has revolved around editing the complex specifications that help developers and designers build web browsers. Hickson wasn't alone in ACID3, though, as many other specialists helped build ACID3; for example, developer Erik Dahlström, formerly of Mozilla, contributed by creating subtests for images and rich media standards (Dahlström, 2008).

The work of building ACID3 began in April 2007, and the test was publicly released in March of 2008, although several minor revisions were made shortly afterward to fix minor portions of the test. ACID3 is notable in that it attracted much attention from browser manufacturers, engineers, tech writers, and web designers. Indeed, web designers have been some of its more interested users. Since ACID3 was released, designers have often referenced the page as a gold standard for web browser quality. Browsers that pass ACID3 can be considered high-performance machines. For instance, when Internet Explorer 8 was released, many designers noted that despite many innovative features, it still did not pass ACID3, which for them indicated the browser's weakness as a piece of software (Mackie, 2011). ACID3 became an important benchmark for quality, at least among designer publics.

Despite this, many designers also questioned the validity of the test. One of the most notable critics of ACID3 was Eric Meyer, a seasoned web design consultant, popular author, and standards evangelist (Meyer, 2002, p. 334). Meyer suggested the test was a "precisely defined sequence . . . of cunningly (one might say sadistically) placed

hoops, half of which are on fire and the other half lined with razor wire" (2008). For Meyer, ACID3 wasn't a comprehensive evaluation of standards-compliance; ACID3 was an elaborate trick that provided a semblance of standards compliance without a full test. The ACID3 could be gamed by those who knew its tricks.

Yet despite these criticisms, ACID3 has continued to provide a mark of legitimacy for web browser technologies. Designers continue to reference it as a benchmark for legitimacy. This is significant because the test helps to stabilize a technology and sediment standards into a larger World Wide Web infrastructure. To borrow a term from Science and Technology Studies' Wiebe Bijker and Trevor Pinch, the ACID3 works to provide closure for web browser technologies. In their writing, closure is the point in a technology's development when relevant social groups have decided the technology has reached maturation. Reaching closure is not a simple process. Who is to say that a technology adequately solves a problem and should then achieve closure? Who is to say that the technology should not be adapted for new problems? Closure is as much a rhetorical problem as it is a technical problem. In order for closure to occur, relevant publics need to agree to closure. Conceptual arguments that resemble the ideas of closure have been part of many discussions about web browser technologies. Designers and developers often question what the web browser should be able to do, because for many, when the browser is standardized, it locks in public decisions within the material of a technology. ACID3 works to create closure by motivating standardization of a technology.

Ample rhetorical work legitimizes ACID3 so that the test can further legitimize standards. Various publics argue for its validity, and it also has its own affordances that endorse it as an instrument of standardization. The remainder of this chapter explores the rhetorics and affordances of ACID3. My primary means of doing this is through a rhetorical analysis of ACID3. This involves a close textual-intertextual analysis, which examines both the reception and the text of the test (Ceccarelli, 2001, pp. 6-9). By rhetorically reading the reception and the code of ACID3, I am able to make stronger claims about how it works to legitimize web standards rhetorically. Notably, most of my analysis focuses on the reception rather than the text itself. The reason for this is due to the complexity of the technical language. Although I read the test, I also depend on its translators to make sense of it.

My analysis also takes a cue from critical rhetoricians, however. Understanding the reception of ACID3 required more than analyzing the reception to the test, it also required a better understanding of the publics that were a part of that reception. To understand those publics, I draw from several supplementary texts that speak to the role of arête, or virtue, within designer publics. For the sake of space, my analysis is limited to three supplementary texts: a conference presentation titled "How to be a Web Design Superhero," a web page titled the CSS Zen Garden, and a group of job advertisements from the web company Meebo. These texts were selected because of their ability to highlight arête, and I specifically draw from Debra Hawhee's understanding of that term (2002). Virtue amongst ACID3 publics bleeds into the idea of virtue within ACID3. The analysis of the supplementary texts therefore informs and precedes the textual-intertextual analysis. These texts provided three rhetorical commonplaces for discussing the arête of ACID3: mastery, purity, and infallibility.

This research is significant in that it analyzes ACID3, which is used to help standardize browser technologies central to Web communication, which works toward the closure of a shared public infrastructure. To close an aspect of Web browser technology is to close the ways people interact with and understand each other through the Web. Consider the following example: Recent deliberations about browser technologies have included discussions about best image formats to use for the web. From a technological standpoint, this decision affects the historical work associated with different browsers. That is, some browsers aren't currently capable of rendering some image formats. To define an image format is to shift power balances amongst people who have developed infrastructure to support other technologies. ACID3 provides explicit directions for using the SVG image format while ignoring other formats. To understand the rhetorical work of a test like ACID3 is to better understand the politics of the Web's information infrastructure.

ARGUING ACID3

How to be a Web Design Superhero

"How to Be a Web Design Superhero" was a conference presentation given by Andy Clarke and Andy Budd on several occasions (see Budd & Clarke, 2006). One of these presentations was at the 2006 South by Southwest Interactive (SXSWi) an annual web design festival held in Austin, Texas. SxSWi consists of "compelling presentations from the brightest minds in emerging technology, scores of exciting networking events hosted by industry leaders" (SxSW, 2011). The festival is one of the largest and most prestigious conferences for web design professionals, with web development and design presentations making up a large part of SxSW Interactive. Thousands of developers and designers head to Austin, Texas every year to participate.

Before "How to Be a Web Design Superhero," both Clarke and Budd were known as respected web standardization advocates. Andy Clarke is a supporter for W3C-compliant deployment of web standards in technologies; he mostly focuses on the use of the CSS standard in the work of web design professionals. Clarke has written several books on the topic and regularly gives talks about design and standardization. Budd is likewise an advocate for web standardization and a prolific author, one example of this being his co-authored book, *CSS Mastery: Advanced Web Standards Solutions*, which described practical web design skills using standardized technologies (Budd, Moll, & Collison, 2006). Budd is also a founder of ClearLeft, a web design firm with a solid reputation among designers. One of the reasons for this reputation is that ClearLeft is the home of many industry leaders. ClearLeft's work is seen as important, and they spread their ideas through an education initiative. Budd has become not only an advocate of web standards, but also has created an organization around him with a reputation as an industry leader.

Their 2006 SxSWi presentation used a comparison between web designers and superheroes to describe the web professional's activities. They began the presentation with a short question: "Would you rather be a web designer or superhero?" They answered their own question for the audience by suggesting that most web designers would rather be superheroes, but that their presentation would show the audience how to be a web design superhero. Clarke and Budd then worked through the extended analogy that recharacterized what it meant to be a web designer. The presentation depended heavily on descriptions and images from both Marvel and DC comics: Notable among these was DC's Superman, Batman, and Wonder Woman and Marvel's Incredible Hulk and Fantastic Four.

Through this extended parallelism, Clarke and Budd described how the routine practices of web designers were really like those of superheroes. They said that in order to survive as a profession, designers must be superheroes. Clarke and Budd suggested that not only must they be superheroes, they already are. They suggested that designers have unique origins, strong moral codes, secret double lives, secret hideouts, identifiable logos, and superpowers or unique skills. For example, every designer needs to be particularly good with at least one technical skill set, which they compared to a superhero skill or ability. In their

words, "the best designers are skilled with their tools and have a utility belt full of useful stuff" and they have "a weapon of choice." (Budd & Clarke, 2006).

Their presentation slides pictured prominent designers whose pictures were used to craft visual rhetorics for the comparison. Jeffrey Zeldman, "King of Web Standards," was labeled as Superman (Scanlon, 2007). Molly Holzschlag, invited W3C expert, was compared to Wonder Woman. The various activities of each real designer were used as the basis of a superpower. For instance, Zeldman's leadership and longevity within the design profession made him an apt comparison to Superman. In another visual comparison, the web designer's studio—a desk with a computer on it—was described as a secret hideout and laboratory. Clarke and Budd also suggested that any good designer has a dual identity, which could be clearly seen in the number of side projects that the designer takes on in addition to his or her primary work. Although Clarke and Budd's presentation was comedy, it only made sense because they made comparisons that their audience could easily identify with. As prominent voices within the development community they tapped into a persona and identity that their audience understood and sympathized with. The ovation their presentation received reinforces this idea. More so, survey data seconds that idea. Many of the attributes of designers that Clarke and Budd highlighted were also described as some of the defining aspects of web workers in existing surveys (Batt et al., 2001; "Findings from the Web Design Survey 2007," 2008; "Findings from the Web Design Survey 2008," 2009; "Findings from the Web Design Survey 2009," 2010). For instance, the importance of technical skill sets is one of the foremost concerns of web professionals.

In addition to entertaining the audience, the superhero comparison simultaneously embedded a set of rhetorical commonplaces into what it means to work as a web designer. For instance, the work of the web designer is given a storyline taken from the commonplaces of superhero narratives. These narratives come with their own ideas of climax, linear time, and right/wrong, a point thoroughly covered by rhetorician Shaun Treat (2004). One of the more interesting things that Treat identifies as a consequence of the superhero myth in modern society is its ability to refract good and evil into professions (he writes about business leaders) that are multi-faceted and complex. Given Clarke and Budd's description, the work of the web designer can be understood as type of narrative in which the designer plays the central autonomous character, fighting for good, complete with special abilities.

The CSS Zen Garden

Clarke and Budd's presentation shares an affinity with the CSS Zen Garden, a popular design showcase site for web designers (Shea, n.d.). The site was launched in 2003 as a way to demonstrate how standards-compliant markup could be elegantly and beautifully used in modern web design. The site's founder, Dave Shea, has won several awards for site designs and is a frequent speaker at web design conferences similar to SxSWi. In addition, he is also a member of the Web Standards Project, a group dedicated to promoting "standards that reduce the cost and complexity of development while increasing the accessibility and long-term viability of any site published on the Web" ("Web Standards Project," n.d.). Shea's CSS Zen Garden site has also led to a published book titled *The Zen of CSS Design: Visual Enlightenment for the Web* (Shea & Holzschlag, 2005). At one point, it was considered an honor amongst web designers to have one's design showcased on the CSS Zen Garden. By 2008, over 200 official designs had been added to the collection. Although the site hasn't been updated since 2008, it remains a milestone and touchstone within the design community. Shea and his site represent an important articulation point of standardization rhetoric, largely because of the site's reputation as a piece of high-quality work.

The CSS Zen Garden is devoted to demonstrating the benefits of websites developed with web standards artistically, specifically HTML and CSS standards, which define the organization and visual appearance of a website. The Zen Garden demonstrates these benefits by showcasing the work of professional graphic designers who use standards-compliant code. To do this, each Zen Garden designer redesigns the CSS Zen Garden's home page template files with their own graphic layouts. The underlying standardized code remains the same, but the visual effect is different depending on the work of the designer. By reusing the same template files, each designer helps to prove that standards-compliant code can be used in visually appealing design. Imagine the CSS Zen Garden templates as blank slates. The Zen Garden challenges designers to maintain the purity of the standardized slates while creating a beautiful work of art.

The language used to describe the project reflects the idea that standards-compliant code is clean and pristine. Standards-compliant code works to "clear the mind of past practices" that are non-standard and encourages "time-honored techniques" that "create beauty from structure" (Shea, n.d.). The Zen Garden provides a "road to enlightenment" where the designer will become "one with the web" (Shea, n.d.). Clarity, cleanliness, and wisdom are united as concepts to suggest that standards design is enlightened.

The CSS Zen Garden highlights something about Clarke and Budd's presentation that isn't immediately evident in "How to Be a Web Design Superhero." Clarke and Budd had referenced the unique mission of a superhero: a superhero has special motivations, a cause. Although they don't restrict these motivations, the few that they suggest are important include writing web content with correct use of web standards. The correct choice, the enlightened choice, the choice of right in their presentation, included using web standards.

CSS Zen Garden's language may be very different from the superhero rhetoric of Clarke and

Budd, but the site does emphasize many ideas that were also foregrounded in Clarke and Budd's presentation. For example, the CSS Zen Garden highlights the cultivation of special abilities for developing websites. The Zen Garden suggests that "time-honored techniques" can be cultivated by adhering to "lessons from the masters." It's not just that abilities are valued in both texts, however. It's that these abilities are unique or at the very least, difficult to come by. Indeed, the Zen Garden suggests that designers with unique abilities are needed for web design because it compares designers to other web content creators: "To date, most examples of neat tricks and hacks have been demonstrated by structurists and coders. Designers have yet to make their mark. This needs to change" (Shea, n.d.). Designers are asked to seize a rhetorical exigency and forward the development of web standards.

This rhetorical exigency is marked by a sense of agency that parallels that provided in Clark and Budd's presentation. The Zen Garden suggests that designers can take "complete and total control" by developing websites with standards-compliant code, and the fact that the Zen Garden also suggests that code can be learned from masters strengthens this idea. Masters, of course, is a term that suggests authority, which suggests that there is someone who has authority over others ("master," 2011).1 Both the presentation and the CSS Zen Garden suggest that the designer is a knowing and controlling agent who is able to make a difference over a domain area.

There is one other similarity between the CSS Zen Garden and "How to Be a Web Design Superhero." The Zen Garden highlights the same right/wrong dichotomy that the superhero rhetoric does. The opening description of Zen Garden begins with vivid imagery: "Littering a dark and dreary road lay the past relics of browser-specific tags, incompatible DOMs, and broken CSS support" (Shea, n.d.). Those who design in the CSS Zen Garden way work to create a better world, free of the baggage of past web design. Clark and Budd's

presentation uses this same dichotomy, only it is more specific about the broken down world of wrong that the designer must overcome. For Clark and Budd, the broken web of the past is a nightmare for web users. It's the designer's job to create a safe world for their end user. They protect the web from the evil deeds of clients who would try to attain their own ends rather than that of the users. Although the CSS Zen Garden is not quite as forward as Clarke and Budd, it does second this right/wrong dichotomy by literally suggesting how the coding practices of the past contributed to wrong design practices. Taken together, the "How to Be a Web Design Superhero" presentation and the CSS Zen Garden suggest why the web can be thought of in terms of right/wrong and also provide palpable examples of right/wrong.

Between the "How to Be a Web Design Superhero" presentation and the CSS Zen Garden, there are two recurring themes. The first is the idea of an individual with unique abilities as a type of master. The second is the idea of a binary relationship between right and wrong, or to put it another way, the idea of clarity and clear ideas of right and wrong: purity. The CSS Zen Garden, especially, highlights the idea that web standards are right while alternatives are wrong or at the very least misguided. CSS Zen Garden then correlates this concept of standards to the idea of purity.

Web Design Job Announcements

One final set of texts provides a third commonplace for my later analysis of ACID3. In 2007, the instant messaging software producer Meebo circulated employment notices for a JavaScript developer, an ActionScript programmer, and a server administrator. These positions were branded as a JavaScript Ninja, an ActionScript Assassin, and a Server Samurai. ("meeblog » ActionScript Developer (aka 'ActionScript Assassin')," n.d.; "meeblog » JavaScript Developer (aka 'JavaScript Ninja')," n.d.; "meeblog » Server Side Developer (aka 'Server Samurai')," n.d.). This branding may seem eccentric, but it's not particularly unusual, as it resembles many existing naming conventions among developer communities. For instance, Mozilla developer John Resig (2011) used similar language in his book *Secrets of the JavaScript Ninja*.

While the naming is odd, the primary content of the job announcements is more relevant for ACID3. The announcements included a series of puzzles that job respondents were asked to submit with their application. Notably, these puzzles were reminiscent of a phenomenon that Amy Slaton and Janet Abbate (2001) encountered in their historical research on a central Internet standard: the TCP/IP protocol. In their research, Slaton and Abbate used the term "bake off" to describe an organized set of technology tests that do rhetorical work in addition to their testing work (2001, pp. 129-131). In other words, the tests portrayed a correct type of standardization for its test taker, while simultaneously verifying the functional components of those standards. They explain how, during the early development of TCP/IP, several ARPANET contractors organized a series of bake offs at trade shows. The contactors described the agenda for the meeting as an opportunity for technical collaboration and friendly competition. They instructed participants that their event would allow them to meet, collaborate, and test their systems with colleagues. As a requirement, participants would need to standardize their equipment to join in the competition. The underlying framing of a bake off, then, drives participants to standardize their technologies. At the very least, participants would need to temporarily standardize their technologies to take part in the bake off. In Abbate and Slaton's ARPANET case, the organizers also appealed to participants by suggesting that standardization was beneficial in other ways: it would be an opportunity to learn about new technologies.

The contemporary Meebo job announcements included similar appeals. Each announcement ended with a series of questions requiring that applicants, at the very least, be knowledgeable of standardized design practices. For example, JavaScript applicants needed to answer prompts that included the following:

These questions asked job candidates to describe minor differences in standardized JavaScript. For instance, differentiating between "var x=3;" and "x=3;" is a matter of noticing how a JavaScript variable is initialized in either global scope or local scope. In other words, coding "var x=3" will create a consistent piece of data that is only available within an immediate JavaScript function rather than to a complete program of functions. To be able to make this differentiation assumes fairly good experience with the JavaScript standardized language. Because these tests were published online, job applicants were able to si-

Figure 1. Screen capture of Meebo advertisement for a JavaScript programmer

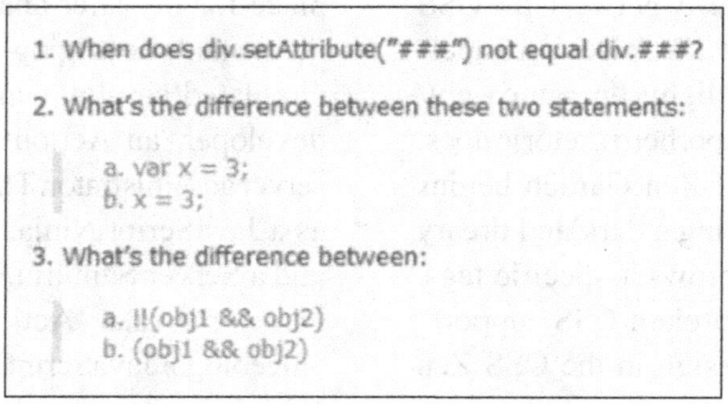

```
1. When does div.setAttribute("###") not equal div.###?

2. What's the difference between these two statements:
     a. var x = 3;
     b. x = 3;

3. What's the difference between:
     a. !!(obj1 && obj2)
     b. (obj1 && obj2)
```

multaneously learn the answers as they answered the test. There were no time constraints; there were no limits on outside resources. Applicants could essentially learn the JavaScript standardized language as they applied for the job.

Notice how this part of the job application is written to look like a formal test, a genre that confers an aura of authoritative certainty. The questions are numbered, and getting "correct" responses would indicate the respondent's conformance to an authority. In this case, these parts of the test genre impel applicants to reinforce or extend their understanding of standards, because the questions already assume a good understanding of the standard. Because correctness is an attribute of the test genre, "learning" the standard becomes an effect of the test. Critical analysis of the standard is subjugated to the act of learning knowledge to repeat verbatim in a test form.

Several web designers wrote online responses about the employment notice, which helped to substantiate the rhetorical effectiveness of these advertisements. Some of these designers provided answers to Meebo's puzzles on their personal websites. Many suggested that any real developer could easily solve these puzzles. One blogger wrote "I would imagine that a true JavaScript Ninja would be subjected to a more rigorous test of skill than these small puzzles" (Stimmerman, 2007). This puzzle-solver not only was familiar enough with the standard to answer the questions, he also suggested further competence by claiming his skills exceed the test. Puzzle solving was equated with professional competence. Simultaneously, the competence is accompanied by a martial arts metaphor that parallels some of the competency discourse from the Clarke and Budd presentation and the CSS Zen Garden.2 The fragments of discourse analyzed here are simply not isolated commonplaces.

Notably, this analysis is not meant to undermine the technical abilities of respondents. As programmers from the field will tell you, programming is hard work. Knowing how to manipulate any

one technical language is usually just one of the many difficult parts of programming. Histories of programmers have shown longstanding power battles among professionals attempting to define the key characteristics of the profession (Ensmenger, 2003). Indeed, battles of what constitutes professional knowledge in related fields like computer programming have been well documented (Ensmenger, 2001). The history of computer programming suggests that the profession is a rich and diverse field of knowledge that includes much more than know-how of a technical language. Yet what emerges in this study of texts is an equation of technical skill with special abilities. The language of standardization is persuasive because it taps into an arête that takes pride in programming abilities. It taps into a concept of arête that understands technical skills as superhuman abilities; this arête understands some types of programming as more enlightened than others, and it assumes a certain material fidelity in technological standardization. Instead, I suggest that these topoi are providing heuristics for ethical behavior in guiding a community.

Commonplaces of Standardization: Mastery, Purity, Infallibility

The previous analysis has highlighted three commonplaces within standardization discourse that I describe as mastery, purity, and infallibility. The first, mastery, is the idea of the designer as a fully autonomous and capable individual without a dependence on others. This can be easily seen in the "How to Be a Web Design Superhero" presentation. The idea of developing or having special abilities is only important for an audience that values a unique sense of self. And when the CSS Zen Garden speaks about learning from masters, it also gestures towards the idea of mastery. When the Meebo job announcements asked for skills that were as specific as a programming language, they reinforced the idea of individuality originally articulated by Clarke and Budd's presentation. This theme of standardization suggests that a recurring idea of techné, or technical ability, is central to legitimizing web standards.

The second theme is purity. The CSS Zen Garden's language best articulated this when it described web design as having a canvas that could be pure and clean. In order for there to be a pure and clean canvas, there needs to be some sort of ideal state of cleanliness. The Clarke and Budd presentation seconded this idea when they spoke to the idea that web design superheroes have strong morals and special motivations, just like the idealistic goals of superheroes. In this rhetoric, their motivations are to create a world of pristine standardization—the ideal clean state of the World Wide Web.

The third theme of standardization is infallibility, which refers to the idea that there is a right and wrong method of standardization. Infallibility also refers to the recurring theme that there is a correct type of design, that there is a correct way of creating for the web, and that the very idea of the standard is correct. Note here that the idea of infallibility abets the belief that everyone already agrees upon the same tenets of right and wrong. Infallibility proceeds with the belief that standards are correct. It deflects the idea that a standard could indeed be corrected to accommodate new ideas and beliefs about web design and writing on the World Wide Web. The examples of infallibility in each of the artifacts were ubiquitous. Superheroes assume a world of right and wrong; the idea of cleanliness assumes the idea of non-standardization as filthy; the advertisements assume a test-like notion of right and wrong answers. The Meebo job advertisements also work towards this idea because they deploy "tests" that draw on a genre that emphasizes clear ideas of right and wrong. These three commonplaces (mastery, purity, and infallibility) work in tandem, providing sensibility for each other, often supporting concepts of one with the premise of another. The idea of a "correct" web and the topoi intermingle. They provide the backing for arguments about what is right and wrong for browsers and the World Wide Web.

Passing ACID3

ACID3 was released as a test for web browser standardization in March 2008. The test drew attention primarily in two ways. First, critics commented on the use of ACID3 as a litmus test for good browser technologies. Various people questioned the motives behind releasing the test. These critics often questioned whether the test's designers adequately considered how the test's goals related to its performance. Second, critics commented on the code of the test. These critics showed interest in the code design of the test. To describe the rhetorical work of ACID3, my analysis follows both those lines of critique.

Responses to ACID3

For many designers, the underlying premise of ACID3 is that a pure web exists that can actually be constructed. ACID3 aids in that construction. In this myth, by standardizing all technologies, the Web will be free, clear, pure, and open. As evidence, consider how the funding organization of ACID3, the Web Standards Project, describes the purpose of ACID3: the job of ACID3 is "to expose flaws in the implementation of mature Web standards in Web browsers" ("Web Standards Project," 2008). If the current Web has flaws, this is understandable because of the commonplace of purity that accompanies the standardization rhetoric. And the idea that the Web is maturing suggests a more perfect future. At the very least, it suggests a more advanced future. And also notice the word implementation, suggesting there is a correct way to begin developing standards.

Notably, the design community has been largely in favor of ACID3. It is referenced frequently as a step forward in the work of designers. The possible reasons for this purity are multiple, and the CSS Zen Garden helps here because it suggested that purity was a "road to enlightenment" (Shea, n.d.). Purity, through the lens of ACID3, is part of the future of the World Wide Web. This sentiment is further reinforced by a central member of the Web Standards Project: "Testing is really important. Without tests that check how well a certain browser follows standards, i.e. applies mark up and displays the result correctly, we can never guarantee an open, fully interoperable web" (Gunther, 2008). The purity commonplace informs the most fundamental motivations of the test, but the other themes of standardization also recur throughout ACID3.

Critics of the test hint at the importance of the mastery and infallibility required for the legitimacy of ACID3. After its release, ACID3 received backlash from several advocates of web standards. Two notable critiques came from Eric Meyer and Mike Shaver. Both of these critics are widely respected and hence, within the discussions of standardization, perform gatekeeper roles (Hammersley & Atkinson, 1995, p. 27). Meyer writes books about the Cascading Style Sheet standard and participates in their development. Shaver was the VP of Engineering at Mozilla Corporation, the manufacturer who makes the popular web browser Mozilla Firefox.

Both of their responses to ACID3 were mixed: they applauded the effort but questioned the value of the test as a method for evaluating web standardization. Meyer (2008), for example, questioned the logic of the ACID3 authors, and wrote:

What I disagree with is the idea that if you cherry-pick enough obscure and difficult corners of a bunch of different specifications and mix them all together into a spicy meatball of difficulty, it constitutes a useful test of the specifications you cherry-picked. Because the one does not automatically follow from the other. (n.p.)

His critique drew attention away from the precision of the test and deployed a testing commonplace that many critics of modern education employ. That is, he questioned the logic of the test to measure what it claimed to measure. He metaphorically revised the test as a spicy meatball, a meal option that is perhaps not even a good idea. In this case, Meyer suggested that ACID3 should be different. It should provide browser manufacturers with a comprehensive series of benchmarks for testing the hundreds of specifications that are built into any one browser. The ACID3 spicy meatball should be redesigned as a real comprehensive taste test. For Meyer, the ACID3 overreached its claims of comprehensiveness: One of the reasons for this was that ACID3 sacrificed thoroughness and even though Meyer (2008) admitted "clever thinking" by "a lot of very smart people" was involved in creating the test, he argued a "more important point had been missed" (n.p.). In this case, the mastery of a few smart people is not enough to persuade Meyer of the test's usefulness.

Meyer is able to point to something that was left unstated by the ACID3 tests. The premise assumed by many of its supporters was that ACID3 was designed by people who could definitively be trusted to create a useful test for standards compliance. The value added by the test-creators rested in leveraging their special abilities as developers. Meyer pointed to a commonplace weakness of the test: If the developers are not considered to be autonomous masters, if they are the sort of people that miss something important, the test's legitimacy suffers.

Recall the first theme of web standardization: mastery. By attacking this commonplace idea, Meyer drew the legitimacy of the standardized test into question. This is to say that Meyer critiqued standardization by questioning the designer's mastery. Meyer raised a fundamental problem of ACID3: If the techniques of the designer are based on the ability to manipulate the technologies of a browser, then how does it follow that the same mastery is appropriate for evaluating those technologies? Meyer raised that issue and then suggested that not enough thought had gone into the appropriateness of the ACID3 tests.

Mike Shaver raised similar concerns about the ACID3 test. As a head of one of the browser vendors that was being tested, he wrote that Mozilla would not put energy into passing the ACID3 test. Shaver (2008) responded to the release of ACID3 by claiming that Mozilla was:

The best browser the web has ever known, and we expect to be putting it in the hands of more than 170 million users in a pretty short period of time. We're still taking fixes for important issues, but virtually none of the issues on the Acid 3 list are important enough for us to take at this stage. (n.p.)

Shaver focused on the work that the browser vendor was doing to continue supporting the users of the World Wide Web. More than that, though, Shaver (2008) suggested that the "improvements" encouraged by the ACID3 test were steps backward for Mozilla, that his company:

Didn't want to be rushing fixes in, or rushing out a release, only to find that they've broken important sites or regressed previous standards support, or worse introduced a security problem. Every API that's exposed to content needs to be tested for compliance and security and reliability, and we already have some tough rows to hoe with respect to conflicts with existing content as it is. We think these remaining late-stage patches are worth the test burden, often because they help make the web platform much more powerful, and reflect real-web compatibility and capability issues. Acid 3's contents, sadly, are not as often of that nature. (n.p.)

Shaver seconded the point made by Meyer, the point that the appropriateness of ACID3 is in question. Shaver questioned how credible it is as a test. He simultaneously questioned the mastery of its designers. He also addressed the purity commonplace by suggesting that ACID3

incorporates an idealistic vision of the web that he believed is not possible. For him, ACID3 remains pure insofar as it constructs an ideal world, but a clearly unrealistic world. Shaver deflected the purity commonplace, thus showing that purity is unrealistic and problematic. For Shaver, and for Meyer, ACID3 did not measure what it claimed to be measuring. It simply consisted of clever tricks with little substance. Shaver exposed the same underlying issue that Meyer found. The authority of mastery cultivated by discourses of the ACID3 test and similarly, those found in Clarke and Budd's "How to Be a Web Design Superhero" and the CSS Zen Garden encourage an orientation toward design that is focused on technological mastery. In the context of ACID3, this was inappropriate.

Perhaps both Meyer and Shaver are seeking a type of arête that would be more appropriately envisioned through the concept of decorum that has been popularized by Michael Leff as a type of practical wisdom. As Leff wrote, decorum "orders the elements of a discourse and rounds them out into a coherent product relevant to the occasion" (1999, p. 62). For Leff, decorum is bound by situation. It is the ability of the rhetor to work with existing fragments of discourse and recompose them with relevance for the situation. Meyer and Shaver seek a different understanding of ACID3's appropriateness, a redefined notion of virtue, shaped not by the technical mastery of its autonomous designers. If mastery is at stake in ACID3, decorum hints at the type of mastery desired.

Both Meyer and Shaver's critiques are particularly useful for pointing out mastery as a key ACID3 commonplace. Both writers took up the idea that was earlier discovered within this study's other texts. That idea deployed through ACID3 is that of the autonomous individual with unique extraordinary abilities to pass the test. But by molding the limits of that concept to the situation of ACID3, the critics were able to show how the idea of this type of individual mastery does not suggest appropriateness. In this case, the critics both do that by showing the inadequacy of the test that was created by web designers and developers. The fact that Meyer and Shaver raised this issue showed how ACID3 depends on the technical authority of its designers as a way to legitimize standardization.

On a side note, it's significant that Meyer and Shaver pointed this out, but the tension they deploy between mastery and decorum seems to be more significant. By balancing back and forth between ideas that seem similar to technical mastery vs. decorum, Meyer, Shaver, and a larger discourse community encourage a richer, more vibrant, and multi-faceted discussion. It's also a discussion that encourages openness to the conclusions of discourse. Imagine the ACID3 discussion without the contributions of Meyer and Shaver. Meyer and Shaver provided a primer for discourse that keeps the discussion from becoming a deterministic race of technological ends.

The Design of ACID3

The above analysis of critical reception of ACID3 has highlighted the use of the mastery theme to encourage standardization discourses. In complement, the design and code of the test reveal how other commonplaces of standardization provide legitimacy for ACID3. Consider ACID3's construction in the abstract: ACID3 is composed of a gamut of 100 tests. These tests are structured in a way than reads much like a coherent narrative marked by forward movement through time. As ACID3 runs, each test checks to see whether the browser conforms to a W3C specification. The tests are independent of each other, but because of the way the browser parses them, they build a tempo similar to a time-dependent storyline. After test one, then test two. After test two, then test three, and so on. Each bit of code leads to a new test. As each test is completed, ACID3 provides further evidence to legitimize it as a coherent bit of programming. By test 100, the code has worked to

fully substantiate its reputation as a potent, nearly omnipotent, type of standards testing.

Notice how this test stacking uses and amplifies rhetorics similar to the bake off discourse identified in the web design job advertisements. Like those job advertisements, the idea of a test discursively builds its own conceptions of right and wrong. For ACID3 and its developers, to pass the test is correct, and thus to standardize in the W3C way is correct. ACID3 extends an open source philosophy on browser development. ACID3 extends this authority through repetition. Piling test after test works to reiterate this concept of the test's infallibility and thus its legitimacy.

The ACID3 idea of infallibility is further reinforced through the enactment of the mastery commonplaces within the code's comments. Comments are descriptions and elaborations written for future audiences. They are included within the actual code itself, and are typically added to explain how the code functions. To use an analogy, if software were theater, the comments would be asides, providing explanation of events unfolding before an audience. Figure 2 demonstrates an example comment from ACID3.

Comments allow developers to describe how related programming code works. In this example, the code describes how visual aspects of the test will be rendered with a web browser. The ability to describe the test in this much detail (down to the level of pixels!) should be read as performing the mastery trope that reinforces the tests of ACID3 and contributes to the idea of parsimonious right and wrong. In this way, the idea of mastery seems to be built into the very artifacts and practices of the programming.

In this case, code builds mastery within the context of the ACID3. In line 211 of the code, a new JavaScript function is initiated that is named kungFuDeathGrip ("The Acid3 Test," n.d.). Comments complement and explain the actual code itself, which works in tandem with code to provide a richer rhetorical text. The comments of this code explain its function a little more: "used to hold

Figure 2. HTML comment from ACID3

```
*
* ...then the line-height matters, otherwise the font does. Note
* that font descent - font-ascent will be in the region of
* 10px-30px (with Ahem, exactly 12px). However, if we make the
* line-height big, then the _positioning_ of the inline-blocks will
* depend on the font descent, since that is what will decide the
* distance from the bottom of the line box to the baseline of the
* block (since the baseline is set by the strut).
*
* However, in Acid2 a dependency on the font metrics was introduced
* and this caused all kinds of problems. And we can't require Ahem
* in the Acid tests, since it's unlikely most people will have it
* installed.
*
* What we want is for the font to not matter, and the baseline to
* be as high as possible. We can do that by saying that the font
* and the line-height are zero.
*
* One word of warning. If your browser has a minimum font size feature
* that forces font sizes up even when there is no text, you will need
* to disable it before running this test.
*
*/
```

things from test to test" ("The Acid3 Test," n.d.). The function (the JavaScript term for a reusable program) is used to hold information about variables from one test to another—from one piece of the narrative to another. This function also serves rhetorical work as a nod to the same types of rhetorics that most clearly were articulated in the "How to Be a Web Design Superhero" presentation. The mastery rhetorics of the designer, specifically through allusion to special abilities, resurface through the actual programming in the test. In this case, the language of the program is built to be read by the same types of audiences that participate in the "How to Be a Web Design Superhero," Meebo job announcements, and CSS Zen Garden texts.

CONCLUSION

Standards are the conceptual specifications of shared public infrastructures. Implementing them builds material resistances into the physical world. In this case, it defines code-writing practices for web designers who depend on browsers for their work. Yet decisions about standardizations are often insulated within a few select groups such as developers, policy-makers, entrepreneurs, and standards bodies. Some of this exclusivity in standards making can be attributed to their invisibility (Star & Ruhleder, 1996). Technical standards simply aren't noticed unless they break down and draw attention themselves (Edwards, 2003). But some of the exclusivity of standards can be attributed to their outward appearance as inevitable or correct. Often, it seems as though a certain standard is necessary. Standards are only created and deployed to support the public aims of infrastructure, though. There is nothing inherently necessary about a standard's existence, and so standards should be looked at as matters of public deliberation.

By examining ACID3, I have worked to reveal techniques that tend to legitimize web standards for publics. This paper suggested three recurring commonplaces that inform web standardization rhetoric: mastery, purity, and infallibility. Recognizing those rhetorics opens rhetorics like ACID3 up for critical scrutiny and debate among publics who might otherwise see ACID3 as a closed test of standardization. The assumptions of standardization rhetorics should be openly challenged, and critiques like Meyer's (2008) and Shaver's (2008) demonstrate how this is possible. Both critics were able to question and refract that concept of mastery. Their critiques demonstrated how being able to challenge tests like ACID3 provides a better position to become involved with standardization issues. The ability to challenge leads to the ability to better advocate infrastructural decisions.

Consider the following as possible issues to explore further. First, at what key points do the commonplaces of this study become involved in other infrastructural decision-making? ACID3 works at the margins of standardization. One of the reasons that it seems reasonable as a test is that many of the standards it includes have already been previously legitimized among standards bodies. The work here suggests that commonplaces like mastery, purity, and infallibility work to distribute an already largely accepted standard amongst other publics that work regularly with standards. It is prudent to investigate when and how those commonplaces do the work of shaping the standards. I have demonstrated one point of intervention here with my analysis of ACID3, but rhetorical critics could also intervene by taking part in standardization-making bodies, online forums, and W3C specification draft writing. It is also significant to think about how the concept of standardization reflexively encourages the rhetorical commonplaces of this study in the conceptual frameworks of professionals. Standards are precise and calculating as a means of building infrastructures that can flatten differences across time and space (Bowker, Baker, Millerand, & Ribes, 2007). How does that practice encourage professionals to incorporate the commonplaces described in this research into their other argumentative structures? In other words, it is a significant task to identify the interfaces of standardization and publics.

Last, it is also prudent to consider the material effects of standards as rhetoric further. For example, which groups are advantaged and disadvantaged as various technologies are privileged through standards specifications? Who benefits from the effect of standardization? Although these questions have been raised among other communities that study new media, a rhetorician's point of view would be a helpful addition. Essentially, these last two points converge on a single line of inquiry. To standardize is to make decisions that should be made publicly. As such, we should be seeking to answer questions of how, why, and who standardizes technical specifications. We should be seeking to answer questions of why that standardization matters for broader groups of people instead of just those within the groups constructing and obeying standards. In this specific case, I would suggest that mastery, in particular, works to interpellate publics that identify with westernized concepts and beliefs. The commonplaces of purity and infallibility work against revisiting old design and implementation practices. If something were pure and infallible, it would not need revision. While tests like ACID3 may continue, it is a significant and important task of the critic to deflect these three commonplaces and work to encourage standards participation amongst greater audiences with an eye on revising past practices rather than building immaculate infrastructures.

REFERENCES

American Council of Learned Societies Commission on Cyberinfrastructure for the Humanities and Social Sciences. (2006). *Our cultural commonwealth: The report of the American Council of Learned Societies Commission on Cyberinfrastructure for the Humanities and Social Sciences*. New York, NY: American Council of Learned Societies Commission. Retrieved from http://www.acls.org/uploadedfiles/publications/programs/our_cultural_commonwealth.pdf

Batt, R., Christopherson, S., Rightor, N., Van Jaarsveld, D., & Economic Policy Institute. (2001). *Networking: Work patterns and workforce policies for the new media industry*. Washington, DC: Economic Policy Institute.

Bowker, G. C., Baker, K., Millerand, F., & Ribes, D. (2007). Towards information infrastructure studies: Ways of knowing in a networked environment. In J. Hunsinger, L. Klastrup, & M. Allen (Eds.), *International handbook of Internet research*. New York, NY: Springer Verlag. Retrieved from http://interoperability.ucsd.edu/docs/07BowkerBaker_InfraStudies.pdf

Bowker, G. C., & Star, S. L. (1999). *Sorting things out: Classification and its consequences*. Cambridge, MA: MIT Press.

Budd, A., & Clarke, A. (2006, March 7). *How to be a web design superhero*. Presented at the South by Southwest Interactive Festival 2006, Austin, TX.

Budd, A., Moll, C., & Collison, S. (2006). *CSS mastery: Advanced web standards solutions*. Berkeley, CA: Springer-Verlag.

Ceccarelli, L. (2001). *Shaping science with rhetoric: The cases of Dobzhansky, Schrodie;] dinger, and Wilson*. Chicago, IL: University of Chicago Press.

Dahlström, E. (2008, January 22). *Getting to the core of the Web*. Retrieved March 27, 2011, from http://my.opera.com/MacDev_ed/blog/2008/01/22/core-web

Ecma International. (n.d.). *What is Ecma International?* Retrieved April 4, 2011, from http://www.ecma-international.org/memento/index.html

Edwards, P. N. (2003). Infrastructure and modernity: Force, time, and social organization in the history of sociotechnical systems. In Misa, T. J., Brey, P., & Feenberg, A. (Eds.), *Modernity and technology* (pp. 185–225). Cambridge, MA: MIT Press.

Edwards, P. N., Jackson, S. J., Bowker, G. C., & Knobel, C. P. (2007). *Understanding infrastructure: Dynamics, tensions, and design*. NSF Grant 0630263.

Ensmenger, N. L. (2001). The "question of professionalism" in the computer fields. *IEEE Annals of the History of Computing, 23*(4), 56. doi:10.1109/85.969964

Ensmenger, N. L. (2003). Letting the "computer boys" take over: Technology and the politics of organizational transformation. *International Review of Social History, 48*(S11), 153. doi:10.1017/S0020859003001305

Feng, P. (2003). Studying standardization: A review of the literature. *The 3rd Conference on Standardization and Innovation in Information Technology* (pp. 99-112).

Findings from the A List Apart Survey, 2008. (2009). New York: A List Apart Magazine. Retrieved from http://aneventapart.com/alasurvey2008/

Findings from the Web Design Survey 2007. (2008). New York: A List Apart Magazine. Retrieved from http://www.alistapart.com/d/2007surveyresults/2007surveyresults.pdf

Findings from the Web Design Survey, 2009. (2010). New York: A List Apart Magazine. Retrieved from http://www.alistapart.com/articles/findings-from-the-web-design-survey-2009/

Gunther, L. (2008, October 2). *Acid3 receptions and misconceptions and do we have a winner?* Retrieved March 28, 2011, from http://www.webstandards.org/2008/10/02/dowehaveawinner/

Hammersley, M., & Atkinson, P. (1995). *Ethnography: Principles in practice*. London, UK: Routledge.

Hawhee, D. (2002). Agonism and aretê. *Philosophy & Rhetoric, 35*(3), 185–207. doi:10.1353/par.2003.0004

Leff, M. C. (1999). *The habitation of rhetoric. Contemporary rhetorical theory: A reader* (pp. 52–64). New York, NY: Guilford Press.

Mackie, S. (2011, March 15). *Internet Explorer 9 released, but should you care?* Retrieved May 12, 2012, from http://gigaom.com/collaboration/internet-explorer-9-released-but-should-you-care/

master, n.1 and adj. (2011, March). *Oxford English Dictionary*. New York, NY: Oxford University Press. Retrieved from http://www.oed.com/view/Entry/114751?rskey=DhchNQ&result=1&isAdvanced=false

meeblog » ActionScript Developer (aka "ActionScript Assassin"). (n.d.). Retrieved from http://web.archive.org/web/20080502081313/http://www.meebo.com/jobs/actionscript/

meeblog » JavaScript Developer (aka "JavaScript Ninja"). (n.d.). Retrieved from http://web.archive.org/web/20080502110640/http://www.meebo.com/jobs/javascript/

meeblog » Server Side Developer (aka "Server Samurai"). (n.d.). Retrieved from http://web.archive.org/web/20080920042629/http://www.meebo.com/jobs/openings/server/

Meyer, E. (2002). *Eric Meyer on CSS: Mastering the language of web design* (1st ed.). New Riders Press.

Meyer, E. (2008, March 27). *Acid redux*. Retrieved March 4, 2011, from http://meyerweb.com/eric/thoughts/2008/03/27/acid-redux/

Miller, K. W. (2005). Web standards: Why so many stray from the narrow path. *Science and Engineering Ethics*, *11*(3), 477–479. doi:10.1007/s11948-005-0017-0

Resig, J. (2011). *Secrets of the JavaScript ninja*. Greenwich, CT: Manning.

Scanlon, J. (2007, August 6). Jeffrey Zeldman: King of web standards. *BusinessWeek Online*. Retrieved March 15, 2011, from http://www.businessweek.com/innovate/content/aug2007/id2007086_670396.htm

Shaver, M. (2008, March 27). *The missed opportunity of Acid 3*. Retrieved March 4, 2011, from http://shaver.off.net/diary/2008/03/27/the-missed-opportunity-of-acid-3/

Shea, D. (n.d.). *CSS Zen Garden: The beauty in CSS design*. Retrieved from http://www.csszengarden.com/

Shea, D., & Holzschlag, M. E. (2005). *The zen of CSS design: Visual enlightenment for the web*. Peachpit Press.

Slaton, A., & Abbate, J. (2001). The hidden lives of standards: Technical prescriptions and the transformation of work in America. In Allen, M. T., & Hecht, G. (Eds.), *Technologies of power: Essays in honor of Thomas Parke Hughes and Agatha Chipley Hughes* (pp. 95–144). Cambridge, MA: MIT Press.

Smart, A., Tutton, R., Martin, P., Ellison, G. T. H., & Ashcroft, R. (2008). The standardization of race and ethnicity in biomedical science editorials and UK biobanks. *Social Studies of Science*, *38*(3), 407–423. doi:10.1177/0306312707083759

Star, S. L., & Ruhleder, K. (1996). Steps toward an ecology of infrastructure: Design and access for large information spaces. *Information Systems Research*, *7*(1), 111–134. doi:10.1287/isre.7.1.111

Stimmerman, B. (2007). *JavaScript ninja : Crisis averted!* Retrieved from http://web.archive.org/web/20081225224424/http://socket7.net/article/javascript-ninja

SxSW. (2011). *South by Southwest interactive*. Retrieved March 27, 2011, from http://sxsw.com/interactive

The Acid3 Test. (n.d.). Retrieved April 20, 2008, from http://acid3.acidtests.org/

Timmermans, S., & Berg, M. (2003). *The gold standard: The challenge of evidence-based medicine* (1st ed.). Philadelphia, PA: Temple University Press.

Treat, S. R. (2004). *The myth of charismatic leadership and fantasy rhetoric of crypto-charismatic memberships*. Louisiana State University.

Web Standards Project. (2008, March 3). *Acid3: Putting browser makers on notice, again*. Retrieved March 28, 2011, from http://www.webstandards.org/press/releases/2008-03-03/

Web Standards Project. (n.d.). *About - The Web standards project*. Retrieved December 6, 2007, from http://www.webstandards.org/about/

World Wide Web Consortium. (2007, June 19). *About W3C*. Retrieved December 6, 2007, from http://www.w3.org/Consortium/

ADDITIONAL READING

Abbate, J. (1999). *Inventing the Internet*. Cambridge, MA: MIT Press.

Abbott, A. D. (1988). *The system of professions: An essay on the division of expert labor*. Chicago, IL: University of Chicago Press.

Allsopp, J. (2000, April). A dao of web design. *A List Apart, 58*. Retrieved from http://www.alistapart.com/articles/dao/

Amman, J., Carpenter, T., & Neff, G. (Eds.). (2007). *Surviving the new economy*. Boulder, CO: Paradigm.

Benner, C. (2003). "Computers in the wild": Guilds and next-generation unionism in the information revolution. *International Review of Social History, 48*(S11), 181–204. doi:10.1017/S0020859003001317

Benoit-Barné, C. (2007). Socio-technical deliberation about free and open source software: Accounting for the status of artifacts in public life. *The Quarterly Journal of Speech, 93*(2), 211–235. doi:10.1080/00335630701426751

Blanchette, J. (2011). A material history of bits. *Journal of the American Society for Information Science and Technology, 62*(6), 1042–1057. doi:10.1002/asi.21542

Bowker, G. C. (2005). *Memory practices in the sciences*. Cambridge, MA: MIT Press.

Clark, D. (2008). Content management and the separation of presentation and content. *Technical Communication Quarterly, 17*(1), 35–60. doi:10.1080/10572250701588624

Downey, G. (2002). *Telegraph messenger boys: Labor, technology, and geography, 1850-1950*. New York, NY: Routledge.

Dunn, E. C. (2009). *Standards without infrastructure*. Ithaca, NY: Cornell University Press.

Edwards, P. N. (2010). *A vast machine: Computer models, climate data, and the politics of global warming*. Cambridge, MA: MIT Press.

Frohmann, B. (1992). The power of images: a discourse analysis of the cognitive viewpoint. *The Journal of Documentation, 48*(4), 365–386. doi:10.1108/eb026904

Gurak, L. J. (1997). *Persuasion and privacy in cyberspace: The online protests over lotus marketplace and clipper chip*. New Haven, CT: Yale University Press.

Haigh, T. (2008). Protocols for profit: Web and email technologies as product and infrastructure. In Aspray, W., & Ceruzzi, P. E. (Eds.), *The Internet and American business* (pp. 105–158). Cambridge, MA: MIT Press.

Hanseth, O., Monteiro, E., & Hatling, M. (1996). developing information infrastructure: The tension between standardization and flexibility. *Science, Technology & Human Values, 11*(4), 407–426. doi:10.1177/016224399602100402

Johnson, N. R. (2012). Information infrastructure as rhetoric: Tools for analysis. *Poroi, 8*(1), 1–3.

McGee, M. C. (1990). Text, context, and the fragmentation of contemporary culture. *Western Journal of Speech Communication, 54*(3), 274. doi:10.1080/10570319009374343

McKerrow, R. E. (1989). Critical rhetoric: Theory and praxis. *Communication Monographs, 56*, 91. doi:10.1080/03637758909390253

Monteiro, E., & Hanseth, O. (1995). Social shaping of information infrastructure: On being specific about the technology. In Orlikowski, W. K., Walsham, G., Jones, M. R., & DeGross, J. I. (Eds.), *Information technology and changes in organizational work* (pp. 325–343). London, UK: Chapman & Hall.

Paling, S. (2004). Classification, rhetoric, and the classificatory horizon. *Library Trends, 52*(3).

Pickering, A. (1995). *The mangle of practice: Time, agency, and science.* Chicago, IL: University of Chicago Press.

Pyati, A. (2005, May). WSIS: Whose vision of aninformation society? *First Monday, 10*(5). Retrieved October 7, 2007, from http://www.first-monday.org/issues/issue10_5/pyati/index.html

Sandvig, C. (2006). Shaping infrastructure and innovation on the internet: The end-to-end network that isn't. In Guston, D. H., & Sarewitz, D. (Eds.), *Shaping science and technology policy: The next generation of research* (pp. 234–255). Madison, WI: University of Wisconsin Press.

Turner, F. (2006). *From counterculture to cyberculture: Stewart Brand, the whole earth network, and the rise of digital utopianism.* Chicago: University of Chicago Press.

Wark, M. (2004). *A hacker manifesto.* Cambridge, MA: Harvard University Press.

Winner, L. (1980). Do artifacts have politics? *Daedalus, 109*(1), 121–136.

KEY TERMS AND DEFINITIONS

ACID Test: A visual benchmark for how well an Internet browser conforms to a set of standards.

CSS: A technology that formats the content of an online document.

Document Object Model (DOM): A standardized technique for modeling online documents. It provides a metalanguage for a document.

Infrastructure: The amalgam of standards, classifications, protocols, and algorithms that provide dependability for foregrounded action.

JavaScript: A standardized technology that allows a browser to interact with the DOM on the fly.

Standards: A technique that enables the interoperability of technologies across time and space by prescribing uniform methods of communication.

SxSW Interactive: A yearly festival located in Austin, Texas that is devoted to current technology trends.

Web Standards: A standard that provides part of the infrastructure for the World Wide Web.

ENDNOTES

[1] It should be noted that the Zen Buddhist version of the word master tends to be different than the Westernized Hegelian (master/slave) version of the word. I would suggest, however, that in this case, the Westernization of the idea of Zen in the CSS Zen Garden has refracted the term master in a way that is more reminiscent of the authoritative master discussed in the previous section.

[2] There is clearly a crude Western concept of Orientalism being deployed here as well. I haven't touched on it, but it would be worth looking at in the future.

Chapter 4
Establishing Ethos on Proprietary and Open Source Software Websites

Kevin Brock
North Carolina State University, USA

ABSTRACT

The increasing prominence and variety of open source software (OSS) threaten to upset conventional approaches to software development and marketing. While a tremendous amount of scholarship has been published on the differences between proprietary and OSS development, little has been discussed regarding the effect of rhetorical appeals used to promote either type of software. This chapter offers just such an examination, focusing its scrutiny on the websites for three pairs of competitors (operating system, Web browser, and image manipulation program). The means by which the OSS websites promote their programs provide a significant set of insights into the potential trajectory of OSS development and its widespread public acceptance, in terms of both its initial philosophy and its perceived alternative nature to traditional software products and models.

INTRODUCTION

In the last decade, the open source software (OSS) movement has garnered considerable attention from the public. Previously, users of computer programs were often restricted to choosing from among a relatively small selection of proprietary products, but today an explosion of freely available alternatives produced by continually reconstituting teams of volunteers has initiated a

radical shift in the ways that computer software (proprietary and OSS alike) are advertised and otherwise promoted to potential users. This shift in self-promotion must be understood in relation to conventional market expansion: Just as the sheer number of individuals who own or have access to computers has increased, so has there been the potential for a wider range of proprietary (as well as OSS) options for specific ends, from the operating system to the word processor.

However, the proprietary model of software development is challenged by OSS philosophy

DOI: 10.4018/978-1-4666-2663-8.ch004

as much as—if not more than—any individual product is challenged by its competitor(s) in a given market. It is this larger challenge to the status quo that is of immediate interest, since the means by which OSS developers present their products to public audiences can tell us about how those developers wish for their projects to be compared to and distinguished from proprietary competition. There has been a tremendous amount of research into the means by which OSS products and communities might be evaluated, often in terms of the social dynamics of a particular community (Aberdour, 2007; Casaló, Cisneros, Flavián, & Guinalíu, 2009; Lemley & Shafir, 2011) or how users determine the value of a specific program (Gallego, Luna, & Bueno, 2008; Grodzinsky, Miller, & Wolf, 2003; Raghu, Sinha, Vinze, & Burton, 2009). However, relatively little has been discussed about the explicit rhetorical appeals and efforts developers make through various media to gain new users/customers. This is potentially problematic given the complex constructions of *ethos* that often underlie particular interactions between users and these media, including appeals to brand loyalty and anticipated recognition of a product's (or media's) functionality. Even those studies focused on users' evaluations of their software do not provide much insight into how those users were made aware of any program qualities or features included in the evaluative criteria of a given study (such as Del Bianco, Lavazza, Morasca, & Taibi, 2011). This text will explore how OSS and proprietary products are marketed to potential users through the content of the products' websites, with special care taken to examine the *ethos* generated through the presentation of OSS philosophy compared with that presented by proprietary developers. Several competitive pairings will be examined (each consisting of one conventional proprietary product and one open source product), with each pair related to a different specific area of computer use: operating system, Internet browser, and image editing and manipulation.

The major difference between these two models of software development—the developmental philosophy of each approach—serves as the primary point from which much of the rhetorical appeals used by each development community are constructed. However, as we will see, these fundamental philosophies are not always presented consistently to their intended audiences. In fact, there appears to be a trend among OSS websites to imitate appeals to *ethos* that are traditionally representative of the websites for proprietary entities and their products. If these imitations are indicative of a larger trend among OSS websites, there could be serious implications for how OSS is perceived in a broader sense by the public. Specifically, the trend may compromise public awareness of OSS as an alternative philosophy to proprietary development and, instead, OSS may come to be viewed as being *fundamentally the same* as proprietary software, with the only significant difference being the market price of a particular product compared to that of its competitors. The potential consequences of such a shift are significant in that many technologies built philosophically, as well as structurally, upon OSS—such as wikis, open-access repositories of scholarly and professional information, and even the web itself—may similarly be reevaluated through this lens, to the potential detriment of these technologies as we currently know them.

BACKGROUND

How Does OSS Differ from a Proprietary Model of Software Development?

Traditionally, software development exists as a centralized process generally closed off from a program's user base until a "finished" version of the program is released into the market, where it may exist until it is either abandoned, updated, or replaced by another product from the owning organization/entity. Raymond (2000) famously referred to this model as the "cathedral" approach to software development (as opposed to the relatively chaotic OSS "bazaar"), in which isolated groups of skilled craftsmen work within a clear hierarchy to release a polished commercial product. Until that product is ready for its release (to maximize profitability), access thereto is often highly restricted, with only its developers—and perhaps a population of beta testers, users given early access to the software in order to help identify bugs and other problems with it—having any knowledge of its features or capabilities (Zittrain, 2004; Grodzinsky, Miller, & Wolf, 2003). Generally located within a single corporate institution—and often existing as one or more teams within a single department—the development community often possesses a highly formalized hierarchy, and this hierarchy is often supported by specialized information management systems to facilitate specialization among individual developers or teams of developers (Hendry, 2008). The development team(s) may or may not be (and usually are not) involved in the marketing or, in some circumstances, technical support for the product, and users may or may not be involved in various types of participatory development of that product in order to bring it to market (Hendry, 2008).

In contrast, OSS development exists within a vast continuum of production models. Many OSS projects are developed by lone individuals who are unable or uninterested in locating assistance for the product under development. Other projects may involve dozens of volunteer developers upon whom certain responsibilities may be delegated or who may provide contributions entirely independent of one another to a central organizer. Still others are "forks," alternative projects run by developers who have split off of existing projects due to creative differences; relatively well-known examples of "root" and fork project pairs include the Debian and Ubuntu distributions of Linux

and also the KHTML and WebKit web browser engines (the latter of which powers the Apple Safari and Google Chrome browsers). In most of these OSS networks, a hierarchy of development contribution value has been established wherein certain developers possess significant influence over a given project and the vast majority of end-users hold very little (Christley & Madey, 2007). A number of studies have attempted to articulate the complex social dynamics of OSS development communities and any hierarchical structures constructed by those communities (Crowston et al., 2005; Crowston & Howison, 2005; Howison, 2006; Howison, Inoue, & Crowston, 2006; Wiggins, Howison, & Crowston, 2008). In all of these cases, there is one unifying factor: The source code for each and every project is publicly available for any interested parties to download, modify, and redistribute as desired, provided that the modifications and redistributions adhere to a communally accepted set of license restrictions designed to protect the "open" nature of OSS ("Open Source Initiative," n.d.).

It is important to note that, while all of the types of OSS projects described above use an open source license, nothing restricts any OSS project from pursuing profitable success, and any developers are free to sell their products to potential users. The major difference between OSS and traditional models in this regard is that a developer cannot prevent *any other* developer from also selling or giving away a modified version of the product ("Open Source Initiative," n.d.). As a result, conventional issues of copyright are complicated through open source licensing in ways that drastically alter the ways that OSS products (whether sold or given away for free) are marketed to potential users as compared to proprietary software products. For us to understand more clearly how these approaches to marketing could alter our comprehension of OSS and proprietary software, we must turn to rhetoric, the study and art of persuasive communication.

UNDERSTANDING ETHOS IN DOCUMENT DESIGN

Over the past two decades, rhetoricians have increasingly turned their attention to the use of rhetorical appeals and devices in visual artifacts and the design of documents in general, and of multimodal documents, those utilizing multiple forms and modes of communication (such as sound, image, video, or text), in particular. Buchanan (2004) defined the fundamental premise of design, in the sense of a critical discipline, as the effort to understand the rhetoric of products and their design, which consists of "how products come to be as vehicles of argument and persuasion about the desirable qualities of private and public life" (p. 230). Similarly, Kress & van Leeuwen (2006) noted that, "[l]ike linguistic structures, visual structures point to particular interpretations of experience and forms of social interaction" (p. 2). These considerations have extended to new media structures and documents as well; Selfe and Selfe (1994) observed that technological *interfaces* reflect multiple systems of cultural value and ideology, including those of race, sex, class, and politics (and not always reflecting those values or ideologies of the user population of a given technology). Carnegie (2009) argued that the multiple levels of interfaces in a new media object serve as examples of rhetorical *exordium*, the component of a conventional oration where an audience is made more receptive to the ensuing argument and thus where a point of interaction between rhetor and audience occurs (pp. 164-165). In short, text is not the only means available for a rhetor to communicate meaningfully to or with an audience, and for those forms of media in which design plays a significant component, the most effective documents are likely to be those whose authors apply a keen understanding of rhetorical appeals to the design, as well as to the content, of those documents.

Of particular interest to the rhetorical scholar is the manner in which the three fundamental rhetorical appeals are used to persuade audiences to engage in some kind of action. In extremely simple terms, *pathos* is an appeal to emotion, *logos* to reason, and *ethos* to character. However, each of these appeals is more complex than the above description suggests, and *ethos* in particular is far more than a reliance on the credibility of a rhetor's character: It aids the rhetor in establishing his or her character in order to demonstrate qualities like goodwill and sense. Such demonstrations serve as means of proving the validity and worth of an argument and for the rhetor to be perceived as someone who, according to Warnick (2004), "has the audience's best interest at heart and also as [...] one who is qualified to speak on the topic at hand" (p. 258). Miller (2001) has suggested that *ethos* acts as "the default appeal, a kind of presumption: an appropriate, trustworthy character lightens the burden of proof" (p. 270). However, as Warnick (2004) argued, *ethos* is difficult to determine in online environments where specific authorship, along with relevant social and characteristic values, are ambiguous if not entirely obscured from audiences. This ambiguity is exponentially increased when one considers the rhetorical contributions from code and related commentary as they both add to the value and function of a software program but are also often impossible to attribute to individual authors. Such determinations have become increasingly difficult to make in recent years with the success of massive (often anonymously) collaboratively constructed online resources such as *Wikipedia* (Hartelius, 2011). How, then, might *ethos* be more fully understood when examining websites for software programs authored by teams of either proprietary or volunteer open source developers? Instead of attributing *ethos* to specific author identification or character, it may well be more worthwhile to follow Warnick's (2004) call to view *ethos* through designers' suggestions and users' perceptions of "site functionality and usefulness" (p. 264). In addition, we can gain substantial insight by exploring how site designs, like Kenney's (2004) definition of the rhetorical quality of visuals, make use of differing perceptions of *ethos* to "cause us to change our beliefs or to act" (p. 326).

Identifying Appeals Made on Software Websites

What sorts of appeals are made on and through the websites for software products? It is important to recognize the set of conventions upon which software developers and marketers may rely so that we can determine how, if at all, OSS projects are actually distinguished from their proprietary counterparts by visitors to their respective websites. For the scope of this text, I am interested specifically in the appeals made *primarily* on the home pages of those websites, following Warnick's (2007) argument that increasing speed of online navigation leads to increasing expectations of shortened searches for information (and usually, an increasing amount of information in those searches' results). The contents of a website's home page is generally the most likely node through which a visitor may be introduced to the website and its purpose and, as such, needs to provide enough information to answer most visitors' initial queries about that site's purpose and, more importantly, per Warnick (2004), the suggested capabilities and usefulness of both the site and the product it supports – that is, what sorts of functions it facilitates (as well as constrains) for a user of either the site or the software program. While the home page of each site is not the sole text under examination for each site, it serves as the likeliest point at which the site's, and the program's, valued capabilities are emphasized in the hopes of gaining a user's interest.

One of the most prominent types of appeals, not surprisingly, relates to a product's ability to meet a user's needs in regards to completing specific tasks, and this is especially important when looking at a purely electronic product. Many software websites combine *ethos* and *logos* both to explain how the product facilitates achieving

certain goals and to persuade potential users to think about achieving goals through the available abilities of the product. Lankes (2007) noted that there are significant concerns with the creation of credibility through electronic (and especially Internet-based) media and that many organizations and individuals alike have yet to come to a consensus on how "best" to do so, resulting in an unreliable set of principles on which *ethos* can be established. While the means through which these appeals are made are thus often extremely different, the objective is almost always the same for software developers: to increase interest in and use of the product advertised on a given website.

While these initial appeals may differ wildly in execution across websites, one component seems almost universally consistent as a means of establishing *ethos* through reliability and developer presence: technical support. Part of the *ethos* establishment undertaken by the vast majority of websites is access to and availability of competent assistance in solving a user's software problems, a reflection of the functionality of a site as *ethos*, especially where recognition of author credibility, or even specific authorship, is scarce or nonexistent (Warnick, 2004). By providing access to technical support through clearly marked channels on a product's website, software developers can demonstrate that an infrastructure to support users of their software exists for accomplishing particular tasks once that product is purchased or (in some cases) freely downloaded.

What often distinguishes a proprietary from an OSS product's website is the emphasis on the *type* of technical support available to the average user. The former often highlights "official" support from employees of the product's developers, while the latter almost exclusively relies on community support from other users and (some) developers of the product. Hocks (2003) argued that the web allows for varying "audience stances," reflective of multiple levels of—and motivations for—user interactivity and agency in order to help establish an author's *ethos*, and these differing types of technical support are certainly representative of Hocks's audience stances. This distinction of support *type* is significant because it suggests at least two very different paradigms of *ethos* establishment that play up the strengths of either the developers (the proprietary model) or the user community and the product itself (the OSS model) to anticipate the types of functionality audiences desire from the product and the site alike. Such a division of *ethos* priorities also suggests very different demonstrations of how developers expect a user to consider his or her own relationship with the software product, and the consequences of this difference will be apparent in the comparisons of three proprietary software and OSS websites.

DISTINCTIONS IN ETHOS BETWEEN COMPETITORS

Operating System: Macintosh OS X vs. Red Hat's Fedora Linux

Given that a user's operating system holds a potentially tremendous influence over that user's interaction with (or accessibility to) specific software programs, it could be argued that the website for an operating system would need to reflect the power, flexibility, and style of the OS. However, this argument assumes that a website seeks to display an accurate and authentic mirror of how the system operates in use; it is just as possible that the website seeks to provide the most accessible (or, perhaps, the "flashiest") means of introducing a system to potential users, even if the actual use varies wildly from the website's presentation thereof. However, such an approach is often complicated by the existing customers of the product who may view the site as a repository for documentation and other technical support, so there may also be a tremendous amount of detailed information intended for (and expected by) this second audience.

The two sites discussed below, which promote Apple's Macintosh OS X and Red Hat's Fedora Linux operating system, attempt to persuade visitors to the site to examine their respective operating systems (and to imagine using each) with very carefully crafted appeals to specific types of *ethos*. While Apple demonstrates a sleek product from a developer dedicated to ease of use by the consumer, Red Hat argues for an operating system that offers a built-in community support system whose users help one another succeed in using the operating system even where it may not have integral functionality.

I. Apple Macintosh OS X

The website for Apple's Macintosh operating system OS X (http://www.apple.com/macosx/) makes an immediate and direct argument for its use in one of several cycling images at the top of the front page, with its claim that it is "[t]he world's most advanced desktop operating system" (Apple, 2011). Directly beneath this claim is an image of an Apple laptop displaying various uses of software running on OS X. The implicit connection made is that the best (if not exclusive) use of the operating system is on Apple computers—along with a further suggestion accompanying this connection that Apple understands how best to meet the needs of computer users through the production of its line of computers. Thus, the constructed *ethos* of OS X is fundamentally contingent upon that of Apple computers as the only appropriate pieces of hardware on which to run OS X successfully, and those most interested in the specifics of OS X are already interested in purchasing an Apple. This could become a problem for individuals interested in learning specifically about the operating system, since the argument for its use is weakened significantly if the user base requires a specific type of computer on which to run it.

Much of this potential problem is mitigated by the repeated argument for ease of use of Apple's products, the suggestion that OS X's accessibility in *application* is preferred to its restrictions—or, at least, the former is assumed to be preferred by the site developers. It is important to note that, while much of this argument may resemble *logos* as its basis, the underlying assumption of the argument is that Apple has the knowledge and expertise to understand, and the ability to implement, the most necessary and preferred features users look for in an operating system. Beneath the major image of the home page, the page is split into several sections, half of which focus on how easy it is to use OS X. One of the sections proclaims, "Better. Faster. Easier." and includes an explanation that OS X is "the Mac you know and love, made even better" (Apple, 2011). While this is primarily another appeal to an established user base, it also suggests that other viewers of the page may be able to easily seek out more information about the operating system in order to learn about and then come to "know and love" it.

The appeal to quality, speed, and ease of use implicitly highlights a potential problem with Apple's moral *ethos*; specifically, the website promotes the efficiency and aesthetic qualities of the company's software and hardware products in such a manner as to promote its mix of clean efficiency and personal character (that, to quote an older Apple ad, leads its consumer to "think different" from most computer users). Such an *ethos* has the potential to backfire against the company given the myriad accusations of human rights violations that have been levied against the company in relation to the China-based Foxconn production facilities where its products, along with those of other computer hardware companies such as Dell, Microsoft, Samsung, and Sony, are manufactured (Chan, 2011; Cheng, Chen, & Yip, 2011; Rulliat, 2010). Katz (1992) has argued that appeals to technological expediency can hide highly impersonal and immoral details beneath their veneer by obscuring the means by which such expediency is achieved. This is not to suggest that Apple's actions are equivalent to those of the Nazis that Katz's article examined, but it is meant to highlight the potential implications of Apple's specific calls to creating a "better, faster, easier" operating system and computer.

Elsewhere on the page are more direct appeals to specific audiences and the uses or functions the OS and the site may be serving for those audiences. For example, near the bottom of the home page, the OS X's built-in integration of Microsoft's Exchange Server into its software is highlighted, which creates a two-pronged argument aimed at business customers that, first, hints further at the overall ease of use of OS X for any unsure corporate visitors to the site and, second, indicates that a current non-user considering a switch to OS X—likely someone who has been using Microsoft products and the Windows operating system—will remain within "familiar" territory (the MS suite of business software) rather than overwhelming him- or herself with the need to learn an entirely new way of using a computer for specific corporate needs. Specific technical support, meanwhile, is provided through a "Support" link at the top of the page, which then offers visitors a list of links related to specific support, primarily in the form of categorized articles detailing installation, troubleshooting, and other issues, for all manner of Apple products ranging from OS X to the iPad to MacBook accessories. While the support pages also link to community forums, the emphasis is on the official documentation provided by Apple.

Just as the *ethos* of the OS is founded upon it being "better, faster, [and] easier" (Apple, 2011) than other OSes, so does the website construct an *ethos* of tech support more upon its ability to provide accessible information to its users than on providing a space for users to assist one another (i.e., the community forum). Accordingly, the website for OS X presents a consistent appeal to expert authority on the part of Apple across its pages, providing a range of information and assistance to its current and potential user bases. This consistency maintains a perceived relationship of provider (Apple) and recipients (OS X users) of both technology and knowledge.

II. Red Hat Fedora Linux

An extremely similar visual design and set of included information exists on the website for Red Hat's OSS Linux operating system Fedora (http://fedoraproject.org/). The front page, like that for Mac OS X, emphasizes a large central content block in which cycles a series of large images emphasizing Fedora's stability, ease of use, wide range of available software programs, and—most significantly—an explicit reference to "a worldwide community of friends" complete with an image of a crowd of individuals wearing Fedora polo shirts (Red Hat, 2011). It is this final image in the cycle that sets the rhetorical arrangement of the website for Fedora apart from that of OS X; where Apple constructs its own authority through the combination of Apple computer and OS X as the superior means of achieving computer-related goals, Fedora establishes its initial *ethos* by implying its worldwide adoption (by a tight-knit community) with a potentially familiar feel through its visual reflection of many of the design elements of the OS X website. Just as the OS X website's images draw the viewer's eye in order to highlight the sleek power of the hardware and software combination, the Fedora home page juxtaposes a presentation of ease-of-use with a variety of customizable and personalized possibilities for the OS.

Positioned as static text just above this series of alternating images is the phrase, "Freedom. Friends. Features. First" (Red Hat, 2011). These terms are repeated below the images as links to further descriptions of the Fedora project (as clarified by each term). Once again, the system's (as well as its developers') credibility is anchored by an appeal for user inclusivity and community. Indeed, the community appears so inclusive that the site suggests, through the central position of

the image and its supporting text, a blurring of the boundaries between developer and user—so much so that, below the image, there is a link for the viewer to "Join us [the developers]! Help make Fedora!" (Red Hat, 2011) For some readers, this may suggest direct contributions to the Fedora code; for others, it may suggest contributions to online documentation and support; for others still, the call may be perceived to refer to use of the software and identification with the goals of the development team.

The site's appeals to a communal-based credibility are supported by the explicit reference to *freedom*, meaning "openness" and user-based choice in employing a given product as much as any financial cost of that product—the backbone of the OSS philosophy. Fedora thus positions itself as a worthwhile operating system in direct opposition to the lock-in to a specific hardware in the way that OS X is locked in to Apple machines; its authority stems from the user's ability to install it on any computer he or she deems appropriate. Because the user makes this decision, his or her own participation is intertwined with the perceived *ethos* of Fedora as a worthwhile operating system. While Apple relies on its control over production (providing the combination of hardware and software for users) and official dissemination of relevant and helpful information, Fedora relies on its inclusion of the user into the production equation: whether the user successfully installs the operating system onto his or her computer plays a huge role in how motivated he or she will be in using it.

The website content that distinguishes Fedora's argument most clearly from that of OS X is its community-centered inclusion of user-related news ("Your Life on Fedora"), the philosophy underpinning its features ("What Makes Fedora Different?") and information about hosting sponsors. Fedora's developers situate the OS within, and as an integral part of, the broader OSS community in two ways. The first is through a highlight of its primary distinction from proprietary software products – its free and open source nature, described as a "fight for software freedom" that positions OSS (and Fedora specifically) as an underdog competing with corporate oppressors (Red Hat, 2011). The second is through a personalization of its user base with stories of how individuals make use of the operating system and its included software programs. As a result, Fedora's potential *ethos* extends beyond the bounds of its developer base (Red Hat and individual volunteer developers) and reaches to the potential authority of all related OSS developer communities and users. This is a profoundly different model of *ethos* than that of the centralized, corporate institution of Apple and its existence as a separate entity from its user base, and its complicated nature may very well reflect the drastic difference in philosophy between OSS and conventional proprietary models.

Internet Browser: Microsoft Internet Explorer vs. Mozilla Firefox

A clearer contrast between developers' *ethos*-based appeals can be seen in the websites for the web browsers Microsoft Internet Explorer and Mozilla Firefox. These sites are especially intriguing for an inquiry into web-based appeals for software use because they are viewed through a web browser (which may or may not be the very browser marketed by each site). The types of authority each site assumes about its browser, and the interests of its potential user base, are telling in regards to the information it presents: Does it expect that the viewer of the site is already using that browser or one of its competitors? What does such an assumption say about how the advertised browser should be perceived, *ethos*-wise, by its potential users?

That these questions are inevitable whenever web browsers are marketed via websites is telling for the presented authority of these specific sites. Based on the content of each browser's home page, Microsoft's site suggests that its viewers are definitely using (and familiar with) Internet

Explorer and thus using some version of the Windows OS, while Mozilla's site appeals to an audience possessing a wider variety of OS and existing web browser choices with lighthearted fun that potentially compromises the OSS philosophy underlying the development of the browser. As a result, Mozilla's effort to brand itself more as a recognizable product than a collaborative process and community (in the sense that Fedora appealed to its website's viewers) may well work against the fundamentals of the broader OSS movement whose members might otherwise be inclined to support the browser.

I. Microsoft Internet Explorer

The website for Microsoft's web browser (http://windows.microsoft.com/en-us/internet-explorer/products/ie/home) uses a particularly intriguing appeal to *ethos*, given that the program is integrated into all copies of the Windows operating systems (Microsoft, 2011). Effectively, Microsoft has a built-in user base that would have to consciously opt out of the browser's use in favor of an alternative product. This could very well be the reason that Microsoft's major content block, compared with those described above for OS X and Fedora, provides an appeal to upgrade an existing installation of Internet Explorer to the newest version. The site serves as a means of deploying the browser software into a Windows computer alongside the "Windows Update" functionality integrated into the operating system. In either case, the website authors assume the reader is familiar with the fundamentals of the browser's features and functions.

Below the image area, several sections of links hint at more information about the software, similarly anticipating preexisting familiarity with IE, such as in the link for "Video: What's Changed." Meanwhile, to the right of the main content block is an image and link to download "Internet Explorer 9 with Bing & MSN" (Microsoft, 2011). There is some slight discontinuity here: The page title promotes one version of IE9, while there is another version made available and unaccompanied by further detail elsewhere (once again, the website authors assume the reader is familiar with both Bing and MSN as already-utilized Microsoft products within a Windows OS). The result is an unclear argument for the use of one of two specific variations of the browser. Visitors to the site are likely unable to gather a full understanding about either software version (and thus an understanding of which they should download and use). The site ends up working against itself by risking a split of its user base between the available products: Which one is "better" or more appropriate for their needs? According to the site, both are equally valuable—and this dual-product promotion can easily work against Microsoft's intent of expanding its user base in a specific direction.

Part of this website-based confusion is surely to extend from Internet Explorer's Windows-based integration: The only individuals who would need to introduce themselves to the browser are Macintosh users that are interested in adopting it. Otherwise, the only audiences Microsoft may need to address are those who use an outdated version of the browser and have turned off their update function, or those who use Windows but prefer an alternative web browser, although this may not be a significant need. After all, if an individual has already paid for the Windows operating system, Microsoft *might* simply not care if its web browser is used. Given Microsoft's presentation of Internet Explorer on its website home page as an unexplained product that allows a user to "experience a more beautiful web," this could be perceived by visitors to the website as a lack of effort on Microsoft's part to promote, or at least justify, Internet Explorer as a major web browser software.

The integration of Internet Explorer into Windows as a fundamental component of the OS is likewise evident in the range of technical support provided on the browser's website, made available from the home page through a link entitled "Help

& How-To" (Microsoft, 2011). There exist two primary databases of information usable through the site itself: the first is a community-based forum, where users can "find solutions and get assistance when [they] need it" from other users, described as "experts and other customers" but never explicitly guaranteeing official Microsoft support on those pages (Microsoft, 2011). Such an approach to establishing functionality through official and expert aid could be viewed as a reflection of the opacity of the software itself: just as users are unable to view explicitly the inner workings of the program, so too are they unable to ensure that Microsoft will provide the implicitly advertised official assistance. This tactic works in contrast to that of many OSS communities, which make the archives for all communications taking place between developers (and between developers and users) publicly available as part of an effort to maintain transparency in the development process, thus forming a vastly different sort of *ethos*.

The other available database is the Microsoft Knowledge Base, a collection of tips, suggestions, and downloadable program updates for various Microsoft products, including Internet Explorer. This information *is* explicitly provided by Microsoft employee authors, but unlike the assistance provided in the forum, it is non-discursive: The information provided on the knowledge base pages is *all* that is provided, and users are asked to rate the quality thereof after making use of particular entries (and these ratings are not made visible to the public). There is also a third method of obtaining technical support in the form of direct correspondence with Microsoft, which is made possible either via web-based chat or outside the scope of the website by email or telephone. The resulting sense of *ethos* demonstrated on the Internet Explorer site is that public discourse related to the browser is primarily for the user community itself, while official and authoritative support exists on a private, one-to-one level. In essence, Microsoft establishes itself as an intimate provider of technical support for each user while *simultaneously* demonstrating its impersonal and corporate character by distancing itself (that is, its official presence) from the community interactions taking place by users to help one another more successfully make use of its software. Users are thus asked to consider their relationship with Microsoft in regards to both of these approaches, accepting the company as a global developer of software but also as a dedicated provider of customized assistance.

Microsoft provides a sense of intimacy through this latter approach, as each user receives help related to his or her specific problem, but this also means that there is not necessarily an immediately available or accessible solution to one's problem without turning to Microsoft's official support staff. Further, the home page's wording in regards to these channels for technical support suggests an implicitly inferior or subjugated position on the part of the user to that of Microsoft, as the former is offered "how-to" information and provided access to "experts" from the company and the community alike with differing means of demonstrating expertise. However, because Microsoft is almost certainly interested in positioning itself as *the* authority on its products, an attempt at demonstrating *ethos* through expertise is, potentially, a very effective method of establishing that authority.

II. Mozilla Firefox

The website for Mozilla Firefox (http://www. mozilla.com/en-US/firefox/) provides a very similar approach to Microsoft towards persuading potential users of its browser, offering little detail on the home page itself (Mozilla, 2011). The Firefox home page distinguishes between the browsers used to view it: If someone looks at the Firefox site while on either an outdated version of Firefox or with a different browser altogether, he or she will see a splash page dedicated to persuading viewers to download the newest version of the software. Taking up the left third of Firefox's home

page is a short block of text that reads, "Made to make the Web a better place. a new look | super speed | even more awesomeness" (Mozilla, 2011). To the right of the text, and taking up the majority of the screen, is a large image of several cartoon creatures holding a Firefox browser window in which an entry from Wikipedia is displayed. Where Microsoft's appeal relies on its corporate authority and operating system integration to promote its browser as an opportunity to view "a more beautiful web," Mozilla constructs its own *ethos* with a prominently-displayed declaration that it will improve the web and make it "even more awesome," just as the newest version of the browser supposedly is improved upon in comparison to its predecessors.

Mozilla provides information on its site to demonstrate the extent to which its Firefox browser differs from Internet Explorer, but these details are hidden behind a small "Tour" link beneath the primary block of content on the page. If Microsoft could be faulted for relying on an expected familiarity of its software among users thanks to the prevalence of the Windows operating systems, Mozilla relies on an initial expectation that visitors to the site are already motivated to use its browser (such as through word-of-mouth recommendations). Unlike Microsoft, Mozilla does not have its brand integrated into a major operating system—even its historical connection to the Netscape browser is not necessarily obvious, or relevant, to users not already knowledgeable about Firefox. Given that there is an extensive, helpful, and "interactive" demonstration of Firefox's capabilities through the "Tour" link, it is surprising that the Mozilla developers do not immediately highlight the browser's power.

However, the corporate tone of the Internet Explorer website is directly combated through the much more lighthearted approach to web browsing undertaken by Mozilla; the use of colorful cartoon characters (and the Wikipedia entry on *Treasure Island* displayed in the image's browser) suggests a level of play associated with the browser and the culture of its users that simply is not presented on Microsoft's website. Supported by the terms to the main image's left—"*super* speed" and "*awesomeness*" (emphasis added)—the Firefox home page generates an *ethos* founded on *fun*, more than the expertise of its developers, as perhaps the defining factor for internet use. At the same time, by obscuring details about the browser or how it is developed, Mozilla arguably compromises the values of openness and collaboration central to the OSS movement: Visitors to the site are clearly perceived as consumers of (what is at least initially presented as) a black-boxed product created by a separate population of programmers.

As a result, Mozilla distinguishes itself from Microsoft as a non-corporate entity interested as much in enjoyment as improvement of the web (although this is not immediately clarified), but Mozilla also resembles Microsoft in its implicit presentation of its developers as a group separate from the consumer base that downloads and uses the browser nonetheless. Rather than demonstrate an argument for product superiority, Mozilla expects the probable viewer of its website to be swayed more by a lack of a business-based atmosphere. However, the level of involvement assumed by many OSS project websites that could highlight the collaborative, volunteer nature of Firefox development is obscured by the developers themselves—it is unclear from the initial set of details provided on the home page how much the development team would actually want further assistance and input (if any) from any other possible contributors. The Firefox website thus sends a message *primarily* of consumption to potential users: The users are to consider themselves uninvolved entertainment or humor-minded consumers using software developed by approved programmers.

This message of consumption is complicated somewhat by the varieties of technical support provided on the Firefox website, linked from the bottom of the home page through the term "Support" (Mozilla, 2011). Like the site for Internet Explorer, Firefox's emphasized means of support is a database of troubleshooting articles with a community forum augmenting the information in the database. Similarly, the articles populating the database exist as one-way communications capable of being "rated" by users in terms of the articles' usefulness to users, and that rating remains invisible to the public. However, unlike Microsoft's knowledge base, the Firefox troubleshooting database identifies the screen names of the individuals who contributed to a particular article. On the one hand, this is not necessarily a guarantee of credible authority, except that contributors can be tracked across both articles and forum posts alike. Where the Firefox and Internet Explorer sites diverge most obviously in regards to site usefulness is the capability to log into the Firefox site in order to compose troubleshooting articles in addition to forum posts. The *ethos* constructed by the collaboration between site developers and article contributors provides both a means for individual demonstrations of expertise (thanks to article authorship attribution) and communal functionality (the implication that an article exists because it is needed by the user population).

Ultimately, Mozilla Firefox's website offers a problematic *ethos*: On the one hand, it positions branding and consumption as the paramount goal for the site and the browser, while on the other hand, it offers—albeit in a comparatively subdued manner—an opportunity for its user community to engage in multiple forms of social interaction and collaborative knowledge-building. For visitors to the site, the initial demonstration of character (on the site's home page) appears to be one of increasingly monolithic identity growing out of a foundation in a distributed, democratic, and volunteer-based community. While the notion of separate developer and consumer identities works well for Microsoft as a corporate producer of software, this is not as true for Mozilla, where its most lauded quality is its open source nature. As a result, Mozilla positions itself in a *potentially* awkward manner where it may be perceived to be neither a concrete proprietary entity nor an entirely distributed, democratic community that fully embraces open source philosophy.

Image Editing and Manipulation: Adobe Photoshop CS5 vs. GIMP

While operating systems and web browsers are relatively universal in terms of computer software used by the average individual, more specific programs, such as image editing software, are geared towards a much smaller user base. Accordingly, the available methods of establishing or maintaining *ethos* for this primary audience, and any appeals for broader audiences that may be unsure about the use of these types of products (or these specific products), play an enormous role in the success of these programs and the size of their user bases. Adobe attempts to persuade visitors to its website by connecting its Photoshop CS5 program to a large suite of design products, demonstrating authority across multiple types of design and media production. GIMP, on the other hand, focuses its efforts on maintaining credibility via demonstrations of its developers' continual improvement of the program's functionality, although it does so at the potential expense of a user base interested in quickly accessible features and tutorials of the program.

I. Adobe Photoshop CS5

As perhaps the most well-known and widely-used image editing program, Adobe Photoshop CS5 (http://www.adobe.com/products/photoshop/) has a website whose home page plays up the program's abilities through the use of image and text placement (Adobe, 2011). The emphasized content area of the page serves as an advertisement for, and a link to, a video tutorial on a feature new to Photoshop CS5. Surrounding that area are clearly visible fields describing and linking to specific pages on which viewers can learn more about the software or purchase it directly.

Below the image, the site explicitly appeals to the quality of the product: "The newest version of Adobe® Photoshop® CS5 software redefines digital imaging with breakthrough tools for photography editing, superior image selections, realistic painting, and more. And now, use it with creativity-boosting mobile apps" (Adobe, 2011). The home page points to specific attributes of the program in order to establish a claim of feature-based superiority over its competitors, thanks to an implicit suggestion that no other program can serve the range of diverse and specific purposes that Photoshop can. The initial appeal to *ethos* established by the page is thus one of corporate credibility and knowledgeable expertise through its assumptions of website visitor needs and interests, both of which are augmented by the site's effort to provide information supporting those visitor assumptions.

In addition, the continued reference to the published version of the software indicates a temporal as well as a professional quality: The user base for CS5 is distinguished from that of previous versions of the program even if its members greatly overlap. The created *ethos* becomes contingent upon the frequency of major updates to the software that could make it obsolete. That is, while CS5 may currently be heralded as the superior product on the site, there is an implicit suggestion that it will be rendered void by the inevitable release of CS6 and future releases, just as the capabilities of CS5 outshine those of previous editions of the software. While the website serves to present the increasingly varied and accessible functions of Photoshop as it is redesigned for each new edition of the Creative Suite of which it is a part, it also implicitly suggests that the existing software will eventually be outdone by its successor.

Intriguingly, the Photoshop website does not demonstrate the program's appearance solely through the use of static screenshot images of the software interface, but instead—and more significantly—it offers a short video demonstrating one of several of the program's newest features in action. This shift in media presentation allows the website to show a more in-depth demonstration of the program in operation than possible through images alone, thus reducing the need for a visitor to infer the "proper" use(s) of the interface. There is a powerful *ethos* generated by this approach because the video both shows the viewer exactly how the program works and also suggests that Adobe has a clear understanding of what a viewer would want (the content of the video and the inclusion of the video itself on the home page).

Adobe also demonstrates its authority through a series of statements relating to Adobe products as defining industry needs and expectations. Within the box to the right of the image where Photoshop is available for purchase, the product's user is described as being able to "[c]reate powerful images with the *professional standard*" (emphasis added). Below that is a link to the CS5.5 suite, given the title "Design Standard." Elsewhere, below the main image on the page is a five-star rating that doubles as a drop-down menu for users to read through other users' reviews of the product. The implicit argument seems to be that work reflecting the quality of the "professional standard" is possible only with the help of Photoshop CS5 and Adobe's other products—the strength of Adobe's appeal to its authority is here stretched to incorporate the company's entire software catalog, and the existence of Photoshop users' reviews verifies its authority.

Finally, Adobe's included set of links for support resources is a clear reflection of its authoritative claims. Official technical support is listed above a set of free video tutorials, related products that are available for purchase, and links to social media sites for public discussion, along with information about user-provided documentation and forums located near the end of the list. This organization allows Adobe to show that it values third-party technical assistance to an extent, but it gives far more weight to its own official set of resources developed by those involved in the creation of the product itself.

II. GIMP

In stark contrast to Adobe's presentation of its product is the website for GIMP, which stands for GNU Image Manipulation Program (http://www.gimp.org/) (GIMP, 2011). The GIMP website is primarily textual, with only the site's relatively small title image (comprised of the product's logo, name, and current version) offering visual distinction from the vast majority of the home page's content, which is almost completely comprised of information about updates to the software. If Adobe establishes its professional credibility through video demonstration of the current iteration of its software, GIMP's developers establish their own *ethos* by demonstrating continuous activity towards the improvement of their product rather than the creation of distinct and separate versions of that product, each of which possessing different capabilities.

However, because the primary emphasis of the home page is on update information, the GIMP site does not offer visitors to the website much introductory material through which they could become familiar with the program. The most information that is initially given is a brief description of the program, located beneath the site's title image, with links to several tutorials and a longer introduction to the software. These details are provided in small text and are not given much of a spotlight; it is as if, as in the case of Mozilla Firefox, the GIMP developers expect visitors to the site to already be familiar with the product. This assumed familiarity is supported by the list of detailed updates made to the program: for example, the text under the headline "GIMP 2.7.3 Released" reads, "The most visible changes in 2.7.3 are the fully working single-window mode, including working session management, and the introduction of a new hybrid spinbutton/scale widget which takes less space in dockable dialogs " (GIMP, 2011). The developers seem to suggest that visitors to the site are knowledgeable of GIMP, of the specific technical changes made to the program, and of how exactly the two work together. If a potential user does not possess this technical information, he or she may easily find him- or herself lost on the site and, presumably, in the program.

The extent of the site's appeal to *ethos* through OSS community collaboration and contribution is a handful of links on the home page. Only one is emphasized: a "Donate" button exists with a unique appearance below the site's navigational menu, which helps it stand out so the developers might receive financial assistance (as opposed to conventional OSS contributions). Along with a repeated "Donating" link (in a textual form mirroring the others on the page), the other relevant links are for "Bug Reports" and "GIMP Goods." The website creates a distinction between developers and users more reflective of a proprietary model than an OSS one; the extent to which most visitors to the site can see their potential contribution to the program's development is in the form of financial donation rather than through any active participation.

GIMP offers an *ethos* that both rejects and conforms to conventional proprietary software company *ethos*. On the one hand, GIMP's website demonstrates the dedication of its developers to the continued improvement of the program; on the other hand, the separation the website creates between those developers and the potential users of the program is extremely clear. Where this can

potentially fail is in regards to the descriptions of those continued improvements: Are the users or the developers the presumed audience? The website does not seem to distinguish between the two here. Instead, the content of the home page suggests that GIMP's developers expect all potential users to have a similar development-oriented interest rather than the design-oriented interest assumed by Photoshop CS5's website.

CONCLUSION AND CONSIDERATIONS FOR THE FUTURE

While there are certainly differences that can be easily seen between proprietary and OSS competitors of various types of computer software, there does not seem to be any broader, shared appeal to *ethos* by the OSS product websites that immediately connects each together philosophically or stylistically. Instead, the OSS websites overall seem to be negotiating a presentation centered on distinguishing themselves from their respective counterparts while, at the same time, remaining accessible and recognizable to potential users that are far more familiar with proprietary approaches and appeals. A general breakdown of the appeals to *ethos* made on each software site can be seen in Table 1.

Table 1. Appeals to ethos made on each website, by development type

Appeal to *Ethos*	Proprietary Website	OSS Website
Aesthetic Quality	Macintosh OSX	Mozilla Firefox
Efficiency	Macintosh OSX	
Features and Functionality	Adobe Photoshop CS5	Mozilla Firefox Red Hat Fedora
Expertise of Developer(s)	Adobe Photoshop CS5 Macintosh OSX Microsoft Internet Explorer	GIMP Mozilla Firefox
Official Support	Microsoft Internet Explorer	GIMP
Community Support	Microsoft Internet Explorer	Mozilla Firefox Red Hat Fedora
Continued Development		GIMP Red Hat Fedora

In only two cases are the OSS product of a compared pair positioned directly against its proprietary competitor, relating first to Mozilla Firefox and the implicit expertise of its development community as compared with Microsoft's Knowledge Base and other technical support, and, second, to the demonstrations of developer expertise provided by the teams for Adobe Photoshop CS5 and GIMP. There is one parallel use of language between a proprietary and OSS pair: The website for Macintosh OSX described its product as "Better. Faster. Easier." (Apple, 2011) and, Red Hat Fedora has a set of values emphasized on its home page through the phrasing "Freedom. Friends. Features. First" (Red Hat, 2011). While the *ethos* suggested by each site is quite different from the other, the similar wording of single-word descriptors suggests there *may* be some recognition by each company of the other's marketing strategies.

The vast majority of comparisons indicate a number of efforts by the OSS products' developers to engage their potential users in ways specifically *distinct* from the strategies employed by their direct competitors, even if some of these appeals are used by different proprietary developers for other products. For example, the website for Adobe Photoshop CS5 highlights the features of its products as much as the websites for Mozilla Firefox and Red Hat Fedora do, but GIMP—the OSS competitor of Adobe Photoshop CS5—takes a markedly different path by displaying an archive of updates from its seemingly constant development on its front page. Such a tactic is potentially employed to distance itself from being viewed as merely a free alternative to Adobe Photoshop CS5. Similarly, while the website for Macintosh OS X highlights the sleek efficiency of the system, the website for Red Hat Fedora avoids addressing a similar appeal—most likely since its focus is on encouraging any and all interested parties to consider becoming stake-holding contributors; the development provided by the volunteer community reflects and responds to the community's

varied requests for programs and features and is definitely not the consistently developed suite of programs specifically crafted by Apple to suit its customers on its hardware. However, there is one unifying factor that *does* exist across all of the OSS websites: the direct and accessible means of freely downloading each OSS product, establishing them as distinct from their competitors by virtue of each product's (absent) purchase price.

However, this accessibility and emphasized lack of financial cost, rather than shared and communal development, may work against the philosophy underlying OSS: Even though only a handful of websites were explored above, their content suggests some initial movement towards the proprietary developer-user dichotomy that separates users from developers and their design and improvement processes (outside of market demand influences), even with the varied set of approaches to explaining and marketing the OSS programs provided on these particular websites. The only site under examination that directly pushes against proprietary development philosophy is Red Hat, through its emphasis on "friends" as an integral component to Fedora's development, although Red Hat itself may already be perceived as too similar to the conventional proprietary institution.

What could be indicated by a growing distinction between developers and non-developers within the OSS community is one of several trajectories. One is that user populations may be growing more and more apathetic towards development, and so there is less reason for developers to reach out to potential volunteers (because they simply aren't there). Another is that a desire exists within development communities to promote OSS products to the widest potential user bases, whether those users are interested in OSS or not, and so imitating the proprietary aesthetic serves to facilitate this promotion. Yet another is that, as Elliot & Scacchi (2008) noted, more and more corporate opportunities have opened for OSS developers, allowing them to enjoy financial stability even while working on OSS projects; it could be that, in turn, promotion of OSS projects can often serve to facilitate career opportunities centered on proprietary software development.

If OSS as a community-centered philosophy is interested in "freeing" users from the dominant proprietary, closed-off software development process, then the OSS movement appears to be in jeopardy—and if OSS developers want to *fundamentally* and *explicitly* distinguish themselves and their products from their proprietary competitors, this (apparently) diminishing philosophy needs to be reintegrated and reemphasized within the appeals to *ethos* provided to potential users of the developers' OSS products through channels such as their websites. Otherwise, OSS may be perceived by potential users as a subset of proprietary programs, and open source as a broadly embraced philosophy could fall out of favor from all but those developers and users already interested and invested therein.

REFERENCES

Aberdour, M. (2007). Achieving quality in open source software. *IEEE Software, 24*(1), 58–64. doi:10.1109/MS.2007.2

Adobe. (2011). *Adobe Photoshop CS5, our latest picture and image editor software*. Retrieved November 6, 2011, from http://www.adobe.com/products/photoshop/photoshop/

Apple. (2011). *Mac OS X Lion – The world's most advanced OS*. Retrieved November 6, 2011, from http://www.apple.com/macosx/

Buchanan, R. (2004). Rhetoric, humanism, and design. In Handa, C. (Ed.), *Visual rhetoric in a digital world: A critical sourcebook* (pp. 228–259). Boston, MA: Bedford/St. Martin's.

Carnegie, T. A. M. (2009). Interface as exordium: The rhetoric of interactivity. *Computers and Composition, 26*(3), 164–173. doi:10.1016/j.compcom.2009.05.005

Casaló, L. V., Cisneros, J., Flavián, C., & Guinalíu, M. (2009). Determinants of success in open source software networks. *Industrial Management & Data Systems, 109*(4), 532–549. doi:10.1108/02635570910948650

Chan, D. (2011). Activist perspective: The social cost hidden in the Apple products. *Journal of Workplace Rights, 15*(3/4), 363–365.

Cheng, Q., Chen, F., & Yip, P. S. F. (2011). The Foxconn suicides and their media prominence: Is the Werther effect applicable in China? *BMC Public Health, 11*, 841–851. doi:10.1186/1471-2458-11-841

Christley, S., & Madey, G. (2007). Analysis of activity in the open source software development community. *Proceedings of the 40th Hawaii International Conference on System Sciences*. Waikoloa, HI.

Crowston, K., Heckman, R., Annabi, H., & Massango, C. (2005). A structurational perspective on leadership in free/libre open source software development teams. *Proceedings of the 1st Conference on Open Source Systems (OSS)*, Genova, Italy. Retrieved from http://floss.syr.edu

Crowston, K., & Howison, J. (2005). The social structure of free and open source software development. [from http://floss.syr.edu]. *First Monday, 10*, 1–27. Retrieved November 6, 2011

Del Bianco, V., Lavazza, L., Morasca, S., & Taibi, D. (2011). A survey on open source software trustworthiness. *IEEE Software, 28*(5), 67–75. doi:10.1109/MS.2011.93

Elliott, M., & Scacchi, W. (2008). Mobilization of software developers: The free software movement. *Information Technology & People, 21*(1), 4–33. doi:10.1108/09593840810860315

Gallego, M. D., Luna, P., & Bueno, S. (2008). User acceptance model of open source software. *Computers in Human Behavior, 24*(5), 2199–2216. doi:10.1016/j.chb.2007.10.006

GIMP Team. (2011). *GIMP – The GNU image manipulation program.* Retrieved November 6, 2011, from http://www.gimp.org

Grodzinsky, F. S., Miller, K., & Wolf, M. J. (2003). Ethical issues in open source software. *Information. Communication & Ethics in Society, 1*(4), 193–205. doi:10.1108/14779960380000235

Hartelius, E. J. (2011). *The rhetoric of expertise.* Lanham, MD: Lexington.

Hendry, D. G. (2008). Public participation in proprietary software development through user roles and discourse. *International Journal of Human-Computer Studies, 66*(7), 545–557. doi:10.1016/j.ijhcs.2007.12.002

Hocks, M. (2003). Understanding visual rhetoric in digital writing environments. *College Composition and Communication, 54*(4), 629–656. doi:10.2307/3594188

Howison, J. (2006). *Coordinating and motivating open source contributors.* Retrieved December 13, 2011, from http://floss.syr.edu

Howison, J., Inoue, K., & Crowston, K. (2006). Social dynamics of free and open source team communications. *IFIP 2nd International Conference on Open Source Software.* Lake Como, Italy. Retrieved from http://floss.syr.edu

Katz, S. B. (1992). The ethic of expediency: Classical rhetoric, technology, and the Holocaust. *College English, 54*(3), 255–275. doi:10.2307/378062

Kenney, K. (2004). Borrowing visual communication theory by borrowing from rhetoric. In Handa, C. (Ed.), *Visual rhetoric in a digital world: A critical sourcebook* (pp. 321–343). Boston, MA: Bedford/St. Martin's.

Kress, G., & van Leeuwen, T. (2006). *Reading images: The grammar of visual design* (2nd ed.). New York, NY: Routledge. doi:10.1016/S8755-4615(01)00042-1

Lankes, R. D. (2007). Credibility on the internet: Shifting from authority to reliability. *The Journal of Documentation, 64*(5), 667–686. doi:10.1108/00220410810899709

Lemley, M. A., & Shafir, Z. (2011). Who chooses open-source software? *The University of Chicago Law Review. University of Chicago. Law School, 78*(1), 139–164.

Microsoft. (2011). *Internet Explorer - Web browser for Microsoft Windows.* Retrieved November 6, 2011, from http://windows.microsoft.com/en-us/internet-explorer/products/ie/home

Miller, C. R. (2001). Writing in a culture of simulation: Ethos online. In Coppock, P. (Ed.), *The semiotics of writing: Transdiciplinary perspectives on the technology of writing* (pp. 253–279).

Mozilla. (2011). *Firefox Web browser – Free download.* Retrieved November 6, 2011, from http://www.mozilla.com/en-US/firefox

Open Source Initiative. (n.d.). *The open source definition.* Open Source Initiative. Retrieved November 6, 2011, from http://opensource.org/docs/osd

Raghu, T. S., Sinha, R., Vinze, A., & Burton, O. (2009). Willingness to pay in an open source software environment. *Information Systems Research, 20*(2), 218–236. doi:10.1287/isre.1080.0176

Raymond, E. S. (2000). *The cathedral and the bazaar.* Retrieved November 6, 2011, from http://www.catb.org/~esr/writings/homesteading/cathedral-bazaar/index.html

Red Hat. (2011). *Fedora Project homepage.* Retrieved from http://fedoraproject.org/

Rulliat, A. (2010). The wave of suicides among Foxconn workers and the vacuity of Chinese trade unionism. *China Perspectives, 2010*(3), 135-137.

Selfe, C., & Selfe, R. (1994). The politics of the interface: Power and its exercise in electronic contact zones. *College Composition and Communication, 45*(4), 480–504. doi:10.2307/358761

Warnick, B. (2004). Online ethos: Source credibility in an "authorless" environment. *The American Behavioral Scientist, 48*(2), 256–265. doi:10.1177/0002764204267273

Warnick, B. (2007). *Rhetoric online: Persuasion and politics on the World Wide Web*. New York, NY: Peter Lang.

Wiggins, A., Howison, J., & Crowston, K. (2008). Social dynamics of FLOSS team communication across channels. *Proceedings of the IFIP 2.13 Working Conference on Open Source Software (OSS)*, Milan, Italy, (pp. 131-142). Retrieved November 6, 2011, from http://floss.syr.edu

Zittrain, J. (2004). Normative principles for evaluating free and proprietary software. *The University of Chicago Law Review. University of Chicago. Law School, 71*(1), 265–287.

ADDITIONAL READING

Aristotle. (2007). *On rhetoric: A theory of civic discourse* (Kennedy, G. A., Trans.). Oxford, UK: Oxford UP.

Arola, K. (2010). The design of Web 2.0: The rise of the template, the fall of design. *Computers and Composition, 27*(1), 4–14. doi:10.1016/j.compcom.2009.11.004

Bitzer, L. F. (1968). The rhetorical situation. *Philosophy and Rhetoric, 1*(1), 1–14.

Bolter, J. D., & Grusin, R. (2000). *Remediation: Understanding new media*. Cambridge, MA: MIT Press.

Brooke, C. G. (2009). *Lingua fracta: Toward a rhetoric of new media*. Cresskill, NJ: Hampton Press.

Brummett, B. (2008). *A rhetoric of style*. Carbondale, IL: Southern Illinois UP.

Bucy, E. P. (2004). Interactivity in society: Locating an elusive concept. *Interactivity in Society, 20*(5), 373–383. doi:10.1080/01972240490508063

Carpenter, R. (2009). Boundary negotiations: Electronic environments as interface. *Computers and Composition, 26*(3), 138–148. doi:10.1016/j.compcom.2009.05.001

Crowston, K., & Howison, J. (2006). Assessing the health of open source communities. *IEEE Computer, 39*, 113-115. Retrieved from http://floss.syr.edu

Enos, T., & Borrowman, S. (2001). Authority and credibility: Classical rhetoric, the Internet, and teaching of techno-ethos. In Rosendale, L. G., & Gruber, S. (Eds.), *Alternative rhetorics: Challenges to the rhetorical tradition* (pp. 93–109). Albany, NY: SUNY Press.

Fagerjord, A. (2003). Rhetorical convergence: Studying Web media. In Liestøl, G., Morrison, A., & Rasmussen, T. (Eds.), *Digital media revisited: Theoretical and conceptual innovations in digital domains* (pp. 293–325). Cambridge, MA: The MIT Press.

Frobish, T. (2004). Sexual profiteering and rhetorical assuagement: Examining *ethos* and identity at Playboy.com. *Journal of Computer-Mediated Communication, 9*(3). Retrieved from http://www.wiley.com

Fuller, M. (2003). *Behind the blip: Essays on the culture of software*. Brooklyn, NY: Autonomedia.

Gaines, R. N. (2000). Aristotle's *rhetoric* and the contemporary arts of practical discourse. In Gross, A. G., & Walzer, A. E. (Eds.), *Rereading Aristotle's Rhetoric* (pp. 3–23). Carbondale: Southern Illinois UP.

Gurak, L. (1997). *Persuasion and privacy in cyberspace*. New Haven, CT: Yale UP.

Hamerly, J., Paquin, T., & Walton, S. (1999). Freeing the source: The story of Mozilla. In C. DiBona & S. Ockman (Eds.), *Open sources: Voices from the revolution*. Retrieved from http://oreilly.com/openbook/opensources/book/netrev.html

Hawk, B. (2004). Toward a rhetoric of network (media) culture: Notes on polarities and potentiality. *JAC: Special Issue on Mark C. Taylor and Emerging Network Culture, 24*(4), 831–850.

Kostelnick, C. (1996). Supra-textual design: The visual rhetoric of whole documents. *Technical Communication Quarterly, 5*(1), 9–33. doi:10.1207/s15427625tcq0501_2

Lanham, R. A. (1993). *The electronic word: Democracy, technology, and the arts*. Chicago, IL: U of Chicago Press.

Mackiewicz, J. (2010). The co-construction of credibility in online product reviews. *Technical Communication Quarterly, 19*(4), 403–426. doi:10.1080/10572252.2010.502091

Manovich, L. (2001). *The language of new media*. Cambridge, MA: MIT Press.

Mara, A. (2008). Ethos as market maker: The creative role of technical marketing communication in an aviation start-up. *Journal of Business and Technical Communication, 22*(4), 429–453. doi:10.1177/1050651908320379

Miller, C. R. (1994). Opportunity, opportunism, and progress: *Kairos* in the rhetoric of technology. *Argumentation, 8*(1), 81–96. doi:10.1007/BF00710705

Miller, C. R. (2010). Should we name the tools? Concealing and revealing the art of rhetoric. In Ackerman, J. M., & Coogan, D. J. (Eds.), *The public work of rhetoric: Citizen-scholars and civic engagement* (pp. 19–38). Columbia, SC: University of South Carolina Press.

Mitcham, C. (2009). Convivial software: An end-user perspective on free and open software. *Ethics and Information Technology, 11*(4), 299–310. doi:10.1007/s10676-009-9209-7

Porter, J. (2009). Recovering delivery for digital rhetoric. *Computers and Composition, 26*(4), 207–224. doi:10.1016/j.compcom.2009.09.004

Selber, S. (2004). *Multiliteracies for a digital age*. Carbondale, IL: Southern Illinois UP.

Shipka, J. (2011). *Towards a composition made whole*. Pittsburgh, PA: University of Pittsburgh.

Spoel, P. (2008). Communicating values, valuing community through health-care websites: Midwifery's online ethos and public communication in Ontario. *Technical Communication Quarterly, 17*(3), 264–288. doi:10.1080/10572250802100360

Yancey, K. B. (2004). Made not only in words: Composition in a new key. *College Composition and Communication, 56*(2), 297–328. doi:10.2307/4140651

Zappen, J. (2005). Digital rhetoric: Toward an integrated theory. *Technical Communication Quarterly, 14*(3), 319–325. doi:10.1207/s15427625tcq1403_10

KEY TERMS AND DEFINITIONS

Ethos: One of the three fundamental rhetorical appeals (along with *logos* and *pathos*), it is used as a means of appealing to a speaker's character, credibility, and expertise rather than to logic or emotion. For texts in which no clear speaker can be identified, the perceived function or potential use(s) of those texts can be evaluated as a component of *ethos*.

Exordium: An introductory component of the classical oration which is used to lay out the purpose of the oration and to make the audience more receptive to the ensuing argument.

Fork: An offshoot of an existing software project whose scope and code base have diverged in potentially significant ways from its root, the project from which it separated. Some forks are eventually merged back with their roots, while others become stand-alone projects (e.g. Ubuntu, which is a well-known fork of the Debian distribution of the Linux operating system).

Interface: A site or point of contact between two or more parties or entities that serves to mediate and facilitate interaction between those parties. For digital technologies, there are multiple levels of interface, including but not limited to: the physical tools used to manipulate a program (e.g. screen, mouse, keyboard, haptic touchscreen), the graphical user interface of the program, the compiled code of the program as it interacts with the operating system, the language(s) of the code used by developers, and the machine code into which developers' languages are compiled for use by the computer hardware in order to run.

Logos: One of the three fundamental rhetorical appeals (along with ethos and *pathos*), it is used as a means of appealing to reason, i.e. rational thought as opposed to moral character or emotion.

Multimodality: The use of multiple modes of communication (e.g. text, spoken word, image) as a means of engaging in meaningful discourse with a given audience.

Open Source Software: A form and practice of software in which the source code of a program is freely distributed alongside the compiled program so that other interested parties may, in turn, modify and freely re-distribute the program and its source code *ad infinitum*. Many open source programs incorporate contributions from the user community to help develop that software.

Pathos: One of the three fundamental rhetorical appeals (along with *ethos* and *logos*), it is used as a means of appealing to emotion rather than to moral character or logic.

Proprietary Software: A form and practice of software in which development entities (such as corporate institutions) are separated from the customer/user base for that software (that is, the users do not contribute to the development of a given software program) and generally do not distribute the source code for, or with, their software.

Rhetor: An individual engaging in an act of meaningful communication in some form (that is, one making use of rhetoric) with one or more audiences.

Section 2
Perceptions of Online Information

Chapter 5
The Ethos of Online Publishing:
Building and Sustaining an Inclusive Future for Digital Scholarship

Joe Erickson
Angelo State University, USA

Kristine Blair
Bowling Green State University, USA

ABSTRACT

This chapter argues that online academic journals are not only a legitimate venue and sustainable source of disciplinary inquiry, but an important professional development opportunity for graduate students as future faculty and are therefore crucial to maintaining a discipline's ethos. The authors begin by reviewing the ethos of individually produced print publications in the humanities, paying particular attention to the value such publications hold in helping scholars earn tenure and promotion. The authors then posit that efforts within the rhetoric and writing scholarly community to recognize the collaborative nature of multimodal digital texts and to advocate for the collaborative production of such digital texts, which has helped such scholarship achieve a higher level of ethos over the past two decades. Emphasizing the role of graduate students in these ongoing efforts, the authors conclude by recommending three benchmarks developing and advanced scholars should implement to increase their own professional development and thus the ethos of online academic publishing: curricular development, team development, and dissertation research.

INTRODUCTION

As Hawisher and Selfe noted (1997), editorial roles on scholarly journals are exceptionally helpful to graduate student professional development in that these future faculty are more successfully able to dialogue with both new and established colleagues in a process that is reciprocal and

supportive, enabling a mentoring approach in the development of digital scholarship. Although Hawisher and Selfe were writing at a time when online journals were few and far between, in many ways, the advent of these digital scholarly models have helped to achieve this professional goal. This chapter will address the need to provide meaningful opportunities for mentoring graduate students for roles as faculty who foster multi-

DOI: 10.4018/978-1-4666-2663-8.ch005

modal composing in both teaching and research; we argue that developing these opportunities is a necessity for all disciplines, but we will focus on existing models of scholarship found in rhetoric and composition to develop our points. In the spirit of Lisa Ede and Andrea Lunsford's (1990) collaborative scholarship over the years, we will also argue that collaborative journal administration moves away from isolating hierarchical models of mentoring and publishing within the academy toward a community of scholar-teachers.

As the Modern Language Association Taskforce on Evaluating Scholarship for Tenure and Promotion (2006) initially suggested, despite the recognition of the potential of various digital tools to produce and distribute collaborative scholarship in the field, very often our incentive and reward systems still privilege single-authored alphabetic literacy models, and tenure and promotion committees, as well as department chairs, have little to no experience in assessing online scholarship. Indeed, Purdy and Walker (2010) argued that "While both faculty members using digital tools and committees charged with evaluating tenure-and-promotion cases have tried to create appropriate categories for digital scholarship, their success remains partial" (p. 177). Our chapter will argue that the scholarly collective necessary to sustain online publishing reacts against traditional evaluative models of scholarship that have limited the voices that can speak and the modalities in which they can be heard. Through our perspective as two editors of the journal *Computers and Composition Online*, we will advocate for the ethos of digital scholarship as a legitimate venue and sustainable source of disciplinary inquiry, and thus a viable professional development opportunity for graduate students as future faculty.

The Ethos of Print

In his influential book *Scholarship Reconsidered*, Ernest Boyer (1990) called upon the academy to rethink its definition of scholarship in ways that not only break down binaries between research and teaching but also between the individual and collaborative. As Boyer concluded, "professors, to be fully effective, cannot work continuously in isolation. It is toward a shared vision of intellectual and social possibilities—a community of scholars—that the … dimensions of academic endeavor should lead. In the end, scholarship at its best, should bring faculty together" (p. 80). Despite Boyer's emphasis upon the collaborative and upon a broadened range of scholarly activity that values pedagogy, engagement with the community, and integration as opposed to separation of the disciplines, more than two decades later, Boyer's ideas are, for many, more rhetoric than reality. While certainly some disciplines in both the sciences and social sciences recognize and value collaborative research, it is not likely to be pedagogical, and still others may continue to privilege single-authored scholarship in print form.

For instance, the 2006 Taskforce of the Modern Language Association's Report on Tenure and Promotion hailed Boyer's emphasis on the scholarship of teaching and stated that the "goals of scholarship in the fields represented by the MLA would be better served if the monographic book were not so broadly required or considered the gold standard for tenure and promotion" (p. 26).

The Taskforce Report acknowledges the lack of progress when reviewing statistical data that suggests the single-authored monograph is considered "a reasonable demand" at a range of colleges and universities with varying missions rather than just the Carnegie doctorate-granting institutions (see "About Carnegie Classification" for details about this classification system). But whether it be a book or an article, individually written or collaborative, what trumps the "tyranny of the monograph" (Waters, 2001) is the "ethos of print." Such an ethos is built upon the presumption that print is somehow more rigorously imbued with an innate quality that results from a blind peer review process in which rates of acceptance are almost as important to a publication as the contribution to the field.

A significant part of that print ethos is ascribed to the author as creator or, as Martha Woodmansee (1994) suggested, "master." For Woodmansee, "The notion that the writer is a special participant in the production process—the only one worthy of attention—is a recent provenience. It is a by-product of the Romantic notion that significant writers break with tradition to produce something utterly new, unique—in a word, 'original'" (p. 16). Certainly, the emphasis on the construct of individual creativity has long been a subject of critique in rhetoric and composition (Brodkey, 1987), in part as a legitimation of both collaborative writing and postmodern emphases on authorship in order to establish ethos of our academic labor as rhetoric and writing specialists. Woodmansee documented the historical and ideological shifts away from what were clearly collaborative production processes of print books in which the author was one contributor among several, and a precursor, we would stress, to the frequently collaborative production processes for digital scholarship as well.

More recently, the critique against privileging print literacy in a digital world has been tied to the increasing role of technology in the teaching of writing and the resulting desire for teachers to establish a recursive relationship between theories of multimodal composing (New London Group, 1996) and the pedagogical practices that sustain digital writing across the curriculum. This exigency has led to a range of online journals and other forums such as the WAC Clearinghouse to *Kairos* and *Computers and Composition Online*, not to mention forums in digital literary studies and humanities as well, from Romantic Circles to the Princeton Dante Project and *Text Technology*.

Yet from the onset, such online scholarly forums have been subject to suspicion; in much the same way that teachers have discouraged students from overreliance on online sources, departmental

colleagues and the tenure and promotion procedures they enforce all too often position the online publication as inferior, lacking the ethos of its print counterpart. Citing Middlebury College's ban on Wikipedia as a research source, Kathleen Fitzpatrick (2010) observed that despite the scholarly analysis of "convergence culture" (Jenkins, 2006), "scholars have resisted exploring a similar sense in which intellectual authority might likewise be shifting in the contemporary world," concluding that "The production of knowledge is the academy's very reason for being, and if we cling to an outdated system of establishing and measuring authority while the nature of authority is shifting around us, we run the risk of being increasingly irrelevant to contemporary culture's dominant ways of knowing" (p. 17).

In other words, through our efforts to maintain our "traditional" scholarly ethos, we ironically lose that ethos among a larger culture that includes undergraduate students who see a gap between how they read and compose and how we define and value such literate activities. The emphasis on students is an important one because they not only represent the social future of literacy in the 21st century but also the academic future among those who choose careers as college-level faculty. Given the gap between our study of digital texts as opposed to our production of such texts in our knowledge-making and distribution role as academics, it is important to consider the ways these processes must maintain their currency and relevance, and ultimately, their ethos. Fitzpatrick pointed to various social media, such as Media-Commons and Philica, as a way to move from a scholarly press model to a scholarly network model, questioning the extent to which Web 2.0 systems foster sustainable conversations among scholars.

It is this paradigm shift away from hierarchical, anonymous feedback between an "established" scholar in the field qualified to comment on the submissions of both established and novice authors that calls into question our traditional concepts of authority and expertise. Undoubtedly, some online journals have upheld, rather than subverted, more hierarchical peer review processes, doing little more than reviewing and publishing print on screen and thus meeting the expectations and maintaining the more traditional model of ethos within their academic discourse communities. And while those journals that emphasize the affordances of new media by calling for multimodal (as opposed to purely alphabetic) contributions break new ground, their trailblazing efforts are not without challenges for editors and authors alike.

Recognizing the gap between advocating for digital scholarship and acknowledging the inability to produce and evaluate such research activity, the MLA taskforce report made it clear that the possibilities and constraints of establishing digital ethos are a significant challenge in English studies, and by extension, rhetoric and writing. Nevertheless, as Tulley and Blair (2009) noted in their discussion of the need for sustainable opportunities for collaborative digital composing across the English curriculum, "Although some faculty within English departments may continue to believe that ... emphasis on the consumption of literary texts exempts them from teaching functional, critical, and rhetorical digital literacy, to shirk this responsibility risks reinscribing a curriculum in which future faculty in English studies fail to bridge the gap between the literate practices of the academy and the actual literacy practices of twenty-first century students" (p. 465). Because of the recursive relationship between teaching and research across the disciplines, the need for venues in which multimodal composing pedagogies, and the student artifacts resulting from them, can be both theorized and showcased is an editorial imperative within our rhetoric and composition.

In their 2009 article in *The Journal of Scholarly Publishing*, authors Joan Cheverie, Jennifer Boetcher and John Buschman observed that humanities scholars have traditionally been solitary individuals who place tremendous value in the scholarly monograph, both for the purposes

of conducting their own library research and, of course, for the purposes of earning tenure and gaining promotion (p. 224). In addition to the privilege given to individually produced ideas, privilege has also traditionally been given to the stability suggested by the materiality of printed pages tightly bound in physical volumes. Traditionally speaking, the most valued scholarly contributions came from single authors in the material form of the scholarly monograph.

This is in sharp contrast to the fluid and multimodal nature of information produced for consumption in digital environments. Cheverie, Boetcher, and Buschman (2009) rightly observed that "the digital humanities are collaborative: Aesthetic design, that is, the visual look and feel and auditory components, is as important as the information design and architecture" (p. 225). Just as the expressive modes of digital scholarship operate in concert to impart meaning, so, too, is digital scholarship often collectively produced. Cheverie, Boetcher, and Buschman also pointed out just how collaborative the production of multimodal texts can be:

Putting the work in the proper structure requires technical designers who program and arrange data structures, and a crucial role is played by the project manager, who wears many hats and who must align the vision across all the design and thought processes and visualize the end product. Grant writers may participate at the beginning of the project or may be a part of the final design or management, or both. Graduate students who work to populate the Web site or database are the workhorses of such projects. Without any of these people, the project will be deeply flawed or fail. (p. 225)

The scenario described by Cheverie, Boetcher, and Buschman above is certainly within the realm of possibility for the production of academic scholarship, particularly with the rise of new online journals like *Vectors*, described as a "journal that brings together visionary scholars with cutting-edge designers and technologists to propose a thorough rethinking of the dynamic relationship of form to content in academic research" ("Dynamic Backend Generator"). However, as editors of an online journal in the field of rhetoric and composition, our experience suggests that this level of collaboration is not typical of the production of an article length piece for many other digital journal in this field such as *Kairos: A Journal of Rhetoric, Technology, and Pedagogy* or *Computers and Composition Online*. While we commonly see two or more authors collaborate on a piece, we don't see pieces that were created by several different people, each with specialized training in graphics, coding, web design, and grant writing working as a production team, per se. Rather, when such pieces are collaboratively produced, which is not always the case, it typically involves scholars from one field pooling their ideas about a particular scholarly topic and working together to arrive at a dialogically produced position or idea and, sometimes with the assistance of journal editors, a workable interface for presenting that idea in a web environment. As a collaborative means of expression, multimodality would seem to practically invite collaborative authorship. The value gained from intellectual products that emerge from a dialogic exchange of ideas via collaborative authorship could be further enhanced when our understandings of collaborative authorship also include the development of a customized expressive framework for a piece that organically supports the scholarly ideas of a piece.

There is no doubt that composing pieces featuring multiple expressive modes demands multiple sets of authorial literacies. In addition to researching and writing, for example, such pieces often involve the labor of creating and arranging static graphics, audio files, animations, and video files, and all of this must occur via the use of increasingly sophisticated web languages like HTML5 in combination with CSS, Javascript, PHP, or Adobe Flash, among so many other possibilities. Perhaps

more importantly, if authors want their work to be taken seriously by a particular journal and it is readers, they must be able to synthesize all of the expressive modes they use in ways that meaningfully support, present, and even add meaning to the content of the piece. In short, producing credible online scholarship involves a lot of often unseen and unrewarded intellectual labor. Authors must select modes of expression from a wide array of possibilities. They must utilize a number of digital technologies to translate ideas into textual, aural, and/or visual expressions. They must then layer those expressions on top of the complex languages and structures of the web, which, if done with care, results in expressive forms that should *add* meaning and intellectual depth to the content of a scholarly piece, not simply distribute it.

Neither the production of digital scholarship or the meaning expressed via digital scholarship adheres to the traditionally valued hallmarks of an individually produced "big idea" expressed in a textual mode and archived in a whole and tangible material artifact. Given these fundamental differences between print and digital forms of scholarship, it might seem that academe has arrived at an ideological impasse. If academe insists upon assigning ethos to scholarship based upon values established from a print dominated age, then, of course, print-based scholarship will necessarily be more highly valued within academic institutions.

For example, in a recent opinion piece for CNN.com, Fitzpatrick (2012), who also serves as director of scholarly communication for the Modern Language Association, argued "that 'writing' is still seen by many as 'producing words on paper,' just as 'reading' may still be associated primarily with 'books.' Such equations run the risk of blinding us to what's actually happening in contemporary culture." Her words underscore the important fact that academic institutions are situated in broader cultural contexts in which millions of people are accepting, even privileging, digital forms of information. For example, Amazon.com, one of the world's largest retailers of books, magazines, and other media, reported that in 2010 sales of ebooks out paced those of printed books for the first time (Kincaid, 2011). Additionally, e-reader devices, such as Amazon's Kindle, Barnes and Noble's Nook, and Apple's iPad, have been selling in the millions of units per quarter in recent years, and the Kindle frequently occupies the top selling product spot on Amazon. com. (McDougall, 2012).

It is important to note, however, that most ebooks are currently just printed books in digital form, thus not necessarily multimodal texts simultaneously deploying a range of other semiotic modes, but it's important to remember, too, that the ebook format is still in its infancy. We might think of them as the incunabula of our age. As Jay Bolter and Richard Grusin (2000) have noted, even the Gutenberg Bible of the fifteenth century, widely accepted as the first press-printed book, could "hardly be distinguished from the work of a good scribe, except perhaps that the spacing and hyphenation are more regular than a scribe could achieve"; it took "generations for printers to realize that they could create a new writing space with thinner letters, fewer abbreviations, and less ink" (p. 8). We have already seen some evolutionary movement in the ebook format with such enterprises as the Vook ebook platform, which lets users create, distribute, and experience "true enhanced multi-media eBooks" ("What's vook?", 2012). Additionally, Apple has recently introduced its new iBook textbook platform, which it described as a "gorgeous full-screen experience full of interactive diagrams, photos, and videos" ("Apple in Education," 2012). Apple's description emphasized the interactive and multimodal enhancements afforded by their new ebook platform, stating that students are "no longer limited to static pictures to illustrate the text" and that "students can dive into an image with interactive captions, rotate a 3D object, or have the answer spring to life in a chapter review" ("Apple in Education," 2012). If history is any indicator of future trends, and if current sales figures are indicative of a cultural

appetite for digital content, than we have clear evidence that digital and increasingly multimodal content has established a broader cultural value, a value that academe would be unwise to ignore considering its position within that broader culture.

How, then, does the academy begin to bridge this apparent gap between the print-based values held by certain influential members of the academy and the emerging values others are assigning to digital scholarship?

The Ethos of Collaboration

Writing of the rise of collaboration in the "hard disciplines," Bayer and Smart (1991) noted that the greatest degree of collaboration tends to occur in those fields that are more "financially supported" and the "soft" sciences, "particularly the humanities, generally have markedly lower rates of collaborative scholarship" (p. 613). Although Bayer and Smart were writing in 1991, and we can certainly point to applied disciplines such as education that promote collaboration, the MLA's taskforce report confirms this presumption about the humanities almost two decades later. As a result, those working in rhetoric and composition in general and digital rhetoric and writing in particular, most of whom are housed in departments of English, can face institutional and ideological obstacles when hoping to collaborate on both print and digital projects and the presumption that there is less work on the part of individual contributors to such projects.

Despite the MLA Taskforce Report endorsement of Boyer's original call for a broadened notion of scholarship that includes teaching and engagement with the community, the irony exists that while we may foster collaborative projects in classroom settings, when we engage in such work ourselves, it is suspect and of lesser ethos. A review of tenure and promotion in English studies, especially from those institutions with the Carnegie designation of "very high research activity," reveals this bias. For instance, Texas

A&M's English Department tenure and promotion document (2012) stated that "In English, promotion to full professor has been based on the publication of a single-authored monograph or substantive scholarly edition from a reputable scholarly press to represent high-quality research." While the document acknowledged flexibility in determining reputation and quality, and does acknowledge co-authored projects, the typical emphasis for both tenure and promotion to associate and full professor is on a completed, presumably single-authored book.

Yet, on Texas A&M's Initiative for the Digital Humanities, Media, and Culture's Commentpress Blog, Laura Mandell (2011) asserted that the definition of digital scholarship must presume that the relationship between content and delivery is collaborative: "In effective digital research, digital media are not incidental but integral to the scholarly work," and that even the basic technical distribution of a scholarly project means that "digital scholarship by its very nature requires collaboration, and so we must have peer-reviewing mechanisms that take that into account," and support the increasing numbers of humanities faculty going the collaborative digital route, not unlike their scientific counterparts. Certainly such dialogues within English studies are productive ones for their impact on tenure and promotion procedures that dismiss online work for its collaborative elements that move away from perceived rigors of individual academic labor and rigorous blind review. However, even institutions like Texas A&M are acknowledging the need to better inform themselves about the production, distribution, and evaluation processes surrounding digital scholarship, as evidenced by the fact that Laura Mandell's post on the IDHMC blog is introduced as "An Open Letter to the Promotion and Tenure Committee at Texas A&M University, Department of English, upon their request for information about how to evaluate digital work for promotion and tenure."

This example certainly represents a move forward to sustainable partnerships among faculty, and potentially among faculty and students. But for digital scholars working in Rhetoric and Composition, there remain concerns that regardless of the groundwork that has been accomplished through digital journals such as *Kairos: A Journal of Rhetoric, Technology, and Pedagogy* (founded in 1996), the voices of digital rhetoric specialists will be ignored by the larger field of English studies, and we will neither be able to shape emerging tenure evaluation guidelines nor have these newer models apply to our work; instead, it remains possible that the traditional model based on print and the monograph, the model that established the preferred ethos of literature scholars within English departments in the 20th century, will continue to hold sway, much the same way scholarship in composition studies that blends theory and pedagogy has traditionally been devalued.

In her response to the MLA *Profession* 2011 special section "Evaluating Digital Scholarship," *Kairos* Editor Cheryl Ball (2012) lamented that "this issue of *Profession* came out and the section on evaluating digital scholarship contained **no** digital writing or rhetoric scholars, and the MLA was repeatedly touted as having 'taken the lead in encouraging the recognition of digital scholarship in promotion and tenure cases' (Schreibman, Mandell, & Olsen, p. 126)" (n.p.). Ironically, in attempting to support teaching and research innovation in the field of English studies, the MLA reinforces divides between literary studies and writing studies, neglecting much of the ongoing work on peer review, multimodal composition, and tenure and promotion policies. This is likely tied to a presumption that digital scholarship in rhetoric and composition is inherently pedagogical, and although we resist the binary that such a presumption establishes between theory and practice, numerous examples of theory-driven digital scholarship *are* published in *Kairos, Computers and Composition Online* and elsewhere.

For instance, the Fall 2011 issue of *Computers and Composition Online* features a collaborative webtext by Dánielle DeVoss and Julie Platt, "Image Manipulation and Ethics in a Digital-Visual World," which analyzes more than twenty popular media images in ways that align with the work of media theorists such as Henry Jenkins (2006). A similar example occurs in Paul Muhlhauser's (2011) "Teaching Moms and Dads to Perform the Family: Rhetoric and Assisted Reproductive Technology Websites," which, although single-authored, provides a powerful example of the type of rhetorical analysis and cultural critique that has its roots in both communication and literary studies.

As with Ball's significant work on establishing mentoring models of peer review in which section editors and editorial board members are assigned to mentor potential contributors on both form and content prior to submission, in our own collaborative work on *Computers and Composition Online*, we have striven to provide an editorial collective in which both new and established authors of digital scholarship receive support in readying their webtexts for peer review and publication. A recent model of this process involves the collaboration between Dirk Remley and Joe Erickson (2010), who initially queried *Computers and Composition Online* about a piece about the role of Second Life in the teaching of professional writing. While Remley had media artifacts and an initial .html formatted text, the first version was not yet ready for publication; thus, Erickson, as a graduate student, worked on the design efforts in consultation with Remley to produce an interface that made stronger use of the affordances of web publishing to integrate the visual and the textual into a rhetorically effective argument and a webtext the editors were ultimately pleased to publish with both Remley and Erickson receiving credit on the piece. Specific elements of the redesign process included making use of cascading style sheets (CSS), embedding video

into .html documents, and creating a design theme and navigation structure that fit the content. Thus, this process was a significant learning opportunity for the author, and since that time, Remley has submitted additional work that he has designed and coded individually but has also served as a reviewer for other multimodal webtexts submitted to the journal, suggesting the important role of mentoring in the design and evaluation of digital scholarly. Also, since Erickson was a graduate student at the time, this process demonstrates the importance of graduate students engaging in editorial collectives and being rewarded for them, and it also chronicles the important role of collaboration in both the digital composing process and the preparation of future faculty.

The differing audience needs within, and expectations for, multimodal scholarship such as Remley's call for the types of team development models we have employed at *Computers and Composition Online* in ways that differ from online journals that publish work in primarily print formats. As is often the case, faculty new to digital composing do not necessarily have the technological skills to develop a high-end webtext, and thus the process of developing both the skill-set and the resulting scholarly artifacts is very time-intensive, much more time-intensive than continuing to hone an already established base of alphabetic writing for print-based production processes. And, just as the MLA Taskforce Report acknowledged that all too many scholars, including faculty working in rhetoric and composition, do not know how to produce digital scholarship, it's clear that an even larger number don't know how to evaluate it either, which provides challenges in establishing viable models of peer review. As Cheryl Ball's efforts at *Kairos: A Journal of Rhetoric, Technology, and Pedagogy* suggested, a range of strategies are needed to educate not only authors but also peer reviewers and students as future producers themselves. In her recent article, "Assessing Scholarly Multimedia: A Rhetorical

Genre Studies Approach" (2012a), Ball stressed that criteria for assessing digital scholarship can—and should be—collaboratively generated by demystifying peer review by documenting the types of criteria used by editors and reviewers at *Kairos: A Journal of Rhetoric, Technology, and Pedagogy*, and then sharing such criteria with students so that they reflect upon and update the criteria for their own rhetorical purposes. Because Ball (2012a) overviewed a series of digital assessment heuristics published and in use at *Kairos*, she concluded that

I hope readers keep in mind that each of these interpretive and evaluative verbs (reading, grading, assessing, evaluating) indicates a different audience—randomly and overlapping: pleasure readers, students, scholars, hiring committees, tenure committees, teachers, and authors—each of which has different needs from, and comes to the reading experience with different value expectations of, such a piece of scholarship. (p. 63).

Inevitably, establishing shared and criteria for assessing digital scholarship not only maintains scholarly ethos but also sustains digital scholarly production. This sustainability manifests itself in several ways: 1) online journals are able to document peer review processes for purposes of tenure and promotion; 2) editors, reviewers, and faculty are able to help educate current and future faculty about the expectations involved in developing scholarly webtexts that go beyond mere print on screen for publication in online journals, and 3) the increased emphasis on multimodality in both writing studies and English studies will lead to even more online venues for publication for faculty and even students. This latter point is evidenced by recent journals such as *The Journal for Undergraduate Multimedia Projects* (JUMP) at the University of Texas at Austin; *Present Tense: A Journal of Rhetoric in Society,* founded by graduate students at Purdue

University; and *Harlot: A Revealing Look at the Arts of Persuasion*, founded by graduate students at The Ohio State University. Given that *Kairos* was itself developed by graduate students back in 1996 (most of whom are now themselves faculty), online journals provide a significant opportunity for current and future faculty to immerse themselves in the theories and practices of digital scholarly production, thereby establishing a long line of powerful digital experience that can positively affect existing and emerging graduate students and faculty, which can have an obvious effect on the field as a whole as well.

The Ethos of Design

Although we use the term "digital scholarship," the definition is a nebulous one given the continuum between alphabetic texts published online and webtexts that integrate a range of visual, textual, and aural modalities. In her 2004 *Computers and Composition* article, Ball also made a persuasive case for the importance of defining exactly what the concept of new media scholarship signifies. For many, this particular phrasing has been synonymous with online scholarship, digital scholarship, or even multimodal scholarship. Based upon her work at *Kairos* and her ongoing engagement with the computers and writing community, Ball observed that there are important distinctions in the form of scholarly articles published online. For example, many online articles are primarily text-based pieces, with perhaps a few images or video clips to augment their text-centered meaning. Ball referred to this format as "online scholarship." She went on to differentiate online scholarship from "new media scholarship," which she described as pieces that "juxtapose semiotic modes in new and aesthetically pleasing ways and, in doing so, break away from print traditions so that written text is not the primary rhetorical means" (p. 403). She argued that distinguishing "online scholarship" from "new media scholarship" can provide readers with an interpretive framework for evaluating the scholarly meaning and value of new media scholarship.

Ball's article proposed an important solution to the well-understood yet lingering problem of applying print evaluation standards to digital texts. According to Jay David Bolter and Richard Grusin (2000), it is logical that people tend to graft their understandings of older media onto new media before the new media evolve and take on new defining conventions that might be harder to grasp; they refer to this as the process of "remediation" (p. 45). While Bolter and Grusin argued that a dialogic relationship persists between old media and new media as they each respond to each other's presence in culture over time, at some point the new media does develop its own defining characteristics, even if it never fully severs its ties to the older media from which it evolved. But just as the new media itself evolves, so, too, must the literacies of those who make use of and evaluate new media.

Media can mature faster than our ability to recognize and embrace its maturity. In their 2005 *Computers and Composition Online* article, Shawn Apostel and Moe Folk referred to this as an example of "First Phase Information Literacy," which they described as adapting evaluation methodologies designed for books and applying them to websites ("First Phase"). Drawing on Bolter and Grusin (2000), Apostel and Folk acknowledged that early websites did in fact look quite similar to print media, for they utilized alphabetic text as their primary expressive mode. They refer to such websites as "first generation." However, by the time their article was published in 2005, the web had already begun to mature to the point where it could feature Flash animations along with video and audio content, features they assign to "fourth generation" websites. With the launch of Facebook in 2004, the web was on the verge of transforming into the interactive and social web 2.0 we know so well today in 2012. Yet, despite reaching its "fourth generation" of maturity, Apostel and Folk still had significant cause in 2005 to claim that,

"our approach to evaluating websites has changed little since they were first introduced." In 2011, Cheverie, Boetcher, and Buschman echoed the same concerns, as did Ball (2012a) in her recent "Assessing Scholarly Multimedia" article.

Given the arguments and observations offered by scholars in the fields of rhetoric and composition, media studies, and digital publishing, it's clear that there exists a cross disciplinary understanding of the distinct nature and value of digital media. The scholarship of Bolter and Grusin (2000), Apostel and Folk (2005), and Ball (2004, 2012a) recognized the relationship between print and digital media but emphasized that the unique, multimodal value of digital media will, with time and collective influence, serve to distinguish digital media in the minds of enough people within the academy to overcome the inertia of the print perspective. Additionally, peer-reviewed and multimodal scholarship featured in online journals like *Kairos: A Journal of Rhetoric, Technology, and Pedagogy* and *Computers and Composition Online* will, as all scholarship should, provoke new ideas and new responses that will be best expressed in multimodal forms. Indeed, the collaboration between the multiple modes of expression that give online scholarship its unique value harmonizes with the work of online journal editors like Ball and Blair to model and encourage collaboration in the production and editorial management of online scholarship. By emphasizing multiplicity and collaboration within each layer of online scholarship – at the level of a scholarly piece itself, but also including the processes of it production, publication, and management – attributes such as "multiple," "collaborative," and "democratic" will eventually overcome the tenure of deeply rooted attributes like "print" and "mono" in collective mindset of the academy, resulting in a more highly recognized ethos as these qualities garner more respect and practice in other disciplines and even outside the academy as well.

In the opening to his 2011 book, *How to Design and Write Web Pages Today*, Karl Stolley acknowledged that "writing for the web is a community activity" (p. xiv), which echoes our emphasis on collaboration in the production of online scholarship. Given the "how to" purpose of his book, however, Stolley also pointed out that,

in order to join or even simply benefit from the knowledge of any community—whether photographers, football fans, carpenters, knitters, poker players, medical doctors, or Web designers—you have to know or be willing to learn the words that that community uses in addition to engaging in photography, carpentry, poker, or whatever activity the community is known for. (p. xiv)

Stolley's book is part of the *Writing Today* series of books published by Greenwood Press that aims to "help students, professionals, and general readers write effectively for a range of audiences and purposes" (p. ix). In other words, Stolley's audience is wide ranging, consisting of students and professionals within and outside of the academy. This diverse target audience is an appropriate fit for Stolley, whose work interests audiences both within the academy and outside of the academy in the web development and open source communities. On his personal website, Stolley (2011a) described himself as a "Chicagoan. Professor. Researcher, practitioner, and evangelist of open standards and open source for Web development and information architecture"; while his "educational background is in classical and contemporary rhetoric," he is also known for his technical expertise as the lead web designer for *Kairos,* as well as the heavily trafficked Purdue Online Writing Lab. In the web design courses Stolley teaches at the Illinois Institute of Technology, he asks his communications students to learn web mark up languages including XHTML and CSS (2011a). For Stolley, learning the markup languages of the web is no different than learning the "words" of any community to which one wants to become a part of or interact with. In other words, Stolley would likely argue that writing web markup is not just for self-proclaimed "techies."

The web has become such a prolific presence in contemporary culture that basic hypterext markup should no longer be considered a specialized knowledge; rather, it should be a widely adopted literacy.

Stolley is not alone in this notion. Bradley Dilger and Jeff Rice's edited collection *From A to <A>: Keywords of Markup* (2010) offered a series of voices that collectively argue that, "In the age of new media, there is no way to avoid markup. Markup is text. Markup is communication. Markup is writing" (p. ix). It should be no surprise that such a book was awarded the 2011 Computers and Composition Distinguished Book Award, for it explicitly recognized the need for widely held understandings of not just how to read and write markup, which are important literacies themselves, but also to understand and appreciate the broader cultural influence of markup, hidden as it is from view for most people. In other words, markup language, like any language, is worthy of—in need of—functional, critical, and rhetorical study and understanding, and therefore members of the humanities should be amply equipped to engage markup as part of their broad concern for language and literacy.

It's no secret that computers and software have long been studied from multiple standpoints within rhetoric and composition as a field, and other fields as well. As early as 1995, for example, Gail Hawisher, Paul LeBlanc, Charles Moran, and Cynthia Selfe published *Computers and the Teaching of Writing in American Higher Education*, in which they outlined the emergence and eventual influence of the computer on the nature of the writing process and the teaching of writing in college settings. Disciplinary journals like *Computers and Composition* and *Kairos,* along with the annual Computers and Writing conference, also serve as manifestations of disciplinary interest in the connections between computer technologies and writing broadly speaking. These and other similar disciplinary enterprises have collectively served to bolster the ethos of online publishing within rhetoric and composition as an academic discipline and, arguably, within the academy as a whole. Until recently, though, many within the umbrella field of English Studies have chosen not to substantially engage the code and markup underlying the digital interfaces that allow us to interact with computer technologies. As Stolley, Dilger, and Rice have argued in recent years, markup is a language foundational to the value of digital publishing, and, as such, if we choose to leave it sequestered to the domains of other disciplines, we forfeit opportunities to fully explore the potentials it affords for composing rich, multimodal documents capable of expressing ideas in ways not possible via alphabetic text alone.

Beyond expressive potential, we must also consider the ubiquity of markup in our world and its influence on the millions of people who consume digital information every day. When we conceptualize web markup as a literacy and study it as such, we adopt a line of research that has value across disciplinary lines both within and outside of academe. In other words, the expressive potential and global reach of markup make it a valuable line of inquiry because of the potentially broad cultural impact such research might produce, and, more specifically, its obvious relevance and importance to the citizens and governments that fund academic research. By both studying markup and publishing our scholarship in venues that rely upon careful and thoughtful uses of markup, we can potentially bring to light the value of the intellectual labor involved in producing online texts and thereby increase the ethos of online publishing within academe. We might also draw greater public attention to the important work of literacy studies and the digital humanities broadly speaking.

CONCLUSION

The value of digital scholarship in academe rests in its alignment with Boyer's original goals of fostering intellectual community and scholarly dialogue as opposed to monologue. Digital scholarship is also, in light of the material conditions and the overall access to information, the present and future of an academy in which both teaching and research are conducted online (and economic realities, among other circumstances, continue to shut down many university print presses). Rather than ignore the important work within digital rhetoric and writing studies in the shift from print to digital, English studies and the digital humanities can benefit from longstanding efforts to professionalize students in rhetoric and composition about the rhetorics and realities of scholarly publishing and perhaps provide a model to other disciplines looking to expand in this area as well. Addressing those rhetorics and realities should involve educating colleagues across the disciplines about the academic labor involved in developing digital scholarship, as well as the need to counter presumptions that digital scholarship in rhetoric and composition is purely pedagogical and not as theoretical as work in literary studies. In short, the work to elevate the ethos of digital scholarship must be undertaken by those who are actively involved in producing it, not relying on those who are involved in consuming digital scholarship separately from its means of production and equating it to print paradigms. Situated within such institutional ecologies, how can digital rhetoricians establish and sustain ethos within the English Departments in which they reside, and how can more graduate programs prepare future faculty to negotiate and advocate for the ethos of digital scholarly and editorial collectives? Below we present a number of benchmarks for fostering these important goals within English departments that we feel could be adopted by many other disciplines as well:

Curriculum: English Departments must revisit curriculum development at the graduate level more frequently to determine how discussions (and instantiations) of research, teaching, and scholarship are remediated in the digital age. For instance, the doctoral program in Rhetoric and Writing at Bowling Green State University offers several courses with this goal in mind. The first is a studio course in multimodal composing, which relies on range of web-authoring and digital image and video editing tools, ultimately leading to an electronic portfolio that students can use on the academic job market. The second is a course in scholarly publishing in which students not only read about the tensions between print and digital scholarship but also develop a scholarly identity through the development of a revised paper for submission to a peer-reviewed journal, which many students revise as multimodal webtexts for submission to online journals. Being able to have students and faculty constantly question digital effects on research, scholarship, and teaching within their own disciplines while also having the opportunity to produce cutting-edge digital texts will be key to academic and professional development.

Team Development Models: Not unlike online course development models in which faculty work with instructional designers to migrate a face-to-face course for online delivery, and not unlike collaborative authorship models in the sciences, online journals and academic departments should foster team relationships in producing digital scholarship in which graduate students can play a significant role as designers and co-authors and receive credit for their efforts. This can occur in courses, as well as with conference presentations and eventual publication in the growing number of online journals in both literary studies and writing studies. Because *Computers and Composition Online* is administered by graduate student section editors, for example, and these editors have the opportunity to review incoming submissions and queries to their sections and provide feedback in a team-based way to prospective authors, similar to Erickson's work with Remley, while other members of the editorial staff have a role in commenting on content and assisting with design prior to and after the peer-review process. In many disciplines,

collaboration is often underwritten by grants and exists so long as there is money to support it, but we argue such a team-based approach should be present regardless of financial support—opportunities should be extended to students and faculty outside of any such grant-supported roles in order to foster a deeper level of sophistication to any final products and to cultivate the professional ethos of more individuals, which ultimately raises the ethos of the field as well.

Dissertation Research: Despite the emphasis we have placed on digital scholarship, it is a sad irony that the culminating requirement for today's scholars continues to be the print-based dissertation. However, many institutions, including Bowling Green State University, are now requiring an online submission component in which doctoral candidates upload their dissertation, in our case in .pdf form, to a statewide repository. Rather than have that be a mere "Save as .pdf" model, graduate programs across the humanities should help student understand the affordances of new media such as video and audio in both the data collection and data representation process, and the inevitable relationship between the visual and the textual in making a significant scholarly argument. Such discussions can and should begin in research methods courses (Graupner, Nickoson-Massey, & Blair, 2009) as well as in the development of the dissertation prospectus. Transcending traditional print-based dissertations across the disciplines would provide dramatically different scholarship that would not only allow a more diverse group of thinkers to attempt doctorates and connect more readily with people outside academe but tackle new problems in new ways.

Undoubtedly, regardless of discipline, the education graduate students receive must be a political education as well, understanding that technology progresses much more quickly than either policy or ideology, and that they are obligated both to learn the values of the departments in which they reside and to educate their faculty colleagues and administrative leaders about the emerging ethos of digital work in the field. Admittedly, given the cultural ecologies that we have outlined throughout this chapter, this may be easier said than done, but, fortunately, organizations such as the MLA, the Conference on College Composition and Communication (CCCC), and the Council of Editors of Learned Journals (CELJ) have provided guidelines for candidates, department chairs, tenure and promotion committees, and journal editors, and thereby enable such discussions to move from the national to the local level, making the rhetoric of digital scholarship a reality for current and future faculty in rhetoric and composition. Our hope is that heeding such calls from official organizations in other disciplines, coupled with taking up the advice we offer in this chapter, can help more graduate students, faculty, departments, and fields alter their ethos in ways congruent with shifting ideas of ethos in a digital age.

REFERENCES

Apostel, S., & Folk, M. (2005). First phase information literacy on a fourth generation website: An argument for a new approach to website evaluation criteria. *Computers and Composition Online*, Spring. Retrieved February 5, 2012, from http://www.bgsu.edu/cconline/apostelfolk/c_and_c_online_apostel_folk/apostel_folk.htm

Apple. (2012). *Apple in education*. Retrieved May 15, 2012, from http://www.apple.com/education/ibooks-textbooks/

Ball, C. (2004). Show, not tell: The value of new media scholarship. *Computers and Composition*, *21*, 403–425.

Ball, C. (2012). Logging on: Review of *Profession 2011* section on "Evaluating Digital Scholarship." *Kairos: A Journal of Rhetoric, Technology, and Pedagogy, 16*(2). Retrieved February 4, 2012 from http://www.technorhetoric.net/16.2/loggingon/lo-profession.html

Ball, C. (2012a). Assessing scholarly multimedia: A rhetorical genre studies approach. *Technical Communication Quarterly*, *21*(1), 61–77. doi:10.1080/10572252.2012.626390

Bayer, A. E., & Smart, J. C. (1991). Career publication patterns and collaborative "styles" in American academic science. *The Journal of Higher Education*, *62*(6), 613–636. doi:10.2307/1982193

Bolter, J. D., & Grusin, R. (2000). *Remediation: Understanding new media*. Cambridge, MA: MIT Press.

Boyer, E. (1990). *Scholarship reconsidered: Priorities of the professoriate*. Washington, DC: Carnegie Foundation for the Advancement of Teaching.

Brodkey, L. (1987). Modernism and the scene of writing. *College English*, *49*(4), 396–418. doi:10.2307/377850

Carnegie Foundation for the Advancement of Teaching. (n.d.). *About Carnegie classification*. Retrieved May 15, 2012 from http://classifications.carnegiefoundation.org/index.php

Cheverie, J., Boetcher, J., & Buschman, J. (2009, April). Digital scholarship in the university tenure and promotion process: A report on the Sixth Scholarly Communication Symposium at Georgetown University Library. *Journal of Scholarly Publishing*, *40*(3), 219–230. doi:10.3138/jsp.40.3.219

DeVoss, D., & Platt, J. (2011). Image manipulation and ethics in a digital–visual world. *Computers and Composition Online*. Retrieved May 12, 2012, from http://www.bgsu.edu/cconline/ethics_special_issue/DEVOSS_PLATT/

Dilger, B., & Rice, J. (Eds.). (2010). *From A to <A>: Keywords of markup*. Minneapolis, MN: U of M Press.

Dynamic backend generator (DBG): A scholarly middleware tool. (2006). *Vectors*. Retrieved May 7, 2012 from http://vectorsjournal.org/journal/blog/dbg-overview/

Ede, L., & Lunsford, A. (1990). *Single texts/plural authors: Perspectives on collaborative writing*. Carbondale, IL: Southern Illinois University Press.

Fitzpatrick, K. (2011). *Planned obsolescence: Publishing, technology, and the future of the academy*. New York, NY: NYU Press.

Fitzpatrick, K. (2012). *My view: Are electronic media making us less (or more) literate?* Retrieved from http://schoolsofthought.blogs.cnn.com/2012/02/01/my-view-are-electronic-media-making-us-less-or-more-literate/

Graupner, M., Nickoson-Massey, L., & Blair, K. (2009). Remediating knowledge-making spaces in the graduate curriculum: Developing and sustaining multimodal teaching and research. *Computers and Composition, 26*(1), 13–23. doi:10.1016/j.compcom.2008.11.005

Hawisher, G., LeBlanc, P., Moran, C., & Selfe, C. (1995). *Computers and the teaching of writing in American higher education, 1979-1994: A history*. Norwood, NJ: Ablex Publishing. doi:10.2307/358464

Hawisher, G., & Selfe, C. (1997). *Scholarly publishing in rhetoric and writing- The edited collection: A scholarly contribution and more.*

Jenkins, H. (2006). *Convergence culture: Where new and old media collide*. New York, NY: NYU Press.

Kincaid, J. (2011). *That was fast: Amazon's Kindle ebook sales surpass print (it only took four years)*. techcrunch.com. Retrieved from May 12, 2012, from http://techcrunch.com/2011/05/19/that-was-fast-amazons-kindle-ebook-sales-surpass-print-it-only-took-four-years/

Mandell, L. (2011). *Promotion and tenure for digital scholarship*. IDHMC Blog. Retrieved from http://idhmc.tamu.edu/commentpress/promotion-and-tenure/

McDougall, P. (2012). *Kindle Fire sales hit 6 million in Q4*. Informationweek.com. Retrieved May 10, 2012, from http://www.informationweek.com/news/hardware/handheld/232500684

Modern Language Association. (2007). Report of the taskforce on tenure and promotion. *Profession*, 9–71.

Muhlhauser, P. (2011). Teaching moms and dads to perform the family: Rhetoric and assisted reproductive technology websites. *Computers and Composition Online*. Retrieved May 12, 2012, from http://www.bgsu.edu/cconline/paul-muhlhauser/

New London Group. (1996). A pedagogy of multiliteracies: Designing social futures. *Harvard Educational Review, 66*(1), 60–93.

Purdy, J., & Walker, J. (2010). Valuing digital scholarship: Exploring the changing realities of intellectual work. *Profession, 19*, 77–195.

Remley, D., & Erickson, J. (2010). Second Life literacies: Critiquing writing technologies of Second Life. *Computers and Composition Online*. Retrieved May 12, 2012, from http://www.bgsu.edu/cconline/Remley/

Stolley, K. (2011). *How to design and write web pages today*. Santa Barbra, CA: Greenwood Press.

Stolley, K. (2011a) *About me. Dr. Karl Stolley*. Retrieved May 11, 2012 from http://karlstolley.com/about/#more

Texas A&M University Department of English. (2012). *Department of English tenure and promotion guidelines*. Retrieved from http://dof.tamu.edu/sites/defaults/files/tenure_promotion/Tenure_and_Promotions_Guidlines_English.pdf

Tulley, C., & Blair, K. (2009). Remediating the book review: Toward collaboration and multimodality across the English Curriculum. *Pedagogy, 9*(3), 441–469.

Vook. (2012). *What's Vook?* Retrieved May 15, 2012 from http://vook.com/whats-vook/

Waters, L. (2001, April 20). Rescue tenure from the tyranny of the monograph. *The Chronicle of Higher Education*, B7–B9.

Woodmansee, M. (1994). On the author effect: Recovering collectivity. In Woodmansee, M., & Jaszi, P. (Eds.), *The construction of authorship: Textual appropriation in law and literature* (pp. 15–28). Durham, NC: Duke University Press.

ADDITIONAL READING

Applen, J. D., & McDaniel, R. (2009). *The rhetorical nature of XML: Constructing knowledge in networked environments*. New York, NY: Routledge.

Ball, C., & Kalmbach, J. (Eds.). (2010). *RAW: (Reading and writing) new media*. Cresskill, NJ: Hampton Press.

Borgman, C. L. (2007). *Scholarship in the digital age: Information, infrastructure, and the Internet*. Cambridge, MA: MIT Press.

Carr, N. (2010). *The shallows: What the Internet is doing to our brains*. New York, NY: W. W. Norton.

Damrosch, D. (1995). *We scholars: Changing the culture of the university*. Cambridge, MA: Harvard University Press.

Davidson, C. (2008). Humanities 2.0: Promises, perils, predictions. *PMLA, 123*(3), 707–717. doi:10.1632/pmla.2008.123.3.707

Davidson, C. (2011). *Now you see it: How the brain science of attention will transform the way we live, work, and learn*. New York, NY: Viking.

Kress, G. (2003). *Literacy in the new media age*. New York, NY: Routledge. doi:10.4324/9780203164754

Kress, G. (2010). *Multimodality: A social semiotic approach to contemporary communication*. New York, NY: Routledge.

Lauer, J. M. (1993). Rhetoric and composition as a multimodal discipline. In Enos, T., & Brown, S. C. (Eds.), *Defining the new rhetorics* (pp. 44–54). London, UK: Sage Publications.

Lessig, L. (2008). *Remix: Making art and commerce thrive in the hybrid economy*. New York, NY: Penguin. doi:10.5040/9781849662505

Manovich, L. (2001). *The language of new media*. Cambridge, MA: MIT Press.

McCarty, W. (2005). *Humanities computing*. New York, NY: Palgrave. doi:10.1057/9780230504219

McCorkle, B. (2012). *Rhetorical delivery as technological discourse: A cross-historical study*. Carbondale, IL: Southern Illinois University Press.

Modern Language Association. (2002). *Guidelines for evaluating work with digital media in the modern languages*. Retrieved from http://www.mla.org/guidelines_evaluation_digital

Moore, C., & Miller, H. (2006). *A guide to professional development for graduate students in English*. Urbana, IL: National Council of Teachers of English.

National Council of Teachers of English. (1998). *CCCC promotion and tenure guidelines for work with technology*. Retrieved from http://www.ncte.org/cccc/resources/positions/promotionandtenure

Prensky, M. (2010). *Teaching digital natives: Partnering for real learning*. Thousand Oaks, CA: Corwin.

Selber, S. (2004). *Multiliteracies for a digital age*. Carbondale, IL: University of Southern Illinois Press.

Selber, S. (Ed.). (2010). *Rhetoric and technologies: New directions in writing and communication*. Columbia, SC: University of South Carolina Press.

Shipka, J. (2009). Negotiating rhetorical, material, methodological, and technological difference: Evaluating multimodal designs. *College Composition and Communication, 61*(1), 343–366.

Westbrook, S. (2006). Visual rhetoric in a culture of fear: Impediments to multimedia production. *College English, 68*(5), 457–480. doi:10.2307/25472166

Wysocki, A. F., Eilola, J. J., Selfe, C., & Sirc, G. (2004). *Writing new media: Theory and applications for expanding the teaching of composition*. Logan, UT: Utah State University Press.

KEY TERMS AND DEFINITIONS

Digital Humanities: An area of study focused on the role of computing in fields such as literary studies, philosophy, and linguistics and on integrating technology into scholarly activity.

HTML: Refers to Hypertext Markup Language, the original underlying language of website development founded by Tim Berners-Lee that allows content to be read online through a process of coded elements.

Modern Language Association: Founded in 1883, the MLA is the major professional organization for language and literature scholar-teachers in the United States, with over 30,000 members.

Multiliteracies: An emphasis on the ways in which knowledge-making and meaning are made through differing cultural, linguistic, and multimodal contexts.

Multimodality: The use of a range of technologies of literacy to make meaning, including the visual, the verbal, and the aural. While it often refers to the digital there are other material modalities as well.

New Media: Refers to a range of digital interfaces: video, audio, gaming, websites, and other forms of social networking.

Remediation: The process by which the newer media of the digital age repurpose and refashion older media genres such as photography, film, and television.

Rhetoric and Composition: A field of English studies related to research about writing instruction, primarily at the college level.

Scholarship of Teaching and Learning: An emphasis on research into excellence in undergraduate instruction and the need for incentive and reward for the publication of pedagogical scholarship.

Webtext: This term refers to articles and other material designed specifically for delivery on the World Wide Web and that rely upon the affordances of that delivery system, integrating multimodal content in .html or other web format.

Chapter 6
The Query is Just the Beginning:
Exploring Search-Related Decision-Making of Young Adults

Veronica Maidel
Syracuse University, USA

Dmitry Epstein
Cornell University, USA

ABSTRACT

Web search has become an integral part of everyday online activity. Existing research on search behavior offers an extensive and detailed account of what searchers do when they encounter the search results pages. Yet, there is limited inquiry into what drives the particular search decisions that are being made and what contextual factors drive this behavior. This study provides a user-centric inquiry focused on in-depth detailed investigation of search-related decision-making processes. It builds on data collected through analysis of structured observations of young adults performing searches on their personal laptops. It focuses explicitly on the decisions the users make after completing a query and facing a list of search results. The study reveals a pattern of sophisticated use of a variety of explicit cues, tacit and contextual knowledge, as well as employment of an incremental search strategy.

INTRODUCTION

The continuously growing volumes of information pose significant challenges in terms of evaluating credibility, quality, and relevance of online content. In recent years, search has emerged as a major tool through which people find information and experience the online world. Search engines

have been referred to as the gatekeepers of online information because they apply algorithms that make decisions about which content to present to the user out of the millions of available options. They have been criticized for enabling the reproduction of the traditional media landscape where a handful of large, influential websites are accounting for the majority of web content and traffic (Granka, 2010). Yet search engines are not a purely technical phenomenon. Hargittai (2007)

DOI: 10.4018/978-1-4666-2663-8.ch006

described them and their uses as "embedded in a myriad of social processes that are important for social scientists to consider in their research in order to understand the social implications of these important tools of our time" (p. 775). Moreover, she emphasized that "[g]iven their popularity, search engines are important brokers of information, and knowing more about how they represent content and how they are used is vital to understanding patterns of information access in a digital age" (p. 775-776).

There are a variety of studies focused on search engines and information seeking behavior. Depending on the home discipline, they range from system-focused studies (development and improvement of search algorithms and indexing techniques) to the human-focused studies (dealing with users' information needs and their information seeking behavior) (Kelly, 2009). Studies that acknowledge human agency typically focus on user behavior, with an emphasis on identifying patterns of how searchers interact with search engines. Thus far, there is limited inquiry into what facilitates the searcher's decision-making as to which search results to follow.

A variety of studies have dealt with the questions of credibility and quality of online information. Some of the early scholarship in this area focused on evaluating the elements of web pages and their content (Flanagin & Metzger, 2000; Fogg et al., 2001, 2003). Another line of scholarship focused on the cognitive processes involved in evaluation of information (Hilligoss & Rieh, 2008; Metzger, 2007). Recently, there is a growing focus on the social aspects of credibility judgments (Metzger, Flanagin, & Medders, 2010). Although on its face useful, there is limited application of this body of knowledge for furthering the understanding of the searcher's decision-making process when she or he encounters a page with a set of results.

Bridging the two areas of inquiry, this study aims to shed light on what users do know about search engines and the search process, and how they integrate this knowledge with their percep-

tions of credibility in their use of search systems. It builds on data collected through analysis of structured observations of young adults in the U.S. while they searched on their personal laptops. It focuses explicitly on the decisions users make after they complete a query and are facing a list of search results. In other words, this paper asks not only what elements of the search results people pay attention to or what search results they actually follow, but also (1) how they interpret the various elements of the search results and (2) what aspects of context of their search activity influence their decision about which result to click on. Our goal is to advance a more holistic view of the Web as a longitudinal social experience.

BACKGROUND

Understanding how people search, particularly how they navigate through their search results, is important for those designing search engines, teaching digital literacy skills, as well as for those concerned with the social implications of search. Yet there is limited research into what facilitates people's decisions to choose certain search results over others. There is extensive literature on what decisions people make when faced with a list of search results or on the information needs that underlie their search behaviors and how to predict them (Agichtein, Brill, & Dumais, 2006; Agichtein, Brill, Dumais, & Ragno, 2006; Downey, Dumais, Liebling, & Horvitz, 2008; Jansen, 2006; Jansen, Booth, & Spink, 2008; Jansen & Spink, 2006; Rieh & Xie, 2006; Silverstein, Marais, Henzinger, & Moricz, 1999). There are numerous studies about the path users take through search results until the moment they decide to click on a particular link (Cutrell & Guan, 2007; Dumais, Buscher, & Cutrell, 2010; Granka, Joachims, & Gay, 2004; Guan & Cutrell, 2007; Lorigo et al., 2008), in other words, how people physically reach a link on a search engine results page (SERP). But there is limited inquiry into what facilitates particular decisions and what contextual factors drive this behavior (Hargittai, 2002; Hargittai, Fullerton, Menchen-Trevino, & Thomas, 2010; Teevan, Alvarado, Ackerman, & Karger, 2004). We identify the main barriers to furthering this line of research as residing in the area of methodology and limited conceptual repertoire.

The Methodological Challenge

Some approaches to studying search behavior, such as transaction log analysis, are predominantly search-data-driven. This type of inquiry is well suited for identifying patterns of search behavior. It provides information about the length of the query, frequency of its reformulation, number of results consulted on the SERP (Silverstein et al., 1999), length of the search session, percentage of single-term queries, and the use of query operators (Jansen & Spink, 2006). It also allows identifying categories of query reformulations (Jansen, Booth, & Spink, 2009; Rieh & Xie, 2006) and stability of search behavior over time and across search platforms (Jansen & Spink, 2006). The search-data-driven approach, however, is inherently limited in explaining the decision making process behind that behavior—a transaction log does not record the users' motivation and does not track their cognitive processes, such as reasons for the search or the decision-making process when it comes to selecting a search result (Jansen, 2006).

Other approaches to studying search behavior are more problem-driven, which in turn requires more controlled research environments. Particularly interesting in the context of decision-making process about search results is the more recent and more sophisticated method of eye tracking. Eye tracking is used to obtain a deeper understanding of where people invest attention on the SERP, for how long, and in what order, before they click on a search result (Hornof & Halverson, 2003). The controlled nature of eye tracking analysis allows researchers to draw a rather detailed map of the users' search behavior, including: the order in which the search results are viewed (Lorigo et al., 2006); styles of search results evaluation as a function of the amount of information consulted on a SERP (Aula, Majaranta, & Räihä, 2005); differences in search results' scanning behavior across different platforms (Lorigo et al., 2008); the influence of the snippet length on search performance (Cutrell & Guan, 2007); and identifying which search results get the most attention (Dumais et al., 2010; Granka et al., 2004; Guan & Cutrell, 2007; Pan et al., 2007).

The experimental nature of eye tracking studies provides more data on information behavior of users when compared to transaction log analysis, but those studies still use behavior elicited through eye tracking analysis as a proxy for the user's cognitive processes. Gaining a deeper understanding of the decision-making processes using eye tracking is problematic because of the difficulty in analyzing and interpreting eye tracking data and the difficulty in integrating eye tracking methods with other usability testing techniques, such as think-aloud (Jacob & Karn, 2003; Poole & Ball, 2005).

The Conceptual Challenge

One of the main research areas in the field of Information Retrieval (IR) pertaining to the decision-making activity of a searcher on the SERP is the study of relevance. There is an ongoing effort to understand relevance and its manifestations. Saracevic (2007) classified a series of criteria used by the searchers when making judgments about the degree of relevancy. Those included: content (topic, quality, depth, scope, currency, treatment, and clarity), object (characteristics of the document, such as representation, availability, costs, etc.), validity (accuracy, authority, trustworthiness of sources, verifiability), situational match (appropriateness to situation, usability, urgency), cognitive match (novelty, mental effort), affective match (emotional responses, frustration), and belief match (personal credence given to information, confidence). He stressed that "user perception of topicality seems still to be the major criterion, but clearly not the only one in relevance inferences" (p. 2130). The growing recognition of the social factors influencing search behavior places more weight on criteria other than content. Focusing on criteria such as validity and belief match offers an avenue to connect research on search decision-making to research on credibility of online content.

Research on credibility judgments of online content focuses primarily on the user's interaction with the content of target web pages. Early research in this area suggested that people judge online content mostly based on appearance. Fogg et al. (2003), for example found that "design look" was by far the most frequently attributed feature in credibility judgment across different types of websites, and Flanagin and Metzger (2000) found that web users do not invest much effort in verifying online information. There are numerous models explaining the process of credibility evaluation. Hilligoss and Rieh (2008), for example, offered a credibility judging model consisting of three layers (construct, heuristics, and interaction) where credibility judgment is built through continuous and repeated engagement with online content. Metzger (2007) proposed a dual processing model of credibility assessment under which more motivated and able users will engage in a more thorough evaluation of credibility compared to the less motivated and less capable users. Similarly, Pirolli's (2005) information foraging theory suggested that people "tend to optimize the utility of information gained as a function of interaction cost" (p. 351). Taken together, these models suggest that web users strive to optimize their cognitive effort and time to the perceived quality of the outcome ratio.

There is a growing focus on the contextual and social aspects of credibility judgments. Flanagin

and Metzger (2000), as well as Hilligoss and Rieh (2008), for example, found that the type of task influences how much people are willing to invest in verifying the credibility of information. Metzger et al. (2010) investigated how social affordances of the Web are integrated in credibility judgments; they identified four main strategies: social information pooling, social confirmation of personal opinion, enthusiast endorsements, and resource sharing via interpersonal exchange. Although the way scholars think about and study credibility has significantly changed over time, and has grown into a more sophisticated and nuanced approach, most of this research still focuses on evaluating online content on a particular web page.

We argue that when people rely on search engines as their primary gate into the Web, credibility judgments often take place before users reach a particular web page—they happen on the SERP. Unlike target pages, however, SERPs contain limited cues that searchers can utilize in their assessments. Limited knowledge of users with regard to how search engines work (i.e., how the ranking system works, including the prioritizing of sponsored and non-sponsored links) further complicates credibility judgments at this preliminary stage (see Hargittai, 2007; Jansen & Spink, 2007). Arguably, this limitation prevents optimal use of search systems, but the ubiquity of search in everyday online activities suggests that users still derive utility from their imperfect use of the systems. In other words, users make sense of search systems based on what they do know, their past experience with search engines, and numerous contextual factors.

Overcoming the Challenges

This study belongs to a line of research trying to address the limitations of the primarily data-driven and highly controlled approaches. This line of research builds on the studies of relevance in Information Retrieval (IR) (Cooper, 1971; Park,

1993; Saracevic, 1975) and attempts to establish links between the real information needs of users and their subsequent search behavior (Kelly, 2009). This line of research ventures beyond the realm of search platform and involves contextual factors and tacit knowledge. Teevan et al. (2004) is a good example of a study focused on people's search behavior in their natural settings. Following the logic of diary studies, the authors conducted semi-structured interviews in which participants reported their most recent search activity. Teevan et al. observed that instead of jumping directly to their information targets using keywords, the participants navigated to their targets with small, incremental steps, using their contextual knowledge as a guide, even when they knew exactly what they were looking for in advance.

Hargittai (2002) used a similarly contextually-rich, yet more controlled, approach where she conducted structured observations of how people find information online. Her study suggested that people's ability to find information on the Web is a function of a complex set of contextual factors that includes their technical and informational environments and the level of their relevant skills (e.g., ability to use a browser's navigational features, ability to enter valid search terms, etc.). In a more recent study, Hargittai et al. (2010) combined structured observations with elements of the think-aloud technique to study the entire process of information seeking of young adults: from search engine selection, through the evaluation of search results, and all the way to the final destination. The researchers found that "the process of information-seeking is often as important as verifying the results when it comes to assessing the credibility of online content" (p. 479), specifically that the participants tend to rely extensively on the search engine rankings. Moreover, the researchers suggested that lower levels of information literacy are associated with higher trust in search engines. Another finding of the study suggested that the participants have in-formation seeking routines and that those routines are built primarily around brands such as specific search engines or information repositories (e.g. Wikipedia).

While studies of information-seeking behavior in electronic environments and in context shed some light on what facilitates particular search choices, there are still more questions than there are answers in this domain. One factor that limits the ability of current research to unpack the search-related decision-making process is the fact that these studies do not actually aspire to do that. For example, Teevan et al. (2004) aimed to explore the range of orienteering behaviors, Hargittai (2002) explored the question of internet skills, and Hargittai et al. (2010) focused on online content evaluation. Another limiting aspect of the existing inquiry lies in methodology. While studies like Teevan et al. pay a lot of attention to the actual information environment of their participants and real-life information seeking tasks, they rely primarily on user-reported behavior. At the same time, studies like Hargittai et al. do collect both user-reported and actual behavior of their participants, but this comes at the cost of context—first, the participants are forced to operate in an alien information environment and second, as much as the tasks are similar to real-life situations, they are still artificial for the individual participants. In this study we aspire to overcome some of these limitations.

METHOD AND DATA

Conceptually, we focus specifically on instances of decision-making where users decide which search results to click on. Methodologically, our study is based on recalling actual activities the participants have gone through and their enactment of those activities in their own technical environment. While this approach has its own limitations and drawbacks, which we will discuss later, it allows investigating real-life situations in information environments our participants are familiar with and that they trust.

The data discussed in this paper were collected in May-July 2010 as part of a larger international study in an urban private university in the Northeastern U.S. The participants were recruited among undergraduate students who met a series of criteria, namely: 25 years old or younger, native English speakers, and owned a laptop, which they were willing to use in the study. Earlier research found different variations of these factors to be related to people's abilities to navigate online information and search (Eamon, 2004; Hargittai et al., 2010; Kralisch & Berendt, 2004; Zhang & Chignell, 2001).

The participants were recruited through campus advertisements and classroom announcements. They were offered a $15 compensation payment for their participation in the study. While we are aware of the limitations of our sampling, particularly when it comes to generalizability, we believe it captures a rather diverse group. A total of 32 participants took part in the study, split evenly between male and female participants. The range of ages spanned from 18 to 25 years old with an average age of 20. Almost half of the participants (47%) were Caucasian, 20% African-American, 17% Asian-American, and 13% Hispanic. Thirty-eight per cent of our participants came from households where at least one of the parents had completed at least some college education, 28% came from households where at least one of the parents held a graduate degree, and 34% from households where none of the parents had any college education. Sixty-six per cent of the participants lived in dorms at the time of the study, 22% rented, and the rest had a different living arrangement. As mentioned above, all of the participants owned a laptop. The oldest laptop was purchased in 2005, the newest in 2010; on average the participants used machines that were about one year old. Among the participants,

56% were Mac users and 44% were Windows users. None of the participants used Linux or other operating systems.

An average data collection session lasted between 45 and 60 minutes and included seven parts. Part one consisted of a structured questionnaire about the technical environment of the participant, including the kind of equipment used in the session, operating system, browser, email client, etc. The questionnaire was administered by the researcher. Part two consisted of a structured questionnaire about internet use and focused on time spent online, frequency of visiting various types of websites and engaging in different online activities. This part was also administered by the researcher and focused on activities related to school work, religion, hobbies, politics, health, and more. Part three of the session included a structured, self-administered questionnaire aimed to survey the participants' online skills. The questionnaire included knowledge questions as well as questions found to be good proxies of digital literacy (Hargittai, 2005, 2009).

Parts four and five of each session included a semi-structured observation of the participant's behavior (Hargittai, 2002) based on their recall of past activities (Teevan et al., 2004). Our method draws on what Kelly (2009) calls "spontaneous and prompted self-report" (p.89), which is a technique of collecting data from subjects while they engage in search and the observer tries to elicit feedback about their search behavior. Subjects are not required to continuously verbalize their thoughts (as with think-aloud) but are instead asked to provide feedback at fixed intervals or when they think it is appropriate. The purpose of this technique is to get more refined feedback about the search that can be associated with particular events instead of summative feedback at the end of the search.

Part four of the data collection session focused on the participants' recall of their online routines. The participants were asked to walk the researcher through their daily online routines, enacting their browsing behaviors and explaining them out loud. Part five utilized the participants' responses to the questionnaire in part three to prompt them about various instances of visiting websites and engaging in different online activities. As before, the participants were asked to think-aloud as they enacted their online behavior. In regard to the users' search activities, the researcher asked the users about their decisions to follow links from the SERPs they reached when executing a query as part of their routine or as part of recalling their past online activities. Following the observation, the participants were asked to answer a few more questions related to their digital literacy. These questions were organized in a structured questionnaire administered by the researcher. Finally, the participants filled out a self-administered questionnaire about their demographics.

Parts four and five of the session were captured on both video and audio, including the participants' laptop screen. These steps later allowed for extracting both quantitative and qualitative data about the participants' behavior and their explanation thereof. This paper builds on the qualitative analysis of over 30 hours of recorded material. On average, each participant had seven instances of search behavior during an observation session, which provided us with a rich account, not only of their behavior, but of their rationale of search-related decision-making. While we did not request the use of a particular search engine, all of our participants used Google as their search engine of choice.

USER DECISION-MAKING ON SERP

As mentioned above, this study focuses explicitly on the decisions users make after completing a query and they are facing a list of search results. While our goal was to capture a wide range of search-related decision-making patters, some behaviors stood out in light of existing literature. For example, we repeatedly observed our participants focusing on the top results and rarely venturing beyond the first SERP. This observation is consistent with numerous eye-tracking studies such as Lorigo et al. (2008) and in-person observations such as Hargittai et al. (2010). One participant openly stated, "I would go to the first hit, rarely I would look at the title or the URL, I just use Google's decision making." But other participants, based on their commentary and responses to the question of what helped them make the decision to follow through with a particular search result, seemed to utilize ranking order as only one piece of information to be considered, as opposed to trusting it blindly. We found that our participants took into account a variety of elements on the SERP before making a decision, especially when they dealt with the more complex informational queries. Moreover, we observed an extensive use of contextual and tacit knowledge when it came to evaluating search results.

Interpretation of SERP Elements

Among our participants, the visual focus on the top search results was consistent across the board, but the decision to follow through with a result included a number of additional elements. When analyzing a SERP, our participants would often factor in ranking, but they also paid attention to discrepancies between their information needs and the title of the result or the short description accompanying it (the snippet). Specifically, they would expect to find their exact search string emphasized in the title of the result or in the snippet. For example, a male participant searching for flu symptoms stated, "some of the titles seem retarded… they don't have the things I'm looking for… I'm judging it by what the title says… I'm not interested in swine flu, only in cold, so I wouldn't go to that one." Similarly, when searching for information about privacy, a female participant pointed at the title of a result and stated, "I'm going to choose the first one because they have my exact [query] that I typed in there." In other words, search string formulation and assessment of the results are tightly intertwined and are not seen as two separate steps.

While title and snippet of the results are usually examined together, titles seem to attract the initial attention. A male participant, looking for names of two scientists, explained his decision-making processes in the following way, "You look at the title and the few sentences underneath to try and decide." When asked explicitly why he did not follow the top search result, the participant explained, "Because the title on the second one contained my research topic and the first one had 'Newman' in it – I don't know who 'Newman' is and 1999 is a bit old." Similarly, another male participant suggested that he will read the snippet only "if the title isn't bluntly obvious." A female participant, who was looking for differences between mosquito and bed bug bites, explained her decision to follow the fifth result on the SERP by pointing at the title and saying, "It's exactly what I Googled – 'mosquito bites vs. bed bug bites'." As before, the participants were looking for an exact match between their search string and terms in the search results.

When the participants explored the snippets, they also looked for a close match to their query. For example, a male participant conducting research about presidential campaigns explained his decision-making process:

I look for my search, bolded. I look at the bolded [words] for my search title, for my search criteria...I look at the links what's bolded and what's not... I'll see a [term] over here a [term] over there... they are not actually what I am looking for [hand-gesturing that the words are not adjacent]... So I'll click on links that have the actual phrase that I was searching for.

Similarly, the absence of relevant terms in the snippet leads to a decision to reformulate a query. The following example of a female participant who decided not to click on any of the search results for information about an artist illustrates this point well:

I don't think any of this stuff does [have any relevant information about the artist]. This is why I am not clicking on it. What's on the top line... [pointing at the titles] None of this has [the name of the artist] in it.

This explanation was followed by a query reformulation.

While an expectation to find their exact search string in the title of the result or in the snippet was a relatively common occurrence, the decision-making process on SERP was not always as straightforward. In many cases, our participants would pay attention to the position of their search terms in the snippet and the immediately adjacent words. For example when a male participant searched for an option to watch free movies online, he explained:

I would go through the results and see which one says "free" because not all of them say "free" in the snippet... I won't click on this one [pointing at a result] because it says "free trailer." So I will go to the one that says "watch movies online for free."

In other words, the term "trailer" adjacent to the term "free" in this case served as a cue for lack of relevance of a particular search result.

In other instances, when examining the words adjacent to the query terms, participants brought in additional, relevant to the search, knowledge in order to interpret the results. They would pay attention to terms that were not originally part of their query, but when those terms appeared in the snippet they were interpreted as an indication of relevance or lack thereof. For example, a male participant searching for the summary of a book explained why he felt confident in following through with a particular search result:

I knew things about the book, but I didn't know anything in the book. But I knew it was a novel. I knew the plot. And the work was set in 1959 [pointing at the year on the snippet]. And the setting of the area is Chesapeake Bay [pointing at the name of the area on the snippet].

Another male participant, who was looking for information about Buddhism, offered the following explanation for his decision to click on a lower ranked result: "It says 'tolerance'… so… it means how to deal with them, and that's not what I was looking for… I was looking for their belief system and what they stand for." A female participant who searched for an answer to a question about relationships clarified her decision-making process by explaining, "It just sounds more informative. I want something that's more social, like kids around my age… it's like 'my' and it's not 'your' it's not, so formal. I like informal information." In this case, she interpreted the words "my" and "your" as a qualifying characteristic of expected content. Since she wanted peer advice, she chose to opt for a website that told individual stories as opposed to providing advice from an adult or a professional. We will address the function of contextual knowledge in the decision-making process later.

Another element of the search results presented on SERP and utilized by our participants was the URL. One participant in fact referred to the URL as the first attribute of the search results she tends to examine. Other participants referred to a number of ways in which they utilized the URLs in their search results evaluation process. On one hand, they looked for simplicity in the URL. For example, a female participant, who searched for information about psychology, explained her behavior in these exact words: "I am mostly looking at things that have a simple URL." On the other hand, when looking for a particular type of information, the participants interpreted the actual or the top level domain as a proxy for the quality of the content. For example, a female student looking for statistics on immigration visas explained her intent to use a government website, "I usually want a government statistic… The URL is a part of my decision… I don't want just .com [or] .net." In another case, a female participant looked for a story about a horse that broke its leg. She preferred to click on a particular result because she was convinced it was a newspaper: "I will look both at the URL and what they say here [pointing at the snippet]… so this is a newspaper." We can see how the participants use additional cues, not just ranking, to make sure the results make sense in the particular context of their search.

In our observations, what may appear as "blind" trust in search engine rankings happened in cases where the participants lacked or did not understand cues described above. For example, a male participant who searched for information about a medical condition explained:

I read the short summary [snippet] down here and it looked like it was going to explain what [name of disease] is. [after clicking on it] Turns out it does nothing except it gives me a different name, so I went back and still flipped through a couple [of search results].

Following a clarifying question regarding whether he checked the results in the order of their appearance, he explained, "As long as they looked relevant… [based on] the short description." In other words, to state that the participants were mindlessly trusting the ranking system would not always be correct. Instead, systematic examination of the websites in the order of their appearance, while utilizing other available cues, becomes part of the evaluation process of the search results. For example a female participant looking for cheap clothes online, who came to a list of unfamiliar websites explained, "I just go through them all because I don't know any of these names" in order to familiarize herself with the available options. In another case, a male participant, who looked for material for a philosophy paper, explained how he deals with the lack of familiar cues on SERP:

I would go to the first one and read through it and if I still don't understand what he's saying then I would go to [a] different website [in the order of their appearance] until I'm satisfied and until I have a clear idea what he's saying and I will write it in a paper.

These findings are consistent with earlier eye tracking studies, which suggested that when lacking clarity about the relevance of results, the users employ more scrutiny examining the SERP (Pan et al., 2007). They also allude to earlier findings in the field of IR, such as those described in Bates's (1989) "Berrypicking" model, which we will attend to later in our discussion. One may also choose to view these findings through the lens of credibility research. In this sense, they are consistent with the expectancy violation and the consistency heuristics (Metzger et al., 2010), whereby the match between the searcher's expectations and the search results trumps the ranking suggestion by the search algorithm and triggers a more thorough review of the results or a query reformulation.

Although usually our participants made very quick decisions regarding which search results to follow on a SERP, when prompted, they exposed a complex system of decision making. A typical decision to follow through with a search result would include a combination of cues and strategies, which incorporates both explicit cues available on the SERP and what we can view as tacit and contextual knowledge, which we discuss in greater detail in the next subsection.

Tacit and Contextual Knowledge

As the last observations suggests, in making their decisions regarding a set of search results, our participants relied not only on the explicit cues, but also on implicit factors, which we describe as tacit and contextual knowledge. By tacit knowledge we mean the participants' prior experience with various websites and their expectations regarding the type and quality of the content they can expect based on that experience or other explicit cues. By contextual knowledge we mean the participants' awareness of the purpose of a particular search and its expected outcome. This classification, in fact, resonates with Park's (1993) categories of variables affecting relevance assessments. Our definition of tacit knowledge draws on Park's idea of "internal context," which "indicates various sources deeply rooted in an individual's previous experience with content in the field and perceptions or beliefs about the problem area" (p. 333). The definition of contextual knowledge, on the other hand, draws from Park's notion of "external context," which "indicates factors that stem from an individual's search and current research," which in turn "tend to originate from the individual's view about the search goal, search process, research stage, or research product" (p. 336).

Wikipedia emerged as one of the more prominent examples of our participants' use of both tacit and contextual knowledge. Even though there are numerous studies suggesting that the quality and reliability of Wikipedia information is reasonably good (Chesney, 2006), there is still prevailing opposition to the use of this resource in education (Denning, Horning, Parnas, & Weinstein, 2005). College students are known to use Wikipedia for academic purposes, but when asked, they report skepticism about the quality of its content (Lim, 2009). As one of our participants explained: "The only .org I don't trust is Wikipedia. That's because they let people edit the stuff."

Consistent with previous research, most of the participants in our sample had experience with Wikipedia, were familiar with the type of content available on the site, and could recognize it easily when it appeared on a SERP. At the same time, their view of how useful this website can be varied by the context of their search. Thus, for example, a male participant who was looking for a technical solution to throttle bandwidth, explained:

If it's Wikipedia I might go into it if I want to learn more about it [process of throttling bandwidth], but if I'm looking for a product or tool – I won't… If it's someone asking a question on a forum, this is what I usually like to go to.

In this case, the participant had expectations about content he could find on Wikipedia and employed this knowledge in the context of a particular task. A number of participants explained under what conditions they would choose Wikipedia among other search results. As one female participant explained it, "I trust Wikipedia, not to cite, but for a general idea of what I am looking for." Another female participant, searching for information on bioluminescence for a research paper, said, "Since this is a research report I probably won't go to Wikipedia." Alternatively, she explained:

If I want to quote something for my paper… this one looks like a good one: JSTOR I know JSTOR … This to me looks a lot more official… If it's more of a common knowledge type thing that I wouldn't have to quote from, then Wikipedia is a good quick thing.

Similarly, another female participant, who was writing a paper on a local theatre, explained: "Wikipedia just to get general overview what it is… and [website of the theater] is their website because I figured I would get accurate information from the source." A male participant, also searching for information for a course paper, presented his justification, "Sometimes I go to Wikipedia and sometimes I don't… Cause Wikipedia – teachers don't like Wikipedia… I use it when I need

biography, but don't use it for school." At the same time, another male participant, who was generally interested in Buddhism, explained, "Buddhism – I would go to Wikipedia because Wikipedia outlines everything for you even before you read the article." From the point of view of credibility analysis, our participants appear to rely heavily on predictive judgment (Rieh, 2002) and reputation heuristic (Metzger et al., 2010).

Our observations about Wikipedia resonate with the findings of Hargittai et al. (2010) about young adults placing trust in brands when they search for and evaluate information online. At the same time, our observations suggest that while the participants recognize brands, they make instrumental use of them based on their prior experience and in the context of a particular search task. For example, a female participant looking for an answer to a relationship-related question explained, "I like stuff from answerbag, yahoo answers… because other questions that I've asked brought me to this website." Similarly, a male participant, who searched on how to deal with a particular medical condition, stated, "Obviously I would go to ehow.com before I will go to dizziness-and-balance.com because I know ehow.com and I know it's organized well and I'm probably gonna find the information quickly." Similar brand recognition of websites such as AllRecipes, Amazon, eBay, New York Times, About.com, etc. was continuously utilized in the decision-making process regarding the results presented on a SERP. Here again, our participants enacted predictive judgment and utilized reputation heuristics (Metzger et al., 2010; Rieh, 2002).

It is important to emphasize that brand recognition was used not only in cases where participants chose to go to a particular website, but also in cases where they chose to avoid a particular link. For example, a female participant who was looking for company information for her internship explained why she avoided one of the top ranked results by saying, "...and then the second one is LinkedIn and I know what LinkedIn is." Another female participant, who looked for a recipe, explained:

There are certain cooking websites that I trust more than others, for example I would be much more apt to go to the JoyOfBaking website because I know it's a cookbook, a well known cookbook. Whereas AllRecipes.com... I know everyone can submit their recipes and I don't know how good they are going to be.

In some cases, the participants seemed to be aware of the low credibility of the source they were referring to, but they chose to proceed because of the particular information need they were seeking to fulfill. For example, a female participant explained her decision to use answers.com: "It is not a credible source, but I just like to get opinionated answers." In a way, such behavior is consistent with what Metzger et al. (2010) referred to as social information pooling. The same participant later also explained why she decided not to follow the first result on the SERP "Because I want something that's more social. This probably won't be considered credible until actual research got over there, but sometimes I like to get firsthand accounts."

As we can see from the quotes above, the participants make instrumental decisions depending on their contextual knowledge (the purpose of their search) and their tacit knowledge (prior experience with the particular site in question or other websites). This is consistent with previous research such as Park (1993), who found that an individual researcher's perceptions and knowledge about journals, as well as about authors and their previous works and affiliated academic programs or institutions, can influence a decision on which bibliographic citation is more relevant. This also suggests that the iterative view of credibility judgments (Hilligoss & Rieh, 2008; Rieh, 2002) can be applied not only to the evaluation of the content of the web pages themselves, but also to the evaluation of SERPs at the point of decision making about which result to follow.

Jumping-Off Points

Another search behavior that we observed is similar to the "Berrypicking" model proposed by Bates (1989) more than 20 years ago. This model suggests that "the query is satisfied not by a single final retrieved set, but by a series of selections of individual references and bits of information at each stage of the ever-modifying search" (p. 409). We touched on this behavior when we discussed cases where our participants did not have enough cues to commit to a single search result, and instead resorted to systematic examination of a number of them in sequence. In addition to that observation, we noticed that for certain searches, when the participants had a gap in their knowledge, they would make use of what a number of them referred to as a "jumping off point." In these cases, the participants would visit a website to gather initial terms to be used in subsequent queries. This would usually be at an initial stage of their research, where they still did not have enough substantive knowledge on the topic in order to perform an effective search that fully satisfied their information need. For example, one male participant noted:

It's just the beginning of my research... I'm studying presidential campaigns [typing in query 'presidential campaigns'] so I'll just go here and this is just to get background information on what's going on and see where I can go from here... it's kind of like a funnel."

Contextual and tacit knowledge come into play in this type of search as well, when our participants knew what type of websites would be able to assist them in establishing a "jumping off point." Once again, Wikipedia was mentioned as a website that can serve this purpose. A female participant searching for biology-related information said:

[A]nd even though you know not to always trust what you read on Wikipedia it's a good jumping off point... I would come here and find out... two different types of beetle families... and so now I had this term, I would probably copy and paste that and then do a Google search on that to get better and more specific info.

Another participant, who was writing a psychology paper, noted, "Wikipedia I can't use for a paper, but I'll read about it and then if it tells me something that I could search, then I'll search that..." The same participant, while performing a different search on foreign policies, went to Wikipedia first and explained: "... just see if they have general information because I don't know enough and I'm just looking for keywords to take me to the next step." The participants seemed to be cognizant of the popular critique of Wikipedia, yet given the ease of use and the great utility offered by this resource, they found a way to integrate it in their search practices (see Lim, 2009, for an extended discussion).

Being able to verify search results through consistency checks also came up as an important aspect, especially in searches related to health. Metzger et al. (2010) referred to this strategy as the consistency heuristics for judging credibility. When our participants were not certain about the relevance or credibility of answers they were getting from the search, they would look for consistency across several websites. For example, a male participant, who was looking to resolve a problem of excessive cough, said:

I start with the first [result] and try to look for consistency... if I see the same thing over on each page then it must be... I look up 15 websites and then 10 say the same thing and 5 are all different from themselves, I think that the other 10 are legitimate who have a doctor writing these things.

A female participant, who wanted to look for side effects of a certain medication, searched "headaches" and explained:

I've had headaches, so I'll search "headaches" or I'll search side effects... I'll go to Wikipedia, I'll go to a couple of blog sites, drugs.com, and then the things I see all across the board... If I only see it on one of the websites I won't take into account, but just the consistency through the websites... just because... if Wikipedia has the same as everything else it's just gonna help me keep searching for different things.

As we can see from the examples in this and the previous sections, the process of checking for consistency would usually be manifested in serial clicking on links, one after another in the order of their appearance on the SERP, especially if the users are unfamiliar with the websites that came up in the search. This finding is in line with previous research by O'Day and Jeffries (1993), who characterized the information-seeking process by presenting "triggers" and "stop conditions" that guide people's search behaviors and who found that people often perform comparisons between the results in order to find consistency.

CONCLUSION AND FUTURE RESEARCH

In this chapter we reported on a study of how college students interact with SERPs and what drives their decision-making process while this interaction occurs. We found that there are various elements in the snippet and the title that the participants are taking into account in order to decide which search results to follow. They had certain expectations for the results such as finding their exact search string in the title of the result or in the snippet. Looking at the position of their search terms in the snippet and the immediately adjacent words was also a rather common practice.

Their decisions were also influenced by terms that did not appear in the query but did appear in the snippets or the titles. In addition to these elements, an aspect of tacit and contextual knowledge guided the decision-making process. We found that the participants' familiarity with the nature of the websites that appeared in the SERP affected their decisions based on the expectations that they had from the content on these websites. Some of these websites were considered by the participants to be good "jumping off points" for further exploration and reformulation of their queries.

As noted in the methods section, the data for this study was collected as part of a larger study on the young adults' online routines. As such, it encompasses a number of limitations that offer concrete avenues for future research and additional analysis of existing data, which did not fit the scope of this paper. Since this study was based on recalling actual past activities, we had limited control over the type of queries that were executed by the participants. Therefore, many of the queries were simple navigational queries, as opposed to more complex informational ones. On one hand, this behavior reflects the participants' real life search behavior, thus adding to the external validity of our findings. On the other hand, we need to understand what cues the participants incorporate into their decision-making process for more complex queries that require substantive effort and multiple reformulations. It could be beneficial to investigate how these cues assist in guiding the reformulation process. Also, the fact that the data collection was based on recall forced the participants to try to remember the exact query they used, which may have introduced bias and may not adequately reflect their first-time interactions with the query and the search results. Incorporating these lessons in future studies will help to produce a more robust inquiry. In addition, future research should look into the difference in the elements that users pay attention to as a function of their expertise in search engines and their online skills.

This project offers detailed insight into the decision-making processes of young English-speaking searchers. It highlights that credibility judgments do not happen solely on the content web pages, but they are employed already at the stage of sifting through information sources via a search engine. The study demonstrated that many of the credibility judgment models and heuristics identified as suitable for evaluation of online content can be also applied to the evaluation of the search results on a SERP. Given the limited information about the target websites provided on the SERP, the reputation heuristic seems to be the dominant strategy, but searchers also incorporate other approaches in their search decision-making. It is important to view this decision-making process as part of a broader online experience where searchers form impressions about the reputation of websites (or online brands) and develop knowledge about the variety of needs that can be met by different websites. The decision-making process on the SERP is a manifestation of that accumulated experience and the ability to assess credibility based on a limited number of cues provided on the SERP. As such, we are asking to advance an experiential view of the Web as an environment one needs to engage with in order to make smart choices in navigating it.

We hope our study will serve as a stepping stone for better understanding search behavior and the related decision-making and credibility judgment processes, thus reinforcing human-focused research in IR. At the same time, we hope our findings can also be used to inform future studies of better SERP design and improvement of the search experience. Particularly, we want to highlight the need for elements that offer signals for the identity and credibility of the source on the other end of the link, since people already engage in credibility judgments on the SERP. For educators, this study challenges the commonly sounded claim that young adults have an almost blind belief in the ranking algorithms of search engines, particularly that of Google. Instead, it draws a picture of rapid—yet complex—decision-making, which utilizes numerous cues and invokes tacit and contextual knowledge. When planning and thinking about educating for critical and thoughtful use of search engines, emphasis should be on creating opportunities for users to accumulate search experience in environments where they can receive feedback on interpretating cues. Those seeking to promote thoughtful engagement with the web should avoid categorically dismissing websites or categories of websites such as Wikipedia or user-generated content and instead point to the various uses different sites can serve in the process of information discovery and query reformulation.

ACKNOWLEDGMENT

The authors are grateful to the former Information+Innovation Policy Research Institute at the Lee Kuan Yew School of Public Policy, National University of Singapore, and especially to Professor Viktor Mayer-Schönberger, for their support and funding of this research. The authors are also grateful to Mary Grace Flaherty for her insightful comments.

REFERENCES

Agichtein, E., Brill, E., & Dumais, S. (2006). Improving web search ranking by incorporating user behavior information. *Proceedings of the 29th Annual International ACM SIGIR Conference on Research and Development in Information Retrieval* (pp. 19–26). Seattle, WA: ACM. DOI:10.1145/1148170.1148177

Agichtein, E., Brill, E., Dumais, S., & Ragno, R. (2006). Learning user interaction models for predicting web search result preferences. *Proceedings of the 29th Annual International ACM SIGIR Conference on Research and Development in Information Retrieval* (pp. 3–10).

Aula, A., Majaranta, P., & Räihä, K. J. (2005). *Eye-tracking reveals the personal styles for search result evaluation* (pp. 1058–1061). Human-Computer Interaction-INTERACT. doi:10.1007/11555261_104

Bates, M. J. (1989). The design of browsing and berrypicking techniques for the online search interface. *Online Information Review, 13*(5), 407–424. doi:10.1108/eb024320

Chesney, T. (2006). An empirical examination of Wikipedia's credibility. *First Monday, 11*(11). Retrieved from http://firstmonday.org/htbin/cgiwrap/bin/ojs/index.php/fm/article/view/1413/1331

Cooper, W. S. (1971). A definition of relevance for information retrieval. *Information Storage and Retrieval, 7*(1), 19–37. doi:10.1016/0020-0271(71)90024-6

Cutrell, E., & Guan, Z. (2007). What are you looking for? An eye-tracking study of information usage in web search. *Proceedings of the SIGCHI Conference on Human Factors in Computing Systems* (pp. 407 – 416).

Denning, P., Horning, J., Parnas, D., & Weinstein, L. (2005). Wikipedia risks. *Communications of the ACM, 48*(12), 152. doi:10.1145/1101779.1101804

Downey, D., Dumais, S., Liebling, D., & Horvitz, E. (2008). Understanding the relationship between searchers' queries and information goals. *Proceeding of the 17th ACM Conference on Information and Knowledge Management* (pp. 449–458).

Dumais, S., Buscher, G., & Cutrell, E. (2010). *Individual differences in gaze patterns for web search*. Presented at the IIiX, New Brunswick, New Jersey.

Eamon, M. K. (2004). Digital divide in computer access and use between poor and non-poor youth. *Journal of Sociology and Social Welfare, 31*(2), 91–113.

Flanagin, A. J., & Metzger, M. J. (2000). Perceptions of Internet information credibility. *Journalism & Mass Communication Quarterly*, *77*, 515–540. doi:10.1177/107769900007700304

Fogg, B. J., Marshall, J., Laraki, O., Osipovich, A., Varma, C., Fang, N., et al. (2001). What makes web sites credible? A report on a large quantitative study. *Proceedings of the SIGCHI Conference on Human Factors in Computing Systems* (pp. 61–68). Seattle, WA: ACM. DOI:10.1145/365024.365037

Fogg, B. J., Soohoo, C., Danielson, D. R., Marable, L., Stanford, J., & Tauber, E. R. (2003). How do users evaluate the credibility of web sites? A study with over 2,500 participants. *Proceedings of the 2003 Conference on Designing for User Experiences* (pp. 1–15). San Francisco, CA: ACM. DOI:10.1145/997078.997097

Granka, L. (2010). The politics of search: A decade retrospective. *The Information Society*, *26*(5), 364–374. doi:10.1080/01972243.2010.511560

Granka, L., Joachims, T., & Gay, G. (2004). Eye-tracking analysis of user behavior in WWW search. *Proceedings of the 27th Annual International ACM SIGIR Conference on Research and Development in Information Retrieval* (pp. 25–29).

Guan, Z., & Cutrell, E. (2007). An eye tracking study of the effect of target rank on web search. *Proceedings of the SIGCHI Conference on Human Factors in Computing Systems* (pp. 417–420).

Hargittai, E. (2002). Beyond logs and surveys: In-depth measures of people's web use skills. *Journal of the American Society for Information Science and Technology*, *53*(14), 1239–1244. doi:10.1002/asi.10166

Hargittai, E. (2005). Survey measures of web-oriented digital literacy. *Social Science Computer Review*, *23*(3), 371–379. doi:10.1177/0894439305275911

Hargittai, E. (2007). The social, political, economic, and cultural dimensions of search engines: An introduction. *Journal of Computer-Mediated Communication*, *12*(3), 769–777. doi:10.1111/j.1083-6101.2007.00349.x

Hargittai, E. (2009). An update on survey measures of web-oriented digital literacy. *Social Science Computer Review*, *27*(1), 130–137. doi:10.1177/0894439308318213

Hargittai, E., Fullerton, L., Menchen-Trevino, E., & Thomas, K. Y. (2010). Trust online: Young Adults' evaluation of Web content. *International Journal of Communication*, *4*, 468–494.

Hilligoss, B., & Rieh, S. Y. (2008). Developing a unifying framework of credibility assessment: Construct, heuristics, and interaction in context. *Information Processing & Management*, *44*(4), 1467–1484. doi:10.1016/j.ipm.2007.10.001

Hornof, A. J., & Halverson, T. (2003). Cognitive strategies and eye movements for searching hierarchical computer displays. *ACM CHI '03 Human Factors in Computing Systems* (pp. 249–256).

Jacob, R. J., & Karn, K. S. (2003). Eye tracking in human-computer interaction and usability research: Ready to deliver the promises. In Hyona, J., Radach, R., & Deubel, H. (Eds.), *The mind's eye: Cognitive and applied aspects of eye movement research* (pp. 573–605). Elsevier Science.

Jansen, B. J. (2006). Search log analysis: What it is, what's been done, how to do it. *Library & Information Science Research*, *28*(3), 407–432. doi:10.1016/j.lisr.2006.06.005

Jansen, B. J., Booth, D., & Spink, A. (2008). Determining the informational, navigational, and transactional intent of Web queries. *Information Processing & Management*, *44*(3), 1251–1266. doi:10.1016/j.ipm.2007.07.015

Jansen, B. J., Booth, D., & Spink, A. (2009). Patterns of query reformulation during web searching. *Journal of the American Society for Information Science and Technology*, 60(7), 1358–1371. doi:10.1002/asi.21071

Jansen, B. J., & Spink, A. (2006). How are we searching the World Wide Web? A comparison of nine search engine transaction logs. *Information Processing & Management*, 42(1), 248–263. doi:10.1016/j.ipm.2004.10.007

Jansen, B. J., & Spink, A. (2007). The effect on click-through of combining sponsored and non-sponsored search engine results in a single listing. *Proceedings of the 2007 Workshop on Sponsored Search Auctions*. Presented at the WWW Conference.

Kelly, D. (2009). Methods for evaluating interactive information retrieval systems with users. *Foundations and Trends in Information Retrieval*, 3(1-2), 1–224.

Kralisch, A., & Berendt, B. (2004). Linguistic determinants of search behaviour on websites. *Proceedings of the Fourth International Conference on Cultural Attitudes towards Technology and Communication, Karlstad, Sweden* (pp. 599–613).

Lim, S. (2009). How and why do college students use Wikipedia? *Journal of the American Society for Information Science and Technology*, 60(11), 2189–2202. doi:10.1002/asi.21142

Lorigo, L., Haridasan, M., Brynjarsdóttir, H., Xia, L., Joachims, T., & Gay, G. (2008). Eye tracking and online search: Lessons learned and challenges ahead. *Journal of the American Society for Information Science and Technology*, 59(7), 1041–1052. doi:10.1002/asi.20794

Lorigo, L., Pan, B., Hembrooke, H., Joachims, T., Granka, L., & Gay, G. (2006). The influence of task and gender on search and evaluation behavior using Google. *Information Processing & Management*, 42(4), 1123–1131. doi:10.1016/j.ipm.2005.10.001

Metzger, M. J. (2007). Making sense of credibility on the web: Models for evaluating online information and recommendations for future research. *Journal of the American Society for Information Science and Technology*, 58(13), 2078–2091. doi:10.1002/asi.20672

Metzger, M. J., Flanagin, A. J., & Medders, R. B. (2010). Social and heuristic approaches to credibility evaluation online. *The Journal of Communication*, 60(3), 413–439. doi:10.1111/j.1460-2466.2010.01488.x

O'Day, V., & Jeffries, R. (1993). Orienteering in an information landscape: how information seekers get from here to there. *CHI '93: Proceedings of the INTERACT '93 and CHI '93 Conference on Human Factors in Computing Systems* (pp. 438–445). ACM.

Pan, B., Hembrooke, H., Joachims, T., Lorigo, L., Gay, G., & Granka, L. (2007). In Google we trust: Users' decisions on rank, position, and relevance. *Journal of Computer-Mediated Communication*, 12(3), 801–823. doi:10.1111/j.1083-6101.2007.00351.x

Park, T. K. (1993). The nature of relevance in information retrieval: An empirical study. *The Library Quarterly*, 63(3), 318–351. doi:10.1086/602592

Pirolli, P. (2005). Rational analyses of information foraging on the Web. *Cognitive Science*, 29(3), 343–373. doi:10.1207/s15516709cog0000_20

Poole, A., & Ball, L. J. (2005). Eye tracking in human-computer interaction and usability research: current status and future prospects. In Ghaoui, C. (Ed.), *Encyclopedia of human computer interaction* (pp. 211–219). Hershey, PA: IGI Global. doi:10.4018/978-1-59140-562-7.ch034

Rieh, S. Y. (2002). Judgment of information quality and cognitive authority in the Web. *Journal of the American Society for Information Science and Technology*, 53(2), 145–161. doi:10.1002/asi.10017

Rieh, S. Y., & Xie, I. (2006). Analysis of multiple query reformulations on the Web: The interactive information retrieval context. *Information Processing & Management, 42*(3), 751–768. doi:10.1016/j.ipm.2005.05.005

Saracevic, T. (1975). Relevance: A review of and a framework for the thinking on the notion in information science. *Journal of the American Society for Information Science American Society for Information Science, 26*(6), 321–343. doi:10.1002/asi.4630260604

Saracevic, T. (2007). Relevance: A review of the literature and a framework for thinking on the notion in information science-Part III: Behavior and effects of relevance. *Journal of the American Society for Information Science and Technology, 58*(13), 2126–2144. doi:10.1002/asi.20681

Silverstein, C., Marais, H., Henzinger, M., & Moricz, M. (1999). Analysis of a very large web search engine query log. *ACM SIGIR Forum* (Vol. 33, pp. 6–12).

Teevan, J., Alvarado, C., Ackerman, M. S., & Karger, D. R. (2004). The perfect search engine is not enough: A study of orienteering behavior in directed search. *Proceedings of the SIGCHI Conference on Human Factors in Computing Systems* (pp. 415–422). Vienna, Austria: ACM. DOI:10.1145/985692.985745

Zhang, X., & Chignell, M. (2001). Assessment of the effects of user characteristics on mental models of information retrieval systems. *Journal of the American Society for Information Science and Technology, 52*(6), 445–459. doi:10.1002/1532-2890(2001)9999:9999<::AID-ASI1092>3.0.CO;2-3

ADDITIONAL READING

Aula, A., Majaranta, P., & Räihä, K. J. (2005). *Eye-tracking reveals the personal styles for search result evaluation* (pp. 1058–1061). Human-Computer Interaction-INTERACT. doi:10.1007/11555261_104

Battelle, J. (2006). *The search: How Google and its rivals rewrote the rules of business and transformed our culture.* New York, NY: Portfolio Trade.

Flanagin, A. J., & Metzger, M. J. (2000). Perceptions of Internet information credibility. *Journalism & Mass Communication Quarterly, 77,* 515–540. doi:10.1177/107769900007700304

Flanagin, A. J., & Metzger, M. J. (2010). *Kids and credibility: An empirical examination of youth, digital media use, and information credibility.* Cambridge, MA: MIT Press.

Hearst, M. A. (2009). *Search user interfaces.* Cambridge, UK: Cambridge University Press.

Lorigo, L., Haridasan, M., Brynjarsdóttir, H., Xia, L., Joachims, T., & Gay, G. (2008). Eye tracking and online search: Lessons learned and challenges ahead. *Journal of the American Society for Information Science and Technology, 59*(7), 1041–1052. doi:10.1002/asi.20794

Metzger, M. J. (2007). Making sense of credibility on the Web: Models for evaluating online information and recommendations for future research. *Journal of the American Society for Information Science and Technology, 58,* 2078–2091. doi:10.1002/asi.20672

Metzger, M. J., & Flanagin, A. J. (Eds.). (2008). *Digital media, youth, and credibility.* Cambridge, MA: MIT Press.

Metzger, M. J., Flanagin, A. J., Eyal, K., Lemus, D. R., & McCann, R. (2003). Credibility in the 21st century: Integrating perspectives on source, message, and media credibility in the contemporary media environment. In Kalbfeisch, P. (Ed.), *Communication yearbook 27* (pp. 293–335). Mahwah, NJ: Lawrence Erlbaum. doi:10.1207/s15567419cy2701_10

Metzger, M. J., Flanagin, A. J., & Medders, R. B. (2010). Social and heuristic approaches to credibility evaluation online. *The Journal of Communication*, *60*, 413–439. doi:10.1111/j.1460-2466.2010.01488.x

Metzger, M. J., Flanagin, A. J., Pure, R., Medders, R., Markov, A., Hartsell, E., & Choi, E. (2011). *Adults and credibility: An empirical examination of digital media use and information credibility. Research report prepared for the John D. and Catherine T. MacArthur Foundation*. Santa Barbara: University of California.

Palfrey, J., & Gasser, U. (2008). *Born digital: Understanding the first generation of digital natives*. New York, NY: Perseus Books Group. Retrieved from http://borndigitalbook.com/

Pan, B., Hembrooke, H., Joachims, T., Lorigo, L., Gay, G., & Granka, L. (2007). In Google we trust: Users' decisions on rank, position, and relevance. *Journal of Computer-Mediated Communication*, *12*(3), 801–823. doi:10.1111/j.1083-6101.2007.00351.x

Rieh, S. Y. (2002). Judgment of information quality and cognitive authority in the *Web*. *Journal of the American Society for Information Science and Technology*, *53*(2), 145–161. doi:10.1002/asi.10017

KEY TERMS AND DEFINITIONS

Contextual Knowledge: The participants' awareness of the purpose of a particular search and its expected outcome.

Decision-Making: The process of choosing the next action after examining the search engine results or the target pages.

Search: The process of finding information to satisfy various information needs (e.g. finding the cause for a certain disease symptom).

Structured Observations: A method that allows capturing behaviors of interest to the researcher in a reliable fashion.

Tacit Knowledge: The participants' prior experience with various websites and their expectations regarding the type and quality of the content they can expect based on that experience or other explicit cues.

User Behavior: Patterns of how people make decisions and interact with search engine results.

Web Search Engine: An online tool to perform searches on the World Wide Web.

Chapter 7
Ethos [edit]:
Procedural Rhetoric and the Wikipedia Project

Ryan McGrady
North Carolina State University, USA

ABSTRACT

This chapter examines the credibility of Wikipedia from a rhetorical point of view, using ethos, one of Aristotle's original modes of persuasion, to assess the community behind the content of the site's articles. To do so, the author adapts a newer perspective from video game studies, procedural rhetoric (Bogost 2007), to provide a means with which to analyze the site's community-created rules, which he argues, operates symbiotically with a unified body of editors to shape what the reader sees. By considering Wikipedia within the encyclopedia genre, and by looking beyond the surface content to the archived and easily accessible sets of rules and user data, those who must make decisions about why, how, and to what extent they should use and/or trust the site—or permit it to be used under their purview—may be able to avoid the mire of evaluating constantly evolving pseudonymous documents for factuality.

INTRODUCTION

In 2001 Jimmy Wales and Larry Sanger launched Wikipedia to act as a testing ground, or sandbox, to supplement Nupedia, their online encyclopedia. Nupedia attempted to create a gift economy in which experts would write articles for a free, open resource that would take advantage of the spatial and temporal advantages of the Web.

Wikipedia exploded with activity, quickly overshadowing and then obsoleting its predecessor. In the years that followed, Wikipedia has grown to be one of the most popular websites in the world, surpassing all other nonprofit organizations, encyclopedias, news sites, educational institutions, and other traditional sources of information. In fact the only names above Wikipedia at the time of writing, as reported by Alexa Internet's list of most visited websites, are Google, FaceBook, YouTube, Yahoo, and Baidu ("Wikipedia.org,"

DOI: 10.4018/978-1-4666-2663-8.ch007

2012). Clearly a special case, Wikipedia has had a dramatic influence not just on popular culture and the Internet, but on the ways in which we think about how knowledge is produced and consumed.

Despite its weight as a cultural force—and, of course, because of it—Wikipedia has also attracted sharp criticism from diverse groups of stakeholders in many sectors of the information economy. Journalists scoff at it as an accurate source, educators ban it outright from classrooms, and late night talk show hosts use it as fodder for their opening monologues. An easy way to indirectly question someone's claim today has become to ask if he or she "read it on Wikipedia."

The issue of credibility is central to critical discourse around Wikipedia, as well as to the nature of the site itself. What is an encyclopedia, after all, without trustworthy entries? Certainly the Encyclopaedia Britannica would not have been able to maintain its place as the gold standard for summarized knowledge for so long without the reverence earned through a long history of strict standards for content (Kafker & Loveland, 2009; Kogan, 1958). When the common comparison is made between Britannica and Wikipedia, the former is often treated as truth itself, as if it, unlike Wikipedia, had not been written by real and fallible humans.[1]

A good deal of research has been conducted about the factual accuracy of the content of Wikipedia's articles, with mostly positive results (Andrews, 2007; Booth, 2007; Matthews, 2005; Read, 2006; Rosenzweig, 2006). In fact, a 2005 study conducted for *Nature* (Giles, 2005) found that it was very close in accuracy to *Britannica*—especially impressive findings considering Wikipedia's exponentially larger database and ability to rapidly improve (all errors noted in the study were fixed within days of its publication). Useful as these assessments are, they nonetheless conceal, omit, or otherwise marginalize what is both the site's greatest strength and most glaring weakness: Wikipedia is not static and is not gated;

anybody can change almost anything at any time. The text analyzed for a study at one moment could the next day have been improved upon or deleted and replaced with misinformation or random expletives. Edits like the latter ("vandalism") are typically obvious and quickly removed, but the volatility such changes demonstrate clearly problematizes methods traditionally used to evaluate websites as sources of information, such as those discussed in the next section.

The question then becomes how one might assess the credibility of a dynamic, open access encyclopedia without relying on traditional, empirical evaluations of content. The exigence this creates is clear: Wikipedia is ubiquitous, accessible, and popular, and there are diverse groups of stakeholders in many sectors of the information economy who must make decisions about its use. To err on the side of caution by abstaining or forbidding its use without proper investigation foolishly writes off a tremendous potential for learning from its unquestionably vast stores of information.

Above all else, an encyclopedia must persuade its potential user that it is trustworthy. Just as a speaker communicates credibility through more than just the words contained in his or her speech, so does a wiki. A rhetorical perspective thus seems a fitting approach. In particular, I will focus on ethos, one of Aristotle's original modes of persuasion, to look at the pseudonymous community behind the content. I will also adapt a newer perspective, procedural rhetoric (Bogost 2007), to provide means with which to analyze the site's community-created rules as well as the coded, technological rules that together shape what the reader sees. By looking through these lenses, we can understand Wikipedia's credibility through its community of editors and the complex system of norms and rules that guide how the content comes to be, rather than by wrestling with the messy uncertainty inherent in evaluations of articles themselves.

BACKGROUND

Ethos

In *Rhetoric*, perhaps the single most important treatise on the art and study of persuasion, Aristotle laid out three modes of persuasion: logos, ethos, and pathos. In the millennia since, these terms have been analyzed from countless perspectives by people in diverse cultures at many points through history, resulting in a good deal of complexity when addressing any of them today.

In the strictest Aristotelian sense, ethos can only be found in the content of a speech itself and takes the form of good sense, moral character, and good will (Johnson qtd. in Enos, 1996, p 243). This character-in-speech version of ethos stands in contrast to another notable perspective, that for which it is commonly used today: source credibility, a relatively new, author-centric view that assumes an intelligible, autonomous individual behind each message (Jasinski, 2001). Seen this way, speeches and other works exist as an extension of the author, copying the traits he or she embodies, and morality is subordinated to a belief in a more objective truth.

But postmodernism, with its disruptions of self, truth, and absolute meaning, thoroughly rejected the idea that messages—especially communication artifacts—have a singular, comprehensible author with a mind acting apart from the rest of society (Baumlin qtd. in Sloane, 2001). Reflecting these sensibilities, there is a third ethos, also dating back to antiquity but obscured by Modernist rationalism, which describes the character of a culture or community. Extending beyond the speech itself, ethos becomes contingent upon the roles, habits, and conventions that construct the author. Ethos can thus also be thought of in terms of community habits and customs.[2]

For the purposes of the Web, use of the modern, Cartesian version may prove problematic. Two sets of criteria for evaluating Web content which have been popular at the college level are those of Jim Kapoun (1998) and Hope Tillman (1995/2003). Kapoun suggested five kinds of features to scrutinize: accuracy, authority, objectivity, currency, and coverage. Tillman similarly offered a checklist of quality indicators to look for: criteria the site uses for inclusion of information; identity and authority of the authors; currency and records of updates; stability and reliability for future reference; and ease of use. Writing a decade later, Barbara Warnick offered a useful update with *Rhetoric Online*, in which she pointed out five elements of Web-based communication that require new consideration by rhetorical critics: reception, source, message, time, and space (2007). Most important for this discussion is source, which she pointed out we typically evaluate based on some observable credentials that speak to the trustworthiness of an author independent of the text itself. This connection between perceived expertise or reputation and the accuracy of a message has loosened on the Web, she argued, as we see a declining importance of source (p. 34). A significant percentage of websites don't make the identity of the author clear, and when an identity is offered, misrepresentation is easy. Such was evident on Wikipedia when, in a well-publicized controversy in 2007, a veteran editor who claimed to be a college professor with two doctoral degrees was exposed as a 24-year old unemployed college dropout ("Essjay," 2008).[3] Additionally, there is a trend, typically attributed to the mid-2000s phenomenon of "Web 2.0," for websites to be collaboratively created. With the blurring of the lines between consumers and producers on the Internet, many sites rely on people's inherent desire to express, create, and contribute, leading to the near ubiquity of terms like "multiplayer," "social," and "collaborative."

Assuming, for the moment, a clear attribution to "the author" of a document, there are still other "authors" at work. A typical Web page is written in a combination of HTML, CSS, PHP, ASP, JavaScript, Cold Fusion, and XML, created by a programmer to dynamically select a collection of media (image, video, audio, interactive text, video games, etc.), which were, in turn, created by designers and are frequently customized for the user or pulled dynamically from databases, accompanied by advertisements tailored to each reader's browsing habits, displayed in one of many customizable browser windows on the user's computer, which processes the code at varying speeds depending on connection quality (Warnick, 2007). Two people could load the same URL and have vastly different experiences. This instability is certainly true on Wikipedia, as comedian and talk show host Stephen Colbert demonstrated on the air when he urged his viewers to change the Wikipedia entry on elephants to include the statement "the number of elephants has tripled in the last six months" (Brumm, et al., 2006). Ever since the broadcast, and especially in the weeks after it aired, other editors have had to remain vigilant in patrolling the article on elephants, frequently locking the page to restrict editing access because Colbert's misinformation is repeatedly added.[4]

Roland Barthes's distinction between Work and Text may be useful here. A "Work" for him is a textual object that has reached a final state and will not be changed, with an identifiable source who created it from his or her own imagination. But the author, he said, is dead. In our postmodern world we instead have "Texts," remixes of other writings wherein the reader creates the meaning, produced not by authors but by "Scriptors," who cannot simply create, but rather combine what is already created (Barthes, 1978).

In a world of Scriptors and Texts, when someone produces a painting, mystery novel, cookie recipe, rock song, sitcom, academic essay, or stand-up comedy routine, he or she does not do so as an author up in a tower, secluded and uninfluenced; the final product necessarily consists of bits and pieces of other people's ideas (Brodkey, 1987). We cannot help but draw from our vast stores of memory, creating in accordance or discordance with precedent, convention, and experience, whether consciously or not, so each cultural

product, when placed under proper scrutiny, will reveal tightly woven genealogies of influence.

Most people nonetheless find difficulty fully divorcing themselves from the paradigm of Works and Authors. Working to draw attention to this discrepancy, Lawrence Lessig became one of the more prominent defenders of "remix culture" as a popular and valuable mode of creativity rather than a mindless fetishizing of reckless appropriation. He takes specific issue with intellectual property laws, which, all over the world but especially in America, trend toward increasingly tighter control, stricter penalties, and longer periods of ownership. Lessig argued that the strengthening of these laws, which has largely been in reaction to, or catalyzed by, the popularization of file-sharing on the Internet, prevents important raw creative materials from entering the public domain. The alternative he proposed is a "free culture" in which a person can freely build upon the ideas and creations of others, drawing from and contributing to a vast pool of cultural resources (Lessig, 2004; McLeod, 2001). Lessig made a similar—but less severe and more pragmatic—case to Barthes's that refashioning, combining, and appropriating are the dominant forms of cultural production. The Internet, as Lessig noted, makes these activities easier than they ever have been, offering countless means to manipulate, duplicate, collaborate, share, attribute, update, assemble, and publish, exemplifying the constructivist, postmodern attributes of decentralized meaning and destabilized persons.

WikiEthos

Use of the simplified version of Aristotle's ethos-in-the-speech is complicated by the fact that some of the central tenets of writing on Wikipedia preclude authors, original research, and everything that falls outside a "neutral point of view." Contributors are referred to as "editors" and are required to draw from other sources for their content. They are archetypal "Scriptors" working on the archetypal "Text"—an unstable, changeable pastiche. However, if we consider Aristotle's three components of ethos it is certainly possible an audience would see good sense, good moral character, and good will in the words of an article written for a general audience, well-organized and drawing from multiple points of view. Nonetheless, I consider ethos-in-the-speech as least helpful for coproduced messages online.

Wikipedia is most frequently seen through the lens of modernist ethos, treated as though it has a distinct, singular voice. Setting aside the cautions of the previous section for a moment, the best starting point from which to look at Wikipedia may be through the ethos of encyclopedias in general—a long history that frames public expectations.

Encyclopedias have always had lofty goals. One of the earliest known examples is the *Naturalis Historia*, Pliny the Elder's (1601) ambitious attempt at collecting all knowledge having to do with the natural world. The vast, systematic volumes we see today, however, had their root in the Renaissance, where they took on universal and sometimes even transcendent qualities. The details of their history extend beyond the scope of this chapter, but a few specific references may be useful. Diderot, writing in his *Encyclopedie,* argued that "the purpose of an encyclopedia is to bring together the knowledge scattered over the surface of the earth to expose the general system for men with whom we live" (France, 1998).[5] *Britannica*, published only a few years later, carried on Diderot's vision of encapsulating human knowledge, but placed greater emphasis on its own appearance as a unified work. Whereas Diderot would simply "expose the general system" of human knowledge, *Britannica's* editors would concisely mirror it in book form, insulting one of its competitors, Chambers's *Cyclopaedia,* for being diffuse: "a book of shreds and patches" (France 1998).[6]

Peter France (1998), a Professor of French at the University of Edinburgh, treated the "encyclopedia as an organism" in order to aid his historical work: Far from diffuse, "[they] reflect in some way an ordered body of knowledge, a unified science

which may in turn be seen as expressing the unity of nature" (p. 62). He further incorporated the text and the readers into this organism, showing how encyclopedists sought to involve their readerships by answering questions, using the second person in prefaces, asking for advice, and even using subscription as a way to help the noble goal that is the encyclopedia. Writing before Wikipedia, however, he downplayed the role of the content writers and readers alike, perhaps assuming that the presentation of facts, when done accurately, ends the conversation about the ethos of the writer, while he was able to see the readers only in their superficial relationship with encyclopedists. Especially ironic given the privilege of time and the reader/writer blur of Wikipedia, he ended his paper with this observation: "Clearly, however …the idea of an organic grouping of readers... remains a metaphor or a myth. It is nonetheless a fruitful myth in that it helps to retain the notion of the encyclopedia as attractive living organism rather than inert database" (p. 73).

The most ambitious encyclopedia project was perhaps that of Gottfried Leibniz, whose conceptual, global encyclopedia was one of three parts comprising his vision for real-life utopia, the other two being his *characteristica universialis*, a universal language; and *calculus ratiocinator*, a mathematic system that would calculate real human problems when represented by the *characteristica*. For Leibniz, with Neo-Platonist ideals, an encyclopedia of perfect knowledge was plausible (Yates, 1966; Couturat, 1901/1997). Encyclopedias would go on to play a role in other utopian visions like H.G. Wells's (1938) *World Brain*, which called for a new democratic knowledge store available to everyone, constantly updated by people in all cultures. In both Leibniz's and Wells's dreams, encyclopedias were key to world peace and happiness, an ideal Reagle (2010) argued is present in every encyclopedia and perhaps most effectively realized in Wikipedia.

With such a long history of ambitious epistemic prestige and near synonymy with knowledge itself, encyclopedias carry with them a strong ethos as a source. Even as recently as the time of France's writing in 1998, what reason did someone have not to trust an encyclopedia once its creator's respectable, even Platonic, motivation was made clear or once impressed by the amount and quality of information it presents? That he doesn't address the factual accuracy of the books' contents is telling of this assumed infallibility.

Wikipedia, however, is attributed fallibility as a trait that supersedes all others. Even given the repeated demonstrations of its accuracy, faring well in formal comparisons with the old standards, it breaks the unspoken—and naive—expectation through which encyclopedias gained our good graces: It can be wrong.[7] The organism of Wikipedia has evolved to envelop readers, systematizing community practices such that a 24-year-old college dropout is functionally indistinguishable from a 50-year-old professor with multiple doctorates. The ethos of the source has been polluted. And it is through this contaminated view of encyclopedia ethos that we see the wholesale condemnations and denigration of Wikipedia.

Whether Wikipedia should be considered an encyclopedia at all has been the subject of considerable debate. But while it is not clear whether its identification within the encyclopedia genre is more of a help or a hindrance, it is nonetheless an explicit self-identification. The project's founders describe it as an encyclopedia, it began as an encyclopedia, its most fundamental policies include statements about it being an encyclopedia, and, of course, the word "Wikipedia" is a portmanteau of "wiki" and "encyclopedia."

In writing about the role of Wikipedia in educational contexts, James Purdy made a point of putting "encyclopedia" in quotes to make clear that it should not be thought of or evaluated based on the standards of its print-based predecessors (2009; 2010). Indeed, as Warnick and Barthes described in the previous section, Texts online cannot be considered in the same way we have with those of older media. Freed from the tyranny of Authors and Works, the polluted encyclopedia ethos of Wikipedia becomes strength and credibility when viewed through the perspective of postmodern, community-constructed ethos.

Scriptor Ethos

Wikipedia is exemplary of "commons-based peer production" (Benkler, 2006, p. 60), the creation of meaningful goods via coordination and collaboration of large groups of people such that both inputs and outputs remain available for reuse. It works because of a dedicated community of editors, some of whom spend as much time contributing as a full-time job. Why would they do this? Why do they deserve trust? Such questions are essential to determining their ethos—or, more accurately, how they fit into the community's ethos.

An initial assumption about motivation to volunteer for a cause like this would be to see your name in print, or get some kind of recognition. But that which takes place on Wikipedia is hardly the same type of credit that exists in, for example, scientific communities. Latour and Woolgar (1986) found that the "cycle of credit" is the primary incentive linked to publication among scientists due to its direct connection with power, efficacy, and resources (Forte & Bruckman, 2005). On Wikipedia, by contrast, no page is attributed to an author, editors get no credit in terms of what is presented to the public, and the rules and procedures that dictate how content is to be created are nuanced to prevent the expression of self or opinion. There is a small element of credit behind the scenes wherein people are recognized not for their scholarly work (original research is not allowed), but for their sheer amount of time and effort put into the project. The most common way an editor might reward another is to place one of a variety of "barnstar" images on his or her user page, community-created tokens to recognize various kinds of contributions, including writing articles about the television show 24, writing articles about women and science, organizing a broad-scoped biology project, fighting vandalism, designing templates, helping with dispute resolution, attaining a milestone number of total edits, editing by an ambitious new user, and many others. Barnstars, however, have limited practical value outside of building a pseudonymous online identity; they entitle nothing and cannot be used to win arguments, but they do build pseudonymous reputations within the community. When a user is applying for an administrator position, such items on his or her user page might be helpful to demonstrate a level of commitment to the project and social competency, but they are largely unpoliced and require someone take initiative to bestow one upon the deserving.

Wikipedia attracts people who are passionate about knowledge, learning, and/or teaching, and want to see the project succeed. Bryant, Forte, and Bruckman (2005) studied how editors change with experience in an article appropriately titled *Becoming Wikipedian*. They found that new editors tend to edit what they know, making small corrections to articles about subjects that interest them or addressing errors found while reading. Eventually, they work up to more substantial revisions. With experience typically comes a greater sense of the whole, leading to more significant macro-level edits and more meta-work (discussions about organizational structure, debates over rules, and peer-reviewing articles for which peer review has been requested, for example) (Bryant, et al., 2005). As editors begin to see the project's moral character, good will, and good sense (Aristotle's aspects of ethos), the good of the Wikipedia mission becomes more than a collection of articles, and the community of editors becomes more than a large group of information-adders. People put so much work into Wikipedia because of a shared feeling that it has value.

The evolution of a Wikipedian offered by Bryant, Forte, and Bruckman might also be understood as a process of assimilation through which a user moves from initially viewing Wikipedia as an information source or medium for teaching and learning to deeper and deeper levels of understanding of the various processes by which Wikipedia was formed and operates. These norms are so fundamental to the project that Purdy considered Wikipedia to be not just a product or source, but also a "representation of process" (2010, p. 207). It is perhaps most effective, then, to look at the ethos of Wikipedia through the processes that construct content-creation practices and that train new editors. To do this, I turn to procedural rhetoric.

</Authors><Procedural Rhetoric>

The first sense of procedural rhetoric I'd like to use is Ian Bogost's, introduced in his 2007 book, *Persuasive Games*. He first defines "procedurality" as "a way of creating, explaining, or understanding processes" and rhetoric as "effective and persuasive expression," then combines the two to form procedural rhetoric, the "practice of using processes persuasively" (2007, p. 3). Bogost is a scholar of video game studies who noted a trend of games being designed with a raison d'être not for fun or entertainment, but for persuasion. He devised procedural rhetoric to fill a gap in scholarship highlighted by this new genre of "persuasive games": a tool for the rhetorical analysis of procedurally-driven persuasion, derived from representation written into code rather than authored directly. He was careful to point out that this does not mean the study of a computer's technical properties or how computers act as a medium for otherwise fairly typical verbal rhetoric, both of which he ascribed instead to "digital rhetoric" (pp. 25-6).

Designers of games such as *The McDonald's Game, Ayiti: Cost of Life, Peacemaker, Food Force, Fat World*, and *Darfur is Dying* attempted to foment political or social activism or awareness by using not just audio and visual elements, but interactivity, crafting the rules and mechanics of the game to direct the player toward virtual action that forces consideration of real-world circumstances. This is closer to "learning by doing" than listening to the same message delivered by an orator or reading it in a newspaper is. The programmer, as rhetor, rather than writing a compelling speech, steers the audience toward a particular experience.

For Bogost, "procedural rhetoric is a subdomain of procedural authorship; its arguments are made not through the construction of words or images, but through the authorship of rules of behavior, the construction of dynamic models" (p. 29). From this sentence, we can tell his primary audience is designers, or at the very least that there is an assumed knowable source. The most obvious aspect of Wikipedia to apply procedural rhetoric to, then, would be the wiki software itself, MediaWiki, with its relatively stable code and for which a list of developers' real names is publicly available. MediaWiki, like all wikis, has very specific, basic, and uniform processes that govern user experience at the level of the individual in thoughtful ways. For example, Reagle (2010) highlighted, as a particularly important technical feature in the first wiki, the ability to "talk about and refer to something that did not formally exist yet, hence the famous 'red link' on wikis that points to a page not yet filled with content." Anywhere else on the Web, a page that doesn't exist would appear as a broken link or confusingly forward the user to another location, but on a wiki it becomes an opportunity to meaningfully contribute (Purdy, 2009).

The application of procedural rhetoric to wiki software to help us understand the ethos of the Wikipedia community is problematic for at least two reasons. First, it is not specific to Wikipedia; MediaWiki was developed for and is used by Wikipedia, but is also freely downloadable for people to host their own wikis, making it the most popular wiki option on the Internet. A study of the procedural rhetoric of MediaWiki may be a worthwhile undertaking, but does not sufficiently address the problems of credibility Wikipedia in particular faces. The second problem is customizability: Webmasters and wiki editors have the ability to dramatically change the interface, appearance, and accessibility at any time. As the default settings are altered, so too are the true authors of the wiki and the codes that determine what the user experiences. Some features require access to the server on which the software is hosted, but others stem from style templates created and implemented by regular users. If members of the community decided that certain kinds of pages should not be editable, that the first page visitors should see is the list of Recent Changes, or that the text on articles about birds should be pink, it is simply a matter of creating a template and adding the instruction to the style guide. In short, the use of procedural rhetoric with MediaWiki is too far separated from Wikipedia, and a shift in focus to Wikipedia's implementation of it again falls into the mess of evaluating what is easily changed and authorless.

For my purposes, I broaden procedural rhetoric, keeping Bogost's basic definitional terms, but taking some liberties in divorcing it from video games and intelligible authorship. As my argument is that the ethos of Wikipedia can be found in its community, and their system of rules that lead to the creation of content, rather than the content itself, I turn now to an explanation of the rules. To restate an important distinction: My use of procedural rhetoric is to better understand the ethos of Wikipedia as a community by examining the system of rules constructed, adopted, and adhered to by the community. I argue that an end user should be persuaded of Wikipedia's credibility not through the content of articles as they appear at a particular point in time, but through the ethos conveyed by the community practices and procedures governing the articles' creation. Whereas the ethos discerned from an article might include the recent edits of a self-interested interloper, the system of rules is more durable, and understanding it sheds light on a culture in which things like vandalism and bias are to be corrected and marginalized.

Five Pillar Compliance

Successful peer-production, according to Benkler (2006), can largely be seen as consisting of three functions: creation of content, organization and vetting of content, and distribution of content (p. 68). The last is made effectively irrelevant by the Internet, and while the first two are ever-present on Wikipedia, a study by Kittur, Suh, Pendleton, and Chi (2007) found increasing prominence of the second: Between 2001 and 2006, the "percentage of edits going toward policy, procedure, and other Wikipedia-specific pages has gone from roughly 2% to around 12%" (p. 3). As more and more people become involved with the project, the rules broadened into hundreds—perhaps thousands—of pages, each elaborating on those before it. These rules, their creation, and the community that made them are, I argue, the most important sources of Wikipedia's ethos, and not any of the individual editors or the site itself.

On the surface of Wikipedia's policies are the "Five Pillars" (2010), the central tenets from which nearly all the rest of the governing rules are extrapolated:

- "Wikipedia is an online encyclopedia,"
- "Wikipedia has a neutral point of view,"
- "Wikipedia is free content,"
- "Wikipedians should interact in a respectful and civil manner," and
- "Wikipedia does not have firm rules" (n.p.)

Despite their breadth, Wikipedia policies are not intimidating or imperious to members of the community. Most follow common sense extensions of the Pillars, reinforcing, for example, that Wikipedia is an encyclopedia and not a dictionary or a soapbox. The main reason for their acceptance, however, is that every single one of them was formed through democratic deliberation, not put in place from the top of a hierarchy. And if there was ever doubt, all of the debates, revisions, and attempts to find consensus are archived, available to everyone, and searchable. From this process, Wikipedians have a sense of collective sense of ownership and responsibility as a singular encyclopedic organism that applies to the structure as

well as the content. Finally, built into all of this is the big exception: "Ignore all rules" if they impede progress, a value that reaffirms the ultimate goal of creating a high-quality encyclopedia rather than a bureaucracy.

But what about new users who were not a part of the decision-making processes? To think a first-time editor will seek out, read, and accept the rules, and that his or her resistance to one or more conventions will be overcome by extolling the virtues of collaborative evolution is idealistic at best and finds no precedent in the relationship between humans and laws. Here it might help to think of another kind of "procedural rhetoric." Two years before Bogost released *Persuasive Games*, Richard Fulkerson used the term in his 2005 essay *Composition at the Turn of the 21st Century* to describe a tradition of composition instruction. From a pedagogical and rhetorical perspective, Fulkerson described procedural rhetoric as a method in which the student is taught—and expected to use—a wide range of activities when writing and revising. Further, he described procedural rhetoric as consisting of three parts: composition as argumentation, genre-based composition, and composition as an introduction to an academic discourse community. We can see from this perspective, especially when conflated with that of Bryant, et al. (2005), that the rules already in place on Wikipedia serve to train editors as rhetors capable of writing credible encyclopedia articles according to certain standards. Joseph Reagle (2010) supported this when he pointed to the "assume good faith" policy as one of the most fundamental to the community's mindset and to Wikipedia's success. "Assume good faith" and its behavioral counterparts like "Don't bite the newbies" create an environment in which any editor, experienced or not, will tend toward productive exchanges with even the most glib contributor.

The sparsity of Wikipedia policy from 2001 (the year of its launch) necessitated extrapolation for everyday applications. Where there was gray area, people sometimes formed differing interpretations, resulting in intense disagreement. While some saw these conflicts as destructive or unnecessary, legal scholar and White House "information czar" Cass Sunstein (2005) pointed out the generative and absolutely necessary qualities of dissent in the production of knowledge and in democratic society. Deliberation is still very much a part of Wikipedia's existence and identity, and one of the most frequently contested and revised policies today is "notability," which describes who or what is important enough to have an article written about him, her, or it. Notability is also the policy I will elaborate on to give a better impression of how some of the procedure-oriented thinking actually happens.

The idea of notability stems from the pillar of "Wikipedia is an encyclopedia." An article about my sticker collection, including the individual items in the collection and my storage techniques, probably is not worthy of an encyclopedia entry because few people, if anybody, would find it useful. In addition to general notability guidelines, pages with specific criteria have been developed for academics, books, events, films, music, numbers, organizations and companies, people, sports and athletes, and Web content ("Notability," 2010).

Hundreds of new articles about people, places, or things are created all the time that may not be meaningful to the average reader. While some, like my sticker collection, might be intended as a joke or are due to having not read any of the rules, some are advertising or spam, and still others are good-faith interpretations. Why, for example, should one book be considered "not notable" when another, even more obscure or out-of-print, is deserving? The answer to that valid question is usually that while Wikipedia is not bound by the physicality of paper, it is not "a collection of indiscriminate information." Books that are included are set apart from the rest via verifiable published reviews, an especially notable author, literary awards, adaptation for film, or inclusion in the instruction of multiple schools ("Notability books," 2010).

The book notability article was created and suggested to the community by a user in reaction to a debate over the deletion of an article for a self-published political thriller ("Notability: Books," 2010). The argument in favor of notability consisted of pointing to its "banned" status on Amazon.com, its availability on Google Books, a review, the existence of an ISBN number, and a claim that it satisfied the existing criteria that books could be included if they are "available in a couple dozen libraries" and are notable "above that of an average cookbook or programmers manual" ("Notability: Books," 2012). As dozens of editors gathered to discuss the proposal for deletion, it was discovered that the proof for its being banned from Amazon was simply an Amazon search with no hits, all books from the publish-on-demand service with which it was printed with have a presence on Google Books, the "review" was a Craigslist ad, and no record of it could be found at any library. This left an ISBN number and a claim of greater notability "than an average cookbook or programmers manual." Since "Wikipedia is an encyclopedia" and operates under a neutral point of view, the community decided more detailed rules were needed to make sure articles were not easy marketing for self-published propaganda, autobiography, or other non-notable whimsy. What, they asked, is an "average cookbook or programmers manual" anyway?

The criteria used to determine notability have been continuously developed since Wikipedia's beginning, representing thousands of hours of labor and deliberation. As seen by the subject-specific notability guidelines, the work has extended outward as well, each of its progeny likewise representing great amounts of time. In fact, there is a whole formal system through which you can propose changes or additions or weigh in on others. At the time of writing the most prominent of these proposals is a set of guidelines for "Fatal hull loss civil aviation accidents," intended to determine when airplane disasters are important in the context of an encyclopedia ("Notability fatal," 2011).

While disputes can usually be solved by referencing a particular policy, in the instances when consensus cannot be attained, editors fall back on an intricate dispute resolution process involving the solicitation of third party opinions, a mediation system, forums to complain about rule violators, and, as a last resort, arbitration (judgments made by a committee of editors, elected by editors, who levy verdicts based on best interpretation of existing rules). Firer-Blaess (2011) saw the dispute resolution process as having roots in the tension between Wikipedia's dual identity as wiki and encyclopedia: "While the wiki form was stressing a more 'anarchic' and 'let it be' way of doing things, of allowing people to do what they want and of not applying any written rules, the aim of making an encyclopaedia stressed the need of policies and guidelines" (p. 134).

Wikipedia's administrators and their responsibilities exemplify this tension between anarchy and openness on one side and rules and structure on the other. The administrator title grants no more say in day-to-day writing and discussion activities than possessed by any other editor. The primary role of administrators is not necessarily to make decisions, but to make decisions made by others stick; to do so, they are granted the technical ability to block disruptive users from editing and to temporarily lock pages suffering heavy damage (the vandalism to the elephants article after Stephen Colbert discussed it, for example). These abilities are used relatively infrequently, but remain important functions that would lose their effectiveness if more than a small group had access. The rules also have checks and balances built into them, transparent records of all administrator actions,[8] procedures for how administrators are elected or appointed, and a pervasive mindset that Wikipedia should strive for the flattest possible hierarchies while making sure no one user is able to exert disproportionate influence over its content. Although the concept of a small, more powerful group of users may appear to some to undermine the fundamental principles of a wiki, the community is actually using administrators to buttress the community's interests via the system of rules against individual threats. Ultimately, however, administrators must respond to fewer emergencies than one might think, considering Wikipedia is the fifth most popular website in the world. The reason for this—and key to the site's success—lies in the fact that support for the rules is so strong, those who are out for themselves are hopelessly outnumbered by believers in the Wikipedia ethos.

Wikipedia even has its own style guide, which serves a number of purposes. First and foremost in the context of this chapter, a style guide serves to dissolve individualized voices—not to purport a single, unified author, but to remove the perception that there are unique "Authors" at work at all here. Secondarily, standards for composition and presentation are typical of the encyclopedia genre and any criteria used to evaluate sources on the Web will advise consideration of design professionalism or usability. There are basic structural and organizational elements all articles must follow and hundreds of pages of specifics. For example, there is a long article concerning the formatting of road junction lists: what kind of table, colors, and icons to use, and how to treat concurrences and interchanges with multiple exits ("Manual," 2010). Another page details the preferred symbols in articles about logic, with specifics for truth functional connectives, quantifiers, and metalogical symbols ("WikiProject," 2010). At the same time, nobody is expected to memorize all of the rules. If you make a new road junction list, naively using incorrect icons, one of the editors who is familiar with that guideline will simply fix it.

The procedures behind the scenes on Wikipedia, each of them deliberated upon extensively, are coded to define community practices as constrained to actions and habits that further the overall goodness of the project—most importantly, a particular understanding of how articles are to be written.

CONCLUSION: WIKIPEDIA AS A PROCEDURAL ORGANISM

The Wikipedia organism is a community of like-minded people working toward the shared goal of an open, digital, global encyclopedia. It creates rules that standardize, sanitize, and organize content, and which govern community practices, continuously built upon to shape those practices in conformity with the overarching purpose. New users are trained according to these existing rules, taught to properly craft their contributions, while at the same time becoming indoctrinated, guided procedurally toward the mindset that inspired the creation of the rules in the first place.

This community ethos radiates credibility not just from its underlying motivations, but through the fact that it, as a unified entity, acknowledges the fallibility of its individual parts, requiring editors to act as a body of Scriptors, compiling and citing reliable outside sources and mobilizing masses in quick defense of harmful traitors and interlopers. Conceptualizing Wikipedia in this way allows its dynamism and unidentifiable authorship to shift from the primary targets of criticism to be two of its biggest strengths.

The ethos of the Wikipedia community, which above all else seeks to craft the best encyclopedia possible, is facilitated, mandated, maintained, and generated by its evolving system of procedures that exist in a self-sustained symbiosis with all Wikipedians. It is this ethos conveyed by the community as a whole that should be repaired in the eyes of the audience (potential readers or users) rather than engaging in a perpetual struggle with the instability and ungeneralizability of article evaluations.

Using procedural rhetoric to look at the ethos of a community in the way I have done in this chapter leads me to wonder what other applications this approach might have. Procedural rhetoric, divorced from video games and fixed Authors, could probably be used with ethos to evaluate the output produced by any online community so long as it operates according to strict procedures

developed organically by the community rather than handed down from above. Wikipedia may be a rare case that offers an active user base, culturally significant output, an authorless ethos, and complete transparency to see not just the set of rules, but how they are carried out and how they have evolved. Wikis in general would be logical extensions, as well as some of the more open message boards like 4chan or perhaps even chat rooms, but regardless it seems that this idea of Bogost's (2007) and Fulkerson's (2005) holds potential to provide insight into a much wider range of persuasive texts and discourse than have been applied so far. I look forward to reading about its further developments into a notable approach on Wikipedia.

REFERENCES

Andrews, S. (2007, July 12). Wikipedia vs. the old guard. *PC Pro, 154*. Retrieved November 8, 2010, from http://www.pcpro.co.uk/features/119640/wikipedia-vs-the-old-guard

Barthes, R. (1978). *Image, music, text* (Heath, S., Trans.). New York, NY: Hill and Wang.

Benkler, Y. (2006). *The wealth of networks*. New Haven, CT: Yale University Press.

Bogost, I. (2007). *Persuasive games: The expressive power of videogames*. Cambridge, MA: The MIT Press.

Booth, M. (2007, May 01). Grading Wikipedia. *Denver Post*. Retrieved November 8, 2010, from http://www.denverpost.com/search/ci_5786064

Brodkey, L. (1987). Modernism and the scene(s) of writing. *National Council of Teachers of English, 49*(4), 396-418.

Brumm, M., Colbert, S., Dahm, R., Drysdale, E., Dubbin, R., & Gwinn, P. ... Hoskinson, J. (Director). (2006, July 31). Ned Lamont [Television series episode]. In S. Colbert et al. (Producers), *The Colbert Report*. New York, NY: Comedy Central.

Bryant, S. L., Forte, A., & Bruckman, A. (2005). Becoming Wikipedian: Transformation of participation in a collaborative online encyclopedia. *Proceedings of the 2005 International ACM SIG-GROUP Conference on Supporting Group Work,* USA, (pp. 1-10).

Couturat, L. (2002, March 19). *The logic of Leibniz in accordance with unpublished documents* (D. Rutherford & T. R. Monroe, Trans.). Retrieved December 7, 2011, from http://philosophyfaculty. ucsd.edu/faculty/rutherford/leibniz/contents.htm

Enos, T. (Ed.). (1996). *Encyclopedia of rhetoric and composition: Communication from ancient times to the information age.* New York, NY: Garland Publishing.

Essjay controversy. (2010, November 08). *Wikipedia.* Retrieved November 8, 2010 from http:// en.wikipedia.org/wiki/Essjay

Firer-Blaess, S. (2011). Wikipedia: Example for a future electronic democracy? Decision, discipline and discourse in the collaborative encyclopedia. *Studies in Social and Political Thought, 19*, 131-154. Retrieved April 28, 2012, from http://ssptjournal.files.wordpress.com/2011/08/sspt19b1.pdf

Five pillars. (2010, November 07). *Wikipedia.* Retrieved November 8, 2010 from http:// en.wikipedia.org/wiki/Wikipedia:Five_pillars

Forte, A., & Bruckman, A. (2005). *Why do people write for Wikipedia? Incentives to contribute to open-content publishing.* GROUP 05 Workshop: Sustaining Community: The Role and Design of Incentive Mechanisms in Online Systems, Sanibel Island, Florida.

France, P. (1998, June). The encyclopedia as organism. *European Legacy, 3*(3), 62–75. doi:10.1080/10848779808579889

Fulkerson, R. (2005). Composition at the turn of the twenty-first century. *College Composition and Communication, 56*(4), 654–687.

Giles, J. (2005, December 14). Internet encyclopaedias go head to head. *Nature, 438*, 900-901, Retrieved November 18, 2010 from http:// www.nature.com/nature/journal/v438/n7070/ full/438900a.html.

Jasinski, J. (2001). *Sourcebook on rhetoric: Key concepts in contemporary rhetorical studies.* Thousand Oaks, CA: Sage Publications.

Kafker, F., & Loveland, J. (Eds.). (2009). *The early Britannica: The growth of an outstanding encyclopedia.* Oxford, UK: The Voltaire Foundation.

Kapoun, J. (1998, July). Teaching undergraduates Web evaluation. *College & Research Libraries News, 59*(7), 522–523.

Kittur, A., Suh, B., Pendleton, B. A., & Chi, E. (2007). He says, she says: Conflict and coordination in Wikipedia. *CHI 2007: Proceedings of the ACM Conference on Human-factors in Computing Systems,* (pp. 453-462). San Jose, CA: ACM Press.

Kogan, H. (1958). *The great EB: The story of the Encyclopaedia Britannica.* Chicago, IL: University of Chicago Press.

Latour, B., & Woolgar, S. (1986). *Laboratory life: The construction of scientific facts.* Princeton, NJ: Princeton University Press.

Lessig, L. (2004). *Free culture: How big media uses technology and the law to lock down culture and control creativity.* New York, NY: Penguin Press.

Manual of style (road junction lists). (2010, November 08). *Wikipedia.* Retrieved December 10, 2010 from http://en.wikipedia.org/wiki/ Wikipedia:Manual_of_Style_%28road_junction_lists%29

Matthews, R. (2005, December 23). Wikipedia's search for the truth—The online encyclopedia may soon be as credible as it is popular. *Financial Times.* Retrieved November 18, 2010, from http://www.ebusinessforum.com/index. asp?layout=printer_friendly&doc_id=7931

McLeod, K. (2001). *Owning culture: Authorship, ownership, and intellectual property law*. New York, NY: Peter Lang.

Nixon, R. (1998). The feather palace. *Transition, 77*, 70–85. doi:10.2307/2903201

Notability. (2012, November 08). *Wikipedia*. Retrieved November 8, 2010 from http://en.wikipedia.org/wiki/Wikipedia:Notability

Notability (books). (2012, July 9). *Wikipedia*. Retrieved November 08, 2010, from http://en.wikipedia.org/wiki/Wikipedia:Notability_(books)

Notability (fatal hull loss civil aviation accidents). (2011, March 5). *Wikipedia*. Retrieved November 08, 2010, from http://en.wikipedia.org/wiki/Wikipedia:Notability_(books)

Pliny the Elder. (1601). *Naturalis historia* (P. Holland, Trans.). Retrieved July 15, 2012, from http://penelope.uchicago.edu/holland/

Purdy, J. P. (2009). When the tenets of composition go public: A study of writing in Wikipedia. *College Composition and Communication, 61*(2), 351–373.

Purdy, J. P. (2010). Wikipedia is good for you!? In C. Lowe & P. Zemliansky (Eds.), *Writing spaces: Readings on writing,* Vol. 1. Retrieved April 26, 2012 from http://writingspaces.org/essays/wikipedia-is-good-for-you

Read, B. (2006, October 27). Can Wikipedia ever make the grade? *The Chronicle of Higher Education.* Retrieved November 28, 2010, from http://chronicle.com/article/Can-Wikipedia-Ever-Make-the/26960#grading

Reagle, J. (2010/2011). *Good faith collaboration: The culture of Wikipedia*. Cambridge, MA: The MIT Press. Retrieved April 26, 2012, from http://reagle.org/joseph/2010/gfc/

Rosenzweig, R. (2006). Can history be open source? Wikipedia and the future of the past. *The Journal of American History, 93*(1), 117–146. doi:10.2307/4486062

Sloane, T. O. (Ed.). (2001). *Encyclopedia of rhetoric*. New York, NY: Oxford University Press.

Sokal, A. (1996, May/June). A physicist experiments with cultural studies. *Lingua Franca.* Retrieved June 27, 2012, from http://www.physics.nyu.edu/faculty/sokal/lingua_franca_v4/lingua_franca_v4.html

Sunstein, C. (2005). *Why societies need dissent*. Cambridge, MA: Harvard University Press.

Tillman, H. (1995/2003). Evaluating quality on the net. *HopeTillman.com.* Retrieved July 5, 2012, from http://www.hopetillman.com/findqual.html

Warnick, B. (2007). *Rhetoric online: Persuasion and politics on the World Wide Web*. New York, NY: Peter Lang.

Wells, H. G. (1938). *World brain*. London, UK: Ayer.

Wikipedia. org – Traffic Details from Alexa. (2012, June 29). *Alexa Internet, Inc.* Retrieved June 29, 2012, from http://www.alexa.com/siteinfo/wikipedia.org

Wikipedia. Wikiality and other tripling elephants: Revision history. (2012, April 13). *Wikipedia.* Retrieved June 29, 2012, from http://en.wikipedia.org/w/index.php?title=Wikipedia:Wikiality_and_Other_Tripling_Elephants&action=history WikiProject Logic/Standards for notation. (2009, December 15). *Wikipedia*. Retrieved 10 December, 2010, from http://en.wikipedia.org/wiki/Wikipedia:WikiProject_Logic/Standards_for_notation

Yates, F. (1966). *The art of memory*. London, UK: Pimlico.

ADDITIONAL READING

Aristotle. (1954). *Rhetoric* (Roberts, W., Trans.). New York, NY: Modern Library.

Barthes, R. (1986). *The rustle of language* (Howard, R., Trans.). Oxford, UK: Blackwell.

Baumlin, J. (Ed.). (1994). *Ethos: New essays in rhetorical and critical theory.* Dallas, TX: Southern Methodist University Press. doi:10.2307/358718

Benkler, Y. (2006). *The wealth of networks.* New Haven, CT: Yale University Press.

Bogost, I. (2007). *Persuasive games: The expressive power of videogames.* Cambridge, MA: MIT Press.

Clyde, J., Hopkins, H., & Wilkinson, G. (2012). Beyond the "historical" simulation: Using theories of history to inform scholarly game design. *Loading.*, *6*(9), 3–16.

Dickson, A., Mejia, J., Zorn, J., & Harkin, P. (2006). Responses to Richard Fulkerson, "composition at the turn of the twenty-first century. *College Composition and Communication, 57*(4), 730–762.

Foucault, M. (1979). What is an author? In Harari, J. (Ed.), *Textual strategies: Perspectives in post-structuralist criticism* (pp. 141–160). Ithaca, NY: Cornell University Press.

Giles, J. (2005, December 14). Internet encyclopedias go head to head. *Nature, 438*, 900–901. doi:10.1038/438900a

Gleig, G. (1803). Advertisement. In Dobson, T. (Ed.), *Supplement to the encyclopaedia* (pp. iii–vi). Philadelphia, PA: Budd and Bartram.

Hariman, R. (1992). Decorum, power, and the courtly style. *The Quarterly Journal of Speech, 78*(2), 149–172. doi:10.1080/00335639209383987

Jameson, F. (1991). *Postmodernism, or, the cultural logic of late capitalism.* Durham, NC: Duke University Press.

Lanier, J. (2006, May 30). Digital Maoism: The hazards of the new online collectivism. *Edge: The Third Culture.* Retrieved from http://edge.org/3rd_culture/lanier06/lanier06_index.html

Lessig, L. (2006). *Code: Version 2.0.* New York, NY: Basic Books.

Miller, C. (2004). Expertise and agency: Transformations of ethos in human-computer interaction. In Hyde, M. (Ed.), *The ethos of rhetoric* (pp. 197–218). Columbia, SC: University of South Carolina Press.

O'Donnell, D. (2007). If I were "you": How academics can stop worrying and learn to love "the encyclopedia that anyone can edit." *The Heroic Age, 10.* Retrieved from http://www.heroicage.org/issues/10/em.html

Sanger, L. (2005, April 19). *The early history of Nupedia and Wikipedia: A memoir.* Message posted to Slashdot. Retrieved from http://features.slashdot.org/story/05/04/18/164213/the-early-history-of-nupedia-and-wikipedia-a-memoir

Sattler, W. (1947). Conceptions of ethos in ancient rhetoric. *Speech Monographs, 14*, 15–65. doi:10.1080/03637754709374925

Schiff, S. (2006, July 31). Know it all: Can Wikipedia conquer expertise? *New Yorker (New York, N.Y.), 82*(23). Retrieved from http://www.newyorker.com/archive/2006/07/31/060731fa_fact

Smellie, W. (1771). Preface. In Smellie, W. (Ed.), *Encyclopaedia Britannica* (*Vol. 1*, pp. v–vi). Edinburgh, UK: Colin Macfarquhar.

Sunstein, C. (2006). *Infotopia: How many minds produce knowledge.* New York, NY: Oxford University Press.

Warnick, B. (2007). *Rhetoric online: Persuasion and politics on the World Wide Web*. New York, NY: Peter Lang Publishing.

Yoos, G. (1979). A revision of the concept of ethical appeal. *Philosophy and Rhetoric, 12*(1), 41–58.

Zappen, J. (2009). Digital rhetoric: Toward an integrated theory. *Technical Communication Quarterly, 14*(3), 319–325. doi:10.1207/s15427625tcq1403_10

KEY TERMS AND DEFINITIONS

Cartesianism: Relating to the philosophy of Rene Descartes, but here focused on the "Cartesian subject," a rational, autonomous mind in the world.

Ethe: Plural of ethos.

Exigence: A situated, apparently solvable problem, imperfection, or need that is the reason or urgency for speaking or acting.

Gift Economy: A society in which goods and services are frequently given away without specific expectations of repayment. In such a system, donors continue to give because they benefit from the gifts of others.

Neo-Platonism: A mystical philosophy based on the teachings of Plato and especially his theory of Ideas, perfect forms on which the material world is merely a copy of. Neoplatonists have often been preoccupied by the idea of attaining perfect knowledge of these forms.

Procedure: A relatively independent set of operations or actions that can be carried out repeatedly. Procedures run according to rules until completion, running according to the same logic each time.

Process: A flow of activity in pursuit of a goal. Processes are not based on iterability, but take advantage of the iterability of procedures (see above) in their continuous operation.

Vandalism: In the context of Wikipedia, vandalism is the addition, deletion, or alteration of content in a way that deliberately misinforms, offends, or otherwise undermines the integrity of the site.

Wiki: A kind of software, named after the Hawaiian word for "quick," designed to enable easy collaboration, editing, interlinking, and publication of web pages. Though typically minimal in their design, wikis are often meticulously styled for the sake of usability. The wiki became a seminal figure in what came to be known as "Web 2.0" (see above).

ENDNOTES

1. Original Britannica editor William Smellie (1771) wrote in the preface to the 1st edition, "With regard to errors in general, whether falling under the denomination of mental, typographical or accidental, we are conscious of being able to point out a greater number than any critic whatever. Men who are acquainted with the innumerable difficulties of attending the execution of a work of such an extensive nature will make proper allowances. To these we appeal, and shall rest satisfied with the judgment they pronounce" (p. vi). George Gleig (1803), chief editor for Britannica's 3rd edition (1788-1797), wrote, "perfection seems to be incompatible with the nature of works constructed on such a plan, and embracing such a variety of subjects" (p. iv). More recently, in a response to the *Nature* study, Britannica reiterated its disagreement with some perceptions that it sees itself as error-free.

2. The word "ethos" is actually based on the Greek for "custom, habit, or usage" (Sattler, 1947).

3. The Essjay Controversy attracted a great deal of media attention, frequently accompanied by a damnation of Wikipedia. What was generally left out was Wikipedia's policy that credentials should not come into play

since original research isn't even allowed. Furthermore, while the revisions made by Essjay were generally found to be very good under scrutiny, the same cannot be said for the paper at the center of a similar controversy in the print world, the Sokal hoax. In 1996, NYU physics professor Alan Sokal intentionally submitted a nonsensical article in *Social Text* to test whether or not such a prestigious journal would publish it if "(a) it sounded good and (b) it flattered the editors' ideological preconceptions" (Sokal, 1996).

4. Colbert's fans also took it upon themselves to edit the article for Babar the Elephant to reflect tripled publications, the article for Dumbo to describe the rising population of the protagonist's native family, and the article on Idaho mentioning its growth rate being surpassed only by the population of elephants ("Wikipedia: Wikiality," 2012).

5. France's article, *The Encyclopedia as Organism* (1998) contains a number of quotes from encyclopedias in their original French. Google Translate (http://translate.google.com) was used to translate these passages into English. Without a formal translation, I must advise against citing them as I've presented.

6. Also of note in the context of Wikipedia were *Britannica*'s critiques of the organization of several of their competitors for their use of extensive in-text references to other articles, complaining that following all of the links would be too time-consuming to be useful (France, 1998).

7. That encyclopedias other than Wikipedia have been wrong does not require an example, but the article on ostriches from the millennia-old *Naturalis Historia* is telling of the lasting influence an encyclopedia (and its errors) can have. In it, Pliny offers a fact that historians believe is responsible for popularizing a myth that persists today (Nixon, 1998): "...the veriest fooles they be of all others. For as high as the rest of their bodie is, yet if they thrust their head and necke once into any shrub or bush, and get it hidden, they thinke then they are safe ynough, and that no man seeth them" (Pliny, 1601).

8. The ability to see the IP addresses of registered members is one of the very few features that isn't available to the public. Only the upper levels of administration are allowed "CheckUser" access for privacy reasons.

Chapter 8
Credibility in Web Search Engines

Dirk Lewandowski
Hamburg University of Applied Sciences, Germany

ABSTRACT

Web search engines apply a variety of ranking signals to achieve user satisfaction, i.e., results pages that provide the best-possible results for the user. While these ranking signals implicitly consider credibility (e.g., by measuring popularity), explicit measures of credibility are not applied. In this chapter, credibility in Web search engines is discussed in a broad context: credibility as a measure for including documents in a search engine's index, credibility as a ranking signal, credibility in the context of universal search results, and the possibility of using credibility as an explicit measure for ranking purposes. It is found that while search engines—at least to a certain extent—show credible results to their users, there is no fully integrated credibility framework for Web search engines.

INTRODUCTION

Search engines are used for a wide variety of research purposes, and they are often the first place people go when searching for information. Users select search results and then read Web pages based on their decisions about whether the information presented by the search engine is of value to them. Search engines are so popular that they are, together with e-mail, the most commonly used service on the Internet (Purcell, 2011). Every day, billions of queries are entered into the search boxes of popular search engines such as Google and Bing. Therefore, these engines play

an important role in knowledge acquisition, not only from an individual's point of view, but also from society's point of view. It is astonishing to what degree users trust search engines (Hargittai, Fullerton, Menchen-Trevino, & Thomas, 2010; Pan et al., 2007), and users rely on Web search engines to display the most credible results first.

Search engine rankings, however, do not guarantee that credible pages are ranked first for every topic. The construct underlying the rankings of search engines basically assumes that a page's popularity equals credibility, although other factors also play a role. Popularity, in this case, refers to popularity among all Web authors and readers by measuring the distribution of links, clicks within the search engine results pages, time spent read-

DOI: 10.4018/978-1-4666-2663-8.ch008

ing the results documents, and recommendations in social media. While applying these factors in ranking algorithms often leads to good results, it should be stressed that these popularity measures always rely on the users judging the credibility of the documents, i.e., only people make credible pages popular.

Technical means for finding suitable indicators for credible Web pages are an alternative to human credibility judgments about Web search engine results (Mandl, 2005, 2006). Apart from popularity analyses, page and text properties can be used to estimate the credibility of a document, although such approaches can only provide estimates. Before discussing credibility further, we first need to define the concept in the context of search engine results.

According to the Encyclopedia of Library and Information Sciences (Rieh, 2010), credibility is an intuitive and complex concept that has two key dimensions: trustworthiness and expertise. Both are judged by people consuming information, and therefore, credibility always lies in the eye of the beholder. In discussing credibility in the context of Web search engines, we follow Rieh's (2010) definition: "Credibility is defined as people's assessment of whether information is trustworthy based on their own expertise and knowledge" (p. 1338). However, as search engines rate documents algorithmically, we need to consider "people" not only being users of information, but also designers of search engines and their ranking algorithms, which have certain assumptions about credibility that are then used in the system.

Tseng and Fogg's (1999) four types of credibility provided a deeper understanding of credibility:

1. Presumed credibility, where people have general assumptions about a source of information (e.g., assuming that a friend will tell them the truth, or that articles written by full-time journalist will give credible information)
2. Reputed credibility, where sources of information are seen as credible because third

parties assigned credibility to them in the past. E.g., the title of doctor or professor makes most people believe that this person is a credible source of information.
3. Surface credibility, where credibility is given to a source of information because of surface criteria, such as the jacket design of a book or the layout of a webpage.
4. Experienced credibility, where the person judging credibility has first-hand experience with the source of information.

In information retrieval (IR) evaluations, users judge the relevance of documents linked to a search query or information need in a controlled environment. However, while the concept of relevance somehow incorporates credibility, it also incorporates many other aspects. If expert jurors who are instructed to research the credibility of the documents are asked, then statements about credibility can be made. However, such studies are rare, mainly because expert jurors are expensive and the process of evaluating credibility is time-consuming. Furthermore, there is no tradition of credibility evaluation in IR because in traditional IR systems (e.g., newspaper databases or patent collections), the quality of the documents is controlled in the process of producing the database (Rittberger & Rittberger, 1997), and only documents from collections deemed credible are included.

Information quality frameworks (e.g., Knight & Burn, 2005) are of only limited use in the context of Web search engine because the main problem is that search engine users apply credibility judgments when considering (1) the results descriptions ("snippets") on the search engine results pages, and (2) the results documents themselves. In both cases, they have only limited resources for judging credibility. It is much easier to apply information quality criteria to the inclusion of documents into an information system than applying such criteria to the ranking of documents, or even to rely on the system's users to judge the quality of the documents.

Already from this short introduction, we can see that credibility in Web search engines has multiple dimensions; it is a concept that, while it is inherent in search engine rankings and users' perceptions of search engines, has not yet been fully explored. Therefore, the aim of this chapter is to clarify the meaning of credibility in the context of search engines and to explore where credibility is applied in this context.

This chapter is structured as follows. First, the criteria by which search engines decide upon including documents in their indices are discussed. Then, we consider ranking signals generally applied in ranking Web documents and show that while credibility is a measure implicitly considered in these rankings, it is mainly achieved through measuring popularity. As the positions of search engine results greatly influence the probability of the results being selected, it is of interest to content providers to optimize their documents for ranking in search engines. How such optimization can influence the overall credibility of results will be discussed under a separate heading. Then, we will examine the results' credibility when not only considering the general-purpose ranking of search results, but also the inclusion of so-called universal search results into the search engine results pages, i.e., results from specially curated collections. After that, we will turn to search engine evaluation and how credibility criteria are—and could—be applied to it. Finally, we discuss the search engines' responsibility to provide credible results. The chapter concludes with some suggestions for future research.

SEARCH ENGINES' INCLUSION CRITERIA

Search engines are usually seen as tools that aim to index "the whole of the Web." While this surely is not achievable from a technical, as well as from an economic standpoint (Lewandowski, 2005), there are also credibility reasons for not indexing every document available on the Web. While the Web offers a great variety of high-quality information, one should not forget that a large ratio of the documents offered on the Web is of low quality. While there are no exact numbers on how many pages on the Web can be outright considered as spam, the wide variety of spamming techniques facing search engines, as described in the literature, shows that protecting users against this type of content is a major task for search engines (Gyongyi & Garcia-Molina, 2005; Berendt, 2011). It is also understandable that search engine vendors do not provide much information on that issue, since doing so would invite spammers to flood search engine indices, and because techniques for fighting spam are also trade secrets, they are highly valuable to search engine vendors.

While search engines do not explicitly judge the *contents* of documents on the Web, they do decide against the inclusion of documents from known low-quality sources. Decisions against including content in their indices are always based on the source, not on the individual document. However, even though lots of low-quality documents are included in the search engines' databases, a user may not see any of them because of elaborate ranking functions.

However low the barriers for including contents in Web search engines' databases are, they still exist. Consider, for example, pages with spyware or other types of malware. These pages are either excluded from the index entirely, or are specially marked when displayed on a search engine's results page (McGee, 2008). This "hard" credibility measurement is *explicit* in that such documents are either not included in the index at all, or the users are warned when such a document is displayed in the results list. On the other hand, *implicit* judgments on credibility are only shown through the results ranking, which is not comprehensible to the general user.

To summarize, credibility criteria are implemented in search engines in three different ways:

1. Pages of low credibility are excluded from the search engines' indices.
2. Pages of low credibility are specially marked in the results presentation.
3. Pages of low credibility are ranked lower in the results lists.

Again, it should be stressed that while search engines do apply criteria for including documents in their indices, the barriers are very low, and documents or document collections only built for the purpose of being included in the search engines' indices for supporting other, to-be optimized documents through links are mainly excluded. While in some particular cases, a user might miss relevant documents explicitly searched for because the search engine used simply did not index them, this now occurs rather rarely.

DOES POPULARITY EQUAL QUALITY?

It is a mistake to think that in search engines, credibility does not play a role in ranking. However, while search engines do not measure credibility explicitly, measures of credibility are surely implicit of other measures, such as popularity.

Before discussing the influence of credibility on search engine rankings in detail, we will give a short overview of the ranking factors applied in search engine rankings. Ranking algorithms can be broken down into individual signals, and they can consist of hundreds of these signals. However, these signals can be grouped into a few areas, as follows:

1. **Text-Based Matching:** Simple text-based matching, as applied in all text-based information retrieval systems, matches queries and documents to find documents that fulfill the query. Text-based ranking factors such as term frequency / inverted document frequency (TF/IDF) are based on assumptions about the occurrence of terms in a document (e.g., an ideal "keyword density") and allow for a ranking that differentiates not only between documents containing the keywords entered and ones where the keywords are not present, but also between document weights; this allows for a rank-based list of results. However, such text-based ranking algorithms are designed for collections where all documents are deemed credibly (e.g., newspaper databases where quality control is applied before individual documents are added to the database). They fail in the case of the Web, where content providers are able to manipulate documents for gaming the search engines. Therefore, while text-based matching forms the basis of search engine rankings, additional factors measuring the quality of the documents are required.

2. **Popularity:** The popularity of a document is referenced for its quality evaluation. For example, the number of user accesses and the dwell-time on the document is measured, as well as the linking of a document within the Web graph, which is decisive for the ranking of Web documents. For this purpose, not only are the number of clicks and links crucial, but weighted models are also implemented that enable a differentiated evaluation. These models are well documented in the literature (Culliss, 2003; Dean, Gomes, Bharat, Harik, & Henzinger, 2002; Kleinberg, 1999; Page, Brin, Motwani, & Winograd, 1998) and are still considered most important in search results rankings (Croft, Metzler, & Strohman, 2010). Popularity-based measures can be divided into three groups:

   a. **Link-Based Measures:** The classic method to determine popularity is through links. A link pointing to a Web page can be seen as a vote for that page, and when weighting links according to the authority of the linking page, good measurements can be achieved.

b. **Click-Based Measures:** Using click-based measures to determine quality has the advantage that such measures are available almost immediately, while link-based measures require time to build up. The drawback, however, is that in search engines, most users only click on the results presented first; therefore, click-based measures are heavily biased, as not every document even has the opportunity to be selected.

c. **Social Signals:** In the context of social media, explicit ratings of documents are ubiquitous.[1] These judgments can be exploited for ranking, assuming that the search engine has access to data from a social network.

3. **Freshness:** The evaluation of freshness is important for Web search engines in two respects. Firstly, it is a matter of finding the actual or relative publication and refresh dates (Acharya et al., 2005). Secondly, the question concerns which situations it is useful to display fresh documents preferentially.

4. **Locality:** Knowing the location of an individual user is of great use for providing relevant results. This not only holds true in a mobile context, but also for desktop use.

5. **Personalization:** The aim to provide users with tailored results is referred to as personalization and combines measures from the user's own behavior (through queries entered, results selected, reading time) with measures from other users' behavior (focusing on the one hand on all users, and on the other hand on the users socially connected to the user in question), and with general measures (freshness and locality).

From this short explanation of search engine ranking signals, we can conclude that popularity lies at the heart of these systems, whether such popularity exists with all the Web's content producers (who set links to other pages and thus determine their popularity), with a certain user group (e.g., the contacts of an individual user), or with an individual user (through his clicks and viewing patterns).

The question that arises from the discussion of ranking signals is how search engines are able to show credible results without explicitly considering credibility in the documents. When looking at credibility or, more generally, information quality frameworks (Knight & Burn, 2005; Wang, Xie, & Goh, 1999; Xie, Wang, & Goh, 1998), we can see that the criteria generally mentioned are not easily applicable to algorithms. Therefore, "workarounds" must be found.

In Table 1, measures used to determine the credibility of documents are shown. Credibility

Table 1. Implicit measures of credibility as applied by search engines

Credibility through...	Measures	Based on...
Source	Domain popularity	Link graph
Selection behavior	• Click-through rate, i.e., how often a certain document is selected when shown • Time spent reading when document was selected • Bounce rate, i.e., how often a user "bounces back" to the SERP immediately after selecting the document	• Individual user • User group • User population
Recommendation through links	(Weighted) number of links pointing to a certain document	• Links from all other pages • Links from a group of pages, e.g., from topically relevant pages
Explicit ratings	Number of "likes," i.e., number of users who explicitly clicked on a "like" button such as Facebook's	• User group • User population

can be derived through the analysis of the source, the users' selection behavior, recommendations through links, and explicit ratings in social media. All these can be measured in different ways, and in the cases where users are considered, they can further be differentiated according to the group of users taken into account.

THE ROLE OF SEARCH ENGINE OPTIMIZATION

Search engine optimization (SEO) is the modification of Web page content, external links, and other factors, with the aim of influencing the ranking of the targeted documents. It must be stressed that with the well-known search engines, it is not possible to buy positions directly in the results lists, and advertising is separated from the organic results lists (and clearly labeled for the most part). However, with knowledge about search engines' ranking factors, it is possible to influence the results positions of the targeted documents.

While SEO can serve useful purposes, such as making documents findable in search engines by adding relevant keywords to the document, the techniques can also be used to manipulate search engines' rankings, and in the worst case, to help spam documents achieve top positions in the search engines. So while search engine optimization does not directly affect the credibility of an individual document, in the SEO process, vast amounts of non-credible (i.e., pages of low quality) may be generated and pushed into the search engines' results lists. The problem, therefore, does not lie in search engine optimizers producing non-credible documents but in replacing more credible documents on the top of the results rankings.

Search engine optimization has now become an important business, a part of the online marketing industry. Traditionally, companies applied SEO techniques to boost their products and services in the search engines' rankings, but SEO now goes much deeper in that political parties, lobby organizations, and the like use these techniques as well. Even in the academic sector, there are efforts to optimize papers for academic search engines (so-called Academic SEO; see Beel, Gipp, & Wilde, 2010).

Looking at the credibility of search engines' results, SEO may not be problematic when products and services are concerned. One could argue that it does not matter whether a customer searching for some product will buy the best product at the best-possible price or whether he buys an inferior product. However, in the case of knowledge acquisition, the issue is different. Here, credibility matters, and SEO can severely influence the credibility of the results.

Search engine optimization tries to influence all implicit credibility judgments mentioned in Table 1, although selection behavior is an exception, at least to a certain degree, because it is hard to simulate real user behavior.

The real influence SEO has on areas where knowledge acquisition in concerned has still not been explored. However, there is some research on special sources that appear frequently in search engines' results, such as Wikipedia articles (Lewandowski & Spree, 2011). From research on Wikipedia, we can see that lobbyists and public relations agencies try to influence public opinion by writing and editing articles. Furthermore, social media optimization aims at "optimizing" opinions expressed in social media. Again, there is a wide range of methods, ranging from Facebook popularity campaigns to writing fake reviews.

In summary, SEO and other "optimizing" strategies aim at influencing what users see in the top positions of search engine results pages. Agencies now see the control of the full first search engine results page as critical for success (Höchstötter & Lüderwald, 2011), whereas in the past, the aim was just occupying one top position. We assume that the influence of search engine optimization will even rise in the coming years. However, it is difficult to project how search engines will react to increasing efforts to influence their results lists, when informational content is concerned. On the one hand, SEO can help to make good informational content visible in the search engines, and therefore, search engines such as Google even encourage content providers to use SEO techniques (Google, 2012). On the other hand, search engines need to keep the content displayed within the results lists credible, at least to a certain degree.

UNIVERSAL SEARCH RESULTS

Universal Search is the composition of search engine results pages from multiple sources. While in traditional results presentation, results from just one database (the Web index) are presented in sequential order and the presentation of individual results does not differ considerably, in universal search, the presentation of results from the different collections (such as news, video, and images) is adjusted to the collections' individual properties. See Figure 1 for an example showing news and image results injected into the organic results list.

With universal search results, we see search engines turning away from pure algorithmic results. The idea of the general-purpose Web search engine was based on the facts that: (1) there are only low barriers for documents to be included in the index, and (2) due to the same algorithms

Figure 1. Universal search results presentation in Google

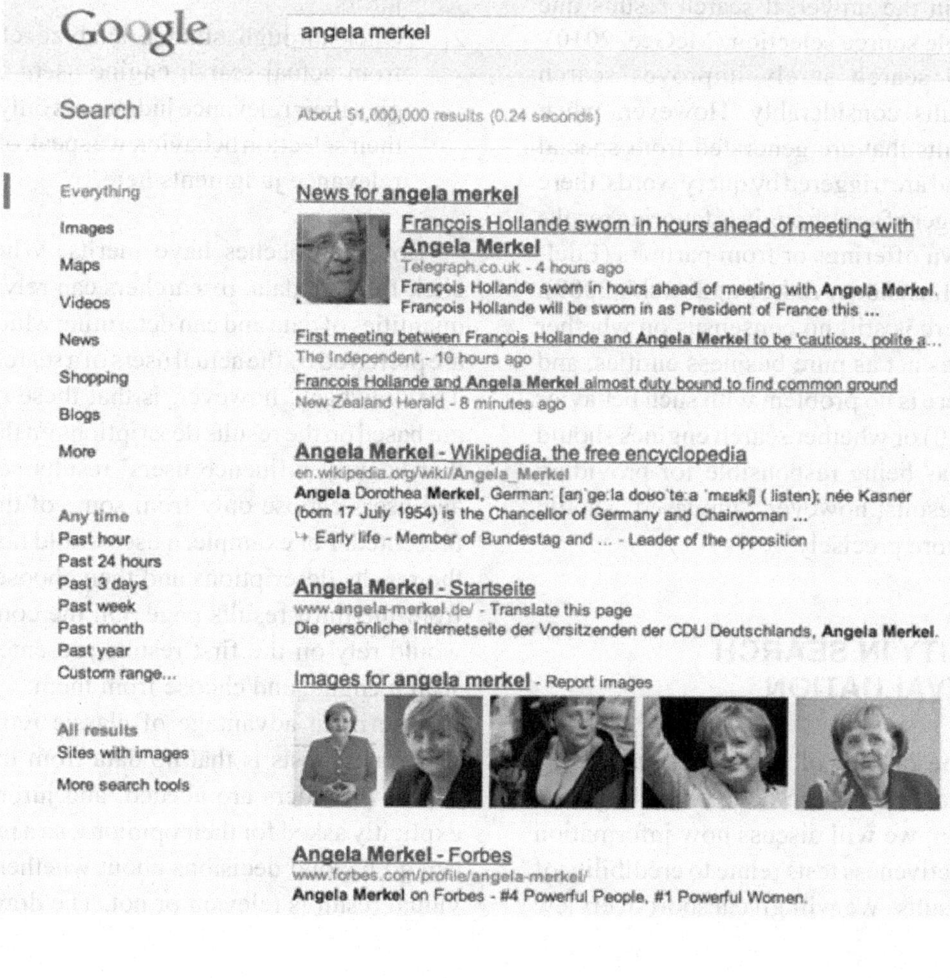

applied to all documents in the index, every document has the same chance of being shown in a results list.

With universal search, however, these conditions do not necessarily hold true anymore. We see that search engines build highly-curated collections where only a very limited number of sources is included; thus, only search results from these sources are displayed. While a search engine's general Web index includes millions of sources (when considering every Web site to be a different source), the news index of a search engine only includes some hundred sources that a search engine considers as news media. While this surely does improve quality in terms of credibility (the documents themselves are considered of equal quality, and credibility judgments are made based on the source), low-quality documents may also be included in the universal search results due to questionable source selection (McGee, 2010).

Universal search surely improves search engines' results considerably. However, when showing results that are generated from special collections and are triggered by query words, there is also the danger of search engines favoring results from their own offerings or from partners (Edelman, 2010; Höchstötter & Lewandowski, 2009). However, there is still no consensus on whether search engines act as pure business entities, and therefore, there is no problem with such behavior (Granka, 2010) or whether search engines should be regarded as being responsible for providing "unbiased" results; however, "unbiased" should be defined more precisely.

CREDIBILITY IN SEARCH ENGINE EVALUATION

Regarding the quality of search results, a vast body of research has been done over the years. In this section, we will discuss how information retrieval effectiveness tests relate to credibility of the search results. We will give a short overview

of the testing methods, and then show how credibility assessments could be added to such tests.

Evaluation has always been an important aspect (and an important research area) of information retrieval (IR). Regarding Web search engines, established IR evaluation methods have been adapted to the context and been modified to suit Web searching. A good overview of newer approaches in Web search engine retrieval effectiveness evaluation is provided by Carterette, Kanoulas, & Yilmaz (2012):

In the retrieval effectiveness evaluation, two approaches need to be differentiated:

1. Retrieval effectiveness tests use a sample of queries and jurors to evaluate the quality of the individual results. These studies employ explicit relevance judgments made by the jurors.
2. Click-through studies analyze click data from actual search engine users. As users give their relevance judgments only through their selection behavior, we speak of implicit relevance judgments here.

Both approaches have merits. When using click-through data, researchers can rely on large quantities of data and can determine which results are preferred by the actual users of a search engine. The drawback, however, is that these decisions are based on the results descriptions on the SERPs that heavily influence users' results selections, and users choose only from some of the results presented. For example, a user would not read all the results descriptions and then choose a result from the third results page. On the contrary, he would rely on the first results presented by the search engine and choose from them.

The main advantage of classic retrieval effectiveness tests is that no data from the search engine providers are needed, and jurors can be explicitly asked for their opinions, so a researcher can go beyond decisions about whether an individual result is relevant or not. The drawback of such tests, however, is that such studies usually must rely on a relatively low number of queries and jurors, and results are seen as independent of one another. This can be illustrated by a user choosing a completely relevant result and who will therefore not need another relevant result that just repeats the information already given.

Regarding the judgment of the results, we speak of implicit relevance judgments in the case of click-through studies, and of explicit relevance judgments in the case of retrieval effectiveness tests. Relevance is a concept central to information science, but there is no agreed-upon definition of relevance in the field (Saracevic, 2007a, 2007b; Borlund, 2003; Mizzaro, 1997). However, this may not even be problematic when considering searches in curated collections and users being experts in their field; thus, the users can also consider credibility when assessing relevance. Tefko Saracevic, after many years of research on relevance, states: "Nobody has to explain to users of IR systems what relevance is, even if they struggle (sometimes in vain) to find relevant stuff. People understand relevance intuitively" (Saracevic, 2006, p. 9).

However, the question remains as to what extent this holds true when speaking of general-purpose Web search engine users. They surely see some results as relevant and others as not, but it is unclear to what extent they are able to discover errors or biases in the documents being examined. Therefore, it is questionable whether judging the relevance of documents is truly sufficient when testing Web search engines.

Research examining the concepts underlying relevance found that many factors influence users' relevance judgments. Mizzaro (1997) reported research by Rees and Schultz, who found 40 variables influencing users' relevance decisions. Cuadra and Katter (1967) found 38 variables, and Barry and Schamber (1998) found 80 variables. Some studies (e.g., Chu, 2011) attempted to identify the most important variables in relevance judgments, but we still lack studies explicitly us-ing individual variables, especially credibility, as criteria in search engine evaluation. Information science seems to regard credibility as just one part of the wider concept of relevance, which is central to the discipline.

One reason for using the more general concept of relevance instead of more specialized, credi-bility-oriented measures lies in the availability of jurors and the workload needed to judge the results' quality. When approaching relevance as stated in the quote by Saracevic above, relevance judgments are relatively easy to obtain and deci-sions can be made relatively quickly.

Most search engine studies use students as jurors, and faculty sometimes also help to judge the documents (Lewandowski, 2008, pp. 918-919). While this is convenient, it is questionable whether these jurors are indeed able to evaluate the credibility of the results, as they may lack deeper understanding of the topics being researched and may invest too little time in checking for credibil-ity. A further concern is that search engine evalu-ations increasingly "crowd-source" relevance judgments, i.e., the tasks are distributed over the Internet to a large group of paid-for jurors. While this allows for larger studies, the jurors often lack commitment to the tasks and are interested in com-pleting them in the least amount of time possible.

When assessing the credibility of the results, it would be best to use expert jurors and to ask them to examine the results thoroughly. However, expert jurors are difficult to contact and must also be paid for their services. Therefore, studies using such experts are, apart from some very topically specialized studies, quite rare.

As can be seen, the main problem with using credibility in search engine evaluations is recruit-ing suitable expert jurors. It may seem simple for experts to judge documents, but often, a rather large amount of time is needed to research every statement within a document. Even when the criteria for judging credibility are unambiguous, the time-consuming process of examining tens or even hundreds of pages (as done in retrieval effectiveness tests) is a major barrier not only for adding credibility to search engine tests, but also to adding other factors.

Keeping this in mind, I suggest taking a four-part approach for incorporating credibility judgments into search engine retrieval effectiveness tests:

1. Use actual search engine users to judge results relevance (as done in retrieval effectiveness studies).
2. Let the same users judge the credibility of the results. Add questions on the type of credibility the jurors assign to the individual result: presumed credibility, reputed credibility, surface credibility, and experienced credibility.
3. Use automatic approaches to judge credibility. These approaches can be used to at least give indications on reputed quality and surface quality. However, it should be noted that search engine rankings themselves could be seen as judgments of reputed quality.
4. Use expert judges to judge the credibility of individual results.

This multi-dimensional approach can be used to compare laypersons' and experts' rating of results credibility. For search engine vendors, information about whether the results presented are seen as credible by both groups is of value for improving their ranking algorithms. For internet researchers, such information would be valuable in determining whether search engines are suitable tools for research in certain areas, e.g. health topics.

In contrast to studies on the credibility of search engines results, there is a vast body of research on frameworks for judging the credibility of Web content (Rieh, 2002; Wathen & Burkell, 2002; Fogg, 2003; Hilligoss & Rieh, 2008; for an overview, see Rieh & Danielson, 2007). A challenge for research on Web search engines lies in combining the advantages of established retrieval effectiveness test methods with judgments on credibility. One can argue that asking jurors only to judge

relevance is realistic, in that users generally do not think much about the credibility of the documents, instead relying on the intermediary (the search engine) and its ranking, but to understand search engines as dominant tools in knowledge acquisition, further research on their overall quality is needed (Lewandowski & Höchstötter, 2008). Research from other disciplines such as journalism, rhetoric, or lexicography, where frameworks for measuring credibility exist, can help to build better credibility measurements for Web search engines (Lewandowski & Spree, 2011).

THE RESPONSIBILITY OF WEB SEARCH ENGINES FOR CREDIBLE INFORMATION

It could not be stressed enough how search engines have become major tools for knowledge acquisition. This can be seen when looking at the number of queries entered into the general-purpose search engines every day. According to research from ComScore (ComScore, 2010), more than 131 billion searches were conducted in December 2009 alone. The question that arises from these figures is what role search engines should play in promoting credible Web documents. While their algorithms surely honor credibility to a certain degree (see above), credibility is still not explicitly a concept applied in search engines. Vertical search engines, i.e., search engines that do not provide a "complete" index of the Web but instead focus on a specific area (such as news, jobs, or scholarly articles), are one step toward source credibility, as the sources are—in many cases—handpicked.

However, this vertical search approach is not applicable to general Web searches, as the Web is simply too large to base search engines on human-curated collections. However, the combination of the general-purpose Web index with vertical indices is a way to achieve higher credibility of the overall results. Problematic with this approach, however, is that neither inclusion criteria for vertical indices nor lists of sources included are currently published by the search engines. Furthermore, such vertical indices exist only for some areas, and these are not necessarily the areas where curated indices are deemed most useful (e.g., health, law, and authoritative information from governments).

The question remains as to whether search engines should be seen as pure business entities that can decide whatever content they want to present and in which way, or whether they have a special responsibility to provide users with credible results free from bias.

FUTURE RESEARCH DIRECTIONS

From the argumentation above, we derive three major areas for further research about credibility in Web search engines:

1. **Applying Credibility in Search Engine Evaluations:** Search engine evaluation traditionally focuses on relevance, which incorporates a multitude of concepts. While relevance is a suitable concept for judging the overall quality of a query result or information need and can easily be used in evaluations using laypersons, it is hard to differentiate between the factors that make a result relevant. Furthermore, jurors often cannot judge the credibility of the results accurately, and may even not notice obvious factual errors. Therefore, search engine evaluations should focus more on the credibility of search engine results. While this surely makes evaluations more complicated, as results must be checked for errors and bias, the results would help a great deal in finding what quality of results users are able to view when using search engines.

2. **Increasing Users' Awareness of Credibility Issues in Search Engines:** It is well known that users do not invest much cognitive effort in the formulation of queries and the examination of search engine results (Höchstötter &

Koch, 2009; Machill, Neuberger, Schweiger, & Wirth, 2004). Furthermore, search engines are often seen as responsible for the content of the results provided, and search engines such as Google are referred to as the source for information, whereas in reality, they are simply intermediaries between the searcher and the content providers. This results in great trust in search engines, especially in Google, the most popular search engine. Further research should focus on how users' information literacy can be improved, especially concerning search engine-based research, where credibility is a major concept to be explored. Furthermore, users' awareness of how search engines rank the results and how search engine optimization can influence (the credibility of) the results should be raised.

3. **Building Collections of Credible Sources:** In vertical search engines, collections of credible sources are already built. These collections can be grouped into automatically generated collections (such as Google Scholar's collection of scholarly articles), and handpicked collections (such as Google News). Especially for critical domains such as health and law, search engines' results could significantly benefit from the latter. Microsoft's Bing search engine shows how health-related information from credible sources can be incorporated into search engines' results. However, the sources deemed credible need to be disclosed in such collections. While Bing uses a limited set of health authorities, in other cases such as news, it remains unclear how the collection is built and what the criteria for inclusion are.

4. **Search Engine Vendors' Responsibility for the Quality of the Results:** It remains unclear how much of a problem (lack of) credibility is in Web search engines. Given examples from such diverse areas as credit-related information, health, law, and politics, we know that search engines at least some-

times present biased and/or non-credible information. Research examining to what extent search engines provide users with such content, and how users react to such results, is needed.

CONCLUSION

Credibility is an important concept relevant to search engine evaluation and for search engine providers as well. However, as we have seen in this chapter, in neither area is credibility explicitly used. Search engine providers could benefit from building more human-curated collections of credible sources, as these would help lead users to the best results available on the Web. Another way of promoting credibility in search engine results pages is to give the users better support for judging credibility. Kammerer and Gerjets (2012) suggested the following three areas: (1) Reducing the prominence of search results ranking, (2) increasing the prominence of quality-related cues on search engine results pages, and (3) automatic classification of search results according to genre categories. While (1) is mainly applied in "alternative" search engines experimenting with results presentation and visualization, we can see that the major Web search engines increasingly give quality-related cues on the SERPs through, e.g., author and freshness information. There are some experiments regarding automatic genre classification, but this has not been applied in the major Web search engines yet.

Search engine evaluation should also focus more on credibility, instead of relying on jurors who are themselves not experts in a field and work under time restrictions. As search engines are employed by many millions of users every day for a multitude of purposes, and these users trust in the results provided by the engines, more effort should be put into researching not only the technical means of search engines, but also their impact on knowledge acquisition.

REFERENCES

Acharya, A., Cutts, M., Dean, J., Haahr, P., Henzinger, M., Hoelzle, U., et al. (2005). *Information retrieval based on historical data.* (US Patent US 7,346,839 B2).

Barry, C. L., & Schamber, L. (1998). Users' criteria for relevance evaluation: A cross-situational comparison. *Information Processing & Management*, 34(2-3), 219–236. doi:10.1016/S0306-4573(97)00078-2

Beel, J., Gipp, B., & Wilde, E. (2010). Academic search engine optimization (ASEO): Optimizing scholarly literature for Google Scholar & Co. *Journal of Scholarly Publishing*, 41(2), 176–190. doi:10.3138/jsp.41.2.176

Berendt, B. (2011). Spam, opinions, and other relationships: Towards a comprehensive view of the Web . In Melucci, M., & Baeza-Yates, R. (Eds.), *Advanced topics in information retrieval* (pp. 51–82). Berlin, Germany: Springer. doi:10.1007/978-3-642-20946-8_3

Borlund, P. (2003). The concept of relevance in IR. *Journal of the American Society for Information*, 54(10), 913–925. doi:10.1002/asi.10286

Carterette, B., Kanoulas, E., & Yilmaz, E. (2012). Evaluating Web retrieval effectiveness . In Lewandowski, D. (Ed.), *Web search engine research* (pp. 105–137). Bingley, UK: Emerald. doi:10.1108/S1876-0562(2012)002012a007

Chu, H. (2011). Factors affecting relevance judgment: A report from TREC legal track. *The Journal of Documentation*, 67(2), 264–278. doi:10.1108/00220411111109467

ComScore. (2010). *comScore reports global search market growth of 46 percent in 2009.* Retrieved from http://comscore.com/Press_Events/Press_Releases/2010/1/Global_Search_Market_Grows_46_Percent_in_2009

Croft, W. B., Metzler, D., & Strohman, T. (2010). *Search engines: Information retrieval in practice.* Boston, MA: Pearson.

Cuadra, C. A., & Katter, R. V. (1967). Opening the black box of "relevance." . *The Journal of Documentation, 291–303.* doi:10.1108/eb026436

Culliss, G. A. (2003). *Personalized search methods.* (US Patent US 6,539,377 B1).

Dean, J. A., Gomes, B., Bharat, K., Harik, G., & Henzinger, M. H. (2002, March 2). *Methods and apparatus for employing usage statistics in document retrieval.* (US Patent App. US 2002/0123988 A1).

Edelman, B. (2010). *Hard-coding bias in Google "algorithmic" search results.* Retrieved April 11, 2011, from http://www.benedelman.org/hardcoding/

Fogg, B. (2003). *Prominence-interpretation theory: Explaining how people assess credibility online. CHI'03 Extended Abstracts on Human Factors in Computing Systems* (pp. 722–723). New York, NY: ACM.

Google. (2012). *Search engine optimization (SEO) - Webmaster tools help.* Retrieved February 21, 2012, from http://support.google.com/webmasters/bin/answer.py?hl=en&answer=35291

Granka, L. (2010). The politics of search: A decade retrospective. *The Information Society, 26*(5), 364–374. doi:10.1080/01972243.2010.511560

Gyongyi, Z., & Garcia-Molina, H. (2005). Web spam taxonomy. *First International Workshop on Adversarial Information Retrieval on the Web (AIRWeb 2005)* (pp. 39-47).

Hargittai, E., Fullerton, L., Menchen-Trevino, E., & Thomas, K. Y. (2010). Trust online: Young adults' evaluation of Web content. *International Journal of Communication, 4,* 468–494.

Hilligoss, B., & Rieh, S. Y. (2008). Developing a unifying framework of credibility assessment: Construct, heuristics, and interaction in context. *Information Processing & Management, 44*(4), 1467–1484. doi:10.1016/j.ipm.2007.10.001

Höchstötter, N., & Koch, M. (2009). Standard parameters for searching behaviour in search engines and their empirical evaluation. *Journal of Information Science, 35*(1), 45–65. doi:10.1177/0165551508091311

Höchstötter, N., & Lewandowski, D. (2009). What users see – Structures in search engine results pages. *Information Sciences, 179*(12), 1796–1812. doi:10.1016/j.ins.2009.01.028

Höchstötter, N., & Lüderwald, K. (2011). Web monitoring . In Lewandowski, D. (Ed.), *Handbuch Internet-Suchmaschinen 2: Neue Entwicklungen in der Web-Suche* (pp. 289–322). Heidelberg, Germany: Akademische Verlagsanstalt AKA.

Kammerer, Y., & Gerjets, P. (2012). How search engine users evaluate and select Web search results: The impact of the search engine interface on credibility assessments . In Lewandowski, D. (Ed.), *Web search engine research* (pp. 251–279). Bingley, UK: Emerald. doi:10.1108/S1876-0562(2012)002012a012

Kleinberg, J. (1999). Authoritative sources in a hyperlinked environment. *Journal of the ACM, 46*(5), 604–632. doi:10.1145/324133.324140

Knight, S. A., & Burn, J. (2005). Developing a framework for assessing information quality on the World Wide Web. *Informing Science Journal, 8,* 159–172.

Lewandowski, D. (2005). Web searching, search engines and Information Retrieval. *Information Services & Use, 25,* 137–147.

Lewandowski, D. (2008). The retrieval effectiveness of Web search engines: Considering results descriptions. *The Journal of Documentation, 64*(6), 915–937. doi:10.1108/00220410810912451

Lewandowski, D., & Höchstötter, N. (2008). Web searching: A quality measurement perspective . In Spink, A., & Zimmer, M. (Eds.), *Web search: Multidisciplinary perspectives* (pp. 309–340). Berlin, Germany: Springer.

Lewandowski, D., & Spree, U. (2011). Ranking of Wikipedia articles in search engines revisited: Fair ranking for reasonable quality? *Journal of the American Society for Information Science and Technology*, *62*(1), 117–132. doi:10.1002/asi.21423

Machill, M., Neuberger, C., Schweiger, W., & Wirth, W. (2004). Navigating the Internet: A study of German-language search engines. *European Journal of Communication*, *19*(3), 321–347. doi:10.1177/0267323104045258

Mandl, T. (2005). The quest to find the best pages on the Web. *Information Services & Use*, *25*(2), 69–76.

Mandl, T. (2006). Implementation and evaluation of a quality-based search engine. *Proceedings of the Seventeenth Conference on Hypertext and Hypermedia* (pp. 73–84). New York, NY: ACM.

McGee, M. (2008). Google explains malware warning policy & how to fix your site. *Search Engine Land*. Retrieved February 20, 2012, from http://searchengineland.com/google-malware-warning-policy-15271

McGee, M. (2010). Google news dropping sites, reviewing inclusion standards. *Search Engine Land*. Retrieved February 21, 2012, from http://searchengineland.com/google-news-dropping-sites-reviewing-inclusion-standards-57673

Mizzaro, S. (1997). Relevance: The whole history. *Journal of the American Society for Information Science American Society for Information Science*, *48*(9), 810–832. doi:10.1002/(SICI)1097-4571(199709)48:9<810::AID-ASI6>3.0.CO;2-U

Page, L., Brin, S., Motwani, R., & Winograd, T. (1998). *The PageRank citation ranking: Bringing order to the Web*. Retrieved from http://ilpubs.stanford.edu:8090/422/1/1999-66.pdf

Pan, B., Hembrooke, H., Joachims, T., Lorigo, L., Gay, G., & Granka, L. (2007). In Google we trust: Users' decisions on rank, position, and relevance. *Journal of Computer-Mediated Communication*, *12*, 801–823. doi:10.1111/j.1083-6101.2007.00351.x

Purcell, K. (2011). *Search and email still top the list of most popular online activities*. Retrieved from http://www.pewinternet.org/~/media//Files/Reports/2011/PIP_Search-and-Email.pdf

Rieh, S. Y. (2002). Judgment of information quality and cognitive authority in the Web. *Journal of the American Society for Information Science and Technology*, *53*(2), 145–161. doi:10.1002/asi.10017

Rieh, S. Y. (2010). Credibility and cognitive authority of information . In Bates, M., & Maack, M. N. (Eds.), *Encyclopedia of library and information sciences* (3rd ed., pp. 1337–1344). New York, NY: Taylor and Francis Group, LLC.

Rieh, S. Y., & Danielson, D. (2007). Credibility: A multidisciplinary framework . In Cronin, B. (Ed.), *Annual review of information science and technology* (*Vol. 41*, pp. 307–364). Medford, NJ: Information Today.

Rittberger, M., & Rittberger, W. (1997). Measuring the quality in the production of databases. *Journal of Information Science*, *23*(1), 25–37. doi:10.1177/016555159702300103

Saracevic, T. (2006). *Relevance: A review of the literature and a framework for thinking on the notion in information science- Part II. Advances in Librarianship* (*Vol. 30*, pp. 3–71). Elsevier.

Saracevic, T. (2007a). Relevance: A review of the literature and a framework for thinking on the notion in information science, part II. *Journal of the American Society for Information Science and Technology, 58*(13), 1915–1933. doi:10.1002/asi.20682

Saracevic, T. (2007b). Relevance: A review of the literature and a framework for thinking on the notion in information science, part III: behavior and effects of relevance. *Journal of the American Society for Information Science and Technology, 58*(13), 2126–2144. doi:10.1002/asi.20681

Tseng, S., & Fogg, B. J. (1999). Credibility and computing technology. *Communications of the ACM, 42*(5), 39–44. doi:10.1145/301353.301402

Wang, H., Xie, M., & Goh, T. N. (1999). Service quality of internet search engines. *Journal of Information Science, 25*(6), 499–507. doi:10.1177/016555159902500606

Wathen, C. N., & Burkell, J. (2002). Believe it or not: Factors influencing credibility on the web. *Journal of the American Society for Information Science and Technology, 53*(2), 134–144. doi:10.1002/asi.10016

Xie, M., Wang, H., & Goh, T. N. (1998). Quality dimensions of Internet search engines. *Journal of Information Science, 24*(5), 365–372. doi:10.1177/016555159802400509

ADDITIONAL READING

Balatsoukas, P., Morris, A., & O'Brien, A. (2009). An evaluation framework of user interaction with metadata surrogates. *Journal of Information Science, 35*(3), 321–339. doi:10.1177/0165551508099090

Bar-Ilan, J. (2004). The use of Web search engines in information science research . In Cronin, B. (Ed.), *Annual review of information science and technology* (*Vol. 38*, pp. 231–288). Medford, NJ: Information Today, Inc. doi:10.1002/aris.1440380106

Battelle, J. (2005). *The search: How Google and its rivals rewrote the rules of business and transformed our culture.* New York, NY: Portfolio.

Buganza, T., & Della Valle, E. (2010). The search engine industry. *Search Computing: Challenges and Directions, 5950*, 45–71.

European Commission. (2010). *Antitrust: Commission probes allegations of antitrust violations by Google.* Retrieved from http://europa.eu/rapid/pressReleasesAction.do?reference=IP/10/1624

Hargittai, E. (2007). The social, political, economic, and cultural dimensions of search engines: An introduction. *Journal of Computer-Mediated Communication, 12*(3), 769–777. doi:10.1111/j.1083-6101.2007.00349.x

Hendry, D., & Efthimiadis, E. (2008). Conceptual models for search engines . In Spink, A., & Zimmer, M. (Eds.), *Web searching: Multidisciplinary perspectives* (pp. 277–308). Berlin, Germany: Springer.

Jansen, B., & Spink, A. (2006). How are we searching the World Wide Web? A comparison of nine search engine transaction logs. *Information Processing & Management, 42*(1), 248–263. doi:10.1016/j.ipm.2004.10.007

Lewandowski, D. (2012). A framework for evaluating the retrieval effectiveness of search engines . In Jouis, C., Biskri, I., Ganascia, J., & Roux, M. (Eds.), *Next generation search engines: Advanced models for information retrieval* (pp. 456–479). Hershey, PA: IGI Global. doi:10.4018/978-1-4666-0330-1.ch020

Lewandowski, D. (2012). New perspectives on Web search engine research . In Lewandowski, D. (Ed.), *Web search engine research* (pp. 1–17). Bingley, UK: Emerald. doi:10.1108/S1876-0562(2012)002012a003

Mintz, A. (Ed.). (2012). *Web of deceit: Misinformation and manipulation in the age of social media.* Medford, NJ: Information Today, Inc.

Poritz, J. (2007). Who searches the searchers? Community privacy in the age of monolithic search engines. *The Information Society, 23*(5), 383–389. doi:10.1080/01972240701572921

Robertson, S. (2008). On the history of evaluation in IR. *Journal of Information Science, 34*(4), 439–456. doi:10.1177/0165551507086989

Sullivan, D. (2011). Study: Google "favors" itself only 19% of the time. *Search Engine Land.* Retrieved from http://searchengineland.com/survey-google-favors-itself-only-19-of-the-time-61675.

Van Couvering, E. (2007). Is relevance relevant? Market, science, and war: Discourses of search engine quality. *Journal of Computer-Mediated Communication, 12*(3), 866–887. doi:10.1111/j.1083-6101.2007.00354.x

Van Couvering, E. (2008). The history of the Internet search engine: Navigational media and traffic commodity . In Spink, A., & Zimmer, M. (Eds.), *Web searching: Multidisciplinary perspectives* (pp. 177–206). Berlin, Germany: Springer.

Zimmer, M. (2010). Web search studies: Multidisciplinary perspectives on Web search engines . In Hunsinger, J., Klastrup, L., & Allen, M. (Eds.), *International handbook of Internet research* (pp. 507–521). Dordrecht, The Netherlands: Springer. doi:10.1007/978-1-4020-9789-8_31

KEY TERMS AND DEFINITIONS

Index: The database on which a search engine is based. A search engine can have multiple indices. The core index is the index of Web documents ("Web index"), where only low inclusion barriers apply and as many documents as possible (and economically maintainable) are included. This index can be accompanied by other (vertical) indices, which are based on a selection of sources.

Ranking Signal: Search engines use many signals that together form the ranking algorithm. A signal refers to a single criterion that can be used in document ranking. Based upon definition, search engines apply some hundreds or even thousands of signals in their rankings.

Search Engine Optimization (SEO): Search engine optimization (SEO) is the modification of Web page content, external links and other factors, aiming to influence the ranking of the targeted documents.

Search Engine Results Page (SERP): A search engine results page is a complete presentation of search engine results; that is, it presents a certain number of results (determined by the search engine). To obtain more results, a user must select the "further results" button, which leads to another SERP.

Universal Search: Universal Search is the composition of search engine results pages from multiple sources. While in traditional results presentation, results from just one database (the Web index) are presented in sequential order and the presentation of individual results does not differ considerably, in universal search, the presentation of results from the different collections (such as news, video, and images) is adjusted to the collections' individual properties.

Vertical Search Engine: Contrary to the general-purpose search engine, a vertical search engine focuses on a special topic. General-purpose search engines often enhance their results with results from vertical search indices, such as news, video, or images.

ENDNOTES

[1] There are also implicit social signals, but as these are also click-based, they fall under c.

Section 3
News/Primary Research

Chapter 9
The Special Case of Youth and Digital Information Credibility

Miriam J. Metzger
University of California Santa Barbara, USA

Rebekah Pure
University of California Santa Barbara, USA

Andrew J. Flanagin
University of California Santa Barbara, USA

Alex Markov
University of California Santa Barbara, USA

Ryan Medders
California Lutheran University, USA

Ethan Hartsell
University of California Santa Barbara, USA

ABSTRACT

The vast amount of information available online makes the origin of information, its quality, and its veracity less clear than ever before, shifting the burden on individual users to assess information credibility. Contemporary youth are a particularly important group to consider with regard to credibility issues because of the tension between their technical and social immersion with digital media, and their relatively limited development and life experience compared to adults (Metzger & Flanagin, 2008). Although children may be highly skilled in their use of digital media, they may be inhibited in terms of their ability to discern quality online information due to their level of cognitive and emotional development, personal experience, or familiarity with the media apparatus compared to adults. This chapter presents the findings of a large-scale survey of children in the U.S. ages 11-18 years examining young people's beliefs about the credibility of information available online, and the strategies they use to evaluate it. Findings from the study inform theoretical, practical, and policy considerations in relation to children's digital literacy skills concerning credibility evaluation.

YOUTH AND DIGITAL INFORMATION CREDIBILITY

With the sudden explosion of digital media content and information access devices in the last two decades, there is now more information available to more people from more sources than at any other time in human history. Most people in the developed world today have ready access to almost inconceivably vast information repositories that are increasingly portable, accessible, and interactive in both delivery and formation. One result of this contemporary media landscape is that there exists incredible opportunities for learning, social connection, and individual enhancement via the vast information resources made available by networked digital media.

DOI: 10.4018/978-1-4666-2663-8.ch009

However, information's origin, quality, and veracity are in many cases less clear than ever before, creating an unparalleled burden on individuals to find appropriate information and assess its meaning and relevance (Metzger & Flanagin, 2008). Access to the tremendous number and range of available sources makes accurately assessing information credibility extremely challenging and laborious. And existing research indicates that there may be reason to fear many individuals are not up to the task of credibility evaluation (Bennett, Maton, & Kervin, 2008; Kuiper & Volman, 2008; Metzger, 2007). Moreover, inaccurate credibility assessments can pose serious social, personal, educational, health, and financial risks (Metzger & Flanagin, 2008).

While this is true for all users of digital media, youth are a particularly intriguing group to consider with regard to information and source credibility, for several reasons. As Livingstone (2009) pointed out, children represent around one-fifth of the population in developed countries and studying the myriad ways that they combine multiple media, multitask, engage with each other online, and blur the boundaries between online and offline socialization could yield more insight into the future of media usage than studying adults alone. Not only is children's digital media use behavior indicative of future trends, it also signals a potentially different relation to information gathering and evaluation in the future. Therefore, it is important to understand children's online information evaluation today.

Children are also of interest due to the tension between their technical and social immersion with digital media and their relatively limited development and lived experience compared to adults (Eastin, 2008; Metzger & Flanagin, 2008). On one hand, as so-called "digital natives," children have grown up in an environment saturated with networked digital media technologies (Palfrey & Glasser, 2008; Prensky, 2001) and thus may be highly skilled in their use of those media to access, consume, and generate information. This

suggests that in light of their special relationship to digital tools, youth are especially well-positioned to successfully navigate the complex contemporary media environment. Indeed, forms of credibility evaluation that rely on information to be spread efficiently through social networks suggest some intriguing advantages for younger populations, who are often more interconnected than adults (Jones & Fox, 2009; Lenhart, Purcell, Smith, & Zickurh, 2010). For example, some argue that older children are better able to embrace networked publics than are adults because adults tend to find the "shifts brought on by networked publics to be confusing and discomforting because they are more acutely aware of the ways in which their experiences with public life are changing" (boyd, 2011, p. 54).

On the other hand, youths can be viewed as limited in their cognitive and emotional development, life experiences, and familiarity with the media apparatus. Although children may be talented and comfortable users of technology, they may lack tools and abilities critical to effectively evaluate information (Eastin, 2008; Rowlands et al., 2008). For example, children have fewer benchmarks than adults to compare against information they find online or to discern the relative reputational cues across sources. In addition, children may not have the same level of experience with, or knowledge about, media institutions, which can make it difficult for them to understand differences in editorial standards across various media channels and outlets (e.g., traditional news media sources versus news blogs) compared to adults who grew up in a world with fewer channels and less media convergence (Metzger & Flanagin, 2008). More generally, some youths may not be as critical of digital media or particular online information sources as adults because these media are not "new" to young people who cannot remember a time without them, and thus they do not apply the same level of skepticism toward digital media as do adults. Finally, many children, and especially very young children, often require assistance from adults to even retrieve information, let alone assess its credibility (Solomon, 1993). Some even argue that information retrieval systems, because they require complicated queries when searching for information, are not well suited to children and often yield "inappropriate results in a format unsuitable for children" (van der Sluis & van Dijk, 2010, p. 9), which may further impede youth from evaluating information adequately.

Although a significant amount of research has explored credibility assessment in the context of digital media with populations over the age of 18 (e.g., Chen & Rieh, 2009; Flanagin & Metzger, 2000, 2007; Fogg, 2003; Hargittai, Fullerton, Menchen-Trevino, & Thomas, 2010; Metzger, Flanagin, & Medders, 2010), there is a paucity of work focused specifically on children of any age. This is surprising, given contemporary youth's unique relationship to media technology. For example, youths are more likely than adults to turn to digital media first when researching a topic for school or personal use; they are more likely to read news on the Internet than in a printed newspaper; and they are more likely to use online social networking tools to meet friends and to find information (Lenhart et al., 2010). Moreover, children's relationship to digital media may impact their approach to learning and research (Ito et al., 2009; Prensky, 2001). As the first generation to grow up with the Internet, young people are comfortable collaborating and sharing information via digital networks, and do so "in ways that allow them to act quickly and without top-down direction" (Rainie, 2006, p. 7). Additionally, the interactivity afforded by networked digital media allows children to play roles of both information source and receiver simultaneously as they critique, alter, remix, and share content in an almost conversational manner using digital tools (Tapscott, 1997). These experiences likely have profound implications for how children both construct and evaluate credibility online.

Despite these realities, discussions of youth and digital media have often been somewhat simplistic, focusing for example on the popular generation gap caricature, which portrays children as either technologically adept compared to adults or as utterly vulnerable and defenseless (Greenfield, 2004). Such considerations miss the most important and enduring byproducts of heavy reliance on digital media: "Growing up digital" (Tapscott, 1997) means that more and more of the information that drives children's daily lives is provided, assembled, filtered, and presented by sources that are largely unknown to them, or known to them in nontraditional ways. Yet research has only begun to explore what this means for younger Internet users, who will be immersed in digital media for the entirety of their lives, and for those who endeavor to teach them the skills they need to evaluate digital sources.

In light of the complex relationship between youth and digital media, coupled with the lack of research on children's understanding of credibility, this chapter seeks to examine a series of fundamental and overarching research questions that explore how children ages 11-18 years old view information and source credibility online, including the extent to which they are aware of, and concerned about, the credibility of information they find online, and how believable they find various *types* of online information to be; the ways in which children evaluate the credibility of information online, as well as how they compare their skill at evaluating credibility compared to that of adults; and the extent that young people's credibility beliefs are influenced by individual differences, such as demographic characteristics, types of Internet usage or skill, and strategies they invoke for evaluating credibility.

To address these questions, we present data from a nationally representative, U.S.-based survey of children living at home, ages 11-18 years. We focus our interpretation of results on how children establish credibility when they evaluate information and sources online. To situate this exploration, we begin by examining young people's basic awareness of credibility as a potential problem in the digital environment by asking how often they think about credibility when they are online, and how concerned are they about the credibility of information they find on the Internet?

Credibility Beliefs across Information Types

Although data on children's general beliefs about credibility are useful, young people's credibility beliefs may vary by the type of information they find. Research has shown that the degree to which adults believe information they find online varies by the type or topic of information for which they search and that assessments of credibility are related to the context under which one finds information (Flanagin & Metzger, 2007; Hargittai et al., 2009). People may put more or less rigor into credibility assessment depending on the type of information in question, and they may be more or less skeptical of information depending on its source. For example, people are less likely to find commercial information or information coming from special interest groups to be credible, presumably because they recognize these sources' strong potential for bias (Flanagin & Metzger, 2000, 2007). Yet, young Internet users, who might not have the same background knowledge or sufficient experience in discerning the underlying motivations of commercial or advocacy sources, may not experience this same skepticism.

Moreover, although the majority of research to date has focused on the credibility of static web sites (e.g., government web sites, ecommerce web sites, health and medical web sites, etc.), the current media environment is composed of a diversity of information alternatives, including user-generated information sources such as Wikipedia and news blogs that operate outside of traditional, top-down models of knowledge generation. As digital natives, young people may be simultaneously less aware of the potential credibility problems associated with user-generated

content, and more comfortable with information produced in this manner than adults, which is likely to impact how they evaluate user-generated information. To examine these issues, we ask: How believable do children find online information to be, and to what extent does this vary by the type of information they find?

Information Evaluation Strategies among Youth

Prior research on credibility specifically, and on decision-making more generally, suggests several cognitive processing strategies people use to evaluate information (e.g., Gigerenzer & Todd, 1999; Metzger, 2007; Scott & Bruce, 1995). People sometimes analyze information and its features carefully; other times they use a more holistic and intuitive approach based on their feelings; and sometimes they may draw upon other people in their social circle for advice and guidance. These three strategies, respectively called "analytic," "heuristic," and "social" information processing strategies, have been examined in terms of their impact on adults' credibility determinations (Metzger, Flanagin, & Medders, 2010), but very little research to date has investigated the use of these strategies among children (the only exception being Flanagin & Metzger, 2010). Given developmental and experiential differences between adults and children suggesting that children may differ from adults in terms of their information processing abilities when assessing the credibility of information online (Eastin, 2008), we examine the degree to which children may invoke similar or different strategies as adults by asking: What kinds of cognitive processing strategies do youth employ to evaluate the credibility of online information?

Predictors of Credibility Concerns and Beliefs

Past research on adults (Metzger et al., 2011) have indicated several factors that predict credibility concerns and beliefs, which may be applicable to children's evaluations as well. For example, demographic and background characteristics, patterns of Internet use and skill, a variety of relevant personality traits, and various strategies for evaluating credibility are all likely to be important in explaining children's evaluation of online information credibility and yet have been largely unexplored in the research literature to date.

For example, previous research on credibility evaluation has paid scant attention to demographic and background factors, although there is reason to believe that children's information evaluation strategies and opportunities may vary developmentally across age (Eastin, 2008), income (van Dijk, 2006), or other demographic groupings such as cognitive or academic abilities. Indeed, although differences in access to, and processing of, online information have been found among people of different demographic backgrounds (Hargittai, 2002a), and among people with different levels of usage, experience, or skill with a medium (van Dijk, 2005), surprisingly little research has focused on what factors influence children's credibility judgments.

Similarly, patterns of Internet usage, access, and past negative experiences with information obtained online are important sources of systematic variation in credibility beliefs among young adults (Hargittai et al., 2010). These patterns may also impact children's perceptions of credibility online by leading them to different types of information and by influencing their level of skepticism. Moreover, it is likely that parental mediation impacts young people's attitudes about, and evaluations of, digital media content in ways similar to how parental mediation is known to affect children's reactions to traditional media content such as television (e.g., Valkenberg, Krcmar, Peeters, & Marseille, 1999). More specifically, the extent to which parents control or restrict children's access to and use of the Internet (i.e., "restrictive mediation") and how often parents talk to their children about the credibility of information online (i.e.,

"informative mediation") is another individual difference that may affect children's credibility evaluations. Receiving formal instruction in evaluating the credibility of Internet information in a school setting may similarly affect young people's credibility perceptions.

In addition, several personality traits, including cognitive dispositions or "thinking styles" that have been shown to influence how adults approach information (Zhang, 2003), may also contribute to young people's credibility beliefs and practices. *Need for cognition*, for example, reflects the degree to which people engage in and enjoy thinking deeply about problems or information and, thus, are willing to exert effort to scrutinize information. *Flexible thinking* measures people's willingness to consider opinions different from their own, which might impact how children process contradictory or contrasting information when judging credibility online. *Faith in intuition* reflects a tendency to trust based on first impressions, instincts, and feelings. And, *social trust*, or the propensity to trust strangers, might also affect the degree to which young people are likely to find information provided by those they do not know online to be trustworthy.

Finally, different strategies or methods for evaluating credibility—that is, the *process* of evaluating information online—may also influence the assessments that young information consumers make. Research in adolescent decision making (Jacobs & Klaczynski, 2005) indicates that adolescents primarily approach information analytically or heuristically when making decisions and evaluations. Analytic processing involves effortful and deliberate consideration, whereas heuristic decisions are made more quickly, with less cognitive effort and scrutiny (Klaczynski, 2001). In addition, the strategy of relying on others to help make decisions may also be relevant online, and especially for youth, given their comfort with social media generally and the recent proliferation of information sources that enable people to see and benefit from each other's experiences (Scott & Bruce, 1995). Given the Internet's vast capacity for social interaction, social approaches to credibility evaluation may thus be a particularly important means of processing digital information for young people (as well as for adults—see Metzger et al., 2010), who may use them in conjunction with or in place of heuristic and analytic means of information evaluation. The fourth research question of this study takes these various characteristics into account by asking: To what extent do demographic and background characteristics, patterns of Internet use and perceived skill, personality traits, and strategies for evaluating credibility affect young people's credibility beliefs?

Relative Skill in Credibility Evaluation

Research on cognition demonstrates that people tend to feel that they are less susceptible to negative influence than others are. This phenomenon is rooted in a cognitive process known as the "optimistic bias" (Weinstein, 1980), which is the tendency to see oneself as less likely than others to experience negative life events. Research on the notion of optimistic bias has examined its impact on the beliefs and behaviors of individuals in many contexts (e.g., Clarke et al., 2000; Weinstein, 1980, 1982), and has demonstrated the stability of this phenomenon across a wide range of demographic variables, including age, sex, and education (Weinstein, 1987). However, little research has focused on the occurrence of the optimistic bias in a digital media environment (for one exception, see Campbell, Greenauer, Macaluso, & End, 2007, who found evidence for optimistic bias on the part of college students across a variety of Internet-related events) and no research to date has examined credibility assessments with regard to this phenomenon.

Given that adolescents demonstrate a greater sense of invulnerability to negative events compared to adults (Alberts, Elkind, & Ginsberg, 2006; Elkind, 1967), the same psychological

processes underpinning the optimistic bias phenomenon might operate in the context of judging the credibility of information online. With regard to children's perceptions of their own ability to evaluate the credibility of information online, we sought to understand how children perceive their own skill at evaluating the credibility of information online compared to others by asking the final research question posed in this study, which is: How do children compare their own perceived skill at evaluating the credibility of information online to that of a "typical" Internet user?

Finally, we examined the extent of developmental differences in children's credibility beliefs and behaviors across all of the research questions posed above by analyzing the extent to which age differences exist in young people's concern about credibility, their credibility beliefs, and the strategies they use to evaluate credibility in the online context. The next section describes the methodology employed to examine each of the research questions.

METHOD

The small amount of empirical research on children and credibility is based almost exclusively on interviews and small, nonrepresentative samples of children and adolescents (e.g., Fidel et al., 1999; Hargittai et al., 2009; Large, 2004). To complement these studies, a large-scale probability-based survey of children in the U.S. with Internet access was conducted in order to be able to generalize the results to the larger population, which is only possible with a study of this magnitude. Indeed, these data comprise the most comprehensive information ever available concerning children's credibility evaluation processes.

Participants

The survey was fielded by the research firm Knowledge Networks to a probability-based panel of participants that is representative of children with Internet access in the United States: 2,747 valid responses were obtained from young people between the ages of 11-18 years, with approximately 340 respondents for each age within the range. Responses were weighted to correct known demographic discrepancies between the U.S. population of Internet households and Knowledge Networks' online panel.[1]

Respondents consisted of 53% males and 47% females, with an average age of 14.33 (*SD* = 2.28). 75% were white; 9% were black, non-Hispanic; 12% were Hispanic; 0.4% were other, non-Hispanic; and 4% reported being mixed race, non-Hispanic. Household annual income ranged from less than $5,000 to more than $175,000, with an average income of between $60,000-$85,000. Most families (88%) had between 3 and 5 members living in the household, and the average number of children living at home was 2.25 (*SD* = 1.39). Participants came from all U.S. geographic areas: the Midwest (31%), Northeast (19%), South (28%), and West (23%).[3]

Materials

The survey instrument used in this study was generated through a multi-step, multi-method process. Initial survey topics were based on a review of past literature and existing surveys on information trust, credibility, and quality. To better understand cognitive and developmental issues relevant to youth information assessment and processing, experts in the fields of Developmental Psychology and Cognitive Psychology were recruited as project consultants, who provided feedback on a draft version of the questionnaire. A focus group was then conducted among children 11-18 years old to help refine survey terminology. Next, 40 children were recruited to undergo an hour-long face-to-face interview, in which they provided feedback on questionnaire content and operationalization of key variables, question wording, and general survey administration. This feedback was used to finalize the questionnaire and to ensure that children as young as 11 years could understand and respond to each of our questions appropriately. The survey was also pilot-tested before being launched in the field.

Measures

The survey measured credibility concern, including how often children think about the credibility of information they find online; the believability of various types of online information and the relative believability of different information delivery channels; strategies employed to evaluate the credibility of online information; Internet use, access, and past experiences; perceived Internet skill; and personality and demographic variables relevant to credibility evaluation. Throughout the survey, credibility was operationalized in terms of "believability," as suggested by past credibility literature (Fogg, 2003) and validated by the children who participated in the focus group and interviews. Details about the measures are presented with the survey results.

RESULTS

Credibility Concern

The first research question asked how often young people think about credibility, as well as how concerned they feel people should be about the credibility of information online. 79% of children said they think about whether they should believe information they find online "sometimes" or more often, and 71% said that people should be "somewhat" to "very" concerned about the believability of online information. Age did not matter much in these findings ($r_{thinkabout}$ = .01, p = .59; $r_{concern}$ = .05, $p < .01$), although 18 year olds felt people should be more concerned about the believability of online information than both 12 and 14 year olds ($F_{concern}$ = 2.51, df = 2, 2732, $p < .05$; post hoc tests showed differences at p

< .05). This may reflect a greater awareness of credibility problems with online information as children grow older.

Credibility Beliefs

Findings for the second research question posed in this study indicate that children found information on the web in general to be relatively believable, with 59% reporting that "some" information was believable and 30% reporting that "a lot" of the information found online was believable. There was also a significant tendency for perceived information believability to increase with age (r = .10, $p < .001$): 18 year olds found more of the information online to be credible than 11 through 14 year olds ($F = 4.28$, $df = 7, 2724$, $p < .001$; post hoc tests at $p < .05$).

Credibility by Information Type

Children were asked how likely they are to believe information on the Internet about a number of topics, including health or medical issues, news, something they may want to buy, entertainment information (e.g., about movies, musicians, celebrities, etc.), other people they meet online, and information they find for school papers or projects. Results show that children varied in their likelihood of believing information across these topics ($F = 928.64$, $df = 5, 8620$, $p < .001$, partial η^2 = .35). Specifically, children were on average most likely to believe information on the Internet about schoolwork, followed by news, then entertainment and health information (which children were equally likely to believe), commercial information, and information about people they meet online. Although there were minor age differences with these findings, the general pattern of findings endured regardless of age.[3] These results are generally consistent with what has been found for adults (Flanagin & Metzger, 2000, 2007).

Credibility Perceptions Across Media

Children were asked which medium, including the Internet, television, books, magazines, newspapers, radio, and someone they talk to in person, provides the most believable information. Consistent with past research (Flanagin & Metzger, 2000), differences emerged across channels depending on the type of information sought: When looking for health or medical information, 39% of children indicated that they would believe someone they talk to in person most, followed by the Internet (21%) and books (20%). Children felt that the most believable news information originated from television (54%), followed by newspapers (24%), and then the Internet (11%). Commercial information was best retrieved from the Internet (41%) or in person (33%), followed by television and magazines (10% each). The most believable entertainment information, according to children, can be found on the Internet (40%), then television (28%), then in magazines (11%). Lastly, 53% of children noted that the most believable information for school paper or projects can be found on the Internet, followed by books (34%), and then people they talk to in person (7%). Overall, children rated the Internet as the most believable source of information for schoolwork, entertainment, and commercial information, as well as second most believable for health information and third most believable for news information.

Some age differences emerged in children's indication of which channels they believe most for specific types of information.[4] Older kids tended to believe entertainment information from the Internet and newspapers more than younger kids, and entertainment information from books and the radio less than younger kids. Additionally, older children believed health information from the Internet, books, and magazines more than younger children, and health information from the radio less than younger kids. With news information, older children believed the Internet, books, and magazines more than younger kids and in-person and radio sources less than them. For school-related information, older children believed books and magazines more than younger kids and in-person sources less than younger kids. Finally, the Internet and newspapers are seen as more credible channels for commercial information for older kids than for younger ones, while television was seen as a less credible source of commercial information by older versus younger children.

Children were next asked (on a 5-point scale where higher values signal greater levels) how much people *should* believe the information they find via particular media channels, including newspapers, television, and the Internet. Children indicated that newspapers should be believed the most ($M = 3.54$, $SD = .87$), followed by television ($M = 3.19$, $SD = .78$), and finally the Internet ($M = 2.94$, $SD = .67$), $F = 701.09$, $df = 2, 5526$, $p < .001$, partial $\eta^2 = .30$. There were significant but very small associations between age and feeling newspapers ($r = -.07$, $p < .001$) and the Internet ($r = .05$, $p < .05$) should be believed, but assessments of how much television should be believed ($r = .00$, $p = .99$) did not vary with children's age. When considered together with the findings above about credible sources by information type, it appears that in some ways children's own use of the Internet may exceed the extent to which they think others should rely on it for credible information.

Credibility of News Blogs and Wikipedia

Overall, kids do not find news blogs to be very credible. 79% say they are either "much less" or "somewhat less" believable than newspaper and television news. This does not vary by age ($r = .03$, $p = .29$). It should be noted, however, that many kids were unsure about the comparative credibility of blogs and mainstream news, with 37% of all kids answering "I don't know" about their relative credibility and 8% of the total sample indicating that they did not know what a blog is.

Nearly all kids (99%) had heard of Wikipedia, and the vast majority of them (84%) had used it to look up information. However, when asked to identify what Wikipedia is from a list of seven possibilities (e.g., whether it is an online encyclopedia where anyone can contribute information, a social networking site, a web site where you can play games, etc.), 9% admitted that they do not know what it is, and only 78% made the correct identification. Moreover, there was a small tendency for older kids (ages 16+) to more accurately understand what Wikipedia is.

Overall, children who understand what Wikipedia actually is find it to be fairly credible. Most believe information from Wikipedia at least "some" (43%) or "a lot" (28%). However, children were slightly more skeptical about how much other people should believe Wikipedia, with 23% saying it should be believed "a little bit," 49% saying it should be believed "some," and 20% saying it should be believed "a lot." Indeed, the extent to which children say people *should* believe Wikipedia ($M = 2.88$, $SD = .87$) is significantly lower than they report believing it themselves ($M = 3.04$, $SD = .93$; $t = 14.31$, $df = 2105$, $p < .001$). There was a significant but very small positive relationship between age and both how much children themselves believed ($r = .05$, $p = .04$) and how much they thought that people in general should believe ($r = .04$, $p = .04$) Wikipedia information.

Methods for Evaluating Credibility

Analyses for the third research question investigated children's strategies for evaluating the credibility of information online. Children were asked the extent to which they based their online credibility assessments on heuristic (e.g., by relying on their gut, making decisions based on feelings, making quick decisions), analytic (by carefully considering the information, double checking facts, gathering a lot of information, and considering all views), or social (by getting advice from others or asking for others' help) criteria. Children reported that they used analytic techniques to carefully evaluate the credibility of information online "sometimes" to "often" (on a 5-point scale with higher values indicating higher levels; $M = 3.45$, $SD = .74$), whereas they used both social ($M = 2.92$, $SD = .71$) and heuristic ($M = 2.96$, $SD = .67$) methods comparatively less often ($F = 575.92$, df $= 2$, 5554, $p < .001$, partial $\eta^2 = .17$).

Although this pattern of using analytic methods most often, followed by heuristic and then social methods, was similar across all age groups, the frequency with which each strategy was used increased with age ($r_{heuristic} = .09$, $p < .001$; $r_{analytic} = .13$, $p < .001$, $r_{social} = .11$, $p < .001$). In other words, there was a general trend that older children reported applying all three methods of credibility evaluation more often than younger kids. Interestingly, these results do not comport with research on adults, who indicate that they often use heuristic methods of credibility evaluation (Metzger et al., 2010). Without further study, however, it is impossible to say whether this difference is due to true differences between kids and adults in their strategies for evaluating credibility, or to the specific question wording or research method used in these two studies (i.e., survey methods versus focus groups). Indeed, the question itself may have prompted kids to think about situations in which knowing the credibility of the information they sought was important, rather than considering how they evaluate credibility across the full range of information-seeking situations (e.g., the question asked how often they used analytic, heuristic, or social-based strategies when deciding *what to believe*, rather than simply asking how often each strategy is used while *looking at* information online).

Predictors of Credibility Concerns and Beliefs

The extent to which demographic and background characteristics, prior experiences with online infor-

mation and its evaluation, patterns of Internet use, personality traits, perceived skill, and strategies for evaluating credibility affect young people's (a) concerns about and (b) beliefs about credibility was investigated using stepwise multiple regression analysis. Measures for each of the predictor variables are discussed first, and are followed by the results of the regression analysis.

The *demographic and background characteristics* examined included children's sex, age, household income, race, and grades in school. Young people's *prior experiences with online information and its evaluation* included whether children had ever had a bad experience using some information they found online that turned out not to be credible, or whether they had ever heard of this happening to others. They were also asked whether they had ever received instruction in how to evaluate the credibility of information, how often their parents talk to them about whether to trust information on the Internet, and how many restrictions the parent of each child imposed in the home, which ranged from 0 (parents set no restrictions on child's use of the Internet) to 4 (parents place the computer in a certain location in the home, limit the sites their child can visit, limit the amount of time their child may go online, and use "other" control mechanisms).

Patterns of Internet use included how much time young people spend with the Internet (per week as well as the number of years they have used it) and their use of the Internet for specific activities (social networking, contributing information online, visiting virtual worlds, and using the web for commercial purposes).[5] Measures for the *personality traits* of need for cognition, flexible thinking, and faith in intuition were adapted from standard measures of these concepts (e.g., Epstein, Pacini, Denes-Raj, & Heiner, 1996; Kokis, MacPherson, Toplak, West, & Stanovich, 2002). The social trust items were taken directly, or adapted from, the General Social Survey (GSS). All measures were pilot tested to ensure that children could comprehend them easily.

Children's *perceived level of Internet skill* included self-assessments of technical skill, search skill, and knowledge of the latest online trends and features on a scale from 0-10, with the midpoint defined as being as skilled as a "typical Internet user" and the endpoints defined as being "much more[less] skilled than other Internet users." Skill was operationalized in this comparative fashion, rather than, as in past studies that advocate self-assessment, via an inventory of specific and demonstrable Internet skills (e.g., Hargattai, 2002b), since an underlying component of concerns and beliefs about credibility is self-efficacy relative to others. Indeed, Gasser, Cortesi, Malik, and Lee (2012) argue that self-assessment of skill is especially important for teachers and educational programs as it broadens the definition of what is considered a skill by users themselves rather than narrowly defining what constitutes skill from the perspective of researchers or teachers.

Of course, there is an inherent danger to self-appraisal of Internet skills as it may lead participants to shortchange or overvalue their abilities. However, the crucial comparison point for this study is children's perceptions of their own Internet skills (rather than their actual skills), and how those perceptions may affect credibility evaluation. Thus, to measure perceived Internet skill, participants were asked to "rate your ability to find what you are looking for on the Internet, compared to other Internet users;" to "rate your technical skill with the Internet (for example, fixing problems, changing computer settings, etc.), compared to other Internet users;" and to "rate your knowledge of the latest Internet trends and features, compared to other Internet users."

Finally, *information evaluation strategies* were measured in two ways. First, the question, "When you decide what to believe on the Internet, do you…[give careful thought to the information, rely on your gut feelings, ask for help from other people, etc.]" was used to gauge the extent to which analytic, heuristic, and social means of information evaluation were used by the respondents. Second,

respondents were asked the degree to which they focused more or less on certain credibility cues or web site elements while evaluating credibility online. Factor analysis showed that these various credibility cues reflect three strategies: evaluating credibility via expert confirmation (e.g., looking to see if information is from expert sources), via information quality (e.g., looking at the currency and completeness of the information), and via web site design (e.g., considering the site's appearance and navigability).

The results of the regression analyses used to examine the influence of the foregoing individual difference factors on credibility concern and credibility beliefs are included in Table 1, and are discussed in turn next.

Credibility Concern

Regression analysis showed that the type of strategies that young people use to evaluate credibility affect their concern about credibility and how often they think about credibility issues while seeking information online. Specifically, children who are more concerned about the credibility of Internet information tend to use a more analytic than heuristic approach when evaluating information and look to expert confirmation to evaluate credibility, but rely less on evaluating credibility by means of looking at the web site design. Kids' online experiences and education matter also: having had a bad experience or even hearing about others who have trusted bad information online, having parents talk to them about the trustworthiness of information found online, and having had formal instruction in credibility evaluation all contribute to greater overall concern about the credibility of information on the Internet and/or how often children think about credibility.

The ways in which children engage with the Internet and participate in content creation also mattered in their concerns about credibility. Specifically, those who use the Internet to immerse themselves in virtual worlds more often (includ-

Table 1. Results of stepwise multiple regression analyses predicting credibility concerns and beliefs

Variable	Concerns about Credibility						Beliefs about Credibility					
	How concerned			How often think about credibility			How much is believable			How likely to believe		
	B	SE B	β	B	SE B	β	B	SE B	β	B	SE B	β
Demographics												
Sex of child										.11	.04	.08**
Age of child										-.02	.01	-.07**
Income							.02	.01	.10***	.02	.01	.09***
Race (white or not)	-.13	.06	-.06*									
Grades in school												
Use, Access, Experience												
Years online	.03	.01	.07**									
Hours online												
Internet skill				.03	.01	.07*	.02	.01	.07*	.04	.01	.11***
Bad experience- self	.13	.06	.06*	.10	.05	.06*	-.08	.03	-.06*	-.09	.04	-.07**
Bad experience- news	.28	.06	.13***	.10	.05	.06*						
Social networking use												
Commercial use												
Virtual use				.06	.02	.08**	.03	.02	.06*	.05	.02	.07**
Info contribution use	-.11	.04	-.08**									
Talk with parents	.13	.03	.11***	.07	.02	.07**						
Credibility training				.17	.05	.09***						
Personality Traits												
Flexible thinking				.12	.05	.07*						
Faith in intuition										.08	.03	.07*
Need for cognition							-.09	.03	-.09***	-.09	.03	-.09**
Internet social trust	-.18	.04	-.12***				.20	.03	.22***	.15	.03	.15***
General social trust							.11	.03	.10***			
Evaluation Method												
Expert confirmation	.21	.04	.16***							.08	.03	.09*
Site design	-.10	.04	-.08**				.10	.02	.13***	.11	.03	.13***
Information quality										.08	.03	.10*
Analytic style	.24	.04	.17***	.23	.03	.21***						
Heuristic style	-.08	.04	-.05*	-.11	.03	-.09***	.05	.03	.05*	.13	.03	.13***
Social style												

Notes: Only significant results are shown. $F_{concern}$ = 25.53, df = 11, 1329, p < .001, adj R^2 = .17; $F_{thinkabout}$ = 24.82, df = 9, 1331, p < .001, adj R^2 = .14; $F_{muchbelieve}$ = 25.30, df = 9, 1329, p < .001, adj R^2 = .14; $F_{likelybelieve}$ = 25.89, df = 13, 1326, p < .001, adj R^2 = .20. *p < .05, **p < .01, ***p < .001

ing playing games such as World of Warcraft) and those who engage in content creation less actively show higher levels of concern about credibility. Also, kids who perceived themselves as more highly skilled and who had been online for a greater number of years thought more about or were more concerned about credibility. These results indicate that as children engage more, and more deeply, with the Internet, they may develop a healthy skepticism toward the believability of online information. This finding refutes fears that kids will become more accepting and less critical of Internet information as they deepen their experience and participation in online activities.

Only two traits, flexible thinking style and Internet social trust, emerged as significantly related to children's level of credibility concern. As kids are more flexible in considering information that runs counter to their own beliefs and are less trusting of others online, they express greater concern about credibility. Attending to contradictory data would naturally raise concern about whose view to trust, as would having little confidence in the trustworthiness of others online.

Interestingly, young people's demographic characteristics did not seem to matter much, with one exception: race made a very small contribution to users' concern about credibility. Children who reported themselves to be minorities expressed slightly greater concern about credibility than did white children, which may reflect subcultural differences found in many surveys for trust of all sorts among minority populations (Alesina & LaFerrara, 2002). It is noteworthy that, overall, age did not impact concern about credibility, despite the fact that older kids have more online experience and more life experience.

Credibility Beliefs

While young people's concern about the credibility of information online seems to be driven to some extent by analytic processes of evaluating information, this is not the case for their actual trust of online information, both in terms of the amount of information on the Internet they feel is credible and their likelihood of trusting information they personally find online.

Indeed, young people's beliefs about credibility appear to be more a function of heuristic processes, as evidenced by the fact that young people who rated online information as more credible tended to use a more heuristic, rather than analytic, approach to evaluating information online. Factors that consistently contribute to young people's actual credibility beliefs are evaluating information based on the web site's design and using heuristic credibility evaluation strategies, such as relying on gut feelings and making quick credibility judgments. Personality traits related to these heuristic strategies also contributed significantly to beliefs about credibility, whereby youth possessing lower need for cognition and higher faith in intuition thinking styles rated information on the Internet as more credible.

Although these results are not surprising in light of what is known from past research on adults that finds people's credibility evaluations are often based on cursory cues rather than thorough examination of online information (Metzger, 2007), the fact that heuristic processes figure so prominently in how much online information children find credible and how likely they are to believe the information they find online is a little disconcerting. This is particularly true for digital literacy advocates who stress the need for kids to apply critical thinking skills to Internet-based information due to the unique characteristics that make discerning credible from non-credible information especially complex and difficult (see Metzger, Flanagin, Eyal, Lemus, & McCann, 2003). Another personality trait that influenced young people's views of the credibility of online information was their trusting nature. Questions that tapped into the degree to which kids felt others could be trusted both generally and online were significant and positive predictors of how much of the information online they felt was believable.

Children's demographic characteristics mattered more for their actual beliefs about the credibility of online information than they did for their concern about credibility. Specifically, young people from families of higher income said they believed more information on the Internet, and both younger kids and girls were more likely to believe the information they find online compared to older kids and boys, respectively.[5] This could be due to differences in girls' and boys' Internet usage or experiences interacting with others online, and to the fact that older children are more likely to have had greater overall exposure to online information generally, and thus perhaps more experiences with bad information, as well as had more formal information literacy training than have younger children.

Indeed, the data show that Internet usage and experiences do also factor into kids' credibility beliefs. In particular, young people who rated themselves as more technically skilled felt Internet information was more credible, as did those who visit virtual worlds more often. Past negative experiences with false or non-credible information also mattered in that having such experiences led kids to say that less Internet information is believable and that they were less likely to believe the information they found online, as one would expect.

Relative Skill in Credibility Evaluation

The final research question sought to examine how young people perceive their own skill at determining the credibility of information online relative to other users. Results showed that, in comparison to a "typical Internet user," even the youngest children saw themselves as equally or slightly better on average in their ability to figure out which information is good and bad online ($M = 3.47$, $SD = 1.37$) and more likely to question information they find on the Internet ($M = 3.55$, $SD = 1.47$), both measured on 7-point scales where scores below the midpoint indicate a favorable

self vs. "typical user" comparison. Children also felt they were equally or slightly less likely than a typical Internet user to believe false information online ($M = 4.61$, $sd = 1.44$) on a 7-point scale where scores above the midpoint indicate a favorable self-other comparison. Interestingly, older kids were more likely than younger kids to report a favorable comparison to typical Internet users when it came to determining good from bad online content ($r = -.11$, $p < .001$), believing false information online ($r = .15$, $p < .001$), and questioning information they find ($r = -.14$, $p < .001$). These results suggest an optimistic bias in kids' perceptions of their own information literacy and credibility evaluation skills.

DISCUSSION

The study detailed in this chapter describes youth who have been using the Internet for much of their lives for a variety of purposes, and adds to the current state of knowledge on children's Internet use and their perceptions of online information credibility specifically. It challenges existing notions about children as information consumers and paints a portrait of how youth establish credibility when they evaluate online sources, which should serve as a springboard for further research.

The data demonstrate that children ages 11-18 show a healthy degree of concern for the believability of online information. They think about the credibility of information they find on the Internet and are fairly concerned about the credibility of online information overall. Interestingly, despite their concern for credibility, kids rely quite heavily on the Internet to find different types of information, they trust information on Wikipedia (even more than they say it should be trusted), and they view certain kinds of information (i.e., entertainment information, commercial information, and information for school projects) on the web as a more credible source of information than books, newspapers, and television.

Children also trust different forms of online information more or less depending on its type. For example, information used for school projects is seen as more believable than information from strangers they meet online. That said, children report an equal likelihood to believe entertainment and health information online, which implies potentially problematic outcomes since these types of information should ideally warrant different levels of skepticism.

As children get older, their Internet use increases both in scope and in time spent online. Older teens trust the Internet more as an information source than do younger kids, but think that people should be more concerned about the quality of information online than do younger children. This might indicate that as kids become more experienced with the Internet, they have a greater appreciation for the potential of deceptive information online as well as greater confidence in their ability to find credible information sources.

The findings pertaining to the methods or strategies by which young people evaluate the credibility of information online are particularly revealing and show that several demographic, usage, and personality characteristics significantly predict children's credibility concerns and perceptions. As mentioned in the results section, race played a minor role, such that nonwhite youth manifested greater concerns about credibility, perhaps reflecting known patterns of greater social skepticism among minorities. Kids who have had negative past experiences with online information, who have had training in credibility evaluation, and whose parents talk to them about the credibility of online information demonstrate a higher degree of apprehension and concern about online information. Children who perceive themselves as more skilled Internet users, who spend time immersed in digital worlds, and who have been using the Internet for longer also report a higher degree of concern for credibility, as do those who experience less social trust. Greater concern

about credibility was also linked to more analytic approaches to evaluating information online.

Demographic and personality characteristics also played a role in how much web-based information children believe. Younger children reported believing more information online than did older kids. Females were more trusting than males, and children from wealthier backgrounds and who were more trusting of others believed more of the information available online. Usage and experience mattered too, as children who spent more time in virtual worlds and who rated themselves to be more skilled users overall believed more information, whereas those who had negative past experiences believed less information online to be credible. Perhaps most interestingly, certain credibility evaluation strategies were associated with credibility beliefs. Children who relied on heuristic methods of credibility assessment reported believing more information than children who reported relying on analytic methods and who had higher need for cognition. This suggests that curriculum to increase children's motivation and ability to engage in critical thinking may help to enhance skepticism toward online information, as manifested in employing more effortful means of information evaluation.

Collectively, the findings on belief of online information, concern for its credibility, and strategies used to assess its credibility indicate that as kids become more experienced with an online environment, they develop a greater concern for the credibility of online information, and employ a greater diversity of strategies when assessing it. This is an encouraging finding for media literacy advocates, and to some extent mitigates fears that kids become more accepting and less critical of online information as their usage increases.

However, not all of this study's findings were so encouraging. For instance, while children's concern about credibility appears to be largely driven by analytic credibility evaluation processes, those who find Internet information most credible tend to rely on heuristic (hasty and feeling-based) processes to evaluate credibility. This, coupled with the fact that most children said that people should be concerned about the credibility of information online, suggests that while children take the issue of credibility seriously, they may not always act diligently when evaluating the information they find online. Moreover, children appear to overestimate their own skill levels and capacity to discern good information from bad information as compared to others. Such overconfidence is troubling, inasmuch as it might imply a correspondingly reduced level of vigilance or attention.

Findings from this study reveal a relationship between youth, the Internet, and credibility that is far more nuanced than most previous research has suggested. This study indicates that a combination of experience using the Internet over time and vital cohort-related changes in youth's cognitive development interact to promote a better awareness of general credibility concerns. This has implications for several domains, including education and the creation of media literacy curricula, children's use of the Internet, and digital media policy formulations. For example, based on our findings, online media literacy programs should emphasize a structured but graduated approach to guiding children's use of the Internet that stresses the accumulation of personal experience online, early parental involvement, and the sharing of positive and negative online experiences at an early age. Additionally, education efforts regarding credibility evaluation should be ongoing at the upper elementary, middle, and secondary education levels, and should stress the importance of critical thinking skills, including analytic methods of credibility assessment over heuristic ones.

Although these findings are generalizable to child Internet users in the U.S., limitations inherent in survey research still color our findings. For example, while our data appear to indicate overconfidence in kids' assessment of their ability to discern good and bad information, experimental designs may better reveal any biases that might result from this overconfidence. In addition, multiple methods would be useful to better capture the rich reality of children's web use and evaluation experiences in some instances, particularly those that are social in nature.

Ultimately, this study appears to underscore a reality we would hope for as citizens, Internet users, and parents: children are for the most part aware of the issues surrounding information veracity on the Internet. Thus, the best strategy to help children become more skillful consumers of information online would appear to be the adoption of a perspective that empowers them and capitalizes on their unique upbringing in an all-digital world. In a future in which the information that drives kids' lives is assembled, transmitted, shared, and processed digitally, children need to develop the skills necessary to navigate that information environment effectively. Perhaps the most encouraging conclusion from these data so far is that, for the most part, children seem to be making inroads toward that goal.

ACKNOWLEDGMENT

The authors thank the John D. and Catherine T. MacArthur Foundation for their support of this work.

REFERENCES

Alberts, A., Elkind, D., & Ginsberg, S. (2006). The personal fable and risk-taking in early adolescence. *Journal of Youth and Adolescence, 36,* 71–76. doi:10.1007/s10964-006-9144-4

Alesina, A., & La Ferrara, E. (2002). Who trusts others? *Journal of Public Economics, 85,* 207–234. doi:10.1016/S0047-2727(01)00084-6

Bennett, S., Maton, K., & Kervin, L. (2008). The 'digital natives' debate: A critical review of the evidence. *British Journal of Educational Technology*, *39*, 775–786. doi:10.1111/j.1467-8535.2007.00793.x

boyd, d. (2011). Social network sites as networked publics: Affordances, dynamics, and implications. In Z. Papacharissi (Ed.), *Networked self: Identity, community, and culture on social network sites* (pp. 39-58). New York, NY: Routledge.

Campbell, J., Greenauer, N., Macaluso, K., & End, C. (2007). Unrealistic optimism in Internet events. *Computers in Human Behavior*, *23*, 1273–1284. doi:10.1016/j.chb.2004.12.005

Chen, S.-Y., & Rieh, S. Y. (2009). Take your time first, time your search later: How college students perceive time in web searching. *Proceedings of the 72nd Annual Meeting of the American Society for Information Science and Technology*.

Clarke, V. A., Lovegrove, H., Williams, A., & Macpherson, M. (2000). Unrealistic optimism and the health belief model. *Journal of Behavioral Medicine*, *23*(4), 367–376. doi:10.1023/A:1005500917875

Eastin, M. (2008). Toward a cognitive developmental approach to youth perceptions of credibility. In M. J. Metzger, & A. J. Flanagin (Eds.), *Digital media, youth, and credibility*, (pp. 28-46). MacArthur Foundation Series on Digital Media and Learning. Cambridge, MA: MIT Press.

Elkind, D. (1967). Egocentrism in adolescence. *Child Development*, *38*, 1025–1034. doi:10.2307/1127100

Epstein, S., Pacini, R., Denes-Raj, V., & Heier, H. (1996). Individual differences in intuitive-experiential and analytical-rational thinking styles. *Journal of Personality and Social Psychology*, *71*, 390–405. doi:10.1037/0022-3514.71.2.390

Fidel, R., Davies, R. K., Douglass, M. H., Holder, J. K., Hopkins, C. J., & Kushner, E. J. (1999). A visit to the information mall: Web searching behavior of high school students. *Journal of the American Society for Information Science American Society for Information Science*, *50*, 24–37. doi:10.1002/(SICI)1097-4571(1999)50:1<24::AID-ASI5>3.0.CO;2-W

Flanagin, A. J., & Metzger, M. J. (2000). Perceptions of Internet information credibility. *Journalism & Mass Communication Quarterly*, *77*, 515–540. doi:10.1177/107769900007700304

Flanagin, A. J., & Metzger, M. J. (2007). The role of site features, user attributes, and information verification behaviors on the perceived credibility of Web-based information. *New Media & Society*, *9*(2), 319–342. doi:10.1177/1461444807075015

Flanagin, A. J., & Metzger, M. J. (2008). Digital media and youth: Unparalleled opportunity and unprecedented responsibility. In Metzger, M. J., & Flanagin, A. J. (Eds.), *Digital media, youth, and credibility* (pp. 5–27). Cambridge, MA: MIT Press.

Flanagin, A. J., & Metzger, M. J. (2010). *Kids and credibility: An empirical examination of youth, digital media use, and information credibility*. Cambridge, MA: MIT Press.

Fogg, B. J. (2003). Computers as persuasive social actors. In B. Fogg's (Ed.), *Persuasive technology: Using computers to change what we think and do* (pp. 31-60). San Francisco, CA: Morgan Kaufmann.

Gasser, U., Cortesi, S., Malik, M., & Lee, A. (2012). *Youth and digital media: From credibility to information quality*. Berkman Center for Internet & Society. Retrieved May 22, 2012, from http://ssrn.com/abstract=2005272

Gigerenzer, G., & Todd, P. M. (1999). *Simple heuristics that make us smart*. New York, NY: Oxford University Press.

Greenfield, P. M. (2004). Developmental considerations for determining appropriate Internet use guidelines for children and adolescents. *Journal of Applied Developmental Psychology, 25*, 751–762. doi:10.1016/j.appdev.2004.09.008

Hargattai, E. (2002a). Second-level digital divide: Differences in people's online skills. *First Monday, 7*(4). Retrieved May 2, 2012, from http://firstmonday.org/htbin/cgiwrap/bin/ojs/index.php/fm/article/view/942/864/

Hargittai, E. (2002b). Beyond logs and surveys: In-depth measures of people's web use skills. *Journal of the American Society for Information Science and Technology, 53*, 1239–1244. doi:10.1002/asi.10166

Hargittai, E., Fullerton, F., Menchen-Trevino, E., & Thomas, K. (2010). Trust online: Young adults' evaluation of Web content. *International Journal of Communication, 4*, 468–494.

Ito, M., Baumer, S., & Bittanti, M. boyd, d., Cody, R., Herr-Stephenson, B., et al. (2009). *Hanging out, messing around, and geeking out: Kids living and learning with new media.* Cambridge, MA: MIT Press.

Jacobs, J. E., & Klaczynski, P. A. (Eds.). (2005). *The development of judgment and decision making in children and adolescents.* Mahwah, NJ: Lawrence Erlbaum.

Jones, S., & Fox, S. (2009, January 28). *Generations online in 2009.* Pew Internet & American Life Project report. Retrieved from http://pewresearch.org/pubs/1093/generations-online

Klaczynski, P. A. (2001). The influence of analytic and heuristic processing on adolescent reasoning and decision making. *Child Development, 72*, 844–861. doi:10.1111/1467-8624.00319

Kokis, J., Macpherson, R., Toplak, M., West, R. F., & Stanovich, K. E. (2002). Heuristic and analytic processing: Age trends and associations with cognitive ability and cognitive styles. *Journal of Experimental Child Psychology, 83*, 26–52. doi:10.1016/S0022-0965(02)00121-2

Kuiper, E., & Volman, M. (2008). The Web as a source of information for students in K–12 education. In Coiro, J., Knobel, M., Lankshear, C., & Leu, D. (Eds.), *Handbook of research on new literacies* (pp. 241–266). New York, NY: Lawrence Erlbaum.

Large, A. (2004). Information seeking on the Web by elementary school students. In Chelton, M. K., & Cool, C. (Eds.), *Youth information-seeking behavior: Theories, models, and issues* (pp. 293–320). Lanham, MD: Scarecrow Press.

Lenhart, A., Purcell, K., Smith, A., & Zickuhr, K. (2010). *Social media and mobile Internet use among teens and young adults.* Retrieved February 18, 2010 from http://pewresearch.org/pubs/1484/social-media-mobile-internet-use-teens-millennials-fewer-blog

Livingstone, S. (2009). *Children and the Internet: Great expectations and challenging realities.* Cambridge, UK: Polity Press.

Metzger, M. J. (2007). Making sense of credibility on the Web: Models for evaluating online information and recommendations for future research. *Journal of the American Society for Information Science and Technology, 58*, 2078–2091. doi:10.1002/asi.20672

Metzger, M. J., & Flanagin, A. J. (Eds.). (2008). *Digital media, youth, and credibility.* Cambridge, MA: MIT Press.

Metzger, M. J., Flanagin, A. J., Eyal, K., Lemus, D. R., & McCann, R. (2003). Credibility in the 21st century: Integrating perspectives on source, message, and media credibility in the contemporary media environment. In Kalbfeisch, P. (Ed.), *Communication yearbook 27* (pp. 293–335). Mahwah, NJ: Lawrence Erlbaum. doi:10.1207/s15567419cy2701_10

Metzger, M. J., Flanagin, A. J., & Medders, R. B. (2010). Social and heuristic approaches to credibility evaluation online. *The Journal of Communication, 60,* 413–439. doi:10.1111/j.1460-2466.2010.01488.x

Metzger, M. J., Flanagin, A. J., Pure, R., Medders, R., Markov, A., Hartsell, E., & Choi, E. (2011). *Adults and credibility: An empirical examination of digital media use and information credibility. Research report prepared for the John D. and Catherine T. MacArthur Foundation.* Santa Barbara: University of California.

Palfrey, J., & Gasser, U. (2008). *Born digital: Understanding the first generation of digital natives.* New York, NY: Perseus Books Group.

Prensky, M. (2001). Digital natives, digital immigrants. *Horizon, 9*(5). doi:10.1108/10748120110424816

Rainie, L. (2006, March 23). *Life online: Teens and technology and the world to come.* Keynote address to the annual conference of the Public Library Association, Boston, MA. Retrieved November 7, 2006, from http://www.pewinternet.org/ppt/Teens%20and%20technology.pdf

Rowlands, I., Nicholas, D., Williams, P., Huntington, P., Fieldhouse, M., & Gunter, B. (2008). The Google generation: The information behaviour of the researcher of the future. *Aslib Proceedings, 60,* 290–310. doi:10.1108/00012530810887953

Scott, S. G., & Bruce, R. A. (1995). Decision-making style: The development and assessment of a new measure. *Educational and Psychological Measurement, 55,* 818–831. doi:10.1177/0013164495055005017

Solomon, P. (1993). Children's information retrieval behavior: A case analysis of an OPAC. *Journal of the American Society for Information Science and Technology, 44,* 245–264. doi:10.1002/(SICI)1097-4571(199306)44:5<245::AID-ASI1>3.0.CO;2-#

Tapscott, D. (1997). *Growing up digital: The rise of the Net generation.* New York, NY: McGraw-Hill.

Valkenburg, P. M., Krcmar, M., Peeters, A. L., & Marseille, N. M. (1999). Developing a scale to assess three styles of television mediation: 'instructive mediation,' 'restrictive mediation,' and 'social co-viewing.'. *Journal of Broadcasting & Electronic Media, 43*(1), 52–66. doi:10.1080/08838159909364474

van der Sluis, F., & van Dijk, E. M. A. G. (2010). A closer look at children's information retrieval usage: Towards child-centered relevance. In *Proceedings of the Workshop on Accessible Search Systems, The 33st Annual International Conference on Research and Development in Information Retrieval* (ACM SIGIR 2010), 23 July, 2010, Geneva, Switzerland, (pp. 3-10).

van Dijk, J. (2005). *The deepening divide: Inequality in the information society.* Thousand Oaks, CA: Sage.

van Dijk, J. (2006). Digital divide research, achievements and shortcomings. *Poetics, 34,* 221–235. doi:10.1016/j.poetic.2006.05.004

Weinstein, N. D. (1980). Unrealistic optimism about future life events. *Journal of Personality and Social Psychology, 39,* 806–820. doi:10.1037/0022-3514.39.5.806

Weinstein, N. D. (1982). Unrealistic optimism about susceptibility to health problems. *Journal of Behavioral Medicine, 5,* 441–460. doi:10.1007/BF00845372

Weinstein, N. D. (1987). Unrealistic optimism about susceptibility to health problems: Conclusions from a community-wide sample. *Journal of Behavioral Medicine, 10*(5), 481–500. doi:10.1007/BF00846146

Zhang, L. F. (2003). Contributions of thinking styles to critical thinking dispositions. *Journal of Psychology (Savannah, Ga.), 137,* 517–544.

ADDITIONAL READING

Agosto, D. E. (2002). Bounded rationality and satisficing in young people's Web-based decision making. *Journal of the American Society for Information Science and Technology, 53*, 16–27. doi:10.1002/asi.10024

Agosto, D. E., & Abbas, J. (2010). High school seniors' social network and other ICT use preferences and concerns. *Proceedings of the American Society for Information Science & Technology, 47*, 1–10. doi:10.1002/meet.14504701025

Bennett, S., Maton, K., & Kervin, L. (2008). The 'digital natives' debate: A critical review of the evidence. *British Journal of Educational Technology, 39*, 775–786. doi:10.1111/j.1467-8535.2007.00793.x

Bowler, L. (2010). The self-regulation of curiosity and interest during the information search process of adolescent students. *Journal of the American Society for Information Science and Technology, 61*, 1332–1344. doi:10.1002/asi.21334

Brem, S. K., Russell, J., & Weems, L. (2001). Science on the web: Student evaluations of scientific arguments. *Discourse Processes, 32*(2-3), 191–213. doi:10.1080/0163853X.2001.9651598

Julien, H., & Barker, S. (2009). How high-school students find and evaluate scientific information: A basis for information literacy skills development. *Library & Information Science Research, 31*, 12–17. doi:10.1016/j.lisr.2008.10.008

Kafai, Y., & Bates, M. (1997). Internet web-searching instruction in the elementary classroom: Building a foundation for information literacy. *School Library Media Quarterly, 25*, 103–111.

Kuhn, D. (2010). What is scientific thinking and how does it develop? In Goswami, U. (Ed.), *Handbook of childhood cognitive development* (2nd ed., pp. 371–393). Oxford, UK: Wiley-Blackwell. doi:10.1002/9781444325485.ch19

Kuiper, E., Volman, M., & Terwel, J. (2005). The Web as an information resource in K-12 education: Strategies for supporting students in searching and processing information. *Review of Educational Research, 73*, 285–328. doi:10.3102/00346543075003285

Large, A. (2005). Children, teenagers, and the Web. *Annual Review of Information Science & Technology, 39*, 347–392. doi:10.1002/aris.1440390116

Nicholas, D., Huntington, P., Jamali, H. R., Rowlands, I., & Fieldhouse, M. (2009). Student digital information-seeking behaviour in context. *The Journal of Documentation, 65*, 106–132. doi:10.1108/00220410910926149

Shenton, A. K., & Dixon, P. (2004). Issues arising from youngsters' information-seeking behavior. *Library & Information Science Research, 26*(2), 177–200. doi:10.1016/j.lisr.2003.12.003

Walraven, A., Brand-Gruwel, S., & Boshuizen, H. (2009). How students evaluate information and sources when searching the World Wide Web for information. *Computers & Education, 52*, 234–246. doi:10.1016/j.compedu.2008.08.003

KEY TERMS AND DEFINITIONS

Analytical Evaluation Strategy: Carefully considering the information, double checking facts, gathering a lot of information, and considering all views to evaluate information.

Credibility: The believability of information.

Digital Natives: People who were born during or after the introduction of digital technology.

Faith in Intuition: Reflects a tendency to trust based on first impressions, instincts, and feelings.

Flexible Thinking: Reflects people's willingness to consider opinions different from their own.

Heuristic Evaluation Strategy: Relying on gut instincts, making decisions based on feelings, making quick decisions to evaluate information.

Information Evaluation Strategies: The ways in which individuals evaluate information for its quality or credibility.

Need for Cognition: Reflects the degree to which people engage in and enjoy thinking deeply about problems or information.

Optimistic Bias: The tendency to see oneself as less likely than others to experience negative events.

Relative Internet Skill: The extent to which individuals believe themselves to be accomplished Internet users compared to an average Internet user.

ENDNOTES

[1] First, a post-stratification adjustment using demographic distributions from the most recent U.S. Census Bureau's Current Population Survey data was used to balance errors due to panel recruitment methods and panel attrition. Demographic variables used for this weighting included gender, age, race, education, and Internet access. This weighting was applied before the selection of the sample was made for this study. In addition, a study-specific post-stratification weight was applied after data collection to adjust for the study's sample design and survey non-response. A weight was calculated for all qualified children to make them comparable to 13 to 18 year olds who have Internet access at home. Household income was also included as a weighting variable since education could not be included (i.e., most of the children in this age range have less than a high school education). The sample design effect for this weight is 1.58.

[2] Data regarding household income and size were reported by parents of the children to the survey research firm, as part of their induction into the Knowledge Networks panel.

[3] F test statistics for these tests are too cumbersome to report here but are available upon request. All tests were significant at $p < .001$. Post hoc tests ($p < .05$) showed that some of the older children did not distinguish between the believability of health and commercial information, and children of some ages did not distinguish between commercial and entertainment information.

[4] As determined by cross-tabulations. All chi square tests were significant at $p < .05$ or $p < .001$.

[5] These uses were derived from factor analyses of Internet usage data collected in another part of the survey. The usage data are not reported in full here due to space limitations.

Chapter 10
The Credibility of Sources 2.0 in Journalism:
Case Study in Portugal

Paulo Serra
Universidade da Beira Interior, Portugal

João Canavilhas
Universidade da Beira Interior, Portugal

ABSTRACT

This chapter addresses the use and credibility of news sources 2.0 in journalism. Starting with traditionally established views about the credibility of news sources in pre-Internet journalism as depicted by Gans (2004) and other authors, this chapter discusses the new situation that the Internet and, in particular, Web 2.0, brought about. More specifically, the authors intend to: i) Characterize the way Portuguese journalists use sources 2.0; ii) Study how Portuguese journalists assess the credibility of sources 2.0; iii) Compare the results obtained among Portuguese journalists with the results of other international studies in this field. To do this, the authors analyze and discuss the main results of a survey administered to Portuguese journalists, which is also compared with results from other international studies, in order to discuss its external validity. According to the data, Portuguese journalists, like journalists in other countries, consider news sources 2.0 to be unreliable, but the Portuguese journalists surveyed still use them, so the authors examine that discrepancy and other findings in light of other research.

INTRODUCTION

This chapter investigates the credibility of news sources 2.0 in journalism. According to the definition proposed by Canavilhas and Ivars-Nicolás (2012, p. 66), news sources 2.0 are all the "in-

formation providers who do it for free and from the author's spontaneous will, being either an individual or a group, using collaborative tools," and whose information journalists use to produce their news stories. Those tools are blogs, social networks, forums, chats, wikis and video/photo/sound sharing websites such as YouTube, Flickr or SoundCloud. We do not include web search

DOI: 10.4018/978-1-4666-2663-8.ch010

engines in this group because they are not a direct source of information but an intermediary to get the original news sources.

The 2.0 sources are commonly known as "social media," which may be defined as "a group of Internet-based applications that build on the ideological and technological foundations of Web 2.0 , and that allow the creation and exchange of User Generated Content" (Kaplan & Haenlein, 2010, p. 61). This definition prioritizes the technological factor; however, as stressed by the *Reuters Handbook of Journalism*, social media "are not sources per se" (Reuters, 2008, p. 541). The sources are the people and/or organizations that produce or transmit the content we can find in social media: texts, photos, videos, etc.

The importance and value of Web 2.0 sources for journalists and news organizations has been emphasized by several authors, on several occasions. To give a single example, Johnson and Kaye (2009) drew attention to the importance taken on during the Iraq War by "obscure war bloggers such as Salam Pax, an Iraqi living in Iraq, and military bloggers such as Lt. Smash, a reservist stationed in the Persian Gulf, as well as stateside armchair political pundits such as Sean Paul Kelley (agonist.org)" (p.1).

With regard to credibility, and applying concepts from Aristotle's *Rhetoric* to mass communication, Hovland and Weiss (1951) and Hovland, Janis, & Kelley (1953) empirically identified expertise and trustworthiness as the main dimensions of the credibility of communicators or sources. Later on, several authors working on persuasion also emphasized expertise and trustworthiness (or trust, or believability) as crucial dimensions of credibility (Fogg, 2003, pp. 121-181; Gass & Seiter, 2003, pp. 74-95; Mills, 2000, 14-37; O'Keefe, 2002, who adds source attractiveness and dynamism; Perloff, 2003: 159-168, who adds goodwill; Larson, 2004, pp. 230-3, 296-7, who adds dynamism). The differences among respective analyses notwithstanding, researchers also generally agree that credibility is a multidimen-sional variable (it has more than one dimension or component), and a perceptual one (it depends on the receiver's or audience's perception about the communicator or source) (Stiff & Mongeau, 2003, pp. 104-5). Some authors identify, as key dimensions or components of credibility, characteristics such as believability, accuracy, trustworthiness, bias, and completeness (Flanagin & Metzger, 2000, p. 522).

The credibility of journalists and news organizations depends, above all, on the credibility of their sources. What is a credible source? How do we evaluate the credibility of a source? These are questions that journalism professionals, their organizations, and their associations have been asking at least since the beginnings of the twentieth century—ASNE's Statement of Principles was adopted early in 1922, under the name "Canons of Journalism." A significant part of journalism and mass communication theory, which includes authors like Lippmann (1922), Tuchman (1972), Wolf (1985), McQuail (1992) or Gans (2004), has been trying to answer those questions.

With the advent of the Internet, and particularly Web 2.0, the credibility problem of news sources took on a new form. The proliferation of user-generated content has transformed each user not only into a receiver but also into a potential source, originating a multiplicity and often anonymity of sources that make their selection and evaluation an increasingly complex task (Levinson, 2009; Asur et al., 2011; Wu et al., 2011). As Gillmor (2004) confessed in his seminal book, "This new media has created, or at least exacerbated, difficult issues of credibility and fairness." Can we be confident, like Gillmor, that "the community, with the assistance of professional journalists and others who care, can sort it all out"? (p. 238).

We hope this chapter will give a contribution to this overall purpose. More specifically, the chapter has three main objectives: i) To characterize the way Portuguese journalists use sources 2.0; ii) To study how Portuguese journalists assess the credibility of sources 2.0; iii) To compare the results obtained among Portuguese journalists with the results of other international studies in this field in order to shed light on how we might approach sources Web 2.0 in the future.

THE CREDIBILITY OF NEWS SOURCES ON THE WEB 2.0

The Credibility of News Sources in Web 2.0

As stressed by Robert Park (1940), news is a form of knowledge that, although different from scientific knowledge, is also based on facts. Knowledge of these facts is rarely direct, generally depending on the use of certain sources. By sources, Gans (2004) meant "the actors whom journalist observe or interview, including interviewees who appear on the air or who are quoted in magazine articles, and those who only supply background information or story suggestions" (p. 80). According to the classification proposed by Kovach and Rosenstiel (2010), we can distinguish seven kinds of news sources: i) Sourceless news: the audience as witness; ii) The journalist as witness; iii) The journalist as credentialed expert; iv) Sources as witnesses: firsthand accounts; v) Participants but not witnesses; vi) Expert sources and analysts; vii) Anonymous sourcing (pp. 74-93).

Whatever the news sources, their credibility is seen both by journalists and news organizations as an essential requisite for (good) journalism (Society of Professional Journalists, 1996; Gillmor, 2004; Reuters, 2008). But how and where can journalists ascertain, among the multiple kinds of potential sources, which are the credible ones? We know that, even in the case of "sources as witnesses," we have no guarantee that the sources are credible ones, hence the need for the "two-source rule" (Kovach & Rosenstiel, 2010, p. 82).

Accordingly to Gans (2004, p. 128), journalists try to select their sources in order "to obtain the most suitable news from the fewest number of sources as quickly and easily as possible, and with the least strain on the organization's budget." To do this, journalists depend on six major considerations: past suitability, productivity, reliability, trustworthiness, authoritativeness, and articulateness. These considerations explain, at least partially, why journalists choose authoritative sources (Gans, 2004, p. 282; Reese, 2009, p. 288; Pew Research Center, 2011, p. 1). It is also the quest for efficiency that explains why, for journalists, the most credible sources are the journalists themselves (Reich, 2011, p. 28).

Unlike what happens off-line, in the Web 2.0 environment the problem is not the scarcity of sources—which implies that journalists actively seek them—but their excess. In fact, in the Web 2.0 every citizen—not only the powerful people or organizations—becomes a potential source for journalists but also a potential "journalist." So, this "networked journalism" (Jarvis, 2007; Beckett & Mansell, 2008) makes a multiplicity and diversity of sources and contents from all parts of the world available to journalists, and opportunities for crowd-sourcing and citizen investigation of facts that allow them to save part of his work and efforts (Singer, 2009; Bradshaw, 2011). Besides the "crisis of choice" (Lankes, 2008, p. 679) that an infinite amount of sources can cause, those sources are often anonymous, which contradicts one of the rules of thumb of journalism, which forbids the use of anonymous sources (Haiman, 2000; Gillmor, 2004; Carlson, 2011, pp. 37-48). Even when many of the anonymous sources convey the same information, this chorus can rarely replace the "reputed credibility" (Fogg, 2003) or the authority of a source for the reliability based on multiple sources (Lankes, 2008, p. 678).

The tendency of mainstream news organizations has been to see the problem of the use, selection and assessment of the credibility of news sources 2.0 in a way similar to traditional off-line sources (Reuters, 2008; Associated Press, 2011; Hohmann, 2011). The strategies recommended by news organizations to their journalists in order to deal with the problems posed by sources 2.0 are, briefly, the following: search for references in other known, reputable sources, apply what Fogg (2003) calls "reputed credibility"; establish direct contacts by e-mail, by phone or personal contacts; assess the importance of the in-bound and out-bound links (what Bowman & Willis, 2003, call "distributed credibility"); verify dates and locations; and verify the authenticity and authorship of the documents (texts, photos, videos, etc.) (Reuters, 2008, pp. 541-3; Associated Press, 2010, pp. 4-5; Hohmann, 2011, p. 11). Furthermore, it is argued that information deriving from sources 2.0 should only be taken as the starting point for news stories and journalists must further investigate the sources and events they report through words and images, an investigation that should be made personally, by phone, or by other means (Reuters, 2008, p. 541; Maurer, 2011, p. 12).

As we may easily conclude, these recommended strategies seek to evaluate credibility in light of the reputation, the authority, the prestige, and other social characteristics of the sources in a manner usually associated with "powerful sources" (Gans, 2004) and other credible sources (other journalists, experts, etc.). However, as stated earlier, it is often impossible to implement these strategies—since, as was also previously stated, news sources 2.0 are multiple and often anonymous.

The Use of Sources 2.0 by Journalists

The special problems posed by the selection and assessment of the credibility of news sources 2.0—basically concerning identity, attribution and verification—have not prevented their widespread use by journalists. This use is evidenced by some recent studies.

The online survey conducted by Cision and Don Bates of George Washington University (2009), from September 1, 2009, to October 13, 2009, among 371 print and Web American journalists,

produced the following answer to the question "How important have social media become for reporting and producing the stories you write?": 56% of the journalists answered that social media were "Important" or "Somewhat important." The study produced other interesting results (see Tables 1 and 2).

The Oriella PR Network Digital Journalism Study (2011), attended by about five hundred industry professionals from 15 countries, asked journalists "about the extent to which they use social media both for sourcing new leads, and verifying stories journalists are already working on." The results are presented in Table 3.

From these studies we can conclude that the first survey confronts us with an apparent paradox: Most journalists say that they do use news sources 2.0; however, they admit that sources 2.0 aren't credible and reliable because they don't fit into the journalistic standards of verification and fact checking. But, as shown by the second survey, the paradox is in fact apparent, i.e., journalists

Table 1. How often do you visit the following types of sites when doing online research for a story?

	%
Corporate Websites	96
Blogs	89
Social Networks	65
Multimedia sharing	58
Microblogs	52
Forums	42

Table 2. Do you think that news and information delivered via social media is more or less reliable/vetted than news delivered via traditional media?

	%
Much less reliable	54
Slightly less reliable	31
About the same	13
Slightly more reliable	2
A lot more reliable	0

Table 3. Please select which of the following you use when sourcing new story leads

	Verifying	Sourcing
PR Agencies	61.0	61.4
Corporate spokespeople	57.0	58.9
Industry insiders	55.9	53.2
Twitter	32.5	46.7
Facebook	25.4	35.2
Other social sites	23.9	31.6
Blogs	27.5	29.7
Other	14.0	15.5
Linkedin	14.0	15.3

use 2.0 sources more for sourcing new leads than to verify stories they are already working on. In other words: In what refers to journalism—by which we mean real, professional journalism—news sources 2.0 seem to be more an heuristic tool than a verification one.

In Portugal there are many studies on how journalists use the Internet, but few address the issue of credibility. When they do address credibility, the result appears as a secondary matter. This is the case of Canavilhas (2004), who sought to examine how Portuguese journalists attached credibility to the information they collected on the Internet, without explaining the type of tool used: 32% considered the information reliable, 21% thought it was hardly credible, and 37% had no opinion.

To study how Portuguese journalists use and assess the credibility of the different sources 2.0 and to compare the results with the international studies cited above, we undertook a survey that intended to answer the following specific research questions (RQ):

- **RQ1:** For which purposes do Portuguese journalists use the Internet?
- **RQ2:** Which 2.0 tools do journalists use more?
- **RQ3:** Do journalists use source 2.0 to produce their news stories? Does this vary with the type of medium they work in?
- **RQ4:** Do journalists find group tools – such as wikis and forums – to be more credible than individual tools such as blogs, micro-blogs and social networks? Does this vary with the type of medium?
- **RQ5:** Is the anonymity of sources 2.0 seen by journalists in the same way as the anonymity of traditional news sources?

New Directions in Assessing the Credibility of Sources 2.0

With regard to the RQ 4, some recent studies have revealed an important shift in the way users in general assess the credibility of the information, a shift that necessarily includes the specific group constituted by journalists.

In fact, up until some years ago, research about web credibility had mainly been focused on websites, comparing, for instance, the websites of news organizations, personal websites, or commercial ones (Flanagin & Metzger, 2007); the objective was to identify and describe the criteria (such as accuracy, authority, objectivity, currency, and coverage or scope) that users in general should use for assessing information, materializing what Metzger calls "the checklist approach" (Metzger, 2007, p. 2079).

Surprisingly—or maybe not—some studies have revealed that there is an important gap between what users *should do* and what they actually *do*; and what they actually do is that they tend to apply the criteria that are easier and only require their opinion (for instance, examining coverage or scope of the content), and not the ones that require more effort and time (such as verifying authors' qualifications or credentials)—"the most worrisome finding" being that "the strategy least practiced (i.e., verifying an author's qualifications) is perhaps the most important for establishing credibility" (Metzger, 2007, p. 2080).

A possible explanation for this tendency, especially in Web 2.0, is that users tend to rely less on the application of the demanding criteria identified by the literature of "information sciences" and more on models of collaborative filtering and peer review that are at stake on sites like Amazon or eBay, on topical discussion forums, or on social networks like MySpace or Facebook, thus enacting their "collective intelligence" (Metzger, 2007; Metzger, Flanagin & Medders, 2010); and also on the "cognitive heuristics" that derive from their experience as web users (Metzger, Flanagin & Medders, 2010).

So, apparently, we are facing an important shift in the way users assess the credibility of web information, "from a model of single authority based on scarcity and hierarchy to a model of multiple distributed authorities based on information abundance and networks of peers" (Metzger, Flanagin & Medders, 2010, p. 415). This shift is also explained by the fact that a great part of the web, namely Web 2.0, does not consist of websites but includes tools such as blogs, wikis, social networking sites, chat groups, etc., with their specific characteristics and credibility problems (Metzger, 2007; Johnson & Kaye, 2009).

Based on this literature, and with regard to RQ 4, we formulated the following hypothesis about the way Portuguese journalists assess the credibility of sources 2.0:

- **H1:** Journalists find group tools—such as wikis and forums—to be more credible than individual tools such as blogs, microblogs and social networks. Nevertheless, the shift in assessing the credibility of sources 2.0 does not solve the problems of anonymity—the fact that a statement is made by a thousand unknown people instead of one doesn't make it more true or believable. So, since the essence of journalism continues to be "a discipline of verification" (Kovach & Rosenstiel 2001, p.12), we formulated the following hypothesis:

- **H2:** Journalists see the anonymity of sources 2.0 in the same way as the anonymity of traditional news sources.

PORTUGUESE JOURNALISTS AND NEWS SOURCES 2.0

Methodology and Results

During the months of July, August and September 2011, Portuguese journalists were asked to complete an online questionnaire. We received answers from the largest circulation newspapers in Portugal, two informative radio stations and three national television broadcasters.

The survey included a first group of questions designed to characterize the sample using multiple choice questions. The next group sought to measure the perception of credibility that journalists have about the information found in news sources 2.0, using a Likert scale from 1 (not credible) to 5 (very credible). The last group of questions sought to identify what kind of news sources journalists had already used in previous works, in order to cross-reference these responses with those of the previous group of questions: for this we used close-ended questions (yes/no).

Sample

In a total of 62 participating journalists, 53.2% were men and 46.8% were women. In the segmentation by media we tried to make the representation proportional to Portuguese reality, with 45.2% journalists from newspapers, 30.6% from radio, 25.8% from television and 16.1% from the online sector (journalists working for native online publications or mostly in the online versions of traditional media). We received answers from the newspapers with the highest circulation in Portugal (daily: *Correio da Manhã, Jornal de Notícias, Diário de Notícias, Público*; weekly:

Expresso and *Sol*), from the two radios specializing in news (Antena 1 and TSF) and from the national broadcast televisions (RTP, SIC and TVI).

The Internet experience survey of our sample reveals that 69.4% have been using it for more than 10 years, 27.4% have used it between 6 and 10 years, and only 3.2% use the Internet for less than 6 years. All reported using the Internet several times a day. Internet experience is an important factor because several studies have observed a positive correlation between experience with the Internet and perceptions of credibility (Flanagin & Metzger, 2007, p. 324; Metzger, 2007, p. 2080).

Results

In regard to the RQ 1 (For which purposes do the Portuguese journalists use the Internet?), we highlight purposes like "document an issue or confirm information" (100%), "receive and send information" (95.2%), "search topics for future work" (87.4%), "contact information sources" (82.3%), "access to press releases" (71%), and "read other media" (71%).

About the RQ2 (Which 2.0 tools do Portuguese journalists use more?), the results (Table 4) show that only social networks have high values (48.4%) and are the most used tool of journalists.

These results may be related to how social networks function because they often imply prior authorization for two people to become connected to each other and, therefore, involve the existence of a prior mutual knowledge. To verify this hypothesis, in a previous question we had asked journalists (the ones with more "friends") if having more acquaintances in common lends credibility to the information provided by a user of social networks; only 25.8% replied positively. In this table we also highlight the blogs, wikis and sharing sites that are used by about a fourth of the journalists.

In what refers to RQ3 (Do Portuguese journalists use source 2.0 to produce their news stories? Does this vary with the type of medium?), we see that, despite the lack of credibility attributed to news sources 2.0, a substantial proportion of journalists has produced news with information gathered with these tools (Table 5). In this field, the highlight is social networks, with 69.4% of the respondents answering that they have already published news with information gathered with these tools. Some of them pointed out that this information was confirmed from other news sources before being published, confirming a tendency identified in previous studies (Oriella PR, 2011).

Blogs are another important source of information and more than half of the Portuguese journalists use data collected from them to write stories. The sharing websites are mentioned by almost a third of journalists, a situation verifiable by the significant numbers of television news reports that include images collected on YouTube. Our analysis by medium (Table 6) confirms that 75% of TV journalists use this source and that these

Table 4. Use of 2.0 tools

	Frequency	%
Social networks	30	48.4
Blogs	18	29.0
Wiki	15	24.2
Sharing websites	15	24.2
Newsgroups	5	8.1
Online Forum	4	6.5
Chat	4	6.5

Table 5. Did you publish some work with information collected in...

	Frequency	%
Social networks	43	69.4
Blogs	34	58.4
Sharing websites	18	29.0
Online Forum	14	22.6
Wiki	14	22.6
Chat	10	16.1

are the journalists who use this tool more frequently. Except in sharing sites and wikis, online journalists (working for native online publications or mostly in the online versions) are the group that published the most news using sources 2.0. Looking at the traditional media it seems that the newspaper professionals publish news relying significantly on information gathered from blogs and social networks.

In Table 7 we try to answer our RQ4 (Do journalists find group tools—such as wikis and forums—to be more credible than individual tools such as blogs, microblogs and social networks? Does this vary with the type of medium?), by comparing the levels of credibility attributed by journalists to the six news sources 2.0 considered in this work. By adding the two negative values (not credible and not very credible) and the two positive values (credible and very credible), we can conclude that wikis are considered the most credible source (40.3%), followed by the sharing websites (38.7%), blogs (30.6%) and social networks (29%).

The amount of credibility that journalists assigned to wikis may be related to the concept of "collective intelligence" (Lévy, 1994) referred to

above. Wikis are the news source 2.0 where this function is more visible because the content is under the surveillance of the wiki administrators and the published texts use citations and references from other resources, a trustworthiness criterion that we have mentioned previously. In the case of sharing websites, the credibility assigned is related with the multimedia nature of contents: Journalists have the notion that it is more difficult to manipulate images than texts. So, we may consider that our H1 (Journalists find group tools—such as wikis and forums—to be more credible than individual tools such as blogs, microblogs and social networks) is only partially confirmed—on what refers to wikis, but not to forums.

Table 8 shows that, in general, newspaper and online journalists present the higher values in what refers to the credibility assigned to the sources 2.0. In the case of newspaper journalists, the numbers are justified because newspapers were the first medium to have online versions, and therefore, there is a greater binding between media. In the case of online journalists, the numbers are justified by their closeness with the environment in which they work. The difference between the first

Table 6. Did you publish some work with information collected in...

	Press	Radio	TV	Internet
Social networks	78.6	63.2	68.8	80.0
Blogs	64.3	52.6	43.8	70.0
Online forums	25.0	21.1	25.0	30.0
Chat	25.0	5.3	6.3	30.0
Wiki	21.4	36.8	12.5	20.0
Sharing websites	17.9	15.8	75.0	30.0

Table 7. How do you consider news sources 2.0?

SN		B	OF	CH	WK	SW
Not Credible	56.5	56.4	62.9	67.8	33.9	38.7
No opinion	14.5	12.9	22.6	19.4	25.8	22.6
Credible	29.0	30.6	16.1	12.9	40.3	38.7

Social Networks (SN), Blogs (B), Online Forum (OF), Chat (CH), Wiki (WK) and Sharing Websites (SW).

Table 8. Perception of news sources 2.0 credibility by medium

Press		Radio	TV	Internet
Social networks	28.6	31.6	31.3	40.0
Blogs	28.6	31.6	18.8	30.0
Online forums	14.3	5.3	12.5	40.0
Chat	17.9	5.3	12.5	20.0
Wiki	60.7	31.6	25.0	70.0
Sharing websites	42.9	26.3	37.5	60.0

Table 9. How do you consider the following news sources...

PAB		PIB	SNNP	SNWP	TW
Not Credible	95.2	56.5	96.8	51.6	35.5
No opinion	1.6	12.9	1.6	14.5	11.3
Credible	3.2	29.0	1.6	33.9	53.2

(PAB) Post in an anonymous blog; (PIB) Post identified in blog; (SNNP) Post in a social network without author photo; (SNWP) Post in a social network with author photo; (TW) Text in a Wiki

Table 10. How do you consider the following news sources...

APC		AE
Not Credible	88.7	89.1
No opinion	8.1	6.5
Credible	3.2	6.5

(APC) Anonymous phone call; (AE) Anonymous email

and the second is also consistent with the result of a survey by Cassidy (2007) that found "online newspaper journalists rated Internet news information as significantly more credible than did print newspaper journalists" (p. 478).

Finally, we tried to answer our RQ5 (Is the anonymity of sources 2.0 seen in the same way as the anonymity of traditional news sources?), to measure the influence of anonymity on the perceptions of 2.0 news sources credibility (Table 9), comparing these results with anonymity in two kinds of common news sources (Table 10).

The results allow us to verify that the identification of the author in a publication increases significantly the perceived credibility of the content. Once again we highlight the texts in a Wiki, which are credible for more than half of the respondents (53.2%). In our view, the credibility assigned to the Wikis is explained by the fact that, despite the fact that their texts are anonymous, they use traditional means of assessing the credibility, such as quotations, references and links to sources deemed as credible and that can be verified—thus, there is a process of transfer of credibility from these sources to the Wikis.

Comparing the importance of anonymity in news sources 2.0 and in the two most common anonymous news sources (Table 5) we can see that the values are similar. So it can be concluded that social media do not have enough strength to balance the negative connotation of anonymity—what confirms our H2, that journalists see anonymity of sources 2.0 in the same way as the anonymity of traditional news sources. In fact, these data confirm previous studies that link the information credibility to the person or institution that uses the 2.0 source and not the device itself (Canavilhas & Ivars-Nicolás, 2012). The exceptions are the tools group (wikis, sharing websites) where the concept of collective intelligence seems to nullify this effect.

DISCUSSION

Much like what happens with journalists all over the world, all the Portuguese journalists surveyed use the Internet in their normal daily work. But depending on the nature of the medium in which journalists work, the Internet takes different roles in their routines of news production. However, contacting information sources is mentioned only in the fourth place of journalists' use of Internet, with documenting themselves or confirming information for contextualizing their stories being mentioned in the first place.

In what specifically refers to sources 2.0, Portuguese journalists say they predominantly use the social networks (which are used by about 50% of the journalists), blogs, wikis and sharing sites (each one of them being used by about a fourth of the journalists). As we have already noted, the reason may be that a social network is a network of "friends" that at the very least know each other and so can place more trust on each other. Moreover, having a group of "friends" may easily allow the journalists the multiple cross-checking of the information they want to use.

About the use of sources 2.0 to produce their stories, the Portuguese journalists also favour the use of social networks (about 70%) and blogs (about 60%). These values are somewhat different from those presented by the international studies we mentioned before, since in the Cision and Bates (2009) study the values are much higher and the positions of blogs and social networks are inverted; and in the The Oriella PR Network Digital Journalism Study (2011), which presents general lower values, the microblogs and blogs also have higher values than social networks. The reason for the difference may be that, in Portugal, there is not a core of journalistic blogs as important and consolidated as in countries like the USA (*The Huffington Post* being a well-known example). Analyzing the results by medium, we confirm the tendency identified in other studies that experience with the Internet is a key factor

in the use of Internet sources, namely sources 2.0, with the online journalists using them more than the others.

However, when we ask Portuguese journalists about the credibility they attribute to sources 2.0, the results are quite different from those of the previous questions, with the Wikis and the sharing websites occupying the first two places (both with about 40%), followed by the blogs, and the social networks (both with about 30%).

Concerning these results we must note, in the first place, the gap between the higher percentage of journalists that use sources 2.0 to produce their works and the lower percentage of them that consider those very sources as credible—which led us to the hypothesis that a certain percentage of journalists produce stories using sources that they do not consider credible (a hypothesis that would need further verification). Secondly, we must note the difference between the credibility attributed to group tools like wikis and individual tools like blogs, microblogs, and social networks. Even if that difference is not a very significant one, the results seem to partially confirm the new directions in assessing the credibility of web information identified by Metzger (2007) and Metzger, Flanagin and Medders (2010). This tendency is also confirmed when we analyze the results about the anonymity in the sources 2.0. In general, with the notable exception of Wikis, the respondents see the tools where they can identify the author of the content by name or by photo as being more credible than the others where they cannot.

We have said that these results only *partially* confirm the new directions in assessing the credibility of web information identified by Metzger (2007) and Metzger, Flanagin and Medders (2010) because the tendency to consider group tools as more credible is not verified in what pertains to the other group tools like forums and chats; in fact, they always have the lowest ranks in terms of credibility. This tendency is documented in other studies, such as the online survey carried out by Johnson and Kaye (2009) of "politically interested

Internet users during the two weeks before and the two weeks after the 2004 presidential election," where the users ranked chat rooms/instant messaging as the least credible (7.4 in a 4-20 scale) of five Internet tools, with blogs (12.2), issue-oriented websites (11.4), electronic mailing lists/bulletin boards (9.7) and online candidate sites (9.0) occupying the first four places (p. 179).

In our opinion, the main difference lies in the fact that, from the journalists' point of view, what is important in the expression "collective intelligence" (Lévy, 1994) is not only the word "collective" but also —and above all—the word "intelligence", i.e., the fact that what someone says, or writes, or shares is well justified, documented and referenced, even if not identified, something which generally happens in wikis. Even if wikis may be seen as one of the best examples of the "cult of the amateur" (Keen, 2007), and their texts are very asymmetric in regard to quality and deepness, the fact is that in most cases they constitute a good starting point for journalists to get a minimum of information about a topic—information that they may deepen and cross-check using the links and references indicated in the articles of wikis.

In this respect, we may say that wikis combine the more traditional, established means of assessing credibility – based on citations, references, hyperlinks, etc. – with the new, technological means, based on the collaboration between users and the sharing of the information they produce (about the modalities and motivations for information sharing in social media, see Owens, Shaikh & Chaparro, 2011).

This combination is, probably, one of the distinctive signs of the journalism in the age of Web 2.0. In the pre-Internet journalism, the sources should have a name and be the *authors* of their information—the exceptions were handled very carefully, as happened in the Watergate affair—being preferentially treated "powerful sources" (politicians, officials, etc.). One of the effects of this journalism was that it was a kind of "dance" (or

dialogue) between the journalists and the sources, both seeking mutual access and trying to use each other to get some kind of influence (Gans, 2004). In this process, citizens were more or less passive receivers of the news that journalists provided them. With the Internet, and especially Web 2.0, every citizen with a Web 2.0 tool becomes an information provider—although not a "journalistic" one—or, according to our definition, a (potential) "source 2.0." Even when these citizens have a name (i.e., are not anonymous), their names aren't verifiable or are mere pseudonyms. In this context of generalized anonymity or pseudonymity, what matters is not only the name of the author, but also elements like the quality of writing, the cogency of the arguments, the wealth of data, and information relevance. Besides that, tools like wikis allow users to review, denounce and criticize their content, on the basis of authorities and sources different from the indicated—a peer-reviewing process that leads to a gradual increase in the quality of information that is provided. In spite of the difficulties in the selection and assessment of the credibility of its sources, this "new journalism" seems to have positive consequences for the acquisition and consumption of news (Rosenstiel, 2010).

CONCLUSION

Portuguese journalists, like journalists all over the world, use sources 2.0 to look for new sources, to discover new stories, themes or framings, to look for a content that can make a difference—even if they consider those sources as less credible and as lacking verification and fact-checking.

To compensate for this deficit of credibility, journalists rely both on traditional means to assess the credibility of sources 2.0 (reputation and other social characteristics), and the new means based on the "collective intelligence" provided by group authorship and collaborative tools—moving from one form of credibility assessment to another to solve problems of identification, attribution, authorship, and so on.

This study also concludes that the sources 2.0 most used by Portuguese journalists and that ultimately originate the most news are social networks and blogs. However, these are not the sources that journalists consider the most credible sources—those are wikis and sharing sites—which confirms the importance of group tools as a means of constructing credibility online.

To decode this apparent inconsistency—using the less credible sources more often—it would be necessary to use other research methods such as ethnographic studies or focus group.

Nevertheless, we can speculate that this inconsistency has to do with a certain logic of "fast-journalism," since, while social networks and blogs are "push tools"—information reaches the journalist in a fast, easy, and "automatic" manner, wikis and sharing sites are "pull tools," i.e., they imply a process of actively seeking information and participating, and, consequently, necessitate more time (and money).

Regarding the anonymity of the sources, data allow us to conclude that, with the notable exception of Wikis, the identification of the author of the information received from a source 2.0 increases the degree of reliability, as is the case with traditional sources.

Concerning the use of news sources 2.0 in their daily work, Portuguese journalists seem to follow the same rules that they use regarding traditional sources, seeking to contrast the information of different, multiples sources and awarding more credibility to instances where this contrast is easier to perform.

If the use of sources 2.0 apparently made journalists' jobs easier, it, in fact, entailed new complexities and pitfalls.

Probably the main pitfall is the idea that journalists can find their stories only by searching the web, avoiding the legwork and personal fact-checking. There are general economic, social, and political circumstances that favor this idea:

the economic downsizing that affects the overall press; the loss of prestige of the profession of journalist; the concurrence of non-professional, amateur "journalists" and "news organizations"; the attempt of political candidates and parties to bypass the mainstream media in their communication with the citizens; and the loss of the ethical values of journalism.

Another pitfall is the idea that, with the advent of the Web 2.0 tools, journalists are mere information providers, with no need to contextualize, comment upon, and critically analyze the information they provide. When this happens, they give reason to those who argue that there are no essential differences between journalists and the citizens (non-journalists)—that we are all "journalists."

The good news is that sources 2.0 allow journalists and news organizations to access a plurality of voices, perspectives and stories that they hardly could access before Web 2.0. Even not accepting the idea that Web 2.0 provides citizens with its own press, we must admit that, especially in countries and world regions where there is a lack or restriction of public information via the mainstream, traditional press, Web 2.0 constitutes a powerful tool for citizens to make their voice heard through those mainstream media that use them as their sources.

To avoid the pitfalls and keep the potential advantages of sources 2.0, we must insist on the need to improve the professional training and ethical formation of journalists—especially in regard to web-based journalistic work.

As with any study, our study has its own limitations. One of them is its national character. Another one is the small size of the sample we used (62 Portuguese journalists), despite the fact we tried our best to make sure it was a representative one. Both limitations raise the problems of generalization and external validity of the conclusions. However, these limitations can be overcome if we compare and contrast our study with our international studies in the same field—which is exactly what we tried to do. Above and beyond that, given the global character of both journalism and Web 2.0, the problems and issues we discuss here are far from being specifically Portuguese ones and thus will continue to affect hundreds of countries and millions of people.

FUTURE RESEARCH DIRECTIONS

The use of sources 2.0 and the evaluation of their credibility by journalists is a work in progress. In this process, new strategies seem to combine with traditional ones, in a very complex way. It is not a surprise that this is happening; we just have to keep in mind that the consolidation process of using traditional news sources and evaluating their credibility in pre-Internet journalism took more than a century.

To study how the process takes place we need, first of all, more worldwide empirical studies to compare how journalists within different cultures, traditions, and practices use news sources 2.0 and evaluate their credibility, and then we need to ascertain if any pattern emerges. These studies may assume not only the form of surveys, but also the form of ethnographic studies in the newsrooms—the work of Gans (2004) is probably the best example of this method—or focus groups with journalists and editors.

Another direction is, probably, to study the creation and use of technological tools that allow journalists to deal in a faster, safer, and easier way with the multiplicity and anonymity of news sources 2.0. To mention only a few examples in this direction, Canini, Suh and Pirolli (2011) proposed an algorithm, based both on the topical content of the messages and the link structure of social network structure, to allow users to identify and evaluate "topically relevant and credible sources of information in social networks," namely in Twitter; and Diakopoulos, Choudhury and Naaman (2012) developed a system, which they call SRSR ("Seriously Rapid Source Review"), that

aims to help journalists to select and assess social media sources, a model "informed by journalistic practices and knowledge of information production in events."

REFERENCES

Associated Press. (2011). *Social media guidelines for AP employees*. Retrieved December 3, 2011, from http://www.ap.org/pages/about/ pressreleases/documents/ SocialMediaGuidelinesNov.2011.pdf

Asur, S., Huberman, B., Szabo, G., & Wang, C. (2011). *Trends in social media: Persistence and decay*. Retrieved January 8, 2012, from http:// www.hpl.hp.com/research/scl/papers/trends/ trends_web.pdf

Beckett, C., & Mansell, R. (2008). Crossing boundaries: new media and networked journalism. *Communication, Culture & Critique*, *1*(1), 92–104. doi:10.1111/j.1753-9137.2007.00010.x

Bowman, S., & Willis, C. (2003). *We media: How audiences are shaping the future of news and information*. The Media Center at The American Press Institute. Retrieved October 11, 2011, from http://www.hypergene.net/wemedia/download/ we_media.pdf

Bradshaw, P. (2011). *Mapping digital media: Social media and news*. Open Society Foundations. Retrieved January 3, 2012, from http:// www.soros.org/initiatives/media/articles_publications/ publications/mapping-digital-media-social-media-and-news-20120117/ mapping-digital-media-social-media-20120119.pdf

Canavilhas, J. (2004). *Os jornalistas portugueses e a Internet*. Retrieved March3, 2012, from http://www.bocc.ubi.pt/pag/canavilhas-joao-jornalistas-portugueses-internet.pdf

Canavilhas, J., & Ivars-Nicolás, B. (2012). Uso y credibilidad de fuentes periodísticas 2.0 en Portugal y España. *El Profesional de la Información*, *21*(1), 63–69. doi:10.3145/epi.2012.ene.08

Canini, K., Suh, B., & Pirolli, P. L. (2011). *Finding credible information sources in social networks based on content and social structure*. Third IEEE International Conference on Social Computing (SocialCom), October 9-11, Boston, MA. Retrieved January 4, 2012, from http://www.parc.com/content/attachments/finding-credible-information-preprint.pdf

Carlson, M. (2011). Whither anonymity? Journalism and unnamed sources in a changing media environment. In Franklin, B., & Carlson, M. (Eds.), *Journalists, sources and credibility: New perspectives* (pp. 37–48). New York, NY: Routledge.

Cassidy, W. P. (2007). Online news credibility: An examination of the perceptions of newspaper journalists. *Journal of Computer-Mediated Communication*, *7*, 478–498. doi:10.1111/j.1083-6101.2007.00334.x

Cision & Bates. D. (2009). *Social media & online usage study*. George Washington University. Retrieved from www.gwu.edu/~newsctr/10/pdfs/gw_cision_sm_study_09.PDF

Diakopoulos, N., De Choudhury, M., & Naaman, M. (2012). *Finding and assessing social media information sources in the context of journalism*. CHI'12, May 5–10, 2012, Austin, Texas, USA. Retrieved February, 2, 2012, from http://research.microsoft.com/en-us/um/people/munmund/pubs/chi_2012.pdf

Flanagin, A. J., & Metzger, M. J. (2000). Perceptions of internet information credibility. *Journalism & Mass Communication Quarterly*, *77*(3), 515–540. doi:10.1177/107769900007700304

Flanagin, A. J., & Metzger, M. J. (2007). The role of site features, user attributes, and information verification behaviors on the perceived credibility of web-based information. *Medicine and Society*, *9*(2), 319–342.

Fogg, B. J. (2003). *Persuasive technology. Using computers to change what we think and do*. San Francisco, CA: Morgan Kaufmann Publishers.

Gans, H. (2004). *Deciding what's news: A study of CBS Evening News, NBC Nightly News, Newsweek and Time*. Evanston, IL: Northwestern University Press.

Gass, R. H., & Seiter, J. S. (2003). *Persuasion. Social influence and compliance gaining*. Boston, MA: Allyn and Bacon.

Gillmor, D. (2004). *We the media: Grassroots journalism by the people, for the people*. Sebastopol, CA: O'Reilly Media.

Haiman, R. J. (2000). *Best practices for newspaper journalists: A handbook for reporters, editors, photographers and other newspaper professionals on how to be fair to the public*. Arlington, VA: The Freedom Forum. Retrieved December 23, 2011, from http://www.freedomforum.org/publications/diversity/bestpractices/bestpractices.pdf

Hohmann, J. (2011). *10 best practices for social media. Helpful guidelines for news organizations*. ASNE Ethics and Values Committee. Retrieved January, 16, 2012, from http://asne.org/portals/0/publications/public/10_Best_Practices_for_Social_Media.pdf

Hovland, C. I., Janis, I. L., & Kelley, H. H. (1953). *Communication and persuasion*. New Haven, CT: Yale University Press.

Hovland, C. I., & Weiss, W. (1951). The influence of source credibility on communication effectiveness. *Public Opinion Quarterly, 15*, 635–650. doi:10.1086/266350

Jarvis, J. (2007). Networked Journalism. *BuzzMachine*. Retrieved October 11, 2011, from http://www.buzzmachine.com/2006/07/05/networked-journalism

Johnson, T. J., & Kaye, B. K. (2009). In blog we trust? Deciphering credibility of components of the internet among politically interested internet users. *Computers in Human Behavior, 25*(1), 175–182. doi:10.1016/j.chb.2008.08.004

Kaplan, A. M., & Haenlein, M. (2010). Users of the world, unite! The challenges and opportunities of social media. *Business Horizons, 53*, 59–68. doi:10.1016/j.bushor.2009.09.003

Keen, A. (2007). *The cult of the amateur: How today's Internet is killing our culture*. New York, NY: Doubleday.

Kovach, B., & Rosenstiel, T. (2001). *The elements of journalism*. New York, NY: Three Rivers Press.

Kovach, B., & Rosenstiel, T. (2010). *Blur: How to know what's true in the age of information overload*. New York, NY: Bloomsbury.

Lankes, R. D. (2008). Credibility on the internet: shifting from authority to reliability. *The Journal of Documentation, 64*(5), 667–686. doi:10.1108/00220410810899709

Larson, C. U. (2004). *Persuasion: Reception and responsibility*. Belmont, CA: Thomson/Wadsworth.

Levinson, P. (2009). *New new media*. Boston, MA: Allyn & Bacon.

Lévy, P. (1994). *L'Intelligence collective: pour une anthropologie du cyberespace*. Paris, France: La Découverte.

Lippmann, W. (1920). *Liberty and the news*. New York, NY: Harcourt, Brace and Howe.

Mauer, A. (2010). Using social networks as reporting tools. In Society of Professional Journalists' Digital Media Committee (Ed.), *The SPJ digital media handbook, Part I* (pp. 12-13). Retrieved October 21, 2011, from http://blogs.spjnetwork.org/tech/wp-content/uploads/2010/03/SPJDigitalMediaHandbookV3.pdf

McQuail, D. (1992). *Media performance: Mass communication and the public interest*. London, UK: Sage.

Metzger, M. J. (2007). Making sense of credibility on the Web: Models for evaluating online information and recommendations for future research. *Journal of the American Society for Information Science and Technology, 58*(13), 2078–2091. doi:10.1002/asi.20672

Metzger, M. J., Flanagin, A. J., & Medders, R. B. (2010). Social and heuristic approaches to credibility evaluation online. *The Journal of Communication, 60*, 413–439. doi:10.1111/j.1460-2466.2010.01488.x

Mills, H. (2000). *Artful persuasion*. New York, NY: AMACOM.

O'Keefe, D. J. (2002). *Persuasion: Theory and research*. Thousand Oaks, CA: Sage.

Oriella, P. R. Network. (2011). *The state of journalism in 2011*. Retrieved September 27, 2011, from http://www.centroperiodismodigital.org/sitio/sites/default/files/publication.pdf

Owens, J. W., Shaikh, A. D., & Chaparro, B. S. (2011). Patterns of information sharing among inner and outer social circles. *Usability News, 13*(1). Retrieved January 7, 2012, from http://www.surl.org/usabilitynews/131/sharing.asp

Park, R. (1940). News as a form of knowledge: A new chapter in the sociology of knowledge. *American Journal of Sociology, 45*(5), 669–686. doi:10.1086/218445

Perloff, R. M. (2003). *The dynamics of persuasion: Communication and attitudes in the 21st century*. Mahwah, NJ: Lawrence Erlbaum Associates.

Pew Research Center (PEW). (2011, September 22). *Views of the news media: 1985-2011. Press widely criticized, but trusted more than other information sources*. Retrieved November 12, 2011, from http://www.people-press.org/files/legacy-pdf/9-22-011%20Media%20Attitudes%20Release.pdf

Reese, S. (2009). Managing the symbolic arena: The media sociology of Herbert Gans. In Becker, L., Holtz-Bacha, C., & Reust, G. (Eds.), *Festschrift for Klaus Schoenbach* (pp. 279–293). Wiesbaden, Germany: VS Verlag für Sozialwissenschaften. doi:10.1007/978-3-531-91756-6_20

Reich, Z. (2011). Source credibility as a journalistic work tool. In Franklin, B., & Carlson, M. (Eds.), *Journalists, sources and credibility: new perspectives* (pp. 19–36). New York, NY: Routledge.

Reuters. (2008). *Reuters handbook of journalism*. Retrieved May 27, 2011, from http://handbook.reuters.com/extensions/docs/pdf/handbookofjournalism.pdf

Rosenstiel, T. (2010, September 12). A new phase in our digital lives (Commentary). In Pew Research Center (Ed.), *Ideological news sources: Who watches and why. Americans spending more time following the news* (pp. 79 -81). Retrieved November 21, 2011, from http://www.people-press.org/files/legacy-pdf/652.pdf

Singer, J. B. (2009). Barbarians at the gate or liberators in disguise? Journalists, users and a changing media world. In J. Fidalgo, & S. Marinho (Eds.), *Actas do Seminário "Jornalismo: Mudanças na Profissão, Mudanças na Formação"* (pp. 11-32). Universidade do Minho, Braga: Centro de Estudos de Comunicação e Sociedade.

Society of Professional Journalists (SPJ). (1996). *Code of ethics*. Retrieved December 12, 2010, from http://www.spj.org/pdf/ethicscode.pdf

Stiff, J. B., & Mongeau, P. (2003). *Persuasive communication* (2nd ed.). New York, NY: The Guilford Press.

Tuchman, G. (1972). Objectivity as strategic ritual: an examination of newsmen's notion of objectivity. *American Journal of Sociology*, *77*(4), 660–679. doi:10.1086/225193

Wolf, M. (1985). *Teorie delle comunicazioni di massa*. Milano, Italy: Bompiani.

Wu, S., Hofman, J. M., Mason, M. A., & Watts, D. J. (2011). Who says what to whom on Twitter. *Proceedings of the 20th International Conference on World Wide Web*, New York, USA.

ADDITIONAL READING

Center for the Digital Future at the USC Annenberg School. (2011, June 3). *2011 digital future report* (Press Release and Highlights). Retrieved Aril 4, 2012, from http://www.digitalcenter.org./pdf/2011_digital_future_final_release.pdf

Fallows, J. (1996). *Breaking the news: How the media undermine American democracy*. New York, NY: Pantheon Books. doi:10.1002/ncr.4100850113

Franklin, B., & Carlson, M. (Eds.). (2011). *Journalists, sources and credibility: New perspectives*. New York, NY: Routledge.

Gans, H. J. (2003). *Democracy and the news*. Oxford University Press.

Garrison, B. (1999). *Journalists' perceptions of online information-gathering problems*. Paper presented to the Newspaper Division of the Association for Education in Journalism and Mass Communication, Southeast Colloquium, Lexington, Ky., March 5-6.

Jackob, N. G. E. (2010). No alternatives? The relationship between perceived media dependency, use of alternative information sources, and general trust in mass media. *International Journal of Communication, 4,* 589–606.

Manning, P. (2001). *News and news sources: A critical introduction*. London, UK: Sage Publications.

Online News Association. (2001). *Digital journalism credibility study.* Retrieved December, 2, 2011, from http://banners.noticiasdot.com/termometro/boletines/docs/marcom/prensa/ona/2002/ona_credibilitystudy2001report.pdf

Pavlik, J. (2001). *Journalism and new media*. New York, NY: Columbia University Press.

Shoemaker, P., & Reese, S. (1996). *Mediating the message: Theories of influence on mass media content* (2nd ed.). New York, NY: Longman.

KEY TERMS AND DEFINITIONS

Credibility: The perceived quality of a source or a message that makes journalists (or other people) to believe in it.

Journalism: Activity which consists in collecting, processing, and disseminating information on current events of general interest, following professional rules and ethical codes.

News: Journalistic text about current issues or unheard fact of general interest that answers to six basic questions: what, who, when, where, how and why.

Online: Content or social interactions that occur or are available on the Internet.

Social Media: Set of web technologies based on the principles of Web 2.0 that allow the exchange of content, facilitate interaction between users, and promote user-generated content in a typical bidirectional channel.

Sources 2.0: Information providers who do it willingly using the Web 2.0 tools.

Web 2.0: Concept created by O'Reilly Media to describe a new generation of web services where the user is in the center of the system to consume, but also to produce, online contents.

Chapter 11

Whose News Can You Trust?
A Framework for Evaluating the Credibility of Online News Sources for Diaspora Populations

Rick Malleus
Seattle University, USA

ABSTRACT

This chapter proposes a framework for analyzing the credibility of online news sites, allowing diaspora populations to evaluate the credibility of online news about their home countries. A definition of credibility is established as a theoretical framework for analysis, and a framework of seven elements is developed based on the following elements: accuracy, authority, believability, quality of message construction, peer review, comparison, and corroboration. Later, those elements are applied to a variety of online news sources available to the Zimbabwean diaspora that serves as a case study for explaining the framework. The chapter concludes with a discussion of the framework in relation to some contextual circumstances of diaspora populations and presents some limitations of the framework as diaspora populations might actually apply the different elements.

INTRODUCTION

Metzger (2007) suggested people need to "know when and how to exercise" (p. 2089) the skills needed to evaluate credibility of information on the Internet. No matter what framework a user applies to evaluate the credibility of online news, it is important to realize that user motivation is key (Metzger, 2007). As a responsible global citizen,

users of online news need to make an effort to seek out online sources of news that are credible, recognizing the "negative consequences of misinformation online" (Metzger, 2007, p. 2089). Fritch (2003) argued living "in an information-rich, networked world" (p. 327) requires all users to take individual responsibility for making assessments about the credibility of the information they consume.

It can be argued that populations living in a diaspora may have added motivation to seek

DOI: 10.4018/978-1-4666-2663-8.ch011

credible news about their home countries given their concerns about identity maintenance and challenges to that identity maintenance as they live and work in host countries and cultures. In addition, diaspora populations are motivated to seek connections to home, and some of those connections are kept up through the consumption of news. "Under conditions of high motivation, online information seekers will likely pay more attention to information quality cues and perform more rigorous information evaluation than when motivation is lower" (Metzger, Flanagin & Medders, 2010, p. 416).

Kovach and Rosenstiel (2001) argued that the Internet is a medium that increases the need for people to apply judgments to the news that they consume and that a journalist's role is to help the audience make sense of all the information available online. The aim of the digital literacy movement is, in part to, "assist Internet users in developing the skills needed to critically evaluate online information" (Metzger, 2007, p. 2079). Drawing on findings in the literature, this chapter proposes a framework to help diaspora populations evaluate the credibility of online news about their home countries. Online sources available to those in the Zimbabwean diaspora are used as case study examples to which the framework is applied.

The framework suggests a set of criteria that defines credibility that diaspora populations might use to analyze the credibility of online news about their home countries. These questions, based around the framework, will be explored:

- Is this true? (Accuracy)
- Does this source seem qualified? (Authority)
- Does this news report seem like it is reporting something that could happen? (Believability)
- Does the report show evidence of professionalism and is it free of obvious errors in form? (Quality of Message Construction)

- What are other users saying about this site? (Peer Review)
- Does this source live up to the standards set by other credible sites? (Comparison)
- Can this information be found on other sites too? (Corroboration)

The objectives of the chapter are to establish criteria for defining credibility, to apply those variables to online news sources available to those living in the Zimbabwean diaspora, and to explain the utility of those ideas for other diaspora populations in making decisions about the credibility of online news about their home countries.

Defining Credibility Online

Chiagouris, Long and Plank (2008) pointed out that "although news may be reported accurately through a website news source, the news will not automatically be perceived by consumers as credible" (p. 544). Simply reporting news accurately will not ensure that audiences perceive that news as credible. While we speak of something being credible—having or lacking credibility—that evaluation lies in the minds of the audience, and does not actually reside in the thing being evaluated. Of course, various elements of a news source impact an audience's evaluations of credibility, but credibility lies in the mind of the user of an online news source, not in the source itself. As Rottenberg (2003) argued, credibility is the *audience's belief* about the knowledge, dependability and good intention of the source. So, conceptualizing credibility as *existing in the minds of the audience* is an important first step in exploring the concept of credibility.

In order to develop a framework of credibility for applying to online news sites, it is necessary to define what is meant by credibility. An audience's perceptions of credibility "are the result of multiple dimensions of assessment…" (Robins, Holmes & Stansbury, 2010, p. 14). Chiagouris,

Long and Plank (2008) suggested that from an audience perspective "credibility issues arise from two areas—the content of what is reported and how it is reported" (p. 529). It is important to note that not only is what is said in the reporting significant, but how the report is delivered also plays a role in determining credibility. This suggests that the medium through which news is channeled and received must be considered.

Credibility of information can be thought of in part as the "believability, fairness, accuracy, and depth of information" (Johnson and Kaye, 2010, p 6). From a review of the literature, Metzger (2007) found researchers claiming that part of assessing credibility of online information is to apply variables used to determine credibility in *other* communication channels. She suggested that the five main criteria were *accuracy, authority, objectivity, currency and coverage/scope*. The following discussion of assessing credibility of online information is based on what is known as the "checklist approach" to evaluating credibility.

Kovach and Rosenstiel (2001) pointed out that media audiences have a right to expect certain elements in the news they consume and that those expectations are based on the principles that journalists agree are important. Among those elements are that reporting will be truthful, stories will have been verified, and stories should be interesting and relevant to the audience. Journalists who publish online need to use their traditional journalistic skills to build and maintain credibility: "Online journalism ties it all together: the old values with the new technologies, traditional skills with innovative production, and journalists with their audiences" (Kolodzy, 2006, p. 190).

But establishing credibility for an online audience is more than simply adhering to old values. The Internet is the channel that online news is communicated to an audience, and that channel has an important impact on assessing credibility. From her review of eight different research article findings, Metzger (2007) presented factors that

researchers suggest may "play into credibility assessments" (p. 2081). Among those factors are *citing of sources, identification of author(s), professional, attractive and consistent page design, professional-quality writing that is clear, message relevance and tailoring, and the ability to verify claims elsewhere.*

From their research, Chiagouris et al. (2008) reported that the related variables *ease of use* and *design* were credibility factors that influenced audience evaluation of news websites. Metzger (2007) concurred, suggesting that the literature allows this conclusion to be drawn about Internet users: "They place a premium on professional site design in their credibility appraisals" (p. 2089). Further, Mehrabi, Hassan and Ali (2009) reported that factors such as time spent on the Internet, the salience of the stories to the audience, and how much a person relies on a specific medium to get their news are all factors that impact the evaluation of credibility online.

More recent research has also isolated a number of factors that seem to have an impact on evaluating the credibility of online sources. As Dochterman and Stamp (2010b) suggested "research uncovered 12 factors impacting Web credibility judgments: authority, page layout, site motive, URL, cross-checkability, user motive, content, date, professionalism, site familiarity, process, and personal beliefs" (p. 44).

In addition to the checklist approach to assessing credibility discussed above, another approach to evaluating credibility is worthy of consideration in this chapter. Meola (2004) argued for a *contextual approach* to evaluating websites. He suggested that the utility of this contextual approach is that it makes use of information that is "external to the Web site in order to evaluate it" (p. 338). Evaluating a website in "its wider social context" allows for "facilitating reasoned judgments of information quality" (Meola, 2004, p. 338). While Meola's (2004) arguments were directed at how students could be trained to make good judgments about the credibility and quality of the websites they use in their work, there are some ideas embedded in the contextual approach that apply to this chapter.

Meola's (2004) contextual approach uses three techniques: 1) promotion and explanation of reviewed resources, 2) comparison and 3) corroboration. It can be argued that promoting and explaining reviewed resources for people in a diaspora occurs when communities form in the foreign countries they live in and discuss sites where they access online news. Essentially a form of informal peer review occurs and members of the diaspora lead each other to sites that are "peer evaluated" for credibility. Additionally, suggesting sites to each other that have editorial review of content from sources that are known, trusted, and have expertise familiar to the disapora population, will also allow for a more directed use of *quality sources* which is at the heart of Meola's first technique. In other words, online sources that are of low quality and lack credibility will be weeded out and not accessed by the diaspora population.

Comparison is Meola's (2004) second technique for the contextual approach to evaluating websites. By comparison he meant "the examination of the similarities and differences between two or more items...analyzing the similarities and differences in the content of two or more" websites (p. 340). The power of this contextual approach, he argued, lies in the role that comparative thinking plays in the way that we evaluate things. We need something to compare another thing to in order to make evaluative decisions about that thing. How do we know what an interesting movie plot is unless we have some standard by which we know what an interesting movie plot is? By the same token, how can we evaluate the credibility of an online news source without having something of quality and credibility with which to compare that news site?

Finding online news sources that are defined as credible allows for corroboration, Meola's (2004) third technique. "To corroborate information is to verify it against one or more different sources" (Meola, 2004, p. 341.). He argued that since the Internet is a very information rich environment, people should use that resource to check if information that they are getting from one source is reliable. The driving theory behind his notion is that the "more sources that can be found to corroborate the information, there is a greater probability that the information is reliable" (p. 342). So, in addition to elements from the checklist framework discussed earlier, Meola's three techniques for seeking to establish credibility of online sources are also useful tools. Below, I propose an analytical frame based on the preceding discussion of online credibility.

Analytical Framework Proposed

Metzger (2007) made a convincing argument that from a review of the literature, the conclusion can be drawn that "Internet users as a group are not willing to exert a great deal of effort in assessing the credibility of the information they find online..." (p. 2089). From this conclusion, it is reasonable to suggest that any framework for evaluating online news sources for diaspora populations needs to be *complex enough to work, but simple enough that people will be motivated to use it.*

Keeping in mind the discussion of credibility given above and Metzger's last point above, a reasonable framework for analyzing online news credibility should include the following variables: accuracy, authority, believability, and quality of message construction. These elements, which will be clearly defined in the application of the framework below, should be considered tools from the checklist model of assessing credibility of online news sources.

In addition, Meola's (2004) idea of contextual analysis, specifically the techniques of promotion of reviewed resources, comparison, and corroboration, are important elements of the analytical framework to be applied in the rest of the chapter. These techniques add to the power of the framework because they are largely tools that make use of elements outside of the specific site being evaluated.

Collectively, the checklist variables and the contextual approach can provide users with a complex, yet imminently achievable, analytical framework to help make decisions about online news credibility. These have all formed the basis for the framework proposed and applied to online media available to Zimbabweans in the diaspora later in this chapter. The next section of the chapter will discuss diaspora populations to contextualize their circumstances in regards their online news needs.

Diaspora Populations

Brubaker (2005) isolated three core elements that define a diaspora: the notion of dispersion through space, the idea of a homeland that is a source of identity and values, and the notion of boundary maintenance, meaning that people in a diaspora, to one degree or another, keep up elements of their cultural/national identities that are distinct from their host countries/cultures. The impetus to leave home often comes from social, political, and economic upheaval.

According to Clifford (1994), a diaspora "involves dwelling, maintaining communities, having collective homes away from home…" (p. 308). In other words, a diaspora is not simply individuals having left their home country, but involves the building of communities in clusters in different countries, where people are engaged, interact, and communicate with each other as citizens of their home countries. Often this interaction includes maintaining elements of their cultures in their host countries. "One of the impacts of recent globalization is the formation of new offline and online transnational connections among migrants worldwide" (Ghorashi & Boersma, 2009, p. 667). Communication between people in a diaspora takes place both in interpersonal and mediated settings, within a host country and across national boundaries, with others in the same diaspora in other host countries, and with family, friends and media from their home countries.

Issues of identity are important to people in a diaspora because they are "involved in constructing, imagining and changing identity amid a variety of cultures and discourses that are articulated and negotiated in a transnational context" (Ghorashi & Boersma, 2009, p. 670). Using the Eritrean diaspora as an example, Bernal (2006) argued that people "use the Internet as a transnational public sphere where they produce and debate narratives of history, culture, democracy and identity" (p. 162).

Part of negotiating identity in a diaspora involves the transnational connection to home, and one of the ways people stay connected to what is going on at home is through online news. It is this link to home that news provides that is the focus of this chapter. The news people in a diaspora obtain from online news sources about their home country is important because it provides current information, so being able to gauge how credible that information is very important.

The Zimbabwean Diaspora

The diaspora that is being used as a basis for case study analysis in this chapter is the Zimbabwean diaspora. Over the past decade, Zimbabwe experienced "years of mass unemployment, mutant inflation, chronic shortages and state violence…" (Ross, 2010, p. A7). The economy was devastated, and "Zimbabwe's economic output decreased dramatically between 1998 and 2008. Official inflation rose above 200,000,000% in 2008, and although the economy has since stabilized, unemployment remains estimated at more than 90%" (Ploch, 2010). Shortages of food, rampant hyper-inflation, massive levels of unemployment, political violence and intimidation, crumbling education and health infrastructures, high levels of HIV/AIDS, and restrictions of freedoms of the press, assembly and speech all contributed to a climate ripe for emigration. These crises led to enormous numbers of people, especially educated professionals and those involved in manual labor work, leaving the country to seek greener economic pastures and escape political persecution.

There are significant Zimbabwean populations in countries like the United Kingdom, the United States of America, Australia, New Zealand, Botswana, Namibia and South Africa. While there is no agreement about the exact size and composition of this diaspora, estimates of about 4 million people are often given and those estimates include both professional classes and non-elite labor migrants (McGregor and Pasura, 2010). From a total population of Zimbabweans, an estimated 12-13 million people, the diaspora represents a significant proportion of Zimbabweans no longer living at home.

In a study on Zimbabwean populations in South Africa and the United Kingdom Bloch (2005) reports that 48% left Zimbabwe for economic reasons and 26% for political reasons. Of those surveyed, 82% had educational qualifications beyond high school, including university degrees and other forms of tertiary education. Zimbabweans in the diaspora strive to maintain community and links to Zimbabwe. Bloch (2005) further discovered that 81% of Zimbabweans surveyed were involved in activities with other Zimbabweans in their new countries of residence. Social activities, religious activities, and clubs were the most common community building interactions. Forty-eight percent reported being involved in activities in Zimbabwe, with 21% involved in Internet discussion groups. Bloch (2005) found 96% of respondents kept in regular touch with their friends and families in Zimbabwe, often through email and text messaging. The vast majority of them also reported sending money and other forms of help back to Zimbabwe for their friends and families.

People in the Zimbabwean diaspora are motivated by complex sets of reasons for relocating. Many of those reasons were precipitated or necessitated by the crises of the last decade in Zimbabwe:

Zimbabwe has been through a period of extremely intense social and political upheaval in the last decade and all aspects of Zimbabwean life have been affected by that upheaval…A new political party emerged to challenge the entrenched political leadership, an agrarian land reform policy was implemented with national and international consequences, a hyperinflationary economy made living and working in Zimbabwe extremely difficult, and the social fabric of the country was torn. (Malleus, 2011, p. 130)

These diverse reasons for being in a host country play into the motivations that Zimbabweans have for sourcing online news about home. Kanu's (2010) point that it is "vital to recognize that the Diaspora itself is not a homogenous group that holds a single and defined political view" is important.

"The strong nationalist tradition and appeals for national unity reflected in Zimbabwean diaspora associational life coexist uncomfortably with party-political divisions, class, racial, ethnic and gender tensions…" (McGregor, 2009, p. 187). The Zimbabwean diaspora is not divorced from the realities of social and political tensions that exist in Zimbabwe. There are major political differences between those in the diaspora who support the two main political parties: the Zimbabwe African National Union-Patriotic Front (Zanu-PF) and the Movement for Democratic Change (MDC). Until a government of national unity (GNU) was formed after elections in 2008, Zanu-PF, led by President Mugabe, had been the ruling party in Zimbabwe since independence in 1980. Prime Minister Tsvangirai leads the MDC, which came to partial power with the formation of the GNU but has been represented in parliament for the past decade. The divisions between the main parties have been stark in both terms of policy and governing philosophy. These political divisions are also evidenced by those in the diaspora by Zimbabweans who support one party over the other.

In addition to those political partisans, there are Zimbabweans in the diaspora who are not aligned with, or members of, a political party, whose positions on news connected to politics and governing are less partisan, and who look for cues in online news coverage showing that it is balanced. Partisans may be inclined toward online news coverage that is tilted in their party's favor. Zimbabweans in the diaspora will have their individual and collective takes on what policies and which politicians and groups are to blame for Zimbabwe's problems. That discussion is not the focus of this chapter, but that those differing views exist needs to be understood in order to contextualize the diaspora's consumption of online news.

As Kovach and Rosenstiel (2001) claim "journalism exists in a social context" (p. 42) and it is important to understand the media regime that Zimbabweans in the diaspora would have been exposed to at home. The Zimbabwean broadcast media (television and radio) is government controlled. However, print media is either government or privately controlled. The media environment is politically divided, with government controlled media generally supporting politicians, groups and policies of the Zanu-PF wing of government and being critical of the MDC and their policy stances. The private/independent media on the other hand are generally critical of Zanu-PF and their policy positions, and in the main, supportive of the MDC wing of government. Zimbabweans are accustomed to a divided and partisan press whose news coverage can be skewed and partial, especially when related to political coverage. It is in this social context that the Zimbabwean diaspora's consumption of online news should be framed and understood.

Applying the Framework

This section of the chapter will apply the proposed framework to online news sources of media to which the Zimbabwean diaspora has access. A variety of online news sources will be used to illustrate how the proposed framework has utility for that population in evaluating the credibility of the news that they are accessing online. The following will be addressed:

- Accuracy
- Authority
- Believability
- Quality of Message Construction
- Promotion of Review Resources
- Comparison
- Corroboration

Accuracy

Feargal Keane, a well respected foreign correspondent who works for the *British Broadcasting Corporation*, made a good case for defining journalistic accuracy as "'truth-telling' that is 'artful, fearless and intelligent'" (qtd. in Allen and Thorsen, 2011, p. 20). Kovach and Rosenstiel (2001) claimed that journalists have truth as their first obligation as a guiding ethic. The central criteria for evaluating accuracy of an online news story is to answer the question "Is this true?" To be able to do that, journalists need to use material in their reports that has been independently verified (Allen and Thorsen, 2011, Kovach and Rosenstiel, 2001), in other words, to use more than one source of information before publishing it for their online audiences. As Goggin (2011) pointed out, part of the journalistic ethic calls for fact and source checking to ensure the accuracy of news reports. Online users should expect their reports to be accurate as a "key quality of factual information" (Nguyen, 2011, p. 206).

The Herald Online is an Internet version of the Zimbabwean government-controlled daily newspaper *The Herald*. People in the Zimbabwean diaspora know that this online source represents the Zanu-PF element of the Government of National Unity's position on issues of public importance. The crafting of a draft constitution for the country has been a burning issue in recent years, and three examples from one report recently published in *The Herald Online* allows for an analysis of how users might gauge the accuracy of the article. The headline of the article is "New Constitution's principal drafters must be fired." The first paragraph of the article is:

PRINCIPAL drafters of the envisaged new Constitution should be fired and a new team should start rewriting the first draft or the country should go for harmonised elections using the current Constitution to eliminate political compromise, analysts and legal experts have said. ("New Constitution," 2012)

This paragraph is a restatement of an oft-repeated policy and political position of Zanu-PF. The accuracy of the article hinges on an analysis of who the "analysts and legal experts" are that the unidentified reporters consulted as sources for the article. One political analyst quoted extensively throughout the article was Professor Jonathan Moyo. Moyo is a member of the Zimbabwean parliament, was a minister of information for Zanu-PF for the first part of the decade, and is an outspoken and often controversial politician. Zimbabweans know him, including his political history and ties to Zanu-PF. To identify him simply as a "political analyst" is not accurate. This makes the article and the perception of the accuracy of the article questionable.

A second expert quoted in the article is Professor Madhuku, who heads the National Constitutional Assembly (NCA), and who has been vocal for years about what he and his organization see as the flawed, politicized process of gathering views and writing of the proposed new constitution. So, while his views are depicted accurately, his views are well known, and are not new in relation to the just released first draft of the new constitution. NCA's opposition is not new to what they see as a flawed process, he is not calling for the drafters to be fired, but assured the reporters that the NCA will campaign for a "no" vote on any constitution that comes out of this process.

The third paragraph of the article reads: "Zimbabweans from all walks of life jammed *The Herald* switchboard demanding that the document be rejected and referred back to the drafters who should rewrite it with clear instructions…they should incorporate the peoples' views" ("New Constitution," 2012.). There are no specific quotes from these "callers," no evidence provided to the

audience about the veracity of these claims. On its own, people may not question the accuracy of this assertion, but the government-controlled media is often claiming outrage on the part of the people without providing clear evidence of these claims. Therefore, as part of a pattern, this claim in the report has no way to be verified by the audience.

From these examples, from just one article, users of *The Herald Online* could ask legitimate questions about the accuracy of the report, and hence be on the lookout for similar examples of questions of veracity and framing on that website. As Potter (2012) said, the "frame of news stories is constructed by journalists in the way they select certain bits of information while ignoring others and by how they structure their stories to direct attention toward certain facts" (p. 76). *The Herald Online* leaves itself open to claims about inaccuracy of its reporting by the deliberately misleading way it framed the report. Firstly, it did not name Moyo as a member of parliament and connected to Zanu-PF. Secondly, it did not point out the NCA's opposition to the whole constitutional process from very early on, and thirdly, it did not provide further evidence to support the claims about their switchboard being inundated by callers.

A second example that can be used to apply the framework element of accuracy comes in comparing two articles on an official ceremony to celebrate World Media Freedom Day. The *Zimbabwe Broadcasting Corporation's* (*ZBC*) website carried a report with the headline "Govt respects media pluralism: Shamu" ("Govt," 2012.) Covering the same event, another private newspaper, the *Zimbabwe Independent's* website headline for the story was "Shamu threatens media" ("Shamu," 2012). A person seeing both headlines for these articles would be hard-pressed to say which one was accurate given that there are different interpretations of the same event implied. They would have to read each article in order to judge accuracy.

The *Zimbabwe Independent's* article leads with the claim that the Minister of Media, Information and Publicity Mr. Shamu threatened some sections of the Zimbabwean media in a speech he gave at the ceremony. They support this contention by quoting at length from Shamu's speech:

I can also predict that if the clearly anti-African and anti-Zimbabwe frenzy we have experienced through some media outlets and platforms in this country continues, and if the conspiracy of silence within the media industry and journalism profession also persists, the gloves may soon be off here as well…If the last five years of change do not show the media industry and the journalism profession to have fulfilled their promises, then the sovereign people of Zimbabwe have no option but to intervene and protect themselves through the instruments of state, that is to revert to the regulatory regime of 2001-2007. ("Shamu," 2012)

The article then goes on to quote the views of Prime Minister Tsvangirai and quote representatives of six media related organizations: the Voluntary Media Council of Zimbabwe, the Media Alliance of Zimbabwe, MISA Zimbabwe, the Media Monitoring Project of Zimbabwe, the Zimbabwe Union of Journalists, and the Zimbabwe National Editors' Forum. The views expressed by the prime minister and all the representatives were in contrast to the minister's comments and lamented the lack of media freedom, and called for more media freedom.

In contrast, the *ZBC* report's lead was that "Government says it respects media pluralism and diversity which should be applied in a responsible manner so that the sovereignty and national values of the country are protected" ("Govt," 2012). The article interprets and summarizes Minister Shamu's views without quoting him directly. Two other media voices are quoted in the article. The Zimbabwe Media Commission's chairperson was quoted as saying in part "our duty is to ensure that media enjoy its freedoms. However, what is important is to ensure that the freedom is not abused in violation of other people's rights" ("Govt," 2012). While the Zimpapers Group Chief Executive Officer (Zimpapers being the government controlled group of publications) was quoted

as saying "It's not possible to have 100% media freedom because this country has values that need to be protected" ("Govt," 2012).

Both online news sources have framed the reports in very different ways. The *ZBC* article framed Shamu, a government minister, as representing the government's view. The article did not mention nor frame Prime Minister Tsvangirai's view as being an alternate government view of the state of media freedom in Zimbabwe. Shamu is a Zanu-PF member and Tsvangirai the leader of the MDC. The two sources the journalist(s) writing the article selected to quote both supported Shamu's position at least in part. On the other hand, *The Zimbabwe Independent*'s framing of the report was of two divergent interpretations of Zimbabwe's media environment being presented at the ceremony. Shamu represented one view, while Tsvangirai and the six other people quoted in the article represented quite a different view. The journalist(s) did not quote the two people that *ZBC's* report quoted who would have in part bolstered Shamu's position.

Which telling of the events at the ceremony is most accurate? Keeping Potter's (2012) ideas mentioned above in mind, it would seem that the ZBC's framing of the report is the least accurate of the two article. It told only told one side of the story, and characterized one government representative's version of the media landscape as true, while not giving voice to another, more senior government representative's position on the issue, and ignoring six other voices who contradicted Minister Shamu's contentions.

Thinking further about what builds or busts perceptions of accuracy in online news, there are some special elements of production of online news stories that require journalists to go the extra step in building perceptions of credibility for an audience. Included in those elements are the hyperlinks that the medium allows: "Journalists need to think of a link as an element of an online news story. As such, they must often decide if a link is reliable and accurate, just as they do with a quote from an interview or details from a source" (Kolodzy, 2006, p. 196). Hyperlinks can also be politically or critically framed.

So, one way that Zimbabweans might make assessments of accuracy in articles is to consider how the online news sources they consult use hyperlinks in their reports. Is there a consistent quality to the links that are used, are those links to other sites that have reliable information that adds to the quality of the news report? If an online source does not do a good job of assessing the accuracy of the additional information being supplied to their users through the hyperlinks they use, then users have reason to question the accuracy of their reports overall.

Another way accuracy can be evaluated is how hyperlinks are used as what Stepno called "entry points" for a report (quoted in Kolodzy, 2006, p. 196). A user should be able to see an obvious tie to the report they are reading, with clear information that adds to the reliability of the report from the entry points links provide. Remembering that an online news article is not consumed or often constructed in a linear fashion is important, and a user's consumption of all the elements of the report should contribute to an accurate decoding of the journalist's message.

An example of an online news source that people in the Zimbabwean diaspora turn to is *SW Radio Africa*. This online source is also a radio station that broadcasts to Zimbabwe from the United Kingdom on the shortwave band, so their online presence is used as a news source for people in the diaspora who do not have access to their terrestrial shortwave broadcasts. *SW Radio Africa* staff are opposed to the Zanu-PF wing of the Zimbabwean government, and several were broadcasters in Zimbabwe before they left the country to set up and run what Zanu-PF sees as a broadcaster bent on regime change in Zimbabwe, while those operating *SW Radio Africa* bill themselves as the independent voice of Zimbabweans (in opposition to the broadcasters in Zimbabwe that are government controlled entities.) That is

the context is which Zimbabweans in the diaspora know *SW Radio Africa* operates. In addition to their broadcasts, they also provide links to various reports, most of them damning for Zanu-PF.

A particularly controversial report they linked to in 2011 was a list of names, addresses, and telephone numbers of what they purported to be agents of the Zimbabwean Central Intelligence Organization (CIO), essentially the Zimbabwean secret police agency. The report was claimed to be a leaked document that *SW Radio Africa* were given, with CIO agent's names and addresses that were accurate as of April 2001. In July 2011, station manager Gerry Jackson (2011) issued a statement about the report that read in part that *SW Radio Africa* was "serializing the release of this document in the interests of transparency and accountability and in the hope that by exposing these names some of the daily fear Zimbabweans live under will be taken away. We also hope that it might make some of the perpetrators of violence think twice before they commit further human rights abuses." (n.p.). Jackson (2011) also noted that due to responses and concerns from their listeners, they had removed the addresses of the alleged CIO operatives from their report. There is no way that readers can gauge the accuracy of the report, since *SW Radio Africa* is keeping the source of the leaked document a secret in order, presumably, to protect that person or those persons from any form of retribution or legal sanction.

Another example of a link to a report that *SW Radio* has is a list of farms that the site claims have been forcibly taken from owners of full title to their land under Zimbabwe's controversial land reform program that has taken place this decade. The report (available at (www.swradioafrica. com/Documents/farmlist241209.pdf) names the farm(s) and named legal title-holders, the name of the beneficiary of the land (new "owner") and the beneficiary's relationship to the Zimbabwean government; at the bottom of the report is this note:

NB: This list has been compiled through talking to title deed holders. It is by no means complete due to the lack of transparency within the process. None of the above title deed holders have been evicted through an eviction order from a court. The vast majority have been evicted through other, often severe, intimidatory means.

Given the genesis of how *SW Radio Africa* came in to existence, with journalists practicing in exile, and with their strong stance against Zanu-PF, users who go to links such as these reports have no mechanism for evaluating their authenticity and accuracy, and must therefore judge them accordingly.

On the other hand, *SW Radio Africa* also offers links to many other reports that can be evaluated more easily for accuracy. For example, the site offers users a link to a report about healthcare in Zimbabwe compiled by Physicians for Human Rights (PHR). The link takes users to PHR's site where the report "Health in Ruins: A Man-Made Disaster in Zimbabwe" is available to be read in its entirety. Users of *SW Africa Radio* can read not only the report, but also about PHR, what they do, who their experts are, and make judgments about the report based on some of those other factors too. Having hyperlinks like this example will allow Zimbabwean's in the diaspora to be able to evaluate the accuracy of the stories on which *SW Radio Africa* report.

Authority

A second variable to be discussed from the proposed framework is authority, and links can be made between authority and accuracy as discussed above. I am using "authority" in this sense: "Authority refers to the places from which the Web page gains authority on a topic, including the author or the author's credentials..." (Dochterman and Stamp, 2010a, p. 39). Having information about the writer of report on a website can increase the likelihood that users will perceive the writer as having the authority to write the report (Johnson and Wiedenbeck, 2009).

Additionally, from a large quantitative study on what makes a website credible, Fogg et al. (2001, qtd. in Johnson and Wiedenbeck, 2009) suggested that markers of authority or expertise include whether the site is operated by a news organization that are also well respected outside of the Internet, whether reports have citations and references, and whether a site displays any awards it might have won.

Using *The Herald Online*'s report about the Zimbabwean constitution drafting process discussed previously, there was no reporter's name (or names) attached to the report. The article was cited as being written by "Herald Reporters." This leaves the user very little information to work with in evaluating the authority of the reporter/reporters who wrote the article.

Table 1 summarizes the findings of reporter attribution from the articles billed as top reports on the May 19, 2012 website publications of *The Herald Online*, *The Chronicle*, and *The Zimbabwe Independent*, three major online news sources used by Zimbabweans in the diaspora. An examination

Table 1. Reporter attribution

Top Online News Reports: May 19, 2012	The Herald Online (Six articles)	The Chronicle (Six articles)	The Zimbabwe Independent (Seven articles)
Articles With Reporter Named	Four	Four	Six
Articles With Reporter Not Named	Two	Two	One
Articles With Reporter Beat Identified	Two	One	Zero
Articles With Reporter Beat Not Identified	Four	Five	Seven

of those sources' practice of reporter attribution further demonstrates an inconsistency as to when a journalist is identified and when a journalist is not. There is also an inconsistency in whether a reporter's beat is identified or not.

As can be seen in Table 1, *The Zimbabwe Independent*, which provided no reporters' beat, did provide the names of the reporter for six of the seven top news articles of the day, and *The Herald Online* which provided names of reporters four times, but those reporters' beats only twice, the three online news sources provide an uneven pattern of attribution. This uneven pattern makes it difficult for a reader to accurately gauge the authority of a reporter. For example, one report in *The Herald Online* headlined "Zim capable of hosting 2013 UNWTO summit" was attributed to reporter "Isador Guvamombe in Vic Falls." The attribution identifies the reporter as well as her location. Mentioning her location, the town of Victoria Falls, which is the potential site of the conference, lends some authority to her reporting because she was present at those sites in that town where the summit might be held. Providing the reader with her usual reporting beat might further strengthen her authority.

It would be more effective if *The Herald Online*, *The Chronicle*, and *The Zimbabwe Independent* consistently attributed their reports to individual reporters or teams of reporters, then Zimbabweans in the diaspora might be able to evaluate their authority more effectively. If for example, users get to know that one reporter for *The Herald Online* is covering the constitution-drafting story on an ongoing basis, this would provide a way to establish that reporter's expertise or credibility in covering that issue in the mind of a user. Users could see other articles that the same journalist had written on the constitution making process and be more confident in that person's ability. Additionally, if the journalist were clearly identified as being on the legal beat, or assigned to the coverage of parliamentary affairs, then that person's credentials as a beat reporter would add

credibility to such stories. As it is, without naming the reporter(s), Zimbabweans in the diaspora would have no way to know who wrote the article.

The Herald Online seems to be doing a better job with attributing stories to specific journalists on the sports beat. For example, when exploring the sports section of the website, numerous reports can be seen from Eddie Chikamhi, specifically identified as a sports reporter. Several of his stories are on football, so the audience can make some assumptions about his authority on that subject, given that he has some experience of reporting on that sport specifically. Zimbabweans in the diaspora have an interest in Zimbabwean sport, especially football, which is Zimbabwe's most popular sport, and would be better able to trust his authority on that topic because he is consistently named as the author of reports on football.

Applying the Fogg et al. (2001, qtd. in Johnson and Wiedenbeck, 2009) idea of authority being generated in part by audience members connecting the respect they have for a media outlet in another medium outside of the Internet to that source's online reporting, one example to consider in the context of the Zimbabwean diaspora could be the *British Broadcasting Corporation* (*BBC*). Despite the colonial past that the United Kingdom shares with Zimbabwe, with the UK being the former colonial power in Zimbabwe, and the often turbulent political relations that the governments have had over the past decade, many Zimbabweans receive some of their news and information from the *BBC*. They have been exposed for decades to their shortwave radio broadcasts, and more recently to their news broadcasts on satellite television. Anecdotal evidence suggests that having established a solid reputation with many Zimbabweans in those media allows for their authority to translate to their online news coverage of Zimbabwe.

Zimbabweans in the diaspora have access to *BBC's* online news coverage of the country, and Zimbabwe is a fairly regular feature of *BBC's* African coverage. For example, a *BBC* news report on February 17, 2012 focused on the death of retired Zimbabwean General Mujuru, who was a former freedom fighter, highly influential politician, and husband of Vice President Mujuru. General Mujuru died in a fire on his farm in circumstances that led to an official inquiry being held as to the cause of his death, and this story would have been of great interest to Zimbabweans in the diaspora. There is controversy surrounding Mujuru's cause of death, with conspiracy theories racing around the blogosphere and in the online media, so *BBC's* authoritative reporting would be a reliable source for Zimbabweans around the world to turn to. In addition, the Mujuru article had very up-to date hyper-links to other authoritative sources like the newswire service *AFP* and *The Mail and Guardian* newspaper with updates on the Mujuru story. So, the respect the Zimbabweans have for the *BBC* in other media may translate to their lending BBC online news an authority that audience can count on.

An additional set of criteria, it can be argued, are also important to a user's evaluation of authority, and those have to do with the design of the web pages on a site. A web page's visual design is important because it is one criteria that people use to judge the credibility of that site, particularly influencing a user's first impression of that site (Robins, Holmes, & Stansbury, 2010). A poorly designed web page will tend to leave the user with a negative evaluation of that site.

Fogg et al. (qtd. in Metzger, Flanagin & Medders, 2010) found that "online information consumers' predominant credibility consideration was 'site presentation,' or the visual design elements of Web sites…" (p. 416) and that "people typically process Web information in superficial ways, that using peripheral cues is the rule of Web use, not the exception" (p. 15). This claim is supported by Walthen and Burkell's (2002) assertion suggesting online users often base their first assessments of a web site on things like page layout and web page design.

Given this additional set of criteria for evaluating authority, another set of cases of media that the Zimbabwean diaspora are exposed to will be examined here. *Zimbabwe News Online* is an online news source that claims to be "A Great Platform for Zimbabwe News" on its overcrowded masthead on the site's homepage. There are *twenty tabs* that users can click on at the top of the home page, all within a three-and-a-half inch space. The tabs range from "Top News," "General" and "World," to "Strange News" and "Name and Shame ZFFE Haters." There were twenty-two tabs at the bottom of the home page, most of which were the same as those at the top of the page. There were four different places on the homepage where users could click to read the top news. At the time of writing this chapter, the homepage had four empty spaces soliciting advertisers to use those open spots to advertise, and five adverting spots that were taken on the homepage. There were *thirty-two* pictures on the homepage, some of which were shown more than once. The homepage of *Zimbabwe News Online* would not inspire most users to assign the site authority because its layout was cluttered, ill-designed and needlessly repetitive. The fact that the site had almost as many empty advertising spots as spots that were filled is a credibility-busting element of their design, and, in an even more ominous note, the note "This site may harm your computer" shows up under their homepage in Google search results as of July, 2012.

In contrast to *Zimbabwe News Online*'s homepage, the *Zimbabwe Independent's* homepage is much cleaner and more clearly designed to be user friendly. The *Zimbabwe Independent* is an online version of an independently owned weekly newspaper printed in Zimbabwe. Their masthead proclaims the paper to be the "Leading Business Weekly." The nine main news category tabs available to users run down the left hand side of the home page and include "Business," "Local," "International," and "Sport". All nine ad spots are filled, and the ads seem unobtrusive and are for seemingly reputable or known products and services. There are only two large pictures to go with the two headline news stories. Upon encountering this web page, Zimbabweans in the diaspora might be led to assign some authority to the site, as it appears professional, organized, and uncluttered, both logical and user-friendly in its design.

Believability

In addition to accuracy and authority, the third element that might be useful for Zimbabweans in the diaspora to evaluate the credibility of online news is believability. Newhagen and Nass (1989) made a good point when they suggested that when considering media credibility, it equates to the "degree to which an individual judges his or her perceptions to be a valid reflection of reality" (p. 278). In other words, people evaluate whether they find something believable or not based in part on whether they think it reflects reality. As Newhagen and Nass (1989) pointed out, people ask themselves if a news report is "a plausible reflection of the events they depict" (p. 278). So, asking the question "Does this news report seem like something that could happen?" seems to be a reasonable test for evaluating whether a report is believable.

Potter (2012) noted that we "encounter bits of information that either conform to our existing beliefs or challenge them" (p. 148) and that media is one source of information that serves to reaffirm beliefs that we have or sometimes to challenge those beliefs. Related to the believability of a news report is also the frame of that story. Considering how a story is told by an online news source, audiences recognize a theme or a meaning, and those themes or meanings shape the story for the audience (Potter, 2012). Part of the shape of the story will help audiences decide if they find the story believable or not.

Mavhunga (2009) argued that Zimbabweans, both at home and in the diaspora, are now able to use twenty-first century technology on their cell phones and computers to contest "versions of truth and falsehood" (p. 159). Further, Mavhunga (2009) claimed that "the state newspaper stable… specifically the *Herald* and *Sunday Mail,* were trusted with ensuring that only approved information was filtered to the public." (p. 161). The Media Monitoring Project of Zimbabwe (2002) claimed that the Zimbabwean public expects state media coverage to be "shallow and misleading." Along with the state media, private Zimbabwean media have also been indicted, for example in their failure to "present balanced information" (p. 17) during the coverage of the 2002 presidential election campaign. Given this context, it seems important that Zimbabweans in the diaspora have a tool to consider which news in believable.

The *Zimbabwe Broadcasting Corporation's* (*ZBC*) website provides a report for analysis around the concept of believability. The headline is "Govt to curb fertilizer looting" and the first paragraph of the article reads, "Government has introduced tight measures to curb the alleged looting of fertilisers by some few individuals at the expense of farmers who are struggling to access the scarce commodity" ("Govt to curb," 2012).

The frame of the report is that government, in the person of the Minister of Agriculture Made, blames the shortage of fertilizer in Zimbabwe on a few individuals looting the product at the expense of the majority of farmers. The minister is also reported to have made a claim often repeated by Zanu-PF government officials, that Zimbabwe's land reform program had been successful despite droughts and other "production challenges" ("Govt to curb," 2012). The report provided no details of who has been looting the fertilizer, no details on the amount of fertilizer stolen, no details on how the looting was going to be stopped, no details on how the fertilizer was now going to be distributed to deserving farmers, and no interrogation of Minister Made's claims about the success of the land reform program, nor details of causes of the production challenges face in the agricultural sector. In short, the report, as constructed, is not plausible and users in the diaspora could make that evaluation if they applied the believability criteria to this *ZBC* article.

In contrast, *ZBC* posted a news report headlined "Miners engage govt on fees" (2012), with a first paragraph that accurately framed the report: "Zimbabwe Miners Federation is engaging government on the newly gazetted mining fees with a stakeholder conference slated for next week to deliberate on the constraints facing the mining sector" (n.p.). The headline accurately reflected the article, evidence was given in the form of a quote from the head of the mining organization, the government's likely position was made clear, and concrete figures were provided on the government-proposed fee increase levels. A user of *ZBC's* website could evaluate this report as plausible given the structure, details and frame it was given.

The *Voice of America, Studio 7's (VOA)* website (an online and radio source run by Zimbabwean journalists based in Washington D.C.) ran a story headlined "Fresh Cholera Outbreak Worries Residents of Zimbabwe's Chiredzi Town" (Pepukai and Gumbo, 2012). The headline accurately reflected the article. The reporters provided details of the current outbreak, including the number of people sick with cholera in the town and the number of deaths from the illness. Additionally, they discussed water shortages as being the cause of the cholera outbreak, provided readers with a list of symptoms of the illness, and quoted a senior ministry of health official who specializes in epidemiology and also the member of parliament for the region that the town is in. Finally, the reporters included a brief history of a recent cholera outbreak in the country. A user of *VOA-Studio 7's* website could have some reassurance

about how believable the report was given its frame, details, and sources used by the reporters.

Quality of Message Construction

Metzger (2007) suggested that professional-quality writing that is clear is one of the elements of a news story that people use to evaluate the credibility of that story. Additionally, according to Dochterman and Stamp (2010a):

Content is composed of statements that refer to the specific information on the Web page that participants use to make credibility judgments, including language levels, typos, and direct quotes that add or detract from the credibility of the page. Many of these comments were directed toward the competency of the author, the Web designer, or the Web site housing the page. (p. 41)

People make attributions about the credibility of content based on the mistakes they notice in the news reports online. Mistakes in sentence construction, misused words, poorly edited sentences, and unclear organization; all might serve to lower judgments of credibility. These ideas mentioned above from Metzger (2007) and Dochterman and Stamp (2010a) have been labeled *quality of message construction* as an important element of the analytical framework proposed in this section. The label was chosen to distinguish these ideas from other parts of the framework that also consider elements of content like accuracy or believability. What follows is an application of the quality of message construction frame to various online sources available to Zimbabweans in the diaspora. There are three areas in which errors in writing quality can be illustrated: incorrect word usage, confusing or convoluted sentence structure, and poor editing.

Examples of incorrect word usage range from minor misuse of prepositions such as "But economists have also raised doubts on the inflation numbers…" (Sherekete, 2012), instead of "*about* the inflation numbers," to the more significant examples wherein the meaning is altered or unclear. From the *Financial Gazette*: "…likely to leave most villagers poorer and without drought power come the next farming season" ("Food shortages," 2012), wherein "drought" has been substituted for "draught," illustrating an ignorance of vocabulary on the part of the reporter that could potentially be confusing to the reader. A third example of incorrect word usage: "…making sure the figures remain within stated projections cannot be scuffed off" (Sherekete, 2012). The words "scuffed off" do not make sense and perhaps what the writer meant to say was "shrugged off" but a simpler word that would be clear to readers would have been "ignored."

Illustrations of the second category of errors in writing quality relate to poor sentence construction such as the meandering style of Charumbira's (2012) following sentence: "The Gentleman had a hefty budget of US$116 000, which is far ahead of the majority costs for the local productions nowadays" would be better expressed more succinctly. Another example of poor sentence construction can be seen in M. Dube's (2012) "Muzanenhamo said they were selling tables of 10, which cost US$1 500, an amount that translates to US$150 per head for the occasion that will see Oliver Mtukudzi providing entertainment for the night," which could be more effectively expressed in two separate sentences. Of more concern is when a reporter makes an important point that is lost in a poorly crafted sentence. For example Manyukwe's (2012) point about the political implications of voting patterns for the Zimbabwean population becomes lost in a sentence that starts with a conjunction and has an unclear last phrase: "And in the case that the combined votes of the two parties hand them a mandate to rule, is there a possibility that Zimbabweans would again be saddled with a coalition government that would expend its energies on quarreling as the current one between Zanu-PF and the MDC formations?"

A final set of examples illustrating the quality, or lack thereof, in message construction focuses on editing errors. Editing errors can range from misspellings such as Sherekete's (2012) "artficail" instead of "artificial" and grammatical errors such as Mazara's (2012) fragment use "Was convicted of unlawful possession of marijuana and sentenced to community service" to formatting issues as evidenced by Moyo's (2012) "far-reachingnegative ripple effects." In addition, more subtle errors of repetition such as Yikoniko's (2012) "…Roki and Maneta, are safe and will be around for some time…Roki and Maneta are safe for another 24 hours if not more…", and "The Supreme Court invited Advocate Thabani Mpofu as a friend of the court (amicus curiae) to assist the bench in arriving at a decision…Adv Mpofu, a friend of the court…" from Bulawayo 24 News ("Supreme Court," 2012) point to poor editing of an article.

The examples of poor message construction quality from various online news reports cited above provided evidence of places where more rigorous editing should have been implemented. The examples came from different online sites available to the Zimbabwean diaspora and were stories selected at random then examined for errors. The question is how much weight should users place on quality of construction factors like these? Alone, and in isolated instances, perhaps not too much weight; however, errors in language, sentence construction and editing that abound on news websites, that perhaps speak to the qualifications of writers and quality of editorial oversight, should become cause for greater concern for the audience. Why? If message construction quality errors slip by uncorrected, errors of fact might be slipping by too. The accumulation of message construction errors can be noticed by users and criticisms of those lapses in standards passed along from user to user; this is one example of where the idea of peer review becomes important and this notion is discussed in the next section of the chapter.

Promotion of Reviewed Resources

As Metzger, Flanagin and Medders (2010) noted:

Paradoxically, then, although digital media and information abundance may complicate people's confidence in and knowledge of who is an authority, electronic networks and social computing applications make it easier for individuals to harness collective intelligence to help them assess and evaluate information and sources online. (p. 415)

Communities of Zimbabweans in the diaspora are connected to each other through both interpersonal and mediated communication. That connection can be seen, for example, in the organizations that diaspora populations have organized and belong to. As Kanu (2010) reported, a "number of Diaspora organizations has grown over the years" and she suggested that diaspora populations are now "returning home" using communication technology to keep in touch with friends and family in Zimbabwe. Bloch's (2005) research supports Kanu's contentions finding that Zimbabweans in the diaspora reported being connected socially through clubs and other organizations and that the vast majority of them use mediated communication to keep in touch with relatives at home in Zimbabwe.

With these connections in mind, the idea of those Zimbabwean networks promoting online news web sites that they have found to be credible seems appealing. This peer review of online news sites could be in-person given the socially networked diaspora communities around the world. Alternately, this feedback and informal peer review of sites can take place on the site itself, when comment facilities are part of the site's feedback loop for users. This review could also be removed from the online news site, taking place

in other Internet forums like social media sites such as Facebook or Sha (one of the Zimbabwean equivalents to Facebook.) Sha, which bills itself as "Zimbabwe's Social Network," in addition to having personal pages, has a "News" tab that link users to online Zimbabwean news reports from sources like the independently owned *Daily News*, as well as the government controlled *The Sunday Mail*. There are also discussion forum tabs for people to discuss the issues and the media that are covering these issues. ("Sha," "2012)

Metzger, Flanagin and Medders (2010) argued that the "collective approach to credibility assessment also reveals concerns about source identity and authenticity, as well as the importance of information domains and context" (p. 434). It seems that achieving this sort of depth of assessment would be useful in the context of the sometimes politically divided Zimbabwean diaspora that was discussed earlier in the chapter. Zimbabweans would be somewhat informed about the different sources, know their reputations, their political leanings, and can contribute to discussions about how much credibility an online news site might have, given those more nuanced considerations.

Zimbabweans in the diaspora would know that the perspectives represented on government-controlled online media news sites like *The Sunday Mail, ZBC,* or *The Herald Online* would tend to be more highly critical of the MDC wing of government and more supportive of the Zanu-PF wing of the government in their news coverage. Similarly, independent online news sites like the *Zimbabwe Independent,* the *Daily News,* and *SW Radio Africa* would tend to be more supportive of the MDC and highly critical of Zanu-PF. The power of promoting this credible sources technique is that users educate themselves and each other, thereby avoiding sources that the collective(s) with which they identify deem not credible or at least not consistently credible.

One example of this "peer review" of content can be seen in the following reader comment on a report in the *Zimbabwe Independent*:

Tsvangirai have been stumbling around for the last three years as prime minister of Zimbabwe, why not ask him for his views on: constitutionalism, health, education, agriculture, indus trial development, foreign policy and his role as Prime Minister of the country. Sit him down and ask the above questions. Then you can sit Mnangagwa down for the same interview and we all would know the difference between these two men and I am sure, in fact I am certain Mnangagwa will come out on top at the end of this process. Stop playing games with the lives of the people. (Alston Sandiford, May 11, 2012)

The comment was in response to an article headlined "I'm ready to rule, says Mnangagwa" (Mambo & Chitemba, 2012). The focus of the report was on how a prominent Zanu-PF politician had confirmed speculation that if asked to run as Zanu-PF's presidential candidate, he would. The article also explored the factionalism within the political party as people start to envision a world in which President Mugabe is no longer the leader of the party or their presidential candidate. What is interesting in the reader's comments is that they provide a critique of the journalism and writing of the article, suggesting that the reporters should focus on the policy differences between Mnangagwa and the MDC Prime Minister Tsvangirai. Implied in the reader critique is that the journalistic practice that went into the report is lacking.

Another concrete example of this peer review of a source can be seen in user comments posted in response to *The Herald Online's* (February 8, 2012) report on the end of the inquest into General Mujuru's death. The following are a sample of user comments that are critical to *The Herald Online's* approach to the article (Note: The comments are reproduced as written, with errors in spelling and grammar not being corrected.):

The way u report herald am sorry u are one sided… (Kwiti, February 8, 2012)

If I were a journalist, i would never ever have allowed someone to use me to this extend just for a meager salary, that can not even buy me underwear…wat are uu hoping to achieve by this silly article… (Murambiwa, February 8, 2012)

Can someone tell me why it sounds like the Herald does not want people to know the truth and they are happy about it. (Cde tambawoga, Feb 8, 2012)

You wonder why this paper wants the inquest to die. And as usual they quote unnamed lawyers. (Mutengezi, February 8, 2012)

Editor, please take our comments into account, we are also experts!! lol (Gudo, February 8, 2012)

…don't you find it strange that these so-called 'legal experts' are neither named or quoted? Of course you don't! It is what a Herald reporter does when his handlers tell him what to write. (Mizo, February 8, 2012)

The comments posted by readers in response to reports (as the examples above) provide other users some perspectives on the article. These perspectives might confirm a user's opinions, but might also point out perspectives on the source that a user had not previously considered. The comments indicate that users think *The Herald Online* has a bias, has framed the report in a specific way that is not balanced, and that the report fits an unfortunate pattern of journalistic practice (using vague, unnamed or un-credentialed sources.) On their own, in response to one article, a user might not pay such peer comments much attention, but an accumulation of such comments on multiple articles on one source over a period of time would provide a user in the diaspora a solid set of peer reviews that would help make evaluations of credibility. Another contextual tool that users might use is comparison, and that concept is the focus of the next section in the chapter.

Comparison

Applying this contextual element of the proposed framework for credibility analysis rests on comparative thinking. Meola (2004) suggested that comparison can serve multiple purposes, including to "…reveal the depth of information available on a topic…reveal specific areas of a topic that are controversial and that need special attention and verification… recognize bias" (pp. 340-341). Users take an online news source report that focuses on a specific topic or event, and then find other sources that are covering the same topic and make comparisons between elements of the story covered. Once users have, over time, located credible online sources, they will have developed expectations about what standards sources need to meet in order to be judged credible by comparison. In other words, over time, users will now know what they find credible when they see it based on previous example. Using one important story that would be of interest to Zimbabweans in the diaspora, the next section of the chapter will illustrate the utility of comparison.

Three reports from three different online sources about the inquest into General Mujuru's death, which was discussed earlier in this chapter, will serve as a point of comparison. A user in the diaspora interested in finding out information about how the inquest ended would have many online news sources to turn to; one of those sources may be *Bulawayo24 News,* a source specializing in news that is important to residents of Zimbabwe's second city, Bulawayo, but that also focuses on news of importance to Zimbabweans at large. They ran an *Associated Press (AP)* report on the end of the inquest. The *AP* report contained information on all the following elements of the story: the magistrate's rejection of the Mujuru family's request to have the body exhumed for an independent autopsy, brief context for who Mujuru was, and details of expert testimony and witness testimony from the inquest ("Mujuru's death," 2012).

ZimOnline, billed as "Zimbabwe's Independent News Agency," also published a report at the end of the inquest. Written by "Own Correspondent," the report, entitled "Mujuru won't be exhumed" (2012), provided readers with the facts of the story, including the magistrate's rejection of a request by the family to exhume the body, the magistrate's reasoning, the Mujuru family's expert pathologist's perspectives on the quality of the autopsy that had been performed based on testimony he had heard, statements from the Mujuru family lawyer about the steps the family were considering, explanations of the law around inquest proceedings and verdicts, and finally, a comprehensive summary of General Mujuru's life and role in Zimbabwe's military and political affairs.

The Herald Online published a report headlined "Inquest findings final, say experts" (Nemukuyu, 2012). The gist of the article is that the magistrate and the Attorney General had control over the inquest proceedings and findings, and that the Mujuru family did not have the standing to make requests about exhuming the general's body for further forensic examination. The reporter made these claims based on "experts," none of which was named in the report.

Applying Meola's (2004) comparative elements, depth of information, covering controversial parts of a report, and showing a bias, *ZimOnline's* article can be evaluated as the most credible. There is a depth to the article that the other two articles lack, there is a balanced discussion of the controversy about who should be allowed, under the law, to make decisions about exhumation, and the article is not framed to benefit any particular party. The *Bulawayo24 News* report taken from *AP* can be evaluated as being credible, but lacking in the depth that *ZimOnline's* report had, not surprising since it is a newswire article that takes a certain summative format in its reporting. No obvious bias is shown in *Bulawayo24's* posted report. *The Herald Online* can be evaluated as least credible based on a lack of depth in the details, focusing only on one side of the controversy surrounding exhumation, and taking what can be viewed as a biased stance on the issue by intimating that the decision is final. In addition, *The Herald Online's* expert testimony was unnamed, thus providing no way for readers to make evaluations of their expertise or bias.

From these three examples it should be clear how the element of comparison could be useful. Zimbabweans in the diaspora who systematically compared sources over a period of time would then come to see which sources consistently met their tests of credibility and could then continue to seek out those sources, and not seek out their information from other online sources that failed to meet up to their standards. In addition to comparison, corroboration is a third contextual tool that users in a diaspora might consider as an aid to evaluating the credibility of online news sources.

Corroboration

Mukundu (2005) pointed out that free "expression and media rights have been seriously impinged upon through legislation" (p. 4) in Zimbabwe. Given these restrictions, Zimbabweans in the diaspora might have concerns about how credible a piece of online news is, and might seek to have that news corroborated as a test of credibility. It is important that users do not confuse comparison with corroboration: "To corroborate information is to verify it against one or more different sources" (Meola, 2004, p. 341). This piece of the proposed framework for evaluating credibility of online news sources takes some of Meola's (2004) ideas of corroboration, and marries them to Dochterman and Stamp's (2010a) findings of "cross-checkability" as important in verifying information: "Cross-checkability deals with…whether other sites had posted this information, and with how easy it would be to find and check either the direct source, or the other sites, posting this information from the page being considered…" (p. 40). When users seeks to

corroborate information in testing its credibility, they are looking for other sources that report the same information.

One tragic incident from Zimbabwe in 2009 will be used to illustrate this contextual technique. Prime Minister Tsvangirai's wife died in a car accident. This happened at a time when there were simmering tensions between the MDC and Zanu-PF as they tried to make the government of national unity (GNU) function. Immediately upon the breaking news of her death, rumors began to swirl around the country and the diaspora about the cause of her death, with people questioning how and why Mrs. Tsvangirai had died. Conspiracy theories abounded about her being murdered and about the Prime Minister being the real target of the supposed assassins. It would be crucially important at times like these for Zimbabweans in the diaspora to seek multiple sources to cross-check or corroborate any information on the story they found online.

Zimbabweans might have first come across the story in the blogosphere or through social networking sites. From there, they should go to a site they trust, for example those Zimbabweans in the USA might consult the *Voice of America* (*VOA*) site that has *Studio 7* devoted to Zimbabwean shortwave and online programming. A *Studio 7* article (Rusere & Nkomo, 2009) reported her death in an accident, provided the limited information about the crash that was available, and hinted that some had suspicions about the cause of the crash being investigated given the tensions that existed in the GNU at the time between the MDC and Zanu-PF wings of government.

In order to corroborate the news of her death, a user could have gone to *SW Radio Africa's* site and would have found a report that did indeed confirm her death, but also would have increased the speculation that her death might not have been an accident because it included information about someone being arrested for taking photos of the crash scene (Jackson, 2009). For further confirmation of the news, *BBC News* online reported her death, provided a picture of the crash site, details of the place on the road where the crash occurred, provided quotes from numerous sources, and confirmed that Harare was abuzz with rumors surrounding her death (Hungwe, 2009).

From this rather in-depth report, and the reports on *SW Radio Africa* and *VOA,* a Zimbabwean in the diaspora would have been certain that she was dead, that an investigation was ongoing, and that there was still speculation of foul play swirling in the country. (Her death was later shown to be an accident.) Using the corroboration technique, users can be more certain about the credibility of the information they gather online than if they do not use that technique.

DISCUSSION

One imperative that drives the need for the Zimbabwean diaspora to have a relatively easy-to-use framework to analyze the credibility of online news sources about their country is that Zimbabwean journalists, like journalists in many African countries, work under difficult conditions (Schriffin, 2009). That many journalists are not well paid and "poorly trained and working under both political and commercial pressure" (Schriffin, 2009, p. 128) has negatively impacted journalists' performance. Given this reality, people in the diaspora need a tool to be able to distinguish reporting that is credible from journalistic work that is not. From the examples given in this chapter in application of the framework, it should be clear that there is utility in the different elements of that framework. Whether using ideas that belong to the checklist approach or to the contextual approach to evaluating credibility of online news sources, the elements of the framework can serve as a tool for the audience to apply.

There are, however, some considerations that need to be explored when thinking about the proposed analytical frame. One consideration

is to recognize that the framework proposes a way for the audience to assess credibility of online news sources from a normative perspective. Christians et al. (2009) "define normative theory of public communication as the reasoned explanation of how public discourse should be carried out in order for a community or nation to work out solutions to problems" (p. 65). The framework in this chapter is not a theory, but has a series of theoretical constructs that are applied to examples of news that form part of a public discourse on issues important to Zimbabweans through a normative lens. The framework in this chapter suggests what *could* work, what *should* be applied, and how the framework *might* work. These ideas may or may not *actually* be how people in the diaspora are assessing credibility online or how they perceive credibility of news from home. Different users might find different elements of the framework more or less important in their evaluation of credibility. These differences may be based on their individual needs, backgroundsm and life experiences.

In their discussion of normative theory, Christians et al. (2009) suggested fundamental issues, one of which is the notion of "free and equal access to open public debate" (p. 72). It is important to note that in the Zimbabwean context, there is discussion about how free and open access to public debate really is, and many believe it has not yet been achieved in the country. As discussed earlier in the chapter when focusing on the news sources that covered media freedom day, there are polarized views on the extent to which freedom of expression should prevail in Zimbabwe. Some view absolute freedom of expression as a challenge to national sovereignty, while others call for more freedoms than the current legal regime allows. Given this tension, it makes sense that people in the diaspora should view with some skepticism the news reports that are published online, and use the proposed framework to help them assess the credibility of the reports they consume.

A second fundamental issue identified by Christians et al. (2009) that is germane to this discussion is that "what defines public truthfulness is a central problem because cultural reality is constructed and truth is always to some extent a construction" (p. 72). Kovach and Rosenstiel (2001) proposed that, from a journalistic perspective, "journalists must tell the truth. Yet people are befuddled about what 'the truth' means" (p. 37). If journalists are befuddled, then where does that leave the audience? Accuracy was one element of the framework proposed clearly in the chapter, and it is dependent upon an audience member's biases, their perceptions of accuracy, and how their perceptions of what is true differ.

Having a quality dialogue, with people in a diaspora taking part in this dialogue primarily through different forms of media, is important when thinking about public discourse through a normative lens (Christians et al., 2009). Considering elements like the quality of message construction, authority, and believability from the framework, it is clear how important it is that Zimbabweans in the diaspora are provided online sources that contribute to the quality of this dialogue. Audience members need to know, for example, what gives a reporter authority to report on a specific topic, need to be conscious of cues in a story that suggest poor message construction, and need to seek out sources that they can believe if they want to be exposed to and participate in a quality dialogue about issues in Zimbabwe.

When sharp polarizations exist in a country, then the media tend to reflect that polarization and often "media content tends to be more politically charged than dispassionate" (Christians et al., 2009, p. 98). This idea is also an important one to consider in applying the framework. As discussed in the chapter, there are sharp political divisions in sections of the Zimbabwean population, and, as seen in some of the examples examined earlier, the media reflect those divisions and biases. Being aware of these divisions, then, it seems audience members would be well served in applying ele-

ments of the framework, especially corroboration and comparison, to expose those biases.

Another of the elements of the framework that might prove useful in a polarized media setting is to seek peer-reviewed sources. As Kovach and Rosenstiel (2001) noted, journalists should know the difference between the journalism of opinion and the journalism of reporting and that it is important that reporters maintain those distinctions. It seems clear that the audience should also be aware of those distinctions and be on the lookout for any blurring of those lines. As seen earlier in the chapter, the comments left by readers in response to *The Herald Online's* report about the inquest into General Mujuru's death clearly suggested that they perceived a bias and that the article was really not reporting as much as it was a reiteration of one political party's opinion of the issue at hand. As Christians et al. (2009) said, "there are many competing interests and warring parties in society but…the news media do not have to take sides or have any vested interests of their own" (p. 148). McGregor and Pasura (2010) have noted a shift in some Zimbabweans in the diaspora, people who are "keen to move away from divisive party politics, and to come up with new, explicitly non-partisan frameworks" (pp. 688-689). People in the diaspora who feel that they want to move away from being partisan would be more likely to consider partisan and unbalanced news reports as less credible.

An important consideration for anyone seeking to use elements of the framework is the notion of time. There is no substitution for thoroughness in applying the proposed framework for the scrutiny of sources, and that thoroughness is only achievable when enough time is allocated to the task. Given the way that people actually search for news online, it is quite probable that consumers will not take the time to apply all the elements of the proposed framework to the news they find about Zimbabwe. Fogg et al. (2003, as cited in Metzger, Flanagin, & Medders, 2010) have found that users often do not spend a lot of time on one website and that they tend to develop strategies for make assessments of credibility that are quick. Metzger (2007) supported this finding that people tend to use strategies for evaluating credibility that take the least amount of time and mental energy.

Those who might not have time to apply the framework elements to all the online news they consume about home might want to consider applying some of the elements for *specific kinds of news stories*. Schiffrin (2009) suggested that in many African countries, reporters do not receive specialized training necessary to cover specific beats. For example, a reporter covering the economy of Zimbabwe might not have had any training or background in economics, and therefore their reporting on complex financial transactions, complex economic issues, and fiscal policy will be lacking. Readers might therefore want to pay more attention to cues like authority and accuracy when reading specialized news on specialized topics. "The brain drain from journalism is high" (Schiffrin, 2009, p. 132) on the continent, and Zimbabwe is no exception to this pattern. In a study of three African countries' press coverage of the extractive sector (like oil exploration) Schiffrin (2009) found that sector specific coverage was often "short and superficial" (p. 138) and reporters had relied largely on official press releases, had only one or two sources, and did not explain complex material to the reader.

A further idea to consider when thinking about applying the framework is the notion that a "diasporic understanding of homeland signifies not a place of return but a source of shifting and ambivalent attachment" (Ghorashi & Boersma, 2009, p. 669). Zimbabweans in the diaspora will have different and changing needs and desires about what news and information they seek about home. These changing needs and desires will likely influence their levels of motivation for seeking credible information and help determine, in part, whether they are willing to put the time and effort that it would take to apply the elements of the framework to their online news searches.

On the other hand, Bernal's (2006) work on the Eritrean diaspora and new media presents an interesting idea: "emotional citizenship" (p.164). As Bernal (2006) explained:

...Eritreans in diaspora in North America and Europe are for the most part legally citizens of the countries where they reside and earn their living, but they are emotionally pushed and pulled into Eritrea's national politics. Their sense of themselves is intertwined with Eritrea's destiny as a nation even if they have no intention of living there (p. 164).

Considering this point, it can be argued that Zimbabweans in the diaspora might feel a similar emotional citizenship and be driven to seek out information that is credible about Zimbabwe so that they can take part in national politics because their sense of self is intertwined with the country and what goes on there. That being the case, there is an imperative to seek out credible information from multiple sources, and therefore the framework may have applicability and utility.

FURTHER APPLICATION

While the examples used in the chapter were all from news stories and online sources that would appeal to Zimbabweans in the diaspora, the framework might also be useful for other diaspora populations. The specific circumstances that propel different groups to leave their homes and form collectives in other countries might be different, but there is a common pull to be informed about news from home that allows a connection to culture, politics, economics, the arts, sport, music, etc., to be maintained. In addition, as was discussed earlier in the chapter, people in a diaspora are negotiating their identities, balancing their home and host cultures, and part of that balancing requires some connection to home, often maintained in part through online media.

Given these similar needs, it seems logical that the proposed framework would work, say, for people in the Nigerian diaspora who want to keep up with the latest credible information to be found about developments in Nollywood movie distribution channels, or for Eritreans who want to be informed through credible sources about the latest developments in their country's rocky relationship with Ethiopia, or for Haitians who are seeking credible online sources of information about the rebuilding of Port-au-Prince.

People in a diaspora maintain ties to their homelands and have often moved away from home in waves, so they may have different generational pools of national and cultural knowledge about home. They are thus quite well placed to make evaluations about information about home. For example, a Haitian living in the diaspora would know something about the history of politics in Haiti, who has held power, what a politician's governing history might be, and that knowledge can help that person in evaluating political news they find online. They can use elements of the framework proposed in this chapter to detect bias, to make judgments about the depth of political stories, to seek corroboration from multiple sources, and they can use their connections in the diaspora to obtain peer feedback on the sources they consult.

Asking questions about some or all of these elements of an online news report would seem a useful tool for motivated individuals in a diaspora seeking to establish some gauge of credibility of an online news source or report:

- Accuracy
- Authority
- Believability
- Quality of Message Construction
- Peer Review
- Comparison
- Corroboration

CONCLUSION

This chapter has contributed to the digital media literacy domain by suggesting a theoretical framework that users of online news in a diaspora can apply to the information they gather online about their home countries. While the vehicle for analysis of this framework was online news sites available to the population in the Zimbabwean diaspora, an argument was made that the framework could prove useful for other diaspora populations trying to gauge the credibility of online news they read about their home countries.

As of 2011, only about 12% of the Zimbabwean population living inside Zimbabwe had Internet access ("Zimbabwe," 2012). It is not unreasonable, then, to surmise that Zimbabwean media outlets such as those examples discussed in this chapter provide online news content that they know will be consumed by Zimbabweans in the diaspora as a large portion of their online audience. As the *Zimbabwe Independent*'s news editor Dumisani Muleya stated, "We are targeting people in the diaspora; there are a lot of Zimbabweans in the UK, United States, Australia, New Zealand and Canada" (qtd. in Mukundu, p. 51).

Cassidy's (2007) research suggested that journalists' ratings of online news sources as being credible are increasing, indicating an acknowledgment both that there are concerns about online news credibility in the profession, but also that perceptions of credibility are improving. If journalists have these concerns, then it makes sense that an online news audience like those in a diaspora should be encouraged to start making their own evaluations about the credibility of the news they consume on the Internet, and perhaps using frameworks like the one proposed in this chapter will provide audiences tools to make those evaluations of credibility for themselves.

A user of online news from a diaspora is motivated to stay connected to home. That motivation should encourage people in a diaspora to seek and find online news sources that are credible and likely to provide useful information. Applying the concepts of the theoretical framework for assessing credibility proposed in this chapter is one way to approach this task.

REFERENCES

Allan, S., & Thorsen, E. (2011). Journalism, public service and BBC News online. In Meikle, G., & Redden, G. (Eds.), *News online: Transformations and continuities* (pp. 20–37). UK: Palgrave MacMillan.

Bernal, V. (2006). Diaspora, cyberspace and political imagination: The Eritrean diaspora online. *Global Networks*, *6*(2), 161–179. doi:10.1111/j.1471-0374.2006.00139.x

Bloch, A. (2005). The development potential of Zimbabweans in the diaspora: A survey of Zimbabweans living in the UK and South Africa. *IOM Migration Research Series, 17*.

Brubaker, R. (2005). The "diaspora" diaspora. *Ethnic and Racial Studies*, *28*(1), 1–19. doi:10.1080/0141987042000289997

Cassidy, W. P. (2007). Online news credibility: An examination of the perceptions of newspaper journalists. *Journal of Computer-Mediated Communication*, *7*, 478–498. doi:10.1111/j.1083-6101.2007.00334.x

Charumbira, S. (2012, May 12). US-based Chimhina fundraises for arts. *The Standard*. Retrieved from http://www.thestandard.co.zw

Chiagouris, L., Long, M. M., & Plank, R. E. (2008). The consumption of online news: The relationship of attitudes toward the site and credibility. *Journal of Internet Commerce*, *7*(4), 528–549. doi:10.1080/15332860802507396

Christians, C. G., Glasser, T. L., McQuail, D., Nordenstreng, K., & White, R. A. (2009). *Normative theories of the media: Journalism in democratic societies*. Urbana, IL: University of Illinois Press.

Clifford, J. (1994). Diasporas. *Cultural Anthropology*, *9*(3), 302–338. doi:10.1525/can.1994.9.3.02a00040

Dochterman, M. A., & Stamp, G. H. (2010a). Part one: The determination of web credibility: A thematic analysis of web user's judgments. *Qualitative Research Reports in Communication*, *11*(1), 37–43. doi:10.1080/17459430903514791

Dochterman, M. A., & Stamp, G. H. (2010b). Part two: The determination of web credibility: A theoretical model derived from qualitative data. *Qualitative Research Reports in Communication*, *11*(1), 44–50. doi:10.1080/17459430903514809

Dube, M. (2012, May 18). Where will the stars party? *The Sunday Mail*. Retrieved from http://www.sundaymail.co.zw

Food shortages hit Mat South hardest. (2012, May 16). *The Financial Gazette*. Retrieved from http://allafrica.com/stories/201205200073.html

Fritch, J. W. (2003). Heuristics, tools, and systems for evaluating Internet information: Helping users assess a tangled Web. *Online Information Review*, *27*(5), 321–327. doi:10.1108/14684520310502270

Ghorashi, H., & Boersma, K. (2009). The "Iranian diaspora" and the new media: From political action to humanitarian help. *Development and Change*, *40*(4), 667–691. doi:10.1111/j.1467-7660.2009.01567.x

Goggin, G. (2011). The intimate turn of mobile news. In Meikle, G., & Redden, G. (Eds.), *News online: Transformations and continuities* (pp. 99–114). UK: Palgrave MacMillan.

Govt respects media pluralism: Shamu. (2012, May 3). *Zimbabwe Broadcasting Corporation*. Retrieved from http://eu.zbc.co.zw/news-categories/top-stories/19055-govt-respects-media-pluralism-shamu.html

Govt to curb fertiliser looting. (2012, February 13). *Zimbabwe Broadcasting Corporation.* Retrieved from http://eu.zbc.co.zw/news-categories/top-stories/16461-gvt-to-curb-fertiliser-looting.html

Hungwe, B. (2009, March 7). Rumours fly after Tsvangirai crash. *BBC News.* Retrieved from http://news.bbc.co.uk/2/hi/africa/7930694.stm

Jackson, G. (2009, March 7). Photographer arrested at crash scene. *SW Radio Africa.* Retrieved from http://www.swradioafrica.com

Jackson, G. (2011, March 7). SW Radio Africa statement on release of CIO names and details. *SW Radio Africa.* Retrieved from http://www.swradioafrica.com/pages/ciostatement010711.htm

Johnson, K., & Wiedenbeck, S. (2009). Enhancing perceived cedibility of citizen journalism web sites. *Journalism & Mass Communication Quarterly, 86*(2), 342–348. doi:10.1177/107769900908600205

Johnson, T., & Kaye, B. (2010). Choosing is believing? How web gratifications and reliance affect internet credibility among politically interested users. *Atlantic Journal of Communication, 18,* 1–22. doi:10.1080/15456870903340431

Kanu, J. M. (2010, November 25). Zimbabwe: Diaspora-untapped growth zone. *The Herald.* Retrieved from http://allafrica.com/stories/201011250083.html

Kolodzy, K. (2006). *Convergence journalism: Writing and reporting across the news media.* New York, NY: Rowman and Littlefield.

Kovach, B., & Rosenstiel, T. (2001). *The elements of journalism: What newspeople should know and the public expect.* New York, NY: Crown Publishers.

Malleus, R. (2011). Whose TV is it anyway? An examination of the shift towards satellite television in Zimbabwe. In Wachanga, N. D. (Ed.), *Cultural and new communication technologies: Political, ethnic and ideological implications* (pp. 128–143). Hershey, PA: IGI Global. doi:10.4018/978-1-60960-591-9.ch007

Mambo, E., & Chitemba, B. (2012, May 11). I'm ready to rule, says Mnangagwa. *Zimbabwe Independent,* May 11. Retrieved from http://www.theindependent.co.zw/political-zimbabwe-stories/2012/05/11/im-ready-to-rule-says-mnangagwa/

Manyukwe, C. (2012, May 16). Poll alliance faces hurdles. *The Financial Gazette.* Retrieved from http://www.financialgazette.co.zw

Mavhunga, C. (2009). The glass fortress: Zimbabwe's cyber-guerilla warfare. *Journal of International Affairs, 62*(2), 159–173.

Mazara, G. (2012, May 6). Drumbeat: Can Roki shed his bad-boy image in BAA. *The Standard.* Retrieved from http://www.thestandard.co.zw

McGregor, J. (2009). Associational links with home amoung Zimbabweans in the UK: Reflections on long-distance nationalisms. *Global Networks, 9*(2), 185–208. doi:10.1111/j.1471-0374.2009.00250.x

McGregor, J., & Pasura, D. (2010). Diasporic repositioning and the politics of re-engagement: Developmentalising Zimbabwe's diaspora? *The Round Table, 99*(411), 687–703. doi:10.1080/00358533.2010.530413

Media Monitoring Project Zimbabwe. (2002, July 23). Weekly media update. Retrieved from http://www.mmpz.org.zw

Mehrabi, D., Hassan, M. A., & Ali, M. S. S. (2009). News media credibility of the internet and television. *European Journal of Soil Science, 11*(1), 136–148.

Meola, M. (2004). Chucking the checklist: A contextual approach to teaching undergraduates website evaluation. *Libraries and the Academy, 4*(3), 331–344. doi:10.1353/pla.2004.0055

Metzger, M. (2007). Making sense of credibility on the web: Models for evaluating online information and recommendations for future research. *Journal of the American Society for Information Science and Technology, 58*(13), 2078–2091. doi:10.1002/asi.20672

Metzger, M. J., Flanagin, A. J., & Medders, R. B. (2010). Social and heuristic approaches to credibility evaluation online. *The Journal of Communication, 60*, 413–439. doi:10.1111/j.1460-2466.2010.01488.x

Miners engage govt on fees. (2012, February 9). Zimbabwe Broadcasting Corporation. Retrieved from http://www.zbc.co.zw/news-categories/business/16363-miners-engage-govt-on-fees.html

Moyo, H. (2012, May 17). Policy failures render Zim a basket case. *Zimbabwe Independent*. Retrieved from http://www.theindependent.co.zw

Mujuru won't be exhumed. (2012, February 7). *ZimOnline*. Retrieved March 20, 2012, from http://www.zimonline.co.za

Mujuru's death shrouded in suspicion. (2012, February 6). *Bulawayo24 News*. Retrieved from http://bulawayo24.com/index-id-news-sc-national-byo-11859-article-Mujuru%27s+death+shrouded+in+suspicion+.html

Mukundu, R. (2005). *Research findings and conclusions. African Media Development Initiative*. Zimbabwe: BBC World Service Trust.

Nemukuyu, D. (2012, February 8). Inquest findings final, say experts. *The Herald Online*. Retrieved from http://allafrica.com/stories/201202080026.html

New constitution's principal drafters must be fired. (2012, February 13). *The Herald Online* Retrieved from http://www.herald.co.zw/index.php?option=com_content&view=article&id=33817

Newhagen, J. E., & Nass, C. (1989). Differential criteria for evaluating credibility of newspapers and TV news. *The Journalism Quarterly, 66*(2), 277–284. doi:10.1177/107769908906600202

Nguyen, A. (2011). Marrying the professional to the amateur: strategies and implications of the Ohmy News model. In Meikle, G., & Redden, G. (Eds.), *News online: Transformations and continuities* (pp. 195–209). London, UK: Palgrave MacMillan.

Pepukai, O., & Gumbo, T. (2012, May 17). Fresh cholera outbreak worries residents of Zimbabwe's Chiredzi Town. *Voice of America-Studio 7*. Retrieved from http://www.voanews.com/zimbabwe/news/Health-Officials-Confirm-Cholera-Outbreak-In-Chiredzi-Amin-Water-Shortages-151928535.html

Ploch, L. (2010). Zimbabwe background. *Congressional Research Report for Congress*, 7-5700.

Potter, W. J. (2012). *Media effects*. Los Angeles, CA: Sage.

Robins, D., Holmes, J., & Stansbury, M. (2010). Consumer health information on the web: The relationship of visual design and perceptions of credibility. *Journal of the American Society for Information Science and Technology, 61*(1), 13–29. doi:10.1002/asi.21224

Rottenberg, A. T. (2003). *The structure of argument*. New York, NY: Bedford-St Martins.

Rusere, P., & Nkomo, N. (2009, March 6). Zimbabwe mourns death of Susan Tsvangirai, PM's wife, in highway crash. *VOA News Studio 7*. Retrieved from http://www.voanews.com

Schiffrin, A. (2009). Power and pressure; African media and the extractive sector. *Journal of International Affairs*, *62*(2), 127–141.

Sha: The Zimbabwe Social Network. (2012). Retrieved from http://www.sha.co.zw

Shamu threatens media. (2012, May 3). *Zimbabwe Independent.* Retrieved from http://www.theindependent.co.zw/political-zimbabwe-stories/2012/05/03/shamu-threatens-media/

Sherekete, R. (2012, May 17). Zimbabwe: Official inflation figures 'not a true reflection'. *Zimbabwe Independent.* Retrieved from http://allafrica.com/stories/201205181095.html

Supreme Court to decide on Zimdollar labour awards. (May 17, 2012). *Bulawayo24 News.* Retrieved from http://www.bulawayo24.com

Walthen, C. N., & Burkell, J. (2002). Believe it or not: Factors influencing credibility on the Web. *Journal of the American Society for Information Science and Technology*, *53*(2), 134–144. doi:10.1002/asi.10016

Yikoniko, S. (2012, May 18). Evicted Teclar shies away from media. *The Sunday Mail.* Retrieved from http://www.sundaymail.co.zw

Zimbabwe. (2012). *Internet world statistics.* Retrieved from http://www.internetworldstats.com/af/zw.htm

Zimbabwe News Online. (2012). Retrieved from http://www.zimbabwenewsonline.com

ADDITIONAL READING

Banning, S. A., & Sweetster, K. D. (2007). How much do they think it affects them and who do they believe? Comparing the third-person effect and credibility of blogs and traditional media. *Communication Quarterly*, *55*, 451–466. doi:10.1080/01463370701665114

Bowman, S., & Willis, C. (2003). *We media: How audiences are shaping the future of news and information.* Reston, VA: The Media Center at the American Press Institute.

Flanigan, A. J., & Metzger, M. J. (2000). Perceptions of Internet information credibility. *Journalism & Mass Communication Quarterly*, *77*, 515–540. doi:10.1177/107769900007700304

Greer, J. D. (2003). Evaluating the credibility of online information: A test of source and advertising influence. *Mass Communication & Society*, *6*, 11–28. doi:10.1207/S15327825MCS0601_3

Hilligoss, B., & Rieh, S. Y. (2008). Developing a unifying framework of credibility assessment: Construct, heuristics, and interaction in context. *Information Processing & Management*, *44*, 1467–1484. doi:10.1016/j.ipm.2007.10.001

Lazar, J., Meiselwitz, G., & Feng, J. (2007). Understanding Web credibility: A review of the research literature. *Foundations and Trends in Human-Computer Interaction*, *1*(2), 139–202. doi:10.1561/1100000007

Lin, C., Salwen, M. B., & Abdulla, R. A. (2005). Uses and gratifications of online and offline news: New wine in an old bottle? In Salwen, M. B., Garrison, B., & Driscoll, P. D. (Eds.), *Online news and the public* (pp. 221–236). Mahwah, NJ: Erlbaum.

Liu, Z. (2004). Perceptions of credibility of scholarly information on the web. *Information Processing & Management*, *4*, 1027–1038. doi:10.1016/S0306-4573(03)00064-5

Metzger, M. J., Flanagin, A. J., & Zwarun, L. (2003). Student Internet use, perceptions of information credibility, and verification behavior. *Computers & Education*, *41*, 271–290. doi:10.1016/S0360-1315(03)00049-6

Rogers, D. (2010, April 15). Zimbabwe's accidental triumph. *The New York Times*, A7.

KEY TERMS AND DEFINITIONS

Credibility: A multidimensional construct tied to the perception of how accurate, knowledgeable and believable a person thinks a message is.

Diaspora: Populations of people who have left their home countries, settled in other countries, have collectives in those new countries and who maintain ties and identities with home.

Framework: A theoretical tool that has a practical use in analysis when applied as intended.

Harare: The capital city of Zimbabwe.

Hyper-Link: A tool that is often embedded in text on the Internet that allows a user to click on it and be taken directly to another web site or location on the Internet.

Online News: Reported information, analysis and commentary found on the Internet on web sites supported by media companies.

Zimbabwe: An independent African country, located in Southern Africa, and that used to be a British colony.

Chapter 12
Digital *Parrhesia* as a Counterweight to Astroturfing

Nicholas Gilewicz
University of Pennsylvania, USA

François Allard-Huver
Paris-Sorbonne University, France

ABSTRACT

Astroturfing—fake grassroots communications about an issue of public interest—is further problematized in digital space. Because digitally mediated communication easily accommodates pseudonymous and anonymous speech, digital ethos depends upon finding the proper balance between the ability to create pseudonymous and anonymous online presences and the public need for transparency in public speech. Analyzing such content requires analyzing media forms and the honesty of speakers themselves. This chapter applies Michel Foucault's articulation of parrhesia—the ability to speak freely and the concomitant public duties it requires of speakers—to digital communication. It first theorizes digital parrhesia, then outlines a techno-semiotic methodological approach with which researchers—and the public—can consider online advocacy speech. The chapter then analyzes two very different instances of astroturfing using this techno-semiotic method in order to demonstrate the generalizability of the theory of digital parresia, and the utility of the techno-semiotic approach.

INTRODUCTION[1]

Astroturfing—fake grassroots campaigns about matters of public interest—presents a particular problem to researchers, particularly to those researchers interested in studying the content of advocacy speech. Specifically, the content may be true, and even compelling, but if the honesty of the speaker is questionable, the truth may be a house of cards.

In this chapter, we expand Pramad K. Nayar's (2010) application of *parrhesia* to digital space. Relying, as did Nayar, on Foucault's (2001) articulation of this ancient Greek concept, this chapter derives a model for analyzing the credibility of digital advocacy speech, and thus a model for truth-telling in the digital public sphere. *Parrhesia*, or the ability to speak freely, implies three public duties for speakers: to speak the truth, to sincerely believe that truth, and to honestly represent themselves when speaking. Astroturf-

DOI: 10.4018/978-1-4666-2663-8.ch012

ing, which conceals identities in order to reduce the risks of speaking truth to power—or to the public—always fails the latter duty.

In networked space, however, pseudonymous and anonymous speech can work both democratically and propagandistically. This chapter proposes that digital *parrhesia* helps evaluate astroturfing and helps understand why such evaluation matters. By using digital parrhesia to analyze astroturfing online, this chapter's analytic model aims to contribute to the preservation—and maybe the revivification of—a culture of truth-telling.

BACKGROUND

Astroturfing is Not for Free

On May 16, 2010, the French popular science show *E=M6* featured a story about "triple play boxes." A new communication service in France, the boxes allowed users to access the Internet, television, and telephone services at the same time through the same provider. *E=M6* achieved its popularity by mixing the points of view of scientists and technicians with discussions of consumer uses and needs—thus, the boxes were well-suited for a story on this program. Every broadcast of *E=M6*—named for M6, its channel—follows a format similar to this episode, where the broadcast first explains the science behind the broadcast in terms fit for a general audience, and then explores the contextual uses of the boxes.

This episode featured a happy French family, a couple with two children, who discovered the features of the box. When the mother called the children for dinner during their favorite cartoon, the box allowed them to pause the program and store it on the box's hard drive. When the family took a walk, and worried they would miss an evening show, the father programmed the box using his smartphone. In these ways, the family met their entertainment needs thanks to the little box. But shortly after the episode aired, fans and customers, re-watching the episode on M6's Web

site noticed that the box, called a Freebox, was a product available exclusively from the Internet provider Free. Fans began to discuss the show, and the website Freenews.fr, created by an association of Free customers, reported that the family was, in fact, a fake: The "father" was Free's marketing director, and the "mother" was Free's press secretary ("La Freebox," 2011; "Reportage," 2011).

Other Web sites and radio and television news reported the dishonesty; ultimately, the French broadcasting authority, the CSA, warned M6 that its astroturfing attempt contravened articles 20 and 22 from its broadcasting convention: "The company must verify the validity and the sources of information [and] must show honesty and rigor in the presentation and treatment of information" ("Conseil supérieur," 2011). A core concern of the circulation and presentation of information in public space is one central to the question of authorship and credibility: astroturfing. In this case, what began as a simple five-minute report on a new digital media technology ended in astroturfing practices being exposed by both digital and traditional media.

This clear case of astroturfing—and how it was uncovered—allows us to observe the interrelationship between astroturfing, digital media use, and the exposure of astroturfing. In France, much as in the United States, audiences are accustomed to marketing and public relations. The Freebox/M6 case became a scandal not because it was marketing, but because it was misleading—the family that enjoyed the Freebox was parented by employees of Free. These employees violated what we see as a fundamental factor governing digital communication space: *parrhesia*, in which the public duty of speakers is to speak the truth, to sincerely believe that truth, and to honestly represent themselves when speaking.

Building a Theory of Digital *Parrhesia*

The act of astroturfing may be thought of as manufacturing support for an issue or attempting to mislead politicians, news media, or citizens about the origins of such support. The use of the term dates at least to 1985, when United States Senator Lloyd Bentsen said, after receiving letters that promoted insurance companies' interests, that, "A fellow from Texas can tell the difference between grass roots and Astroturf. This is generated mail" (qtd. in Sager, 2009). Astroturfing attempts to leech the legitimacy held by grassroots movements, pretending that it is a response from below to governance from above.

Growing access to the tools of digital media production, from email to website design to video, have created new communication spaces and communities. Citizens, corporations, and governments all have enhanced abilities to engage in public dialogue about their beliefs, products, and intents—and enhanced abilities to conceal their identities while doing so. Thus, digital communication space introduces new problems for ethos; this realm depends on a proper balance between the ability to create pseudonymous or anonymous online presences, and the public need for transparency in public speech.

Pseudonymity and anonymity surely have their places, for they accommodate truthful comments from individuals who may have valid reasons—such as the fear of community disapproval to the fear of being "disappeared" by a government—to conceal their identity. Yet, corporations, governments, and their public relations or advertising companies can exploit that same anonymity. What may be legitimately defensive for an individual becomes a public relations tactic for an organization attempting to reduce the risk of advocacy. But if astroturfing is easier than ever in the digital era, so is learning the true identity of astroturfers, as demonstrated by the Freebox/M6 scandal.

In order to fully understand the role of digital communications in astroturfing, and to develop a method to analyze digital astroturfing, this chapter turns to Foucault's (2001) articulation of the ancient Greek concept *parrhesia*. Commonly translated as "free speech," *parrhesia* implies that when one has the ability to speak freely, one also

has the public duty to speak the truth, to sincerely believe that truth, and to honestly represent oneself when speaking—criteria worth repeating, and to which this chapter will repeatedly return.

This concept was first ported to digital space to make an affirmative argument for the value of the Web site WikiLeaks as a defender of "the agora of information" and a culture of digital truth-telling (Nayar, 2010). The argument is compelling, but the implications of digital *parrhesia* are both wider and deeper than simply defending WikiLeaks, because, according to Nayar himself, digital cultures generate new communities: "Digital cultures create a new communications culture, which generates a new community, the global civil society . . . and the globalisation of conscience. [WikiLeaks] is an embodiment of this new form of communications-leading-to-community, a digital *parrhesia*" (2010, p. 29). Under this view, new communities emerge whose participants may be judged by whether they adhere to the duties implied by *parrhesia*. Discourse under *parrhesia* centers on truth-telling in the service of community. Digital *parrhesia* is then a necessary component of digital communities, like *parrhesia* was a necessity in the Greek agora.

Risk balances the duty to speak truthfully in digital *parrhesia*, and in what Foucault called the "parrhesiastic game," speakers balance the risk to themselves with the duty to speak the truth: "In *parrhesia*, the speaker uses his freedom and chooses frankness instead of persuasion, truth instead of falsehood or silence, the risk of death instead of life and security, criticism instead of flattery, and moral duty instead of self-interest and moral apathy" (2001, p. 19-20). If engaging in the parrhesiastic game is courageous, then undermining and exploiting the game is cowardly. Moreover, doing so suspends or negates the rule of the game, and thus suspends—and threatens—the role of the society as a discursive community as well.

Digital *parrhesia*, then, may be considered a discursive space where a wide range of individuals can engage in truth-telling practices, and a space whose boundaries—the duty to speak the truth, to believe that truth, and to honestly represent oneself, all though online media—also provide the beginnings of a critical framework for assessing the credibility of digital texts. Clearly, identifying digital *parrhesia* as a discursive space and defining the boundaries of that space is useful; it allows us to distinguish between digital actors who seek to reveal the truth or to conceal it. Getting there, however, requires a clear methodology. And the importance of good methods here cannot be overstated; accusing an author of astroturfing, under the aegis of digital *parrhesia*, is tantamount to accusing that author of propagandistic lying.

Digital *parrhesia* lends itself to semiotic analysis because it identifies different levels of speech. At each level, truth-claims hinge on the medium wherein the speech occurs, how the speech is distributed, the content of the speech, and the identity of the speaker herself. People who have the ability to speak freely in digital culture also have the obligation to become Bentsen's "fellow from Texas" who can distinguish between grassroots content that emerges organically from below and content that is covertly astroturfed down from above. Distinguishing between the two is often contingent on questions of authorship and discourse. In order to help researchers make this distinction, the next section operationalizes digital *parrhesia* by integrating the author and the medium into what we call a "techno-semiotic" method of analysis.

Building a Techno-Semiotic Method for Digital *Parrhesia*

The idea that every human construct has different levels of meaning is the basis of semiotics, which itself can be a key that unlocks the structure of communication by revealing patterns of meaning at those levels. Semiotics builds a coherent approach for analyzing units of meaning. The goal of this chapter is not to solve questions asked by generations of semioticians from their foundational work (de Saussure, 1977; Barthes, 1968;

Morris, 1964; Greimas, 1989) to contemporary scholars (Eco, 1976; Klinkenberg, 2000; Veron, 1988), but rather to operationalize their theoretical work into an easily applied method. The different steps of this method have much in common with the analytical skills used in the humanities and literature studies. And the "techno" part of the techno-semiotic method does not require advanced technical knowledge so much as an awareness that a medium is itself a complex object or condition.

In this way, we propose to understand online statements and the systems in which they evolve. Of the object of research—in the case of this chapter, an advocacy statement that may or may not be astroturfing—four questions must be asked: *Where* does the statement occur? *How* is the statement enunciated? *What* does the statement say? And *who* said it? These questions correspond to different levels of meaning: the medium, the document, the text, and the discourse, respectively. In the techno-semiotic method, the levels, while having separate and identifiable characteristics, are not isolated from each other. Rather, each level plays a role and influences, and is also influenced by, the other levels. So each level must be considered through two points of view: looking at properties intrinsic to each specific level of meaning, and looking at how the levels of meaning can and do interact.

First, the practice of semiotics in social science, communication, and media studies has shown that exhaustive analyses must not restrict themselves only to content—the technical apparatus of communication must be considered as well. Davallon (2004), for example, suggested that what makes objects of communication research unique is their "techno-semiotic weight." From the sheet of paper to the PDF document, every document has material features that transform the way we receive and perceive signs, but also influences our research practices and the meanings we assign to objects. This is but one aspect of the method—particularly significant at the level of the medium—that is highly influential but not deterministic, because, as Wright (1986) suggested, a technical apparatus does not determine communicative processes, which are themselves social, not technological, in nature. Thus, the first step in describing an object is to describe the technical apparatus and the system by which it is produced. For example, this method would ask whether an article published in *The New York Times* and on nytimes.com digital newspaper was the same. Similarly, is a 1933 speech by United States President Franklin Delano Roosevelt the same when heard on the radio back then and when read in a history textbook today? These are the types of questions that the techno-semiotic method prompts: *Where* does the statement occur? And how does the medium in which it occurs affect the meaning of the statement? These questions serve to avoid the pitfall of technological determinism—while still insisting that a statement's technological context affects its meaning.

The second step of this method takes us to the level of *how* a statement is enunciated. This is closely related to the *where*, or to the medium, but is distinct. Rather than looking at the medium and its systems—the differences between *New York Times* stories in print or online, or the differences between a contemporaneous radio speech and a textbook—the second step turns to the document itself, and the process by which it came into being. The question of *how* a statement is enunciated regards how statements become text and how those texts are disseminated. For example, authors rarely publish handwritten drafts of their work. Instead, they uses word processing software, then send a copy—sometimes digital, sometimes paper—to their editors, who may send it along for further review by peers and copyeditors, until the document is transformed into a printable version for the actual publishing apparatus. Thus, techno-semiotic analysis requires attention to how documents are produced and distributed, and to how those processes affect and inform the meanings of statements.

Of course, analyses of communication texts are commonly concerned with the content of statements, which is our third level: *what* the state-

ment says, returning to the classic core question of finding meaning in a text. A news story viewed on YouTube will be different than the same news story viewed during a CNN broadcast. Neither will be understood in exactly the same way, nor will they be understood the same way as the script of the broadcast or the audio track heard without the video. The medium informs this level, because audiences receive different media differently. [2] Nonetheless, texts—particularly news and digital advocacy—have claims. Those claims must be identified and evaluated, as well as understood in the context of the previous two levels: to what extent the medium informs those claims, and to what extent how those claims are presented and distributed affects their reception.

The final level of meaning to investigate is the discourse itself. The analysis of discourse can be as complex as the definition of the term. In the techno-semiotic method, research into the content of the message requires gathering some information about the speaker in order to understand his intentions and purposes. When considering a statement, the question of *who* said it is then a more global question about the speaker and her relation to the statement. Analysis at the level of discourse is closely and strongly interrelated with the other levels of meaning. Through analysis at the levels of 1) the technics of the medium, 2) the production and distribution of a text, and 3) the content of a text, a holistic understanding of a statement and its meaning begins to emerge. To paraphrase and expound upon Marshall McLuhan (1994), if the message is the medium, then we can say that the discourse is the medium: analysis of the medium reveals the space in which the discourse can evolve and can be influenced and transformed, and analysis also reveals for whom it was crafted and to what purpose it was deployed. Meaning is conveyed through discourse and its intent; the techno-semiotic model thus treats the author, in a way, as text.

Traditionally, mass media have served to confer status upon certain speakers—news anchors of major television networks, editors of major newspapers, politicians, and so forth—but in digital communication space, traditional status conferral is dramatically weakened. When discussing matters of public interest in digital communication space, we argue, status is conferred by the honesty of the speaker. Her discourse must fulfill her public parrhesiastic duties, which, again, are: to speak the truth, to sincerely believe that truth, and to honestly represent herself when speaking. As we will see in the examples that follow, analyzing the last of these—honest representation—is at the crux of determining whether advocacy speech is astroturfing.

Astroturfing the European Commission: From Public Consultation to Risk Manipulation

In a previous project, Allard-Huver (2011) tried to understand how negotiating the concept of risk in the European public sphere transformed advocacy communications. He analyzed the public deliberation from 2002 to 2009 surrounding the 91/414 European Directive regulating pesticides, finding that some public feedback was surprisingly similar, considering letters were supposedly from individuals writing individually. Using the techno-semiotic method, it quickly became clear that an astroturfing attempt was being made within the European legislative process.

During its public consultation for the report *Thematic strategy on the sustainable use of pesticides*, the European Commission invited pesticide stakeholders to send comments, suggest modifications, and put forward reservations and criticize the commission's first publication, *Towards a thematic strategy on the sustainable use of pesticides* (European Commission, 2009). Some feedback that initially seemed to be from individuals appeared to be a part of a coordinated campaign when seen through the prism of digital *parrhesia* and evaluated by the techno-semiotic method. These questions followed from the method:

First, is the European Commission Web site, as a digital public sphere, more subject to astroturfing attempts? The first level of inquiry focuses on the Web site—the media layer—of the European Commission, its functions, and the ways it created a digital public sphere. The site functioned in three ways: it served as a medium that raised public awareness of the problems of pesticides; it built a digital discussion space for public participation in debates about pesticide use; and now, it serves as a public archive for a completed process. Each function makes clear that the Web site is a mediator between different publics. The site, by enunciating the perspectives of European legislators as well as those of other stakeholders, suggests that the rules of *parrhesia* are at work; in turn, stakeholders, by participating in the process, imply that they accept those rules. But the physical and material distance introduced by Internet communication itself must not be forgotten. This is the second level of meaning, extending from the first—participating in this discussion space created by the European Commission creates a public archive. But on this website, distinguishing speakers can be difficult, and one can easily submit false information, or falsify an identity; this admits the possibility of astroturfing into the process. Nevertheless, because the site also plays the role of an archive, the public—and researchers—can also investigate that advocacy speech, see how *parrhesia* operates in these debates, and see wheter the ethical duties prescribed by *parrhesia* are met, or not, by speakers.

Now, we can look at the third level of meaning: the content of the actual documents. The principal element of our interrogation is that some of the stakeholder texts are remarkably similar. The text of Birgitt Walz-Tylla is almost the same as the text sent by Carlo Lick, B. Birk, or Joseph Haber. For example, all four letters include this text: "As a scientist who has dedicated most of his career to researching and developing crop protection products, I believe there are a number of elements of this strategy that need to be further considered," even that of Birgitt Walz-Tylla, a woman who, rather humorously, has "dedicated most of *his* career" (emphasis ours) ("European Commission," 2009). Here, the content analysis is less the analysis of signs themselves, and more the recognition that the texts are the same. And these seams—like Birgitt Walz-Tylla's apparent claim to manhood and our ability to quickly compare texts—suggest that these letters are part of a coordinated astroturfing campaign.

So, *who* then is the speaker? The person who signed these letters? The person or people who wrote the original text, which was then distributed to these four scientists? These questions go directly to the third duty of a speaker in the realm of digital *parrhesia*: the duty to honestly represent oneself when speaking. These four letters share the same content, but differ slightly in their presentation and the ways in which their authors present themselves publically. All identify themselves as scientists, and some sign their letters with their academic titles, laying a public claim to be experts in their fields. The letters from Birk and Lick clearly state their professional affiliations; both work for BASF, a chemical company with interests in pesticide production. Walz-Tylla and Haber do not provide their professional affiliations. But no matter: a simple Google search reveals that Walz-Tylla is an employee of Bayer CropScience, and Haber is an employee of BASF. Both companies are industry stakeholders.

Thus, what separately seem to be legitimate individual positions of experts are revealed to be the direct participation of industry. This discourse does not arise from the individual concern of scientists, but from what appears to be coordinated industry propaganda. An industrial agent almost certainly wrote the original text, and suggested the campaign to other industrial stakeholders. This actor, in fact, is the true author of the discourse, but stays in the shadows, uses different identities, and ultimately leaves the ultimate intention unclear—is the issue one of good science or good business? In this debate, then, we can say that

these four scientists—and the real author of their letters—do not respect *parrhesia*. While they may have attempted to exploit the ease of submitting digital feedback, the realm of digital *parrhesia* also affords the opportunity to uncover their campaign. Therefore, digital *parrhesia* and the techno-semiotic method reveal what we believe to be a clear case of astroturfing.

Astroturfing Pinellas County, Florida: Secreting Racism

Astroturfing can be professional, well-styled, and coordinated, as seen in the BASF and BayerCropScience employees submitting letters as individual stakeholders, even though the content thereof is so similar as to suggest a coordinated campaign by industrial stakeholders. Astroturfing can also be petty, but still astroturfing, when a public official spreads individual social biases and political accusations under pseudonyms.

In 2010 and 2011, a commenter on the Web site of the *St. Petersburg Times*, a daily newspaper in Florida, posted a number of controversial comments under the pseudonym "Reality." The commenter complained about "race pimps" who would "walk around looking like an idiot thug trying to hold your pants up. Whitie isn't to blame for your ignorance" (DeCamp, 2011b). Reality also criticized what s/he saw as St. Petersburg's outsized number of "thug shootings" and "prostitute beatings," and also attacked two Pinellas County commissioners—in one case alleging that a commissioner helped a "developer friend" access funds from the county (DeCamp, 2011b).

A reporter noticed that Reality often ended comments with the phrase "just say'n," a phrase also used by another Pinellas County commissioner named Norm Roche, and he noticed that Reality announced a new Web site in a comment— a Web site registered to Roche. Initially, this might not seem to be a case of astroturfing; after all, Roche was not manufacturing wide support for racism. When confronted by a reporter, Roche admitted that he posted both as "Reality" and as "Norm Roche," suggesting a desire to distance his public persona from the views of "Reality." And when critiquing elected officials, including his colleagues, he again used a pseudonym to distance Norm Roche from Reality. Even if the Reality persona was consistent and the author of Reality's comments believed them to be true, that one person operated two personae, whose opinions did not fully align (at least in public), suggested an effort to mislead or misdirect readers of those comments.

The word "secreting" has two meanings: concealing in a hiding place, and forming then emanating a substance. On its face, Roche's comments seem to have more in common with trolling—sowing mischief in online discussion forums, usually anonymously—than with astroturfing. However, by leveling anonymous attacks against political colleagues (or perhaps enemies), Roche's statements moved from simple mischief into the realm of political speech—a realm where *parrhesia* operates. Here, Roche used a pseudonym to conceal the origins of his controversial comments, and possibly to conceal his own controversial views. (It must be noted, however, that Roche has publicly denied being a racist or a homophobe.) At the same time, he used a pseudonym to distribute those controversial comments, and to do so, used a medium that permitted pseudonymous comments and integrated them with news stories. In this case, the journalist who uncovered the relationship between Reality and Norm Roche used something akin to the techno-semiotic method to do so, and we argue that the method works very well to analyze speech in this situation.

As per the method, we first address issues related to the medium. Here, Roche's speech required a news product that offered an online commenting system. Such a system permits an exchange of ideas between readers who participate, and sometimes even between readers and journalists, if journalists choose to respond to comments. Immediately, we see that these texts are polyse-

mous—different readers interpret the meaning of news stories differently, including inscribing their own, sometimes divergent, meanings onto those texts.[3] At the same time, we see how these texts become polyvocal—for readers who do not comment, the news product is the story *plus* the comment threads. Within such polyvocal texts, voices that threaten the peace of the community can easily be identified. In this case, a reporter identified outlandish claims by a commenter. These claims could not exist without the newspaper offering a comment thread, which, in turn, introduced polyvocality into its news product. The digital text therefore has the ability to reveal through its medium the plurality of voices that create and recreate new texts. Identifying these voices is the first step in assessing their credibility.

Second, we address questions related to how the speech is distributed. In the case of these comment threads, reader comments are attached to the end of a news story. Online, the *St. Petersburg Times* publishes stories along with the comments; at the end of the story, the reader must click a link reading "Join the discussion: Click to view comments, add yours." While other content exists on the page, ranging from advertisements to copyright information to links to other news stories, only links to the comments, or links that help readers repurpose the story by sharing or printing it, are directly connected to the story itself. When commenting, a reader becomes a reader-author; when sharing a story by email or on a blog, the reader becomes a reader-publisher. In both cases, a participating reader implicates herself in a case of digital *parrhesia*, especially because she must agree to "Comment policy and guidelines" which include, among others, the requirement that "Your comments must be truthful. You may not impersonate another user or a tampabay.com staff member by choosing a similar screen name. You must disclose conflicts of interest" ("Comments Policy," 2012). Finally, other commenters indicated a parrhesiastic situation, because they implicitly interrogated and summoned the criteria of digital *parrhesia*. On the story revealing that Reality was Norm Roche, many of the 137 comments debated whether the publication had violated its own promises of privacy to its commenters, whether the reporter had used honest techniques to uncover this story, and whether a commenter should take responsibility for his comments by posting them under his real name. Here, the debate is about online ethos itself.

Next, we address questions about the content of the speech. The comments by Reality were often incendiary, supporting the biases of some commenters and provoking outrage among others. Reader comments, in fact, operate as at least three different texts. First, comments exist in relation to the news story—expanding it, criticizing it, and opining on it. Second, comments exist in relation to other comments; they respond to previous comments while anticipating future ones. Third, comments exist as part of a complete news product, one that includes news story and all comments but is also served to non-commenting readers as well. The digital text is at the crossroads of the journalist's production of meaning and the public's reception and sometimes re-appropriation of it. The reporter who revealed Reality as Roche did so by understanding the first two content inter-relationships—by identifying commonalities between supposedly different voices and ultimately revealing them to be the same. Here, examining content, and the ease of comparing that content due to its medium, helped this reporter identify political speech that violated our expectations of online ethos.

Finally, we consider the speaker himself. All four levels of the techno-semiotic method inter-relate, but questions of discourse are perhaps the most pervasive of all. Above, we have seen how online commenting systems promote polyvocal texts, and thus create opportunities for deviant speech. We also have seen that by posting comments, readers become reader-authors, and in doing so, implicate themselves in a parrhesiastic

system. Further, even the most cursory look at the content of reader comments reveals that understanding their intertextual and multitextual nature allows us to see the different ways in which content may be deployed. Discourse, then, is overlaid on all of these. The question of who is commenting and why may be the fundamental question of digital ethos in online texts such as these. In this case, once the reporter marshaled his evidence and asked Roche if he was Reality, Roche admitted that his reasons for concealing his identity (at least part of the time) were entirely discursive. He told the reporter, "A lot of it can be rhetoric and rants. Unfortunately it's part of our communication base now, and you have to be part of it, you have to track it" (Decamp, 2011a).

Thus, we see how a reporter used a process much like the techno-semiotic method to break a news story about a politician who concealed his identity while making possibly racist comments about his constituents. And we also see how different layers of meaning generated through the medium, its distribution, its content, and its author are all available to analyze the credibility of online speech.

CONCLUSION

Digital *Parrhesia* and Digital Communication Texts

Clearly, an application of digital *parrhesia* has the potential to evaluate and assess astroturfing that is spread through digital media. Under the *parrhesia* model, truth-claims are reviewed in three ways: whether they are true, whether the speaker believes that they are true, and whether the speaker is honestly representing herself. Again, *parrhesia* accommodates pseudonymous and anonymous speech because honesty does not require mapping a name onto a real speaker, but rather requires that the speaker honestly believes in and argues for her truth claims. The techno-semiotic method accounts for this, but it also has wider implications.

As Nayar (2010) suggested, digital communication constitutes new communities. This is not a new phenomenon—we have seen it before in the old bulletin board systems and chat rooms, and we see it today in online communities ranging from 4chan to Facebook groups. These communities, as all communities do, develop their own behavioral norms and mores. These norms help define the discursive space of digital *parrhesia*; the risks to a speaker for violating those norms—in the digital space, ranging from chastisement to banishment—help determine when and how the speaker will fulfill her duties to speak the truth, to believe that her truth-claim is indeed true, and to honestly represent herself and her belief. For astroturfers, the risk is that a secret propaganda campaign will be revealed, with consequences ranging from public shame to criminal liability.

To operationalize digital *parrhesia*—to make it useable not only for academic critics, but to make a model that can be used to consider digital communication more broadly—we have integrated the medium and the speaker into our techno-semiotic method. Doing so solves a major problem with the sender-receiver model of communications, which manages to persist even when it is not appropriate. Under a sender-receiver model, texts can be recognized as univocal and polysemous—that is, readers can negotiate their own meanings with texts, even meanings that run counter to the preferred reading of a univocal author. But when texts become *polyvocal*, and when the medium itself—for example, an online news story with comments—creates and sustains polyvocality, the sender-receiver model falters.

Polyvocality in digital media permits the exposure of astroturfers. In this chapter alone, we have seen fans of a product, a scholar (Allard-Huver, 2011), and a journalist (DeCamp, 2011a; DeCamp, 2011b) all used observations made through, or use

techniques reliant upon, digital media to expose astroturfing that was at least partially executed through digital media. This suggests that polyvocal media and polyvocal texts, when functioning in a parrhesiastic way (that is to say, when discussing community issues in ways that hinge on acts of truth-telling), are especially appropriate subjects for the techno-semiotic analysis outlined in this chapter.

The astroturfing cases outlined here—audience members uncovering that a popular science television show became a propaganda and advertising tool in France; the distribution of the PDF of a European Commission report compiling the feedback of stakeholders regarding pesticide use; and a journalist revealing that an elected official clandestinely stoked the fires of racism in Florida—suggest the versatility of both digital *parrhesia* as a theory and the techno-semiotic method.

REFERENCES

Allard-Huver, F. (2011). *Transformation and circulation of the notion of "risk" in the European Commission*. Unpublished Master Thesis, University of Paris-Sorbonne, Paris.

Barthes, R. (1968). *Elements of semiology*. New York, NY: Hill and Wang.

Comments policy. (2012). Tampabay.com Retrieved January 22, 2012, from http://www.tampabay.com/universal/comment_guidelines.shtml

Conseil supérieur de l'audiovisuel. (2011, October 18). *Convention de la chaîne M6*. Retrieved February 12, 2012, from http://www.csa.fr/Espace-juridique/Conventions-des-editeurs/Convention-de-la-chaine-M6

Davallon, J. (2004). Objet concret, objet scientifique, objet de recherche. *Hermes, 38*, 30–37. doi:10.4267/2042/9421 de Saussure, F. (1977). *Course in general linguistics*. Glasgow, UK: Fontana/Collins.

DeCamp, D. (2011a, November 17). Pinellas county commissioner Norm Roche has alter ego for online comments. *St. Petersburg Times*. Retrieved January 22, 2012, from http://www.tampabay.com/news/localgovernment/article1202065.ece

DeCamp, D. (2011b, November 18). Norm Roche's anonymous online snark strains city, county relations. *St. Petersburg Times*. Retrieved January 22, 2012, from http://www.tampabay.com/news/politics/local/ norm-roches-anonymous-online-snark-strains-city-county-relations/1202287

Eco, U. (1976). *A theory of semiotics*. Bloomington, IN: Indiania University Press.

European Commission. (2009). *Towards a thematic strategy on the sustainable use of pesticides*. European Commission. *Environment*.

Foucault, M. (2001). *Fearless speech*. Los Angeles, CA: Semiotext(e).

Greimas, J. (1989). *The social sciences. A semiotic view*. Minneapolis, MN: University of Minnesota Press.

Klinkenberg, J.-M. (2000). *Précis de sémiotique générale*. Paris, France: Seuil.

La Freebox à l'honneur dans E=M6 (MàJ). (2011, May 17). Freenews. Retrieved February 12, 2012, from http://www.freenews.fr/spip.php?article8300

McLuhan, M. (1994). *Understanding media: The extensions of man*. Cambridge, MA: MIT Press.

Morris, C. W. (1964). *Signification and significance: A study of the relations of signs and values*. Cambridge, MA: MIT Press.

Nayar, P. K. (2010, December 25). Wikileaks, the new information cultures and digital parrhesia. *Economic and Political Weekly, 45*(52), 27–30.

Reportage bidon sur la Freebox dans E=M6: intervention du CSA. (2011b, October 11). Freenews. Retrieved February 12, 2012, from http://www.freenews.fr/spip.php?article9131

Sager, R. (2009, August 18). Keep off the astroturf. [Electronic version]. *The New York Times*. Retrieved January 22, 2012, from http://www.nytimes.com/2009/08/19/opinion/19sager.html

Veron, E. (1988). *La sémiosis sociale. Fragments d'une theorie de la discursivité*. París, France: Presses Universitaires de Vincennes.

Wright, C. R. (1986). *Mass communication: A sociological perspective*. New York, NY: Random House.

ADDITIONAL READING

Beder, S. (1998). Public relations' role in manufacturing artificial grass roots coalitions. *Public Relations Quarterly, 43*(2), 21–23.

Berry, J. (1993). Citizen groups and the changing nature of interest group politics in America. *The Annals of the American Academy of Political and Social Science, 528*, 30–41. doi:10.1177/0002716293528001003

Cox, J. L. (2008). Blots and the corporation: Managing the risk, reaping the benefits. *The Journal of Business Strategy, 29*(3), 4–12. doi:10.1108/02756660810873164

D'Almeida, N. (2007). *La société du jugement*. Paris, France: Armand Colin.

Dahlberg, L. (2011). Re-constructing digital democracy: An outline of four positions. *New Media & Society, 13*(6), 855–872. doi:10.1177/1461444810389569

Deleuze, G. (2006). *Foucault*. London, UK: Continuum.

Foucault, M. (2001). *The order of things: An archaeology of the human sciences*. London, UK: Routledge.

Foucault, M. (2010). *The birth of biopolitics: Lectures at the College de France, 1978-1979*. New York, Ny: Palgrave Macmillan.

Gillmor, D. (1994). *We the media: Grassroots journalism by the people, for the people*. Sebastopol, CA: O'Reilly Media.

Lankes, R. D. (2007). Credibility on the internet: Shifting from authority to reliability. *The Journal of Documentation, 64*(5), 667–686. doi:10.1108/00220410810899709

Lessig, L. (1999). *Code and other laws of the cyberspace*. New York, NY: Basic Books.

Lyon, T., & Maxwell, J. (2004). Astroturf: Interest group lobbying and corporate strategy. *Journal of Economics & Management Strategy, 13*(4), 561–597. doi:10.1111/j.1430-9134.2004.00023.x

Roberts, A. (2006). *Blacked out: government secrecy in the information age*. New York, NY: Cambridge University Press. doi:10.1017/CBO9781139165518

Russell, A. (2011). *Networked: A contemporary history of news in transition*. Malden, MA: Polity.

Souchier, E., Jeanneret, Y., & Le Marec, J. (Eds.). (2003). *Lire, écrire, récrire: Objets, signes et pratiques des médias informatisés*. Paris, France: Bibliothèque publique d'information.

Vattimo, G. (1992). *The transparent society*. Malden, MA: Polity.

Wyld, D. C. (2008). Management 2.0: A primer on blogging for executives. *Management Research News, 31*(6), 448–483. doi:10.1108/01409170810876044

Yin, S. (2008, August 7). Astroturfing. *Media*. Retrieved June 20, 2012, from http://search.proquest.com/docview/206258286?accountid=14707

KEY TERMS AND DEFINITIONS

Astroturfing: The practice of generating fake grassroots communication, usually about an issue of public interest, to create the false impression of a public advocating for a particular political, social, or corporate agenda. Usually considered to be practiced by corporations and lobbyists, often concealing the true author of the advocacy documents.

Discourse: A sum of proposition and enunciation that creates a body of knowledge. Foucault calls the circulation of this discourse through media and other organizations the "discursive formation."

Grassroots: A type of political movement driven by a community, or emerging from below traditional sites of power and power structures. Often considered an honest representation of community interests.

Parrhesia: A Greek word meaning "frankness" or "speaking freely." Citizens of democratic ancient Greece had the ability to speak frankly about political issues, and in turn had the public duties to speak the truth, to sincerely believe that truth, and to honestly represent their belief in that truth.

Semiotics: Literally, the science of the signs. Semiotics is the study of the meaning in every form. Here the perspective adopted is to study communication as an exchange, a construction and a negotiation of sign—which can be anything from a word to a photograph to how the color red is used in a film—through media itself.

Technological Determinism: A doctrine focused on the technological evolution of information and communication systems rather than on their interaction and their subordination to the society that developed them. Technological determinism holds that the forms of technology themselves determine how society uses those technologies, and that in turn, technology shapes culture and cultural values.

Techno-Semiotic: A way to understand and analyze media and communication phenomena as both constructions of knowledge and the means by which those knowledge and signification are circulated. The construction of knowledge and the technology of its circulation affect each other; both affect the meaning of signs.

Trolling: The practice of making incendiary comments in online discussion forums, with malicious intent, or to sidetrack discussions with which the "troll" disagrees. Usually practiced anonymously.

ENDNOTES

[1] For a brief history of astroturfing and astroturfing-like activities, see Lee (2010); for legal implications of astroturfing, see Kolivos & Kuperman (2012); for how astroturfing problematizes grassroots movements, see Cho et al. (2011).

[2] For two quite different cases, see the New London Group's work on multiple literacies (1996) and Scott McCloud's use of the comic form to explain visual narrative and sequential art (1994).

[3] A phenomenon readily seen in comment threads following all political stories, for example.

Section 4
User–Generated Content

Chapter 13
Establishing Credibility in the Information Jungle:
Blogs, Microblogs, and the CRAAP Test

Dawn Emsellem Wichowski
Salve Regina University, USA

Laura E. Kohl
Bryant University, USA

ABSTRACT

In this chapter, the authors locate blogs and microblogs such as Facebook and Twitter in the information landscape. They explore their diverse habitats and features, as well as the explosion of uses discovered for them by academic and journalistic researchers. The authors describe an approach to evaluating the quality of blogs and microblogs as information sources using the CRAAP test, and they show how a consideration of digital ethos in the application of the CRAAP checklist imbues the test with flexibility and effectiveness, and promotes critical thinking throughout the evaluation process. The chapter demonstrates how the special features of blogs can be leveraged for rigorous assessment. For the purpose of defining examples, it focuses on blogs and microblogs such as Facebook and Twitter, but the authors see their approach as having application across other yet-to-be developed platforms because of its flexibility.

INTRODUCTION

In this chapter, we demonstrate that blogs and microblogs represent a significant source of information for researchers and contribute to scholarly and journalistic discourse. We show how blogs have characteristics that differentiate them from more traditional scholarly sources such as periodicals and monographs. We center the discussion around our assertion that expanding some of the criteria of the CRAAP test to encompass the concept of digital ethos makes the test applicable to social media applications like blogs. While some scholars argue that checklists like the CRAAP test are inappropriate and mechanistic evaluation tools, we refute this assessment, arguing that this checklist is a useful device especially for students new to research or scholars new to social media

DOI: 10.4018/978-1-4666-2663-8.ch013

resources. We demonstrate how application of the CRAAP test can promote critical thinking. At the core of the chapter is the concept of digital ethos, which, as we apply it, contrasts with the model of authorship in traditional scholarly publications.

The concept of "digital ethos" plays a starring role in scholarship surrounding credibility on the Internet, whether in the fields of human-computer interaction, rhetoric, or information science (Flanagin & Metzger, 2007; Warnick, 2004; Fogg & Tseng, 1999; Enos & Borrowman, 2001; St. Amant, 2004; Marsh, 2006). "Digital ethos" diverges from traditional concepts of authorship in several significant ways. The credibility of authors of more traditional publications may be assessed by such measures as institutional affiliations, advanced degrees, and recognition in mainstream and scholarly press. The concept of digital ethos is more fluid. A blogger's true identity and affiliation may be unknown. A blogger may actively hide his/her true identity to make candid observations. Or a blogger may choose to highlight interests in a blog which stray from his/her professional specialization. In traditional evaluation frameworks, sources created by authors with these characteristics would be considered unreliable. However, we leverage the CRAAP test criteria to account for these differences in author ethos and evaluate the sources according to the more progressive concept of digital ethos. In more traditional scholarly sources, proper use of grammar and vocabulary is a significant indicator of credibility. In a blog, authentic use of slang and cultural-specific idiom may be a better indicator of credibility. We address these differences, and how the CRAAP test is well suited to address them.

Our perspective as librarians contributed to our choice of the CRAAP test as a foundational tool for assessing the quality and authority of social media sources. Our positive experiences in the classroom using the CRAAP test to help students navigate the open web made it an obvious choice, and further comparison with other evaluation approaches confirmed this choice for us. This tool has been accepted and used by information literacy professionals for other pedagogical reasons. The most obvious attribute is its name. As a mnemonic device, the CRAAP test is effective. Sharing this tool in the classroom, we are often met with amused laughter. As its creator, Blakeslee, of California State University, Chico pointed out, it is memorable and works contextually when instructing users about evaluating a wide variety of resources. "For every source of information we would now have a handy frame of reference to inquire, 'Is this CRAAP?'" (2004, p. 7). The test also incorporates all the widely accepted criteria for evaluating print and online resources.

A CRAAP Test Overview

The CRAAP test consists of five overarching criteria for evaluation: currency, relevance, authority, accuracy and purpose. The application of checklists such as the CRAAP test are widely taught by professionals in the library and information literacy fields, particularly for evaluation of online resources, research papers, or other multi-step academic projects (Doyle & Hammond, 2006; Blakeslee, 2004; Dinkelman 2010).

Throughout the information literacy literature there are multitudes of lists of evaluation criteria based on similar concepts (Kapoun, 1998; Blakeslee, 2004; Doyle & Hammond, 2006; Burkhardt et al., 2010). Doyle and Hammond (2006) summed up the criteria contained in most tests: "to decide whether something can be trusted, we need to consider who thought it up, who made it accessible, what are their motives and biases, and what features, if any, might reassure us that the influence of these motives and biases are minimized" (p. 58).

We see the CRAAP test criteria as the most concise, flexible, and memorable evaluation tool of the series of checklist tests that have been proposed since the late 1990s (Kapoun, 1998; Blakeslee,

2004; Doyle & Hammond, 2006; Burkhardt et al., 2010, Dinkelman 2010). The checklist format gives beginning researchers a simple way to understand the basic elements that lend a source credibility, while aiding seasoned researchers in developing an assessment schema for approaching new sources such as social media.

The CRAAP Test Evaluation Criteria

Currency: The Timeliness of the Information

- When was the information published or posted?
- Has the information been revised or updated?
- Does your topic require current information, or will older sources work as well?
- Are the links functional?

Relevance: The Importance of the Information for Your Needs

- Does the information relate to your topic or answer your question?
- Who is the intended audience?
- Is the information at an appropriate level (i.e. not too elementary or advanced for your needs)?
- Have you looked at a variety of sources before determining this is one you will use?
- Would you be comfortable citing this source in your research paper?

Authority: The Source of the Information

- Who is the author/publisher/source/sponsor?
- What are the author's credentials or organizational affiliations?
- Is the author qualified to write on the topic?
- Is there contact information, such as a publisher or email address?

- Does the URL reveal anything about the author or source? examples: .com .edu .gov .org .net

Accuracy: The Reliability, Truthfulness, and Correctness of the Content

- Where does the information come from?
- Is the information supported by evidence?
- Has the information been reviewed or refereed?
- Can you verify any of the information in another source or from personal knowledge?
- Does the language or tone seem unbiased and free of emotion?
- Are there spelling, grammar or typographical errors?

Purpose: The Reason the Information Exists

- What is the purpose of the information? Is it to inform, teach, sell, entertain or persuade?
- Do the authors/sponsors make their intentions or purpose clear?
- Is the information fact, opinion or propaganda?
- Does the point of view appear objective and impartial?
- Are there political, ideological, cultural, religious, institutional or personal biases?

(Meriam Library, California State University Chico, 2010).

Issue of Ethos, Authority, and Credibility in Social Media

Social media as an information format is closely tied to the identity of its creator. The value of blogs as information sources is related to this personal orientation, but it follows that the intelligent use of the blog hinges on an accurate assessment of the blogger's authority. Though the phrase "digital ethos" could be defined as the overall spirit of an Internet community, we use the phrase "digital ethos" to encompass the characteristics that compose a blogger's online identity. Digital ethos can be investigated to evaluate a blogger's authority to purvey information and express opinion and the CRAAP test can help in that investigation.

Some define ethos *as* credibility (St. Amant, 2004, p. 318; Enos & Borrowman, 2001, p. 93), but we subscribe to the more neutral definition, of ethos as tied to a blogger's character, as well as the blogger's audience perception of his/her character (Brahnam, 2009, p. 10). Aligning with Aristotle's depiction, a *credible ethos* arises from a blogger's persuasive mastery (Marsh, 2006, p. 338-339) A blogger's *authority* to opine on a subject is tied to his or her credibility. In their 1999 paper defining the place of credibility in human computer interaction, Fogg and Tseng defined credibility as "believability." Fogg and Tseng emphasized that credibility is not innate, but relies on an audience's subjective assessment (p. 80). Authority relies on audience assessment, but connotes something stronger than believability; authority implies an ethos infused with experience and wisdom (Reynolds, 1993, p. 327; Segal and Richardson, 2003, p. 138). An authoritative blogger ethos can also be enhanced by audience perceptions of a subject's reputation, as judged by affiliations and comments about the blogger's work. Thus, assessment of a blogger's digital ethos to determine credibility and authority is a highly subjective process. We believe that the CRAAP test's criteria provides the most effective and concise way to consider the variables that contribute to a credible digital ethos.

Authority can be based on observed persuasive skill over time, as seen from the blogger's chronological posts and his or her trail of activity as evidenced by comments on other blogs or online forums. Authority is a key measure of blog quality, and is also measured in the accuracy and purpose elements of the CRAAP test. A thorough analysis of overall blogger ethos is achieved by employing all the criteria of the CRAAP test.

Applying the CRAAP Test to Blogs

In this section, we lay out our approach to evaluating blogs using the CRAAP test, with an emphasis on blogger ethos. As librarians, our approach to blog evaluation is grounded in the skill-set conveyed in the concept of "information literacy" or "the set of skills needed to find, retrieve, analyze, and use information" (Association of College and Research Libraries, 2012). In customizing the CRAAP test for use in social media applications such as blogs, we include under the umbrella concept of information literacy other literacies, such as media, technological and digital literacy. Though this chapter is not the first to use the CRAAP Test to evaluate blogs, we believe that our integration of the concept of digital ethos within the criteria of the CRAAP test, as well as a systematic utilization of the characteristics specific to blogs in the application of the CRAAP test's evaluative elements is original and effectively tailors the CRAAP test for use with blogs and other social media sources.

Currency

Assessment of the currency criteria in blogs is intuitive. A key feature of blogs is their chronological nature, in which posts are displayed in reverse chronological order. Dates and times are automatically time stamped on posts, easily reviewed by a user. This feature also allows the user to quickly determine if the blog is being maintained or is in disuse. The comments feature in blogs also includes timestamps, allowing the researcher to chart the flow of commenter reactions. By convention, blogs indicate at the top or bottom of a post if there have been any updates to the original writing. Researchers can easily determine the dates of posts and establish their currency. This also facilitates historical research by following commentary surrounding major events or the evolution of scholarly thought on a topic, for example.

In the case of audio or video, currency may be more difficult to establish. When media is embedded into a blog, researchers can use Meola's corroboration and comparison approach, in which they attempt to locate the original source of the media (2004, p. 331) to see it in its original context, with its original timestamp. Visual clues in embedded video or photographs, such as clothing style, and audio clues, such as figures of speech or music, can also be helpful for estimating currency.

Relevance

Beyond the initial question, whether information included in the blog or microblog answers the researcher's information need, researchers can determine relevance by assessing the blog's intended audience. In some cases, a blog may be embedded in a website that implies its subject focus, and therefore, relevance. In other cases, the blogger is clear about their intended audience. For example, the blog "I Blame the Patriarchy" (http://blog.iblamethepatriarchy.com/) includes this text on its homepage: "I Blame the Patriarchy is intended for advanced patriarchy-blamers. It is not a feminist primer." In the absence of such a clear statement, linguistic cues can provide indicators about a blogger's digital ethos and intended audience, and be reviewed to test relevance. Is the blogger using simple, clear wording to introduce a topic to people new to the subject or using complex speech and linguistic shortcuts to speak to peers? Technical jargon or obscure slang connotes that the blogger is an expert and has an intended audience of experts. More conventional language implies that the blog is aimed at a general audience. In the case of podcasts, Austria (2007) found that podcast listeners were able to judge the level of information through several factors, noting in particular the presence or absence of jargon during an interview

with a scholar. When language was free of jargon the listeners were able to surmise that the intended audience was more general. Relevance may also be closely related to the evaluation of authority and accuracy, discussed below.

Authority

Contributing to both the value and risk of blogs as information sources is their "low barrier to entry," in which it is free and easy for anyone to set up shop with a broad variety of social media accounts (Metzger, 2007, p. 2078). This has implications for our suggested evaluation of bloggers' digital ethos. The conception of "identity" on the Internet is fluid, and while this can be acceptable, it is incumbent upon the researcher to determine the authority and purpose of the blogger as well as the accuracy of the information the blogger provides. We believe that this low barrier to entry, while making careful evaluation of a blogger's ethos essential, is also the medium's strongest asset, as it gives a stage to previously unheard voices.

Several authors argue convincingly that authority is elevated in importance above other criteria for evaluation. Fritch and Cromwell (2001) considered "cognitive authority," defined as "authorship and affiliation," to be the most significant criteria for evaluation. In order to conduct an effective assessment, researchers will find it essential to understand the *spirit* of the criterion, "authority." Of the five criteria of the CRAAP test, authority is most obviously tied to the blogger's digital ethos (Enos and Borrowman, 2004, p. 95-96). By "spirit" we mean that authority can be determined not through a series of set questions, but by understanding the qualities that would lend credibility to a blogger's ethos, and having at one's disposal a series of strategies to choose from based on the type of research and the subject matter of the blog.

To determine if a blogger's ethos is credible, users can evaluate language, scope of the information presented, and accuracy of the information presented for *trustworthiness* and *expertise*. Fogg and Tseng (1999) defined trustworthiness as "well-intentioned, truthful, [and] unbiased" (evidence of bias should be identified, but it needn't invalidate a source's appropriateness for research; the existence of bias in sources will be addressed in the "accuracy" criterion). They defined expertise as "knowledgeable, experienced, [and] competent" to describe the worthiness of the source (p. 80). These qualities can be established by evaluating the accuracy of the blogger's statements (described below in the "accuracy" criterion), or by examining a blogger's affiliation.

Affiliation can yield important clues about authority. Though a clear determination of the true identity of a blogger can prove difficult, affiliation can be established with other evaluative techniques specific to blogs. For example, blogs often include a "blogroll," or list of other recommended blogs. Microblogs include links to the blogger's friends and associated groups. Researchers can check these for clues about the author's cultural and political persuasion, and also for what types of information sources the author considers valuable. This leads to more questions, which may yield information about the blogger's digital ethos. Is the blogger affiliated with groups that show evidence of strong political opinions? Do the groups or friends seem to express rational thought or reactionary views? Do the blogger's friends or blog roll give clues about whether the blogger is knowledgeable of others who are key in their area of interest? Blogs also often include links to photos and video. If the photo or video includes links to other sources, researchers can follow these to determine if they link to reputable sources for the blogger's field of interest. To evaluate authority in vlogs (video blogs) or podcasts, users can direct their attention to the vocabulary, language, and temperament of participants. Austria (2007) asserted that for the purpose of evaluation, the interviewee can be considered the "author" of the content and the host can be seen as the "publisher." Vlogs and video podcasts can also be analyzed using visual information beyond the text (emotional cues to measure bias, evidence of the vlogger's affiliations based on surroundings and visible possessions, etc).

To get an idea of the blogger's reputation and standing with his/her audience, the "comments" feature of blogs can be seen as a form of peer review, in which readers offer critiques and corrections (Banning & Sweetster, 2007). Comments also add value to information within the blog by providing an opportunity for users to offer opinions, personal experiences, and other perspectives that give the reader a fuller picture of the issue than the initial blog post.

To establish the extent of a blogger's expertise, answering the question, "Is the blogger qualified to opine on this subject?" the researcher can assess the blogger's use of language, either written or oral (through video or audio). It is not necessary that the blogger use the language of journalism or academe to be considered credible. Much depends on the type of information the researcher is looking for when evaluating authenticity. If a blogger is speaking about being a gang member, does s/he use the slang and phrasing that would be consistent with the vocabulary of a person from his/her region and affiliation? Maybe the course of research has taken the scholar into the computer programming community. Does the programmer appropriately use programmer slang or technical terms? Do people commenting on the blog post seem to respect him/her? With social media research these considerations can be investigated over time by reading through archives of posts. Coming from an oral tradition, Quintilian asserted that an insincere speaker would reveal him/herself through a continuously developing relationship with the audience (qtd. In Enos & Borrowman, p. 96). Today, an Internet user leaves a trail of activity, often linked across multiple social media platforms, through which a researcher can see the development of the blogger's thought, his/

her relationships with the audience, and scope of knowledge.

There are cultural considerations to consider when assessing authority, especially related to the blogger's expertise. Kirk St. Amant wrote that different cultures use different standards to assess credibility. St. Amant (2004) referred to the elements each culture uses to assign credibility as the "ethos conditions" (p. 319). He cited multiple examples, including the writing styles preferred by authors of Japanese or American business memos. Americans prefer concise, direct explanations, while Japanese feel that stating obvious information is rude (p. 320). He also cited differences in writing structure (for example, southern Europeans see long sentences as evidence of a credible presenter ethos), different assignations of credibility based on the use of humor, and different ways of presenting intelligence (p. 320-325). Thus, audiences from disparate cultural backgrounds may come to divergent conclusions when evaluating a blog using the authority criterion of the CRAAP test. This cultural subjectivity can be mitigated while the researcher considers whether the blogger is writing to an audience that does not come from the researcher's cultural background. If so, it is incumbent upon the researcher to understand the cultural mores of that audience in order to accurately assess credibility.

Another cultural concern is addressed by Alvarez-Torres, Mishra, and Zhao, who found that study participants assigned more credibility to native speakers than to fluent foreign speakers, regardless of actual content (2001). These findings could have implications for researchers evaluating blogs created by bloggers for an audience of a different cultural background. Researchers may be well-advised to keep this in mind while evaluating text, video, and audio components of blogs, in keeping with Barzun and Graff's suggestion that researchers nurture the virtue of self-awareness (1992, p.99).

If a blog post clearly lists a creator, authorship (which aids in getting a fuller picture of a blogger's digital ethos) can be confirmed several ways. If the blogger states his/her name, a web search may provide an idea of any traditional credentials and affiliations. Authority or affiliation can also be confirmed in more technical ways. A Whois.com search allows a researcher to simply type in a domain name and view information on the owner of the domain. This can clarify whether a blog is actually owned by a company or is a product of an individual or organization. Additionally, dissecting the domain address can provide clues as to authorship or ownership. The suffix, or top-level domain of the site's URL (.uk, .ae, .fr, .ly, .edu, .gov, mil, .com) can sometimes provide a general idea of affiliation. Sites ending in .fr, for example, are hosted by companies in France. Sites ending in .ly are hosted by Libyan companies. Other suffixes are specific to the type of institution that hosts them. For example, .edu is only provided to accredited post-secondary educational institutions in the United States. Suffixes ending in .gov are assigned only to United States government websites.

The rest of the URL can also provide clues about a blogger's ethos. For example, http://blog.microsoft.com is very different from http://microsoft.blog.com. The first part of the URL http://blog.microsoft.com (http://) references the protocol, or how the page gets to the user and how it functions. The last part of this URL (microsoft.com) is the domain. This is the host site. The second portion of this URL (blog) is the subdomain. This represents a section on the host site. In the case of the URL http://microsoft.blog.com/february/20120215.html, one can quickly deduce that it is either a site unaffiliated with Microsoft that is blogging about Microsoft, or it is a site attempting to trick users into thinking it is affiliated with Microsoft, possibly, for nefarious purposes. /20120215 is a file name. We can deduce that it is a file because of the .html extension. The .html

Figure 1. Parts of a URL

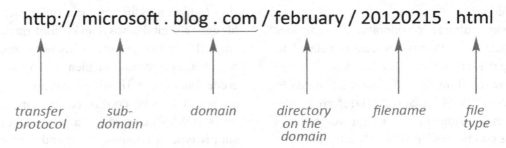

extension tells us that the file is a webpage. If the extension were .doc, we would know that it was a Microsoft Word document. If the extension were .exe, we would know that the file was a program, and would download to our computer if we stayed on the page. /february refers to a directory, like a folder on your computer, in which 20120215. html is housed.

Warnick (2004) asserted that the use of authority is an outdated evaluation criterion in an information landscape where many websites are "authorless" (the actual author cannot be verified). We assert that blogs, as often highly personal modes of online publishing, have a singular preoccupation with authorship if defined more flexibly. The concept of digital ethos in a blog environment may not be compatible with past ideas of an author's ethos. Bloggers may not leave their real names. They may not provide a physical address and may construct identities separate from their physical, real world identities. This may be insignificant though, if their online identity reflects a true aspect of themselves and their expertise in their field of interest is authentic. The strategies for verifying authority detailed above provide an alternative to traditional measures of authority (such as academic credentials and institutional affiliations) in an alternate information landscape.

Accuracy

Accuracy is a crucial element in assessing blogger ethos. The term "blog" is understood to infer subjectivity, immediacy, and less stringent editorial controls (Johnson & Kaye, 2004; Berkman, 2004). The information within the blog is more likely than a major media source to have inaccuracies and errors. Bloggers have varying degrees of concern with their reputation, and therefore may not be as motivated to double-check the information they place online. This is in contrast to a news corporation, which is bound by ethical and professional standards of conduct for journalists and accordingly has incentive to ensure that information it releases is correct (Chung, et al., 2012; Johnson & Kay, 2004). With this in mind, however, researchers can rigorously use evaluation methods imbued with the spirit of the accuracy criteria, including comparison and corroboration and following information to its original context, to judge the reliability of the blog as a source.

Meola's (2004) "contextual approach" to evaluating web resources focused on using comparison and corroboration. This approach fits in nicely with several criteria of the CRAAP test, and is just one example of how the test can lead to higher-order thinking when it is applied to the digital ethos of the online environment. Ideally, a blogger will develop a trustworthy ethos by linking to the original source of the posted information. If s/he doesn't, however, a researcher may look for other sources to corroborate a fact, especially respected or mainstream sources. This is also referred to as verification. In the (1992) *Modern Researcher*, Jacques Barzun and Henry Graff devoted an entire chapter to the importance of verification in research. In the case of blogs, this approach can include following the trail of hypertext links to the original source in order to read and analyze the text in its original context. It can include reading cited sources to confirm that they make the points the author says they do. It can mean double-checking that the story is independently reported in respected news sources. Or, it can even mean contacting the blogger to ask further questions. In certain types of sources, proper use of grammar, vocabulary and punctuation may also give a picture of a blogger's accuracy, and by extension, authority. This issue can become obscured, however, when a blog is not written in the author's home language. Investigation into the author's background may yield clues as to whether they are writing in their native language, which will aid in verification.

The results of not verifying facts in their original context and corroborating them in other news sources can be embarrassing. Viral reposting of satirical news stories on blogs has become so pervasive that a blog, Literally Unbelievable (http://literallyunbelievable.org), was created to chronicle them. Mainstream news sources, such as the Beijing Evening News, which reprinted the story, "Congress Threatens To Leave D.C. Unless New Capitol Is Built" originally from the *The Onion*, have also been humiliated by poor corroboration of facts (Terdiman, 2004).

Throughout the verification process, Barzun and Graff emphasized the importance of skepticism and awareness of how one's personal bias may influence assessment of information quality. Metzger et al. (2010) found that Internet users tend to find sources more credible if they confirm already existing viewpoints. This "bias confirmation effect" is noted in multiple sources, including by Barzun and Graff (1992) in *The Modern Researcher*: "In research as in life one is far more likely to find what one looks for than what one does not care about" (p. 186). This prejudice can create a predisposition to give a source the benefit of the doubt if it agrees with one's own assertions, or to

dismiss a source because it disagrees (Barzun and Graff, 1992, p. 99). Researchers should be aware that their personal opinions may predispose them to be overly critical of information that rebuts their beliefs, and less critical of information that confirms them.

As researchers develop background knowledge of a subject, they can more easily identify information that "doesn't feel right" and should be verified. Barzun and Graff (1992) asserted that successful assessment of a source's accuracy relies heavily on "common sense reasoning, a developed 'feel' for history and chronology, on familiarity with human behavior, and on ever-enlarging stores of information" (p. 99). This intuition is developed over time with exposure to many sources, both traditional and non-traditional.

Purpose

Since the existence of bias may provide important evidence when compiling an accurate picture of a blogger's digital ethos, researchers should consider the reasons a blogger created a blog or post. An "about me" section can provide explicit (though not always completely accurate) information about purpose. Other clues to the purpose, or creator intent, of blogs may be revealed by reader comments and interactions with each other and the blogger. Information about blog purpose may also be revealed through assessment of the kind of advertising on the site. Advertising may be overt, with ads hosted by the blog around the perimeter of a page, or may be more subtle, with posts extolling a particular product or political figure in text or video messages.

Emotional tone and biased or strong language may also give clues to purpose and by extension, the blogger's digital ethos. Bias may also be revealed through the links, photos and videos that the author has included in their post. As addressed earlier, persuasive language or evidence of a biased perspective does not give cause for immediate dismissal of a blog as an unreliable source.

Depending on the topic of research, evidence of strong opinions may make a source more useful.

In terms of evaluating bias, the researcher should vigilantly maintain self-awareness. Banning and Sweetster (2007) explored the "third person effect", or the tendency for people to think that others are more likely to be influenced by media than they themselves are. According to the study, the individual Internet user is likely to believe s/he alone is immune to the wiles of media and advertising, while the rest of the population is more likely to fall prey to such persuasion. The researchers focused on the habits of blog users in particular and found that there were no differences in third person effect when comparing media types (personal blogs, news blogs, online corporate news sources, and newspapers). Though Banning and Sweetster were surprised by the results and urged further study, this could imply that researchers finding information through blogs are susceptible to inflated views of their evaluation abilities.

Meola's (2004) comparison method of source evaluation can also be useful in assessment of purpose. Meola suggests that researchers locate disparate sources (for our purposes, these might include blogs and scholarly or mainstream news sources) with similar subject coverage and compare them. This can reveal bias (as will one-sided coverage of an issue), thoroughness (if the author only discusses economic, as opposed to social implications of a policy, for example) and accuracy.

Blogs and Microblogs as Scholarly and Journalistic Sources

Blogs and microblogs have begun to be recognized as significant sources of scholarly inquiry. In this section, we briefly define blogs and demonstrate how blogs are being utilized in research. A blog (weblog) is technically defined as a series of "frequently modified web pages in which dated entries are listed in reverse chronological sequence." The use of blogs became widespread in mid-1999 and their popularity exploded in the mid-2000s (Her-

ring et al., 2004, p. 1). As of 2008, 33% of Internet users reported that they regularly read blogs, and 12% reported having created a blog, while .5% of Internet users blogged regularly (Smith, 2008).

Microblogs are shorter, often restrict word count in posts, and broadcast updates to other users who choose to subscribe. Platforms include Facebook, Twitter, and Tumblr. Microblogs represent a revolution in social and political communication. As of February 2012, Facebook had 845 million total users (Swift, 2012) and Twitter had 100 million active users with an average of 230 million tweets per day (McMillan, 2011). Hereafter, we will group microblogs under the general heading "blogs," because all of the features, habitats, and uses of blogs described throughout this chapter may also exist for microblogs.

Users choose to follow blogs for news, gossip, editorial opinion, scholarly argument, and personal narrative. In terms of habitat, blogs may be embedded in credible news websites such as those for the *New York Times* and the BBC, freestanding (with their own domain address) or found as a part of a subscription service such as Wordpress. Blogs may incorporate media, such as text, images, video, audio, and hyperlinks to other content. Blogs may include features such as comments by readers, a blogroll (a linked list of recommended blogs), an "about the author page," and a deep history of past writing.

Types and Research Utility of Blogs

Personal narrative blogs represent a significant source of primary, first person information. They may provide unmediated accounts of historical events or "snapshot in time" information. Social scientists and journalists may find personal narrative blogs especially useful because they often provide primary, first person accounts, written and published by the subject which do not pass through the disfiguring lens of an observer or interviewer. These blogs allow researchers and journalists to identify sources who speak at length in their own words, and bring perspectives that might ordinarily be lost or unavailable. Examples include the blog "Baghdad Burning" (http://riverbendblog.blogspot.com/), which represents, as Miriam Cooke detailed in her 2007 article, a platform for civilians trapped by war to reach a global audience while chronicling the details of their daily lives. In the past, the only way such accounts were available was through state-sanctioned and published books or through the eyes of journalists who interviewed people on the ground.

Advertising blogs are created by companies to discuss new products or services and predominately serve as marketing. They represent a source for cultural and historical research related to their host company. This genre has some crossover with the personal narrative genre, as popular bloggers may be paid by advertisers to promote their products. An example of this crossover in purpose is Sony's blog (http://blog.sony.com/), which featured the well known photographer and blogger Ma Ra Koh as a guest blogger.

News blogs take several forms, and each may fill different information needs. An embedded news blog is housed in a traditional news source such as the *Wall Street Journal* or CNN.com. Blogs of this type can be followed for breaking news by staff journalists, or opinions by members of the news organization's editorial staff. Freestanding news blogs, unaffiliated with major news organizations, may aggregate news from a particular beat, cultural or political orientation; conduct original investigations; mine sources to break; provide opinion; or some combination of the above. These blogs have become significant social and political forces in the information landscape. Bloggers, such as Josh Marshall, of "Talking Points Memo" (http://talkingpointsmemo.com/), have broken stories that mainstream news sources disregarded, such as racist statements made by Senate Majority Leader Trent Lott, which resulted in his resignation (Johnson & Kaye, 2004; Marshall, 2002).

News blogs in various permutations can be significant information gathering, dissemination, and story refining tools for journalists and academic researchers. The defining characteristic of all social media news applications, interactivity, encourages more active consumption of information. A blog post about a story may provoke a series of comments and exchanges among those commenting. Reading the story and the associated comments can provide a more complete picture of the issue than the original post alone. Related opinions, personal experiences, and clarifications in the comments can situate the post more clearly for the reader (Chung et al., 2012; Notess, 2010). For example, the *Providence Journal*'s blog coverage of a teen atheist suing to remove a prayer banner from her public high school in Cranston, RI garnered many comments. These comments illustrate the atmosphere of religious controversy in the community where the teen lives, which is not entirely clear from the news blog posting alone (Arditi, 2012). Blog comments by citizen journalists also provide story leads and enrich content, and they have been formalized into discrete news sections by corporate media sources. The *Washington Post*, BBC, and CNN, for example, solicit news, photos and videos from members of the public (Notess, 2010).

Blogs can be a format for journalists to post stories that do not fit in the more formal portions of their publication (Bradshaw, 2008). Examples of this type of blog include the "City Room" blog of *The New York Times* online (http://cityroom.blogs.nytimes.com). Journalists have also established blogs to do more in-depth reporting on a particular beat and have influenced the mainstream media establishment in doing so. An example of this is La Silla Vacía, an investigative journalism blog in Colombia, in which several reporters choose from among of the country's most significant political issues and cover those topics in-depth (Leon, 2010). Blogs may augment their presentation of a story using media formats such as photos, videos, audio, and hyperlinks. This can provide a more complete picture of a story and gives users an opportunity to evaluate the journalist's conclusions themselves (Chung et al., 2012).

Irrespective of subject matter, microblogs such as Twitter and Facebook can play a momentous role in news dissemination. In their working paper "Opening Closed Regimes: What Was the Role of Social Media During the Arab Spring?" Howard and his fellow researchers found, in analyzing the volume, hashtags (folksonomic categorizations, preceded by the "#" sign), and originating location of Tweets over time during the Arab Spring, that a "spike in online revolutionary conversations preceded major events on the ground" (Howard et al., 2011, p. 3). The researchers also tracked the spread of revolutionary topics by hashtag across borders, and noted that protesters in different countries were communicating with one another, spreading news from the ground and also from respected international media outlets (Howard et al., 2011). From this example, it is clear that social media applications like blogs can affect the trajectory of news events, serve as an archive of the events themselves, and can be the subject of study on multiple levels.

Knowledge blogs represent a significant new influence on the scholarly information cycle. Features endemic to blogs facilitate scholarly communication, primarily, their interactive components. Knowledge blogs are publicly accessible; they make new ideas, theories and research available to viewers from outside the creator's discipline and outside of the scholarly community. This opens the floor to unexpected discussions and new conclusions (Kjellberg, 2009). Beyond expanding the base of interested parties, blogs can serve several other purposes in their expansion of scholarly discussion among the various epistemic cultures (defined, in Kjellberg's 2009 article, as differences among scholarly communities in the areas of research practice, knowledge creation, and social characteristics). Kjellberg discussed

Luzon's observations of "strategic linking" among scholars to provoke and engage in "hypertext conversations" (p. 3), which can deepen existing relationships and create new ones. Blogs may also provide context to findings by describing research in practice, and may speed the evolution of ideas by allowing other scholars to build on research that is not completed but is still in process. In this way a blog can be a form of gray literature, allowing researchers to present early results for the express purpose of soliciting feedback (a feature of blogs also significant for journalists, as described by Bradshaw in his 2008 article).

FINDING AND CHOOSING BLOGS

Researchers selecting blogs can use a series of steps similar to the decision-making process for choosing more conventional sources. As Jacques Barzun and Henry Graff suggest, "the researcher must again and again *imagine* the kind of source he would like before he can find it" (1992, p. 47). The process of *articulating the information need* is valuable. It helps to define and clarify the research question, and leads to a consideration of the strengths and weaknesses of the sources at a researcher's disposal. An encyclopedia, for example, will provide an excellent overview of a topic, but is a poor source for in-depth analysis. A research article is a good type of source for in-depth analysis, but one would have to read many articles to begin to see the bird's eye view of a topic.

As a source, a blog post may fulfill any of these needs, ranging from in-depth analysis to overview to breaking news. In making the decision of how to choose a blog as a source, a researcher may also consider which type of blog, among the genres discussed above, would fill their information need. For example, if they are interested in following the evolution of conservative opinion on a topic, they may choose to follow the embedded blog of a conservative newspaper columnist, or a well-respected freestanding news blog whose writers have a conservative slant. They might also follow the knowledge blog of a conservative thinker.

There are several technical approaches to finding blogs. Blogs are sometimes cited and linked to from other news stories or social networking profiles. A politician, scholar, journalist, or other public figure's blog is often listed on their Facebook, or Twitter profile, or on the homepage of their place of work or personal home page. Researchers can also use such finding tools as http://technorati.com/, http://www.google.com/blogsearch, the search functions on such major blog platforms as http://wordpress.com/, or microblog platforms http://twitter.com/ or http://www.facebook.com/. Another way to find blogs is by using the "Search within a site or domain" option within Google's advanced search page. For example, from the advanced search page, entering youtube.com in the "Search within a site or domain" search box will bring up results for your search terms only within YouTube's domain. To keep up with the latest posts or receive posts on a particular topic, a researcher can add blog RSS feeds to their RSS readers. RSS (or Real Simple Syndication readers), such as Google Reader, provide notification of new blog posts as an alternative to regularly checking the blog.

Concerns about Blogs as Information Sources

Brabazon, in her 2006 article, "The Google Effect: Googling, Blogging, Wikis and the Flattening of Expertise," provided a biting indictment of the use of social networking applications in scholarly research. She stated that "'peer production,' … is really peer-less production, where mediocre, banal and often irrelevant facts are given an emphasis and interpretation which extends beyond the credibility of scholarly literature" (p. 157). This condemnation deserves consideration in relation to how researchers *use* social networking applications such as blogs, though we of course

strongly disagree with any characterization that discounts them as endemically invalid sources. We believe that Brabazon is taking a limited view of social media applications in her disparagement. The wide range of subject matter, purpose, and authorship of blogs and bloggers described above, as well as the wide variety of research uses they have already been mined for, is evidence of their value as research sources.

That said, without careful selection, assiduous evaluation, and judicious use of blogs as sources, researchers may find themselves in embarrassing situations. In this section, we describe some of the dangers inherent in blog research, in the belief that an informed researcher is a skeptical researcher. Seemingly factual information may be inaccurate. Authors may not be who they say they are. There are myriad instances of elaborate blog hoaxes; so many, in fact, that a word was coined for these fake blogs: "flog" (Weaver, 2006). There are various categories of flogs. Examples span the spectrum from marketing attempts by major corporations to first-person narrative blogging.

Attempts by major corporations to sew grassroots excitement about their brands are known as "astroturfing." The bloggers "Charlie and Jeremy," for example, were ostensibly two young men who wanted their parents to buy them a Sony PlayStation Portable game console for Christmas and supposedly created "All I want for Xmas is a PSP." However, a domain ownership search revealed that the domain name was registered to Sony's Zipatoni marketing company (Consumerist, 2006). The blog "Walmarting Across America" was created by real people, "Jim and Laura", who traveled across the country in an RV visiting Walmart stores; however, evaluation of the subject matter and the tone of interviews of Walmart employees (described in a businessweek.com article as "relentlessly upbeat") raised questions about bias. Interviews with Jim and Laura revealed that, from the RV to the travel funds, the trip was bankrolled by an advocacy group created by Walmart's public relations firm and funded by Walmart (Gogoi, 2006). In each of these cases, careful assessment of the blogs, using approaches that are also described in the CRAAP test, revealed the hoax.

There are also examples of false identity flogs in the personal narrative genre. The blog "A Gay Girl in Damascus'" garnered a significant and devoted following and was used by journalists to report on events in Syria. Purported author Amina Arraf, a lesbian Syrian-American, was later found to be a 40-year-old American man (Mackey, 2011). This hoax was uncovered by Andy Carvin of National Public Radio, who chronicled the evolution of his opinion on Storify (2012). Carvin detailed the evolution of his suspicions on his social networking account on Storify, and his post provides a fascinating detail of how evaluation techniques described in the CRAAP test can lead to an accurate assessment of blogger ethos. Carvin began to question the true identity of Arraf after he broadcasted a request on Twitter for people who had met Arraf to contact him, and was unable to find anyone who had met her in person. On Storify, Carvin displayed email interviews with multiple skeptical blog readers and contacts in Syria. He showed how blog readers compared pictures purported to be of Amina Arraf, raising questions whether they were of the same person. Blog readers also made a close assessment of the accuracy of Arraf's blog posts:

I can tell you from experience that the post titled my father the hero doesn't make sense whatsoever. They [the secret police] either ask you to come over... yourself to have a chat (usually friendly) or arrest her no matter who her father is. It's as simple as that. (Carvin, 2012)

He also examined the blog's accuracy and purpose by following Arraf's trail of past posts back to an older blog where she explicitly stated that she would be publishing fiction and nonfiction without specifying which was which. Carvin's assessment approach shows how successful utilization of CRAAP criteria can establish an accurate view of

the blogger's digital ethos. He attempted to corroborate the authenticity of the blogger's identity by finding people who had met her in person and by analyzing media embedded in the blog. He compared the accuracy of facts put forth in the blog against those of knowledgeable sources. He used approaches applied in the currency, accuracy, and purpose criteria to follow the thread of past writing and assess the blogger's ethos.

While the proliferation of social media sources can improve access to information, especially breaking news or complicated scientific information, the viral nature of information dissemination in social networking applications increases the need for assiduous evaluation of blogger credibility (Friedman, 2011). In illustration, we can review the case of the viral blog post by "MIT Research Scientist" Josef Oehmen in the wake of the Fukushima Daiichi nuclear accident. The post, apparently originally composed to calm the fears of a cousin in Kawasaki (Jabr, 2011), was picked up by news sources such as the *Telegraph* and *Discover Magazine*, and was also forwarded throughout the web by email and social media posts. The post widely disseminated erroneous information that the accident was not serious. A simple Internet search to confirm the authority of the blogger, however, shows that while Joseph Oehmen is a research scientist at Massachusetts Institute of Technology, his specialty is not nuclear science but "risk management in the value chain" (Massachusetts Institute of Technology, 2012). This example illustrates the need to evaluate a source's authority to opine on the subject in question, as well as the effectiveness of the technique described in the accuracy section above, of following the thread of a viral post back to its original context.

PROBLEMS WITH BLOG USERS' RESEARCH PATTERNS

Several studies substantiate concerns about how researchers use blogs in practice, which have implications for educators and scholarly and journalistic researchers. In this section, we detail some of the problems with how researchers use blogs, in the belief that these risks can be mitigated if researchers and educators are aware of them. According to Johnson and Kaye's 2004 paper, despite well-publicized concerns about authority and accuracy, almost three-quarters of blog readers find blogs to be very credible sources of information, and see little reason to rigorously evaluate online sources of information. However, this motivation to evaluate also relates to the researcher's purpose. According to Metzger et. al., users' motivation to facilitate more methodical evaluation techniques is in proportion to the level of risk associated with inaccuracy (2010). Where risk is determined to be low, motivation to use more rigorous techniques of evaluation is also low. In other words, "people seek to find an optimal balance between cognitive effort and efficient outcomes" (p. 417). In her 2004 article, Warnick reviewed three studies which corroborate the above findings, and also show that the criteria users employ depends on the intent and subject matter of the site (p. 262).

In terms of how Internet users approach credibility assessment, Metzger, et al. found that, in an environment where "source" and creator authority is difficult to assess, users rely on other factors. They evaluate site design, evidence that post authors are "enthusiasts" (apparent experts, based on the volume and thoroughness of their posts), and "social confirmation" (in which users assume that a source is credible because a high number of other users feel that it is credible) (Metzger et al., 2010, p. 416, 424, 435). Though assessing a blogger's commenter perceptions can be an important tool for evaluating a blogger's digital ethos, confirmation bias relates more closely to the *information* in a blog post. This can be risky for researchers choosing whether to use a blog, as it may be popular because it is intentionally controversial rather than because it provides accurate information or thoughtful analysis.

Related to vlogs and podcasts, a study by Lee et al. found that the presence of video in a web environment dramatically affects users' perceptions of credibility, irrespective of how the source rates according to other assessment measures. Participants also asserted that high quality production conveyed more credibility, regardless of the source (in this case, public relations firms vs. news sources) (Lee et al., 2010). This point is also emphasized by Selnow (1998) who contended that users place more stock in primary source media, rather than mediated description of an event or issue (qtd. in English et al., 2011, p. 736). English et al. (2011) also found that researchers use the presence of video to rate sources highly rather than evaluating other elements such as the logic of an argument or a source's attempt to appeal to emotion. In short, the presence of video can influence users to believe a source is credible, regardless of whether the other information therein stands up to additional methods of assessment. With the above concerns in mind, and because the presence of multimedia adds so much value to the information in blogs, we will examine some media-specific approaches to evaluation below.

The above research has implications for educators seeking to inculcate students with a healthy dose of skepticism about web sources, but it is also significant for seasoned researchers using emerging online sources in their research. Barzun and Graff cite self-awareness as one of the "virtues of the researcher." They underscore the precept that to conduct research effectively one must know one's own prejudices. With that in mind, we suggest that the above concerns be at the forefront of every researcher's and instructor's mind while considering the use or teaching the use of social media sources.

MOTIVATION AND ABILITY

Here, we turn to a discussion of how the above problems can be re-envisioned as opportunities for researchers and instructors to instill higher-order critical thinking skills. For example, students are often compelled to be more methodical in their evaluation of sources by the explicit direction of their instructors, but they also have personal motivation to complete a thorough and competent investigation. As Head and Eisenberg (2010) found in their survey of undergraduates:

What mattered most to students while they were working on course related research assignments was passing the course (99%), finishing the assignment (97%), and getting a good grade (97%). Yet, three-quarters of the sample also reported they considered carrying out comprehensive research of a topic (78%) and learning something new (78%) of importance to them, too. (p. 4).

While Metzger (2010) found that "Internet users will use more methodical, systematic evaluation (information processing) approaches when motivation is high and "peripheral" or "heuristic" approaches when motivation is low" (p. 416), our focus is on the highly motivated researcher—one who is looking for information to inform or validate their scholarly work; or in the case of students, one who is researching materials for papers that will be evaluated by experts in the field, i.e. their professors. We posit that it is precisely through completing more methodical, systematic evaluation approaches that Internet users develop good intuitive or heuristic abilities, as well as higher-level critical thinking skills.

Researchers experienced with using more conventional peer-reviewed and news sources may find that they have already developed an intuition that gives them an innate ability to evaluate social media sources in their field of expertise (Metzger, 2007, p. 2088). However, with a strong motivation to ensure that their sources are of high authority, they may also find a systemic framework of evaluation useful for recontextualizing the particular credibility features of social media sources.

We acknowledge that there is a fair amount of controversy over the use of checklists in evaluation, and we aim to address this. We argue, according to Metzger's research on motivation, that users in our context (students with assignments and scholars whose professional reputation depends on quality) will be highly motivated to consider using a clear and concise checklist to address credibility issues of resources, particularly those that are considered nontraditional, like blogs. We also have expanded our application of the CRAAP test to consider the spirit of each of its elements, shifting from a simple checklist to a more flexible assessment framework. Researchers new to evaluating social media need some way to develop their intuition, knowledge and critical eye toward what to look for when determining which resources to incorporate in their work, and our application of the CRAAP test fulfills this need. As Lewis & Smith (1993) pointed out "elaborating the given material, making inferences beyond what is explicitly presented, building adequate representations, analyzing and constructing relationships" are all part and parcel to critical thinking and to our suggested application of the CRAAP test (qtd in King et al., 1998, p. 39).

The CRAAP Test, Heuristics, and Critical Thinking

Higher-order thinking skills are "grounded in lower order skills such as discrimination, simple application and analysis, and cognitive strategies and are linked to prior knowledge of subject matter content" (King et al., 1998, p.1). This implies that the path to higher-order thinking begins with using simple skills and innate knowledge or "intuition." The CRAAP test provides a scaffolded approach to evaluating online resources.

Several sources in the library field have criticized the checklist approach. Meola (2004) asserted that checklists are unrealistically long, do not provide guidance on how to evaluate their criteria, or, when they do, offer unreasonable advice (such as requiring an email address or other contact information). Dahl wrote, "Commonly cited shortcomings of the checklist approach are that it can be difficult and/or cumbersome to implement, it encourages mechanistic rather than critical thinking, and it is not responsive to the varied contexts, needs and motivation levels of students" (2009, p.12). Burkholder asserted, "While convenient, the CRAAP questions imply that high-quality sources are recognizable because they are constructed according to a rigid set of guidelines" (2010, p.5). We believe that our approach to using the CRAAP test, along with Metzger's findings below, negate these concerns.

Metzger, et al. (2010) found that Internet users naturally apply various heuristic methods to web evaluation. In other words, her study showed that users already have several intuitive tools at their disposal. Users tend to turn to "enthusiasts," who are "presumed but noncredentialed experts." They determine whether the enthusiasts are experts by evaluating reputation. This is an approach that is implied in the authority criteria of the CRAAP test. They use other "indicators such as topic mastery, writing style, spelling and grammar, and the extent of details offered" (p. 424). These are methods used in the accuracy and relevance criteria.

That these skills are already intuitive for many searchers considerably lessens the burden of completing a checklist. It also refutes Warnick's (2004) assertion that use of a checklist, which she describes as a "'one-size-fits-all' approach to Web site credibility assessment [,] does not work well because it does not align with what users actually do" (p.262). Instead, the checklist provides signposts to remind researchers what kinds of elements they should be locating and evaluating.

Meola (2004) also asserted that checklist tests do not facilitate higher-order thinking: "The checklist model in practice [...] can serve to promote a mechanical and algorithmic way of evaluation that is at odds with the higher-level judgment and intuition that we presumably seek to cultivate as part of critical thinking" (p. 337). We believe that our approach to using the CRAAP test to evaluate social media sources addresses this concern as well. A researcher need only consider the spirit of each of the criteria and apply them according to the source's specific context and research interest. Critical thinking and reflection arises from this process of developing an understanding of the essence of the CRAAP criteria. It comes from researchers practicing crafting their own questions, which were inspired by the CRAAP test and are dependent on the social media context, to determine whether their source is of high quality.

FUTURE RESEARCH DIRECTIONS

In this chapter, we present the argument that the CRAAP method of website evaluation is an effective approach for assessing the appropriateness of blogs for research. Despite arguments to the contrary in literature about evaluation of online sources, we present evidence that the CRAAP test is an effective tool for evaluation of social media sources such as blogs. The test's effectiveness is boosted by consideration of the essential spirit of its five criteria to craft context-specific questions for evaluation, rather than a more rigid reading of the test elements.

Continuing application of the CRAAP test to emerging online media would be helped by in-depth analysis of new social media formats as they come into use as information resources. This would ideally include a detailed survey of emerging features and research applications, as well as assessment of each feature's potential as a source for information that can be used in evaluation.

Metzger, et al.'s (2010) finding that users employ more rigorous evaluation techniques in proportion to the degree of risk in using inaccurate information suggests that it would be useful to develop a study as to how users evaluate social media formats in practice. With this information, we could further refine evidence-based approaches to credibility analysis and evaluation.

CONCLUSION

The intent of using the CRAAP test is to develop researchers' evaluative skills, eventually endowing them with fine-tuned intuition and enhanced higher-order reasoning in evaluating blogs as a research source. An expanded contemplation of the spirit of the checklist criteria encourages a flexible approach to credibility assessment and is well-adapted to the particular features of blogs. The anarchistic nature of social media demands that a researcher using the CRAAP test account for contextual considerations, employing critical thinking skills. We assert that this contradicts Meola's (2004) claim that checklists foster algorithmic or mechanical thought. Repeated practice in assessing blog credibility with the CRAAP test will develop the reflexive skills that become intuition and knowledge.

Metzger's (2007) depiction of the dual process model of evaluation shows how easily a researcher can leap from having a need for information to making informed judgments when provided with the tools to make the judgment. Other studies (Currie et al., 2010; Hargittai et al., 2010), revealed that Internet users can *verbalize* what they should be looking for to determine credibility, but that they often do not actually take the steps to do so. As Currie, et al. (2010) reported:

Even though the students understood the need to find valid or scholarly information, the authors concluded that the students were not skilled in the application of evaluative criteria. Indeed, these students articulated only three or four specific criteria they would use to evaluate a source, used them repeatedly, and then could not seem to think of any others. They often used the proper terminology in describing their selection process to the authors, but clearly did not understand

the definitions of the terms. For example, while several of the students indicated they wanted to find a "credible" source, they were unable to list many of the specific criteria they could use to determine whether a source was credible or not. (p. 122-123).

With this in mind, we can see how the CRAAP test can be used by those instructing new researchers to lead them to critical evaluation while using exciting and new, but riskier, information sources. Although beginning researchers' heuristic methods of evaluation are a good start, the educator adds a critical thinking component to the research process by discussing expectations, providing tools, and demonstrating examples of rigorous evaluation methods for nontraditional resources such as blogs and other social media applications.

With practice, experienced scholars and beginning researchers alike develop an intuitive, evaluative assessment process, creating their own queries using a multitude of tools and skills. This type of evaluation becomes ingrained, a habit and skill that will aid the researcher in more mundane, low-risk information gathering as well. As Hargittai et al. (2010) pointed out, those using online sources "are not always turning to the most relevant cues to determine credibility of online content. Accordingly, initiatives that help educate people in this domain—whether in formal or informal settings—could play an important role in achieving an informed Internet citizenry" (p. 487). We believe that with the growth of blogs in the online environment, the ability to locate, evaluate and incorporate these rich resources into scholarly research will yield rewards.

REFERENCES

Alvarez-Torres, M., Mishra, P., & Zhao, Y. (2001). Judging a book by its cover: Cultural stereotyping of interactive media and its effect on the recall of text information. *Journal of Educational Multimedia and Hypermedia, 10*(2), 161–183.

Arditi, L. (2012). Cranston West prayer banner was removed Saturday, school chief says. *Providence Journal Breaking News.* Retrieved from http://news.providencejournal.com/breaking-news/2012/03/cranston-west-p-1.html

Association for College and Research Libraries. (2000). *Information literacy competency standards for higher education.* Retrieved from http://www.ala.org/acrl/standards/informationliteracycompetency

Austria, J. L. (2007). Developing evaluation criteria for podcasts. *Libri, 57*(4), 179-207. Retrieved from http://www.librijournal.org/pdf/2007-4pp179-207.pdf

Banning, S. A., & Sweetser, K. D. (2007). How much do they think it affects them and whom do they believe? Comparing the third-person effect and credibility of blogs and traditional media. *Communication Quarterly, 55*(4), 451-466. doi:10.1080/01463370701665114

Barzun, J., & Graff, H. (1992). *The modern researcher.* Boston, MA: Houghton Mifflin Co.

Berkman, R. (2004). *The skeptical business searcher: The information advisor's guide to evaluating web data, sites, and sources.* Medford, NJ: Information Today.

Blakeslee, S. (2004). The CRAAP test. *LOEX Quarterly, 31*(3). Retrieved from http://commons.emich.edu/loexquarterly/vol31/iss3/4

Brabazon, T. (2006). The Google effect: Googling, blogging, wikis and the flattening of expertise. *Libri, 56*(3), 157-167. doi:10.1515/LIBR.2006.157

Bradshaw, P. (2008). When journalists blog: How it changes what they do. *Nieman Reports, 62*(4), 50-52.

Brahnam, S. (2009). Building character for artificial conversational agents: Ethos, ethics, believability, and credibility. *PsychNology Journal, 7*(1), 9-47.

Burkhardt, J. M., MacDonald, M. C., & Rathemacher, A. J. (2010). *Teaching information literacy: 50 standards-based exercises for college students.* Chicago, IL: American Library Association. doi:10.1080/10875301.2011.551069

Burkholder, J. (2010) Redefining sources as social acts: Genre theory in information literacy instruction. *Library Philosophy and Practice (e-journal)*, paper 413. Retrieved from http://digitalcommons.unl.edu/libphilprac/413

Carvin, A. (2012). The gay girl in Damascus that wasn't. *Storify.* Retrieved from http://storify.com/acarvin/the-gay-girl-in-damascus-that-wasnt

Chung, C. J., Nam, Y., & Stefanone, M. A. (2012). Exploring online news credibility: The relative influence of traditional and technological factors. *Journal of Computer-Mediated Communication, 17*(2), 171-186. doi:10.1111/j.1083-6101.2011.01565.x

Consumerist.com. (2006). *Sony's PSP blog flog revealed.* Retrieved from http://consumerist.com/2006/12/sonys-psp-blog-flog-revealed.html

Cooke, M. (2007). Baghdad burning: Women write war in Iraq. *World Literature Today, 81*(6), 23-26.

Currie, L., Devlin, F., Emde, J., & Graves, K. (2010). Undergraduate search strategies and evaluation criteria. *New Library World, 111*(3), 113-124. doi:10.1108/03074801011027628

Dahl, C. (2009). Undergraduate research in the public domain: the evaluation of non-academic sources online. *RSR. Reference Services Review, 37*(2), 155-163. Retrieved from http://ecommons.usask.ca/xmlui/bitstream/handle/10388/281/Undergraduate%20research%20in%20the%20public%20domain-final-Feb.%207.pdf?sequence=1 doi:10.1108/00907320910957198

Dinkelman, A. L. (2010). Using course syllabi to assess research expectations of biology majors: Implications for further development of information literacy skills in the curriculum. *Issues in Science and Technology Librarianship, 60.* Retrieved from http://www.istl.org/10-winter/refereed3.html

Doyle, T., & Hammond, J. L. (2006). Net cred: Evaluating the Internet as a research source. *RSR. Reference Services Review, 34*(1), 56–70. Retrieved from http://search.proquest.com/docview/200503715?accountid=36823 doi:10.1108/00907320610648761

English, K., Sweetser, K. D., & Ancu, M. (2011). YouTube-ification of political talk: An examination of persuasion appeals in viral video. *The American Behavioral Scientist, 55*(6), 733–748. doi:10.1177/0002764211398090

Enos, T., & Borrowman, S. (2001). Authority and credibility: Classical rhetoric, the internet, and the teaching of techno-ethos. In Gray-Rosendale, L., & Gruber, S. (Eds.), *Alternative rhetorics: Challenges to the rhetorical tradition* (pp. 93–110). Albany, NY: State U of New York P.

Flanagin, A. J., & Metzger, M. J. (2007). The role of site features, user attributes, and information verification behaviors on the perceived credibility of web-based information. *New Media & Society, 9*(2), 319–342. doi:10.1177/1461444807075015

Fogg, B. J., & Tseng, S. (1999). The elements of computer credibility. *Proceedings of Computer Human Interface SIG Conference,* (pp. 80–87).

Friedman, S. M. (2011, September/October). Three Mile Island, Chernobyl, and Fukushima: An analysis of traditional and new media coverage of nuclear accidents and radiation. *The Bulletin of the Atomic Scientists, 67*(5), 55–65. doi:10.1177/0096340211421587

Fritch, J. W., & Cromwell, R. L. (2001). Evaluating internet resources: Identity, affiliation, and cognitive authority in a networked world. *Journal of the American Society for Information Science and Technology, 52*(6), 499–507. doi:10.1002/asi.1081

Gogoi, P. (2006). Wal-mart's Jim and Laura: The real story. *Bloomberg Businessweek.* Retrieved from http://www.businessweek.com/bwdaily/dnflash/content/oct2006/db20061009_579137.htm

Hargittai, E., Fullerton, L., Menchen-Trevino, E., & Thomas, K. (2010). Trust online: Young adults' evaluation of web content. *International Journal of Communication, 4*, 168–194.

Head, A. J., & Eisenberg, M. B. (2010). *Truth be told: How college students evaluate and use information in the digital age.* Retrieved from: http://projectinfolit.org/pdfs/PIL_Fall2010_Survey_FullReport1.pdf

Herring, S., Scheidt, L. A., Bonus, S., & Wright, E. (2004). Bridging the gap: A genre analysis of weblogs. *Proceedings of the 37th Hawaii International Conference on System Sciences.* DOI:0-7695-2056-1/04

Howard, P., Duffy, A., Freelon, D., Hussain, M., Mari, W., & Mazaid, M. (2011). *Opening closed regimes: What was the role of social media during the Arab Spring?* Seattle, WA: Project on Information Technology and Political Islam. Retrieved from http://dl.dropbox.com/u/12947477/publications/2011_Howard-Duffy-Freelon-Hussain-Mari-Mazaid_pITPI.pdf

Jabr, F. (March 21, 2011). How Josef Oehmen's advice on Fukushima went viral. *New Scientist.* Retrieved from http://www.newscientist.com/article/dn20266-how-josef-oehmens-advice-on-fukushima-went-viral.html

Johnson, T. J., & Kaye, B. K. (2004). Wag the blog: How reliance on traditional media and the internet influence credibility perceptions of weblogs among blog users. *Journalism & Mass Communication Quarterly, 81*(3), 622–642. doi:10.1177/107769900408100310

Kapoun, J. (1998). Teaching undergrads WEB evaluation: A guide for library instruction. *C&RL News.* (July/August), 522-523.

King, F. J., Goodson, L., & Rohani, F. (1998). *Higher order thinking skills.* Center for Advancement for Learning and Assessment, Florida State University. Retrieved from http://www.cala.fsu.edu/files/higher_order_thinking_skills.pdf

Kjellberg, S. (2009). Scholarly blogging practice as situated genre: An analytical framework based on genre theory. *Information Research: An International Electronic Journal, 14*(3).

Lee, H., Park, S., Lee, Y., & Cameron, G. T. (2010). Assessment of motion media on believability and credibility: An exploratory study. *Public Relations Review, 36*(3), 310–312. doi:10.1016/j.pubrev.2010.04.003

Leon, J. (2010). The blog as beat. *Nieman Reports, 64*(4), 9–11.

Mackey, R. (2011). 'Gay girl in Damascus' blog a hoax, American says. *New York Times.* Retrieved from http://www.nytimes.com/2011/06/13/world/middleeast/13blogger.html

Marsh, C. (2006). Aristotelian ethos and the new orality: Implications for media literacy and media ethics. *Journal of Mass Media Ethics, 21*(4), 338–352. doi:10.1207/s15327728jmme2104_8

Marshall, J. (2002). TPM Editors Blog. *Talking Points Memo.* Retrieved from http://talkingpointsmemo.com/archives/week_2002_12_01.php

Massachusetts Institute of Technology. (2012). *Oehmen, Josef, LAI research scientist.* Retrieved from http://lean.mit.edu/about/lai-structure/faculty-researchers-and-staff/oehmen-josef

McMillan, G. (2011). Twitter reveals active user number, how many actually say something. *Time.* Retrieved from http://techland.time.com/2011/09/09/twitter-reveals-active-user-number-how-many-actually-say-something/

Meola, M. (2004). Chucking the checklist: A contextual approach to teaching undergraduates web-site evaluation. *Libraries and the Academy, 4*(3), 331–342. doi:10.1353/pla.2004.0055

Meriam Library, California State University Chico. (2010). *Evaluating information: Applying theCRAAP test.* Retrieved from http://www.csuchico.edu/lins/handouts/evalsites.html

Metzger, M. J. (2007). Making sense of credibility on the web: Models for evaluating online information and recommendations for future research. *Journal of the American Society for Information Science and Technology, 58*(13), 2078–2091. doi:10.1002/asi.20672

Metzger, M. J., Flanagin, A. J., & Medders, R. B. (2010). Social and heuristic approaches to credibility evaluation online. *The Journal of Communication, 60*(3), 413–439. doi:10.1111/j.1460-2466.2010.01488.x

Notess, G. (2010). The changing information cycle. In P. McCaffrey (Ed.), *The news and its future* (34-39). New York, NY: H.W. Wilson Co.

Reynolds, N. (1993). Ethos as location: New sites for understanding discursive authority. *Rhetoric Review, 11*(2), 325–338. doi:10.1080/07350199309389009

Segal, J., & Richardson, A. (2003). Introduction. Scientific ethos: Authority, authorship, and trust in the sciences. *Configurations, 11*(2), 137–144. doi:10.1353/con.2004.0023

Smith, A. (2008). *New numbers for blogging and blog readership*. Retrieved from http://www.pewinternet.org/Commentary/2008/July/New-numbers-for-blogging-and-blog-readership.aspx

St. Amant, K. (2004). International digital studies: A research approach for examining international online interactions. In Buchanan, E. (Ed.), *Readings in virtual research ethics* (pp. 317–337). Hershey, PA: Information Science Publishing. doi:10.4018/978-1-59140-152-0.ch017

Swift, M. (2012). *Facebook sees its growth slowing down*. The Tennessean.

Terdiman, D. (2004). Onion taken seriously, film at 11. *Wired*. Retrieved from http://www.wired.com/culture/lifestyle/news/2004/04/63048

Warnick, B. (2004). Online ethos: Source credibility in an "authorless" environment. *The American Behavioral Scientist, 48*(2), 256–265. doi:10.1177/0002764204267273

Weaver, N. (December 18, 2006). What we should learn from Sony's fake blog fiasco: A debate over false marketing practices. *AdAge Agency News,* Retrieved from http://adage.com/article/small-agency-diary/learn-sony-s-fake-blog-fiasco/113945/

ADDITIONAL READING

Avery, S. (2007). Media literacy and library instruction: A case study of writing with video. *College & University Media Review, 13*, 77–93.

Baildon, M., & Damico, J. S. (2009). How do we know? Students examine issues of credibility with a complicated multimodal web-based text. *Curriculum Inquiry, 39*(2), 265–285. doi:10.1111/j.1467-873X.2009.00443.x

Castillo, C., Mendoza, M., & Poblete, B. (2011). Information credibility on Twitter. In *Proceedings of the 20th International Conference on World Wide Web, WWW 2011* (pp. 675–684). Retrieved from http://www.ra.ethz.ch/cdstore/www2011/proceedings/p675.pdf

Christian, A. J. (2009). Real vlogs: The rules and meanings of online personal videos. *First Monday, 14*(11). Retrieved from http://frodo.lib.uic.edu/ojsjournals/index.php/fm/article/view/2699

Flanagin, A. J., & Metzger, M. J. (2007). The role of site features, user attributes, and information verification behaviors on the perceived credibility of web-based information. *New Media & Society, 9*(2), 319–342. doi:10.1177/1461444807075015

Gunter, B., Campbell, V., Touri, M., & Gibson, R. (2009). Blogs, news and credibility. *Aslib Proceedings, 61*(2), 185–204. doi:10.1108/00012530910946929

Jin, Y., & Liu, B. (2010). The blog-mediated crisis communication model: Recommendations for responding to influential external blogs. *Journal of Public Relations Research, 22*(4), 429–455. doi:10.1080/10627261003801420

Johnson, T. J., & Kaye, B. K. (2002). We believability: A path model examining how convenience and reliance predict online credibility. *Journalism & Mass Communication Quarterly, 79*(3), 619–642. doi:10.1177/107769900207900306

Jones, T., & Cuthrell, K. (2011). YouTube: Educational potentials and pitfalls. *Computers in the Schools, 28*(1), 75–85. doi:10.1080/07380569.2011.553149

Kang, M. (2010) *Measuring social media credibility: A Study on a measure of blog credibility*. Institute for Public Relations. Retrieved from http://www.instituteforpr.org/

Kim, K. S., & Sin, S. C. J. (2011). Selecting quality sources: Bridging the gap between the perception and use of information sources. *Journal of Information Science, 37*(2), 182–192. doi:10.1177/0165551511400958

Ojala, M. (2005). All generalizations are false, including this one. *Online, 29*(3), 5. Retrieved from http://search.proquest.com/docview/199960071?accountid=36823

Peoples, B., & Tilley, C. (2011). Podcasts as an emerging information resource. *College & Undergraduate Libraries, 18*(1), 44–57. doi:10.1080/10691316.2010.550529

Scale, M. S. (2008). Facebook as a social search engine and the implications for libraries in the twenty-first century. *Library Hi Tech, 26*(4), 540–556. doi:10.1108/07378830810920888

Scherlen, A. (2008). Column people: What's their future in a world of blogs? Part I: Columns and blogs: Making sense of merging worlds. *The Serials Librarian, 54*(1/2), 79–92. doi:10.1080/03615260801973463

Sweetser, K. D., Porter, L. V., Chung, D. S., & Kim, E. (2008). Credibility and the use of blogs among professional in the communication industry. *Journalism & Mass Communication Quarterly, 85*, 169–185. doi:10.1177/107769900808500111

Thorson, K., Vraga, E., & Ekdale, B. (2010). Credibility in context: How uncivil online commentary affects news credibility. *Mass Communication & Society, 13*(3), 289–313. doi:10.1080/15205430903225571

Vraga, E. K., Edgerly, S., Wang, B. M., & Shah, D. V. (2011). Who taught me that? Repurposed news, blog structure, and source identification. *The Journal of Communication, 61*(5), 795–815. doi:10.1111/j.1460-2466.2011.01581.x

Warmbrodt, J., Sheng, H., Hall, R., & Cao, J. (2010). Understanding the video bloggers' community. *International Journal of Virtual Communities and Social Networking, 2*(2), 43–59. doi:10.4018/jvcsn.2010040104

KEY TERMS AND DEFINITIONS

Blog: Often updated web page in which dated posts are listed in reverse chronological order. Usually includes features such as comments and archive of past posts.

Evaluation: Assessment of a source for quality and relevance to a researcher's work.

Flog: Fake blog.

Hashtag: The "#" symbol, placed before a folksonomic keyword. Used in social networking applications to categorize posts and comments.

Hyperlink: Clickable link within an online publication. Takes the user to a new Internet source.

Microblog: A blog with shorter posts. Often the microblog software prescribes a word limit.

Post: A chunk of writing on a particular topic, uploaded to a blog or microblog.

Trackbacks: links from one blog post to another blog's posting.

Tweet: Microblog post on the Twitter platform.

Vlog: Blog in which posts are created using video sometimes with associated text to situate the video.

Chapter 14

Building a Professional Ethos on LinkedIn

Christy Oslund
Michigan Technological University, USA

ABSTRACT

In the face of increasing use of digitally mediated contexts, teachers and students on all levels are expected to be familiar with creating content appropriate for the World Wide Web, and their professional lives are affected by the digital content they create. The professional online networking site LinkedIn, for example, is a group of communities where professionals can create an ethos that will benefit them in both searching for work and maintaining their current working status. In such venues, both students and teachers still need guidance on how to create a profile and presence that will establish a positive, approachable ethos. Specific examples show how the author accomplished this in the $50 billion per year pet industry. These examples clarify both what to do and what to avoid in creating a profile and presence in a professional online community.

INTRODUCTION

LinkedIn began as a professional networking site when launched in May 2003 ("About Us," 2012). Employers, employees, and businesses seeking clients all use the site to connect with represented professionals, who range from dog walkers to Wall Street executives. With 2 million company members, 161 million individual members, including member executives from each Fortune 500 company, LinkedIn is currently the largest online professional networking site ("About Us," 2012). Membership in LinkedIn and participation in one of the more than a million groups on LinkedIn is a free, accessible, and immediate way to begin building a public profile and ethos amongst a professional community. Though paid Premium Subscription membership earnings are up "91 percent year-over-year" ("About Us," 2012), it is still possible to belong to LinkedIn, participate in communities, and submit resumés to companies for no cost to the user. More than 102.5 million unique individual views from around the world

DOI: 10.4018/978-1-4666-2663-8.ch014

were made in the first quarter of 2012; with more than 20 million members who are students or recent college graduates ("About Us," 2012), it is evident to a growing number of people entering the job market that LinkedIn is a tool for helping develop a professional reputation and profile. LinkedIn is rapidly growing, and in one year moved from the 54th most visited web site to the 31st most visited ("About Us," 2012). The popularity of this site continues as the number of page views increases, and the number of members participating in the LinkedIn communities.

By watching the site it is also evident that what is not clear to all users is the difference between building a visible profile and building a visible, *credible* profile that increases personal value to potential employers and clients. LinkedIn provides opportunities to build a global profile; this same site, if used thoughtlessly, can help ruin a reputation in each computer-literate industry and nation. How people build their profile and comment in communities directly affects the ethos and reputation they establish amongst the LinkedIn audience. Being deliberate and careful can make the difference between creating a positive or negative impression on potential professional connections. As Daniel Keller (2007) reminded us, the same rhetorical concepts are at work in digital spaces as are at work in traditional rhetoric, those being "appeals that address an audience's emotions (pathos), rest on a logical argument (logos), or appeal to an understanding of ethical behavior (ethos)" (p. 49). This means that while using LinkedIn can help create a positive ethos, failure to make the right traditional rhetorical moves in LinkedIn communities can create a negative reputation that will follow a person across time and space.

Digitally Compose to Create a Positive Online Ethos

In a college context, teaching writing has traditionally been the work of compositionists. Somewhere between composition and technical communica-

tion classes, most students will be asked to develop a resumé and perhaps learn something about the etiquette of writing and submitting cover letters. Connecting with clients and employers in a digital age, however, means that knowing the formats for paper-based communication is no longer sufficient for a person entering the workforce. As Stuart Selber has pointed out (2004, 2010), teachers in general have not been fully prepared for the new era of digitally mediated composing and writing and could use some support in teaching in this area. Cynthia Selfe's *Multimodal Composition* (2007) was a step in the direction of providing concrete guidance for composition teachers and included ideas for using technology in the classroom. What is still lacking, however, is thick support for teachers and students on how to write and make rhetorical moves in situated technological contexts like LinkedIn, where it is not just the technology that needs to be considered but also the professional implications of the communication moves made within the particular environment. As Selfe (2004) identified, composition teachers often feel "inadequate to the task of teaching students about new media texts and the emerging literacies associated with these texts" (p. 67). The purpose of this chapter is to offer some specific guidance to those seeking to create a positive online ethos by building a profile and presence on LinkedIn, and while those strategies must be specific to that particular context to be effective, the moves I discuss should spur rhetorical ideas that could be adapted in other contexts as well.

I appreciate Moe Folk's (2009) definition of digital composition as "the production of texts that are computer-mediated and draw heavily from the technologies and social practices associated with networked computers" (pp. 2-3). I also appreciate the distinction that Folk made between the more "scholarly tradition" associated with "composing" and what most of us take part in when we are writing in our everyday use of digital mediated contexts, whether that be emailing, texting, tweeting, blogging etc. I will thus follow

Folk's example and refer to less composed work as "digital writing" (p. 3). Digital writing encompasses much of what we take part in when we are working online. Even when writing in what should be a formal context, such as a student emailing a faculty member, there is a tendency to "write" as opposed to "compose." Composing implies a thoughtful process that usually includes revision. Writing is what we all tend to do when firing off a quick message and we sometimes forget that even in "informal activities" like texting, we are nonetheless within digital contexts that might be better served if we took the time to compose our message.

That is what much of the question of this chapter is: How should we compose in a context that, due to its social overtones, subtly encourages informal writing? A broad recommendation would be to treat all digital writing as the semi-permanent record of our thoughts and actions that it is. The line between "composing" and "writing" in digital media is one that we need to realize can quickly erase us from a reader's point of view: a potential future boss or client does not care if a careless, unprofessional comment was meant to be between friends or was made after a night of drinking because such comments still reflect negatively on the responsible person. In "Opening New Media to Writing" (2004), Anne Wysocki reminded us to be "alert" (p. 8) to our own agency when it comes to the content we are responsible for online. Being alert includes realizing that the unfavorable web content we are responsible for can eventually be traced back to us. Such negative occurrences could cost one a job, a promotion, or a good reputation, as people who have made poor choices in using their Facebook accounts have discovered the hard way (Smith, 2012.) Whether someone is posting recipes on the web or showing up in a friend's YouTube clip doing something reckless, each piece of Web content related to an individual will add either negatively or positively to their ethos. This is something to remain mind-ful of when creating all web content or allowing others to post pictures or videos that include you.

Increasingly, faculty are under pressure to be more proactive in introducing and using technology in the classroom (Selber, 2010; Selfe, 2007). Folk (2009), drawing on the New London Group, also reminded us that the argument we face these days in academics is that "educators need to embrace the mulitiliteracies concept because it connects students politically, economically and socially with the worlds they inhabit now, as well as the worlds they will inhabit in the future "(p. 18). Educators, including those who teach rhetoric, are expected to help prepare their students for the use of new media; as Selber (2010) noted, "it is difficult to imagine a rhetorical activity untouched by ongoing developments in writing and communication technologies" (p. 2). The concerns of this chapter are equally immediate; not only is using LinkedIn increasingly likely as a source of elevating professional profiles but how people use LinkedIn can help build or, alternately, help destroy their professional ethos. Using LinkedIn is always a rhetorical activity, and since LinkedIn has such potential as a professional development tool, it is a space where ethical conduct is critical in creating a positive ethos.

One of the underlying assumptions of this chapter, and this collection, is that rhetoric is meant to be used ethically, particularly if one wishes to have and maintain a credible ethos. In the foreword to Selber's *Rhetoric and Technology* (2010), Carolyn Miller observed that, "In seeking to influence the beliefs, feelings, and attitudes of others, we may try too hard to rule, that is to manipulate others for our own purposes…truth and justice, cooperation and disclosure, will suffer" (p. ix). There is an ethical weight and obligation that accompanies our use of persuasion. Digitally mediated communication, with its wide accessibility, requires the rhetor to be even more conscious of the impression her written content has on both her own reputation and on what she is leading her reader to believe about her. In tight economic times, for example, it may be tempting to make yourself look like a stronger candidate for employment relative to the competition by exaggerating your experience. Miller (2010) reminded us that there is a need to be personally vigilant in our use of digital media particularly because technology allows new ways of manipulating the audience. We can combine images with words while approaching our audience in the safety of their own homes or private offices. Ethical rhetors making use of digital media have to be mindful of not letting pathos overtake logos; they have to learn to temper emotional appeals so that they do not promise more than they can actually deliver.

Marilyn Cooper (2010) has suggested that writing is a physical interaction between people and their environment, an evolutionary process that grows out of our responses to the contexts we are in. A writer who does not make a habit of critically reflecting on the implications of her written words may fall into the habit of making claims or implications "naturally"—that is, without thought and consideration—about her aptitudes and abilities that are exaggerations. In a context such as LinkedIn, a place where readers are potential clients and employers, content that is posted without analysis and thought may imply that individuals is capable of offering services or experience that they are not able to deliver. In a digital age, it does not take long for negative comments that reflect disappointing results with someone's work to become attached to names and therefore reputations. The technology allows us to create content that, as Miller (2010) said, "mak[es] some forms of communicative interaction possible or easy and others difficult or impossible"(x); it is possible in a LinkedIn profile, for example, to imply expertise or experience at the same time that it is impossible for a reader to immediately ascertain the truth of these claims. Unlike the more traditional context of a paper resumé, where claims are often associated with contact-people who will act as direct references (and agree to answer specific questions), LinkedIn

allows for the use of reference trading with other people that the profile owner has never actually met or worked with. This leads to a context where references may have no real value. I will return to this final point later in the chapter along with a more complete discussion of why trading false references on LinkedIn is an unethical practice.

When originally introduced into the curriculum, computers, as Selber (2010) has reminded us, were widely considered "more or less neutral, [because it was believed] that technologies and their contexts would not shape or challenge how teachers thought about domain content or about the art and craft of teaching itself" (p. 3). Those of us who have been involved in teaching with computers realize that this stance was a rather utopian perspective. From issues of who had access to technology to how the format of the technology being used shaped the message being received (Banks, 2005; Selfe & Hawisher, 2004) we have discovered that both traditional ethical concerns and new concerns related to the medium itself are definitely part of instructing users about the obligations and implications of new media use.

This is particularly true when using a site such as LinkedIn, where the product that is being advertised and marketed is a person, or a professional profile of a person to be more exact. As Selber (2010) has pointed out, "texts today can record literacy habits, activities, and experiences" (p. 6). These digital records of a person's habits and activities are increasingly easy to access, and thus increasingly likely to be used as a source of making decisions and judgments about a person one has never personally met. When such judgments affect someone's reputation and potential employability, they are particularly critical to keep in mind. The "off-handed" remark made in one context, such as on Facebook or YouTube, can come back to haunt a person, particularly in a job context. While LinkedIn allows a person to create a professional profile, that profile does not exist in a vacuum; other digital content related to a person is just a few keystrokes away. If a potential boss or client is media savvy enough to use LinkedIn, chances are that person will also be able to uncover any other unflattering Web content "out there" about an individual. Building a professional LinkedIn profile, then, is not surefire protection against other careless media content that someone might be related to through Web searches.

Selber (2010) also reminded us that how we as creators use the technology "reflect motivations, needs, and values" that, in turn, can "shape the nature of digital environments" (p. 8). This is something to be particularly mindful of when educating others about the use of digital communication—how are we educating users to shape not just the spaces, but the values held by users of the spaces and contexts they are active in? It not only damages someone's personal reputation to become known in a space and context as having questionable values, but if such actions become more widespread, the negative ethos damages the value of the space and context where the questionable values are distributed. LinkedIn would cease to have value if most of the people using the site were in fact lying about their experiences, abilities, and potential as an employees or service providers.

It is also important for users of LinkedIn to realize that while some community members may be their peers in terms of age or experience, other members of the community are different ages, hail from other cultures, and have different life experiences. For example, something as innocuous as listing "video games" as a hobby may appeal to some audience members, but others will find such a hobby off-putting. A colleague told me that he invited a community businessman into his technical communication class to review student resumés. The students asked this potential employer about listing "video gaming" under their hobbies. The businessman's response: That's a resumé I would toss straight in the garbage. Why? Because in his perception, that person would, at the first and every available opportunity, be gaming at work. What the students had thought would be a fun way to

connect with readers of their resumé was, to a non-gamer, potential employer, a red flag; the older businessman "read" gaming as an indication of workers who would not focus on what they were being paid to do. This is an example of not only differences in generational—and possibly even cultural—experience, it is also a wider reminder that any hobby that is socially perceived to be addictive is probably better left off a resumé—unless one is applying for a job in the field, i.e., when applying for work as a video game tester, gaming would only strengthen your resumé.

Web spaces can create the illusion of equality in professional spaces when, in fact, relationships of employer-employee and client-service provider still lie under the surface. I would not, for example, choose to hire someone who had previously treated me with disrespect while making comments in a community. Being active in LinkedIn, someone will soon discover there is always more than one person active in a field, always more than one potential person to hire. A well-developed reputation is the best way to stand out from the crowd. In *The Brand YU Life*, Hajj E. Flemings (2006) said that in creating a good reputation, "there are no accidents" (p. 28) in becoming known as a responsible, credible, person. Only part of creating a self-brand, or reputation, is "name recognition" (p. 33); a person also needs to create an image that holds the promise of what that person can provide as a professional. Using LinkedIn is a natural extension of what Flemings mentioned: "You," as a professional and an ongoing reputation, are in effect the "product" that you are using LinkedIn to market. The comments you make and the content you post all "say" something about what type of person you are.

Flemings (2006) made another important point—people who choose a profession that is their passion are more likely to be successful because they will be less focused on money as a motivating factor because they actually enjoy the work (pp. 45-46). Whether enjoying work or not, letting employers know you are unhappy reflects negatively on reputation. If someone becomes known for making a large number of negative comments about the work environments she has experienced, for example, she runs the risk of developing a reputation for being a negative person. Given a choice of two qualified individuals to hire, it is reasonable to suppose most employers would not choose the candidate who appeared to have a "bad attitude." How exactly, then, does a person create a positive, ethical profile on LinkedIn, the type of profile that contains a beneficial ethos?

How to Become Appropriately LinkedIn

I first joined LinkedIn in 2009 for one specific reason: to do some background research about a person I was quoting in an article I was writing. I did little, however, to establish my presence on the site —I simply joined, made the connection I wanted to make, and then largely ignored the LinkedIn communities for over a year. I am an academic and a writer who has many interests outside those revolving around academic institutions. Some of my activities outside education have to do with dog rescue and rehabilitation, and I am particularly interested in developing several books related to this other life. In a digital age where the death knell of the traditional book publishing industry has been sounding for years, anyone interested in writing books needs to do a great deal of research on how to break into the field. Publishers are still publishing, but the market is tighter and more nuanced, so writers need to be more aware about what publishers want. Coincidentally, my other areas of interest were the ones reflected in my original LinkedIn profile. In LinkedIn circles, my profile always focused on my background in dog rehabilitation and training, with my academics existing primarily as a footnote.

I belonged to LinkedIn but had been inactive for over a year when I read some advice to writers in *The Essential Guide to Getting Your Book Published* (Eckstart & Sterry, 2010). The authors suggested that in this changing age of digitally mediated contexts, a writer could help herself by creating an online presence and then look for ways to "become actively involved in a community that's relevant to your idea" (Eckstart & Sterry, 2010, p. 19). A light bulb went on for me, and I suddenly realized the real potential of both my LinkedIn profile and the communities available within that space. Before even submitting one of my manuscripts, I first needed to more consciously develop my reputation and connections in the community of dog owners and trainers. I would use LinkedIn to network and also to lead new people to the blog I realized I needed to start.

Although I was not creating an academic profile on LinkedIn, I did benefit from my background in rhetoric, composition, and communication as I created my presence. From the avatar I chose, and the tone I used in making comments in discussions, to the people I linked to, every aspect of my presentation and conduct was done purposefully. I wanted my personal brand to be one of a trustworthy, ethical person in everything from my words to my visual image. The rest of this chapter offers some real-world examples of how to use thoughtful intent to create an ethical LinkedIn professional profile and image. I will also mention at this point that in addition to my other employment, I have also worked in private industry and been part of hiring committees. The comments I make about what employers are looking for is thus also informed by having weeded through many job applicants and selecting out the few from the many.

How and Why to Use an Avatar

In the opening pages of *Writing New Media* (2004), Anne Wysocki argued that writing teachers are exactly what new media contexts require because they "are already practiced with helping others understand how writing…is embedded among the relations of agency and extensive material practices and structures that are our lives" (p.

7). Learning how to develop a LinkedIn profile requires teachers who remind students that each individual writer is responsible for the content they create and post. This is the first step in what Flemings (2006) described as creating a "personal brand" (p. 28). Wysocki (2004) further suggested that we sometimes look at the online content we create as "objects to be seen, to be physically manipulated" (p. 22). When initially composing one's profile, for example, it may seem intuitive to just put a few things (e.g., a picture, a list of achievements, the school you graduated from) on the homepage. Wysocki was right; we need to step back when designing and consider what the "object" (in this case, the profile homepage) we're creating truly looks like and what that object implies about us, how that object will be viewed within its current context.

Before filling out your LinkedIn profile, take time to choose the avatar, the icon or image that will visually represent you. It is important to begin LinkedIn existence with a visual image people can connect to your name and reputation. A profile attached to the default avatar, a generic blank grayscale icon, can say the wrong thing about ethos to viewers. It may suggest that the person lacks attention to detail, or that he does not have sufficient time to take on any new responsibilities—not attractive qualities to either potential employers or clients.

It is also instructive to remember Fleming's (2006) point that part of branding yourself is combining your image with a promise of what you have to offer (p. 33). Creating this emotional and visual impression with the audience requires an image for it to connect with the ethos you are portraying. Consider, for example, how much harder it would be to "picture" the difference between what the largest hamburger chains claim to be offering if the iconic images they had always associated themselves with, such as the complex ideas and values tied to the golden arches and the use of hot colors like red and yellow, were all replaced by the same amorphous, lifeless,

generic grey. Wysocki (2003) reminded us that creating the kind of complex association so that an image (such as a LinkedIn avatar) connects with audiences "everywhere" to promise the same thing takes a great deal of work and concentrated marketing (p. 58). An avatar should come to emblemize ethos, and the ethos should be one that is valued in your profession, e.g., indispensable team worker who gets things done; creative go-to account executive; visionary in the graphic design field, etc. Creating this connection between your image and what you bring to the table should be one of the goals of creating a LinkedIn profile.

When choosing an avatar, I was very conscious of my intended audience—people who work in the pet industry and those who own dogs. I had read Dan Dye and Mark Beckloff's (2003) *Amazing Gracie: A Dog's Tale*, in which they discuss establishing their chain of bakeries for dogs. They associated their successful business not with themselves, but with the images of the three dogs they owned at the time —their Three Dog Bakery openings always featured red carpet events where the dogs would arrive in formal dress. This was a great example of associating an image with a business, and the business's concept of feeding your dog like a member of your family. I also owned three dogs, but for my purposes I thought it best to choose just one of them for my avatar. I made this choice for design and image reasons. Each of my dogs is a member of a recognizable breed, and each breed has a different reputation among "dog people." I wanted to be careful which of those reputations I was drawing on in building my own ethos.

My oldest dog is a rough-haired collie, and while I did have a rather artistic picture that was a close-up of the left side of her face, I decided this image was a little too mysterious. Combine this with the breed's reputation for wandering, and I decided that I might risk implying I was a little *too* artsy and a little less down-to-earth and reliable. My middle dog is an English bull terrier, and in most of her pictures, she is looking away

Figure 1. The default grayscale LinkedIn avatar

Figure 2. Chris Oslund's LinkedIn Avatar

from the camera, giving an impression of aloofness or divided attention. The bully breed also has a reputation for being stubborn and clownish, which is certainly not the best ethos to tie yourself to if wanting to be thought of as professional. My youngest dog, about four months old at the time, was a yellow Labrador retriever, and I had a nice picture of her looking at the camera, her head tilted slightly as if she were interested to see what might happen next. The picture is one that, when associated with comments in the community I was interested in connecting with, would not give an impression of trying to be aggressive, or a "know-it-all," and helps create a friendly tone overall. Labradors also have a reputation for being steadfast, quick to learn, fairly easygoing, and energetic. Amongst dog people, Labs are associated with a good work ethic. Finally, Labs are very popular with the dog-owning public; after becoming more involved with LinkedIn communities, I realized by happenstance that I had nearly a thousand potential business connections/readers just amongst other Labrador owners in the "Labrador Retriever Owners" group (see Wronko, 2008).

Selecting your avatar is a rhetorical choice and heavily dependent on the audience you want to connect with. Are you best served using a headshot of yourself? A picture of you with a diverse group of people? A drawing that you sketched? Each of those images is imbued with a different sense of ethos and thus says something about who you are and implies values to the viewer. There is neither one image that is always appropriate nor one that

will have the same impact on all viewers (Wysocki, 2003). For example, a manager who posts an image of herself in a staid business suit may not seem exciting in certain communities, but that image may be perfectly appropriate to the context and target audience. What might be worth thinking about in this situation is the color of the suit being worn. Are there expectations within the industry you work in about what an appropriate clothing tone is? For example, my cousin once interviewed with a large corporation that for years had an unwritten understanding that all executives would wear navy blue suits. My cousin realized that going to work in any other color of suit would have marked him in a negative way as not "fitting in" with the company's expectations, even if he did his job correctly; read another way, choosing the wrong suit could affect overall ethos because workplace ethos is so much more than just the actual work an employee completes. LinkedIn is a professional website providing access to professional contexts and images should be selected accordingly. People should research other profiles in their targeted LinkedIn communities and field to figure out how others in the field are portraying themselves: Are there any unwritten expectations that can be seen amongst the images being used by others? The avatar sets a tone for the kind of ethos *you are choosing* to portray. If you step too far outside the expectations of the field, you may mark yourself in a negative way.

There is room for personal information in an avatar, if subtle. Some professionals, for example, pose for pictures in suits appropriate to their field, but include a pet in the picture. Others use pictures taken outdoors. Occasionally it is even appropriate to use an avatar that is not the profiled person but an image that is somehow representative or metaphorical; however, this final option requires great care. Cartoon avatars for example, have a very limited application, even for graphic artists. No matter what avatar someone chooses, it needs to be tested before going online. Consider having a friend show your image to a range of people not very familiar with you, and ask that friend to record the viewers' first impressions. Using a friend helps avoid bias from viewers who would self-regulate their comments if, for example, you were to show them a picture of yourself and ask for their comments. After selecting the avatar, make few if any changes to it—if, with time, the image needs to be updated, use a similar color scheme, pose, or other visual element that recalls the essence of the original. Coke and Pepsi, for example, have changed the stylization of their logos over the years, but always stayed with the same basic color patterns since Pepsi switched to red, white, and blue during WW II ("Pepsi," 2005). Constantly changing images shifts the ground under the viewer's feet and risks creating the impression that you are not a stable person or that your values are shifting. It also makes it more challenging for a viewer to follow you over time and space.

Creating Visual Appeal— and a Positive Ethos—by Limiting Information

Visual appeal is also improved or hampered by the blocks of text that appear on your homepage. One of the attractions of digital media is that it offers concise, specific information. The initial information provided in your LinkedIn profile is a hook, a taste of what you have to offer. Giving too much information, whether general or specific, can raise red flags with employers. If I observe that you do not appear to self-sensor in a public format, for example, then I may be concerned about how you will handle private or sensitive information, or whether you will be the person who corners me in the office and talks my ear off. Most employers are looking for evidence that a potential employee is capable of judging what the appropriate length and depth of information for different contexts is. Treating your LinkedIn profile exactly like your Facebook page is never appropriate—these are two different contexts with different purposes and different audiences. Facebook is a personal connection space for family, friends, and associates, and you can afford to bore people there with mundane information or long posts. LinkedIn is a professional space for connecting with those who can potentially impact your financial well-being; if you offend or bore people in this space, your ethos is negatively affected.

Content can be added to over time; however, poorly chosen content will never be "unseen" by those who have viewed it, even if eventually removed. With screen captures and repositories such as YouTube, it is also possible that content only posted briefly can live on to haunt someone for eternity. Additionally, extremely personal information is almost never appropriate in a professional profile. If marketing yourself as a weight loss counselor, then it might be appropriate to disclose how much weight you lost in the last year. Be mindful of the fact that each such specific detail given also removes some viewers from the pool of potential employers. For example, I saw one such profile where the impression created by the posted weight loss details caused me to believe the person was a yo-yo dieter. I would never hire people to guide me nutritionally if they were unable to manage their own weight reliably, just as I would not invest my money with investment consultants who had gone bankrupt. Whatever you post helps build a reputation and reputations are seldom neutral—most people either develop a positive or negative impression when viewing the profile content of others, so a deep understanding of audience and context is necessary if hoping to evoke a positive ethos.

The Usefulness of Group Membership

Your initial content should be a brief introduction to who you are, what professional services and knowledge you have to offer, what experience you have, and if available, some sample references from people you have worked with. When I first joined LinkedIn, I began with the name of my dog business, my education, and a brief response to the formulaic questions about the kind of connections I was interested in. When I came back to LinkedIn as an active member, my most important addition was my avatar and a link to my blog. Since then I have made small tweaks occasionally but only after careful thought. A stable ethos is not created by sudden, dramatic profile changes.

Once a profile is built, the next step is joining community conversations that are always taking part on LinkedIn. The site will suggest possible groups that might interest you based on your responses to a form that must be filled out when creating a profile. The groups you belong to should showcase your interests. As Wysocki advised (2003), step back and look at your profile page as an object; the entire page should provide clues to the viewer about the kind of professional you are and what you value. I did not just join dog groups such as "Dog Lovers" (Garrett), or pet groups like "Pet Friendly Animal Lovers" (Napolitano), I also joined groups that made my concerns for the ethical treatment of animals obvious, such as "The Animal Welfare Group" (Estol) and groups that showed I was interested in business aspects of the animal industry like "Pet Industry International" (Rohrer.) Some groups are open for anyone interested to join, but others are moderated and people need to apply for membership. Applying to a moderated group requires sending the moderator an email with an explanation or reason for want-

ing to join the group. Most moderators (known as group "owners" on LinkedIn) do not immediately respond to requests to join, so be patient. Sending a score of impatient emails asking for a decision is unprofessional and will most likely keep you from gaining access to the group, but more importantly, it will make it that much harder to establish a positive ethos with people who share the same interests as you. Moderators often do not count on the amount of time it will take to view requests, moderate conversations, all while running their own professional and personal lives.

This is why I would advise *against* starting out your LinkedIn life by starting and moderating your own group. Begin by spending time getting to know groups that are already available. It requires time and participation in communities to discover the types of conversations happening in groups. Once you know what is already available, you will be able to recognize whether there is an available niche for a new discussion group. Remember, starting a group that you do not moderate regularly will influence your own ethos in a negative way—your avatar can be displayed on each page of a group you moderate, so you are the most likely person to be thought of negatively if people have long waiting times to get in or have negative interactions once they are in the group. Other professionals are likely to ask themselves, "Why did this person start a group if he/she isn't going to make the time to weed out spam, inappropriate comments, or deal with requests to join in a timely manner?"

Once you begin to join groups, you will be able to send and receive requests to link with other professionals. This is another area where the groups you choose to join can be influential. Some professionals only link with those in a group they belong to; you can also usually send direct requests to others who in a group you are in (although some members restrict their access to those who know their email address.) I was able to send a direct request to link to the founder and CEO of a large national pet business based on the fact that we belonged to several of the same groups. I was somewhat pleasantly surprised when he accepted my request within 24 hours. Once you've been an active member on LinkedIn for a while, other people will regularly approach you, requesting to add you to their professional network. Once you exceed 500 connections, LinkedIn will indicate this status with a blue icon next to your name. This is the only indication regarding the number of contacts one has and is a quick visual indicator of how active a connection is likely to be. More active connections provide more opportunities for you to connect with those you are not directly connected to. Connecting for the sake of connecting, however, is of limited use. While I would never refuse a connection unless I found someone's profile objectionable, I only seek out connections that are related to my primary field and goals. I quit being as active in gaining new contacts once I had slightly over 500 connections; I currently have about 1,100 connections, due in large part to others requesting to be connected with me. Most of these connections are people active in the pet industry and dog owners, although a few have requested the connection because they are also involved in some form of animal rescue work.

Sometimes during this process, you are likely to receive requests for referrals from people you've only recently linked with; the implication, though seldom stated, is that this is a quid pro quo situation—you will provide a referral for another person and based on what you say, they will in turn provide a referral for you. Trading referrals on LinkedIn that suggest you have a working relationship with someone you have not met and/ or not worked with, is just as unethical as writing a letter of recommendation for someone you've never met or worked with. The context does not change the ethics of lying or intentionally misleading others. Wrong is wrong. If, however, someone does not hold the ethical position that lying and misleading others is wrong, then consider this from a practical, rather than ethical perspective. When someone lies regularly, that person is likely to

develop the brand and reputation of an unreliable, untrustworthy person. This is not a reputation that the average employer is seeking in new hires. As satisfying as it might feel to have nice (if false) words about how fine a worker you are posted on your personal profile, such postings are not worth the potential negative impact on your reputation if those falsehoods ever become public knowledge. And in a digital context, all such knowledge can potentially become public. You might even become the target of someone researching the question, "How many people in a given field are willing to trade unethical reviews with others?"

Conversation topics in communities are the other main source of written content that can be used to create reputations on LinkedIn. Some people choose to have little or no interaction in communities and do not post to conversations. Never taking part in conversations is a missed opportunity. LinkedIn is a sea of people, so the best way to catch people's attention is through having your avatar appear in thoughtful topics and comments. Posting topics or comments show your knowledge in an area and indicate whether you are keeping up with the latest developments in your field. Taking part in conversations also demonstrates your ability to have a difference of opinion with others without torching them for the opinions they express. When employers say they are looking for people who can work with diverse groups of people, they can judge whether those qualities and skills are evidenced in comments and conversations, particularly with so many international people taking part in LinkedIn conversations.

Be Positive, Be Moderately Active, Be Polite

One group I belong to is for those specializing in animal rescue, rehabilitation, and policy development for the ethical treatment of animals. A wide range of people belong to this group, everyone from veterinarians at zoos to public figures who shape the policies of governments. Over the course of several weeks, one member of the group started to stand out for posting combative comments that flamed other people's viewpoints, ideas, and opinions. The poster obviously had a set agenda that was not in keeping with the intent of the group—basically, an "outsider" had joined the group just to fight with people. At the same time, this person disclosed so many personal details through the posts that I was able to piece together where the person lived and what their occupation was. With a simple web search, I was able to connect this person to the two businesses they owned and the town they lived in. This person wanted to vent personal views on laws related to animal welfare. A LinkedIn community dedicated to animals was not the appropriate place to do so; the person would have been better off going to Craigslist and the "Rants and Raves" or the general "Pets" discussion groups.

This also leads to an example of how thoughtless comments in a global context can haunt a person. I am connected with people in the pet industry and receive requests for recommendations as well as referrals. I happen to know people in the same geographic location as the combative conversation poster mentioned above. This poster's business and my connections overlap. My own professional standards and ethos keep me from going out of my way to do harm to this careless poster's business; however, at the same time, I can say I have already *not recommended* this person in one particular situation. I will continue not recommending this person as long as they remain in business. Since word of mouth is counted on to help build reputations, having your existence ignored is problematic, just as accruing a negative ethos is. In fact, I wonder if being invisible is worse than being known for negative reasons.

From my own experience I can say that being moderately active in LinkedIn has led primarily to new connections within the LinkedIn communities and very few crossovers from my LinkedIn presence to my blog. The blog is proving to have a

separate life; it has a growing audience that finds the site through searches for images and topics. For example, I get a number of people finding the blog because they are searching online for information about and pictures of bull terriers, people looking for information on dog training, and people wanting information about the characteristics of specific breeds of dogs I have featured. Despite this lack of crossover, I believe that having a positive, professional profile on LinkedIn is part of creating the kind of online ethos I want to establish for myself. Not having a presence on LinkedIn would be like leaving a hole in the brand I am creating for myself. It simply would not make sense to market myself as someone who is part of the current pet professional community and not be present on LinkedIn.

Similarly, if I am reviewing other people's professional credentials, I now expect them to be conversant with LinkedIn communities related to their profession. I can also say I have had one offer of a business connection I was not expecting due to my LinkedIn presence —someone requested I become an import agent for dog supplies from Canada. I have received other offers to market products, and several dog-related books have been shipped to me free in the hopes that I will review them on my blog one day. This suggests that we may not be able to predict what LinkedIn community involvement will bring our way professionally, but opportunities are present that we might not otherwise be aware of, opportunities that arise out of our established ethos.

Will LinkedIn Change Employment Contexts?

In the years to come it will be interesting to see how a professional networking site such as LinkedIn affects hiring practices. Currently, I am employed by a university where the two biannual job fairs attract hundreds of companies that send representatives precisely because they can have face to face interactions with potential employees.

Talented people, however, can be found around the country and internationally, and being able to make initial contact through the internet is an increasingly popular option. It is certainly possible that one day, virtual contact will be the most common way for employers to identify the pool of candidates from which they will make final hiring selections. It is also possible that large national and international corporations will begin to scout junior executives for their own companies through their Web presence, including their professional conduct on LinkedIn.

A colleague also pointed out to me that such technology is allowing me to "reverse-engineer" the way connections have historically been made in the publishing field. It used to be that writers serious about getting published often ended up moving to publishing centers such as New York City in order to better facilitate face-to-face meetings (even chance ones) with agents and publishers. Now, I am able to stay in a rather remote geographical area in the Northern Midwest and continue working my "real" job while spending my other hours creating a reputation and ethos in a field I would be too isolated to take part in otherwise, if taking part can mean being known on a large scale outside my own geographic region.

CONCLUSION

It is clear that people who plan to take part in the job market best serve their interests by becoming familiar with—and present on—LinkedIn. Employers are increasingly likely to look for signs of computer literacy and an awareness of the evolving marketplace by inquiring if one does in fact belong to LinkedIn, not to mention being able to analyze a potential candidate's conduct within LinkedIn to discern rhetorical savvy, ability to work in diverse contexts, and overall ethos. While presence on the site may not be the deciding factor as to whether an employer does

make a job offer, a positive, professional profile will always be a mark in one's favor. It certainly adds appeal to potential employees when employers observe that they have the adequate computer literacy to create a professional ethos online. In a competitive job market, the more someone does to stand out in a positive way, the greater the chances of getting a desirable job offer, and not just any job offer, but a chance at a job whose duties and trajectories are actually appealing. Building a positive ethos through a LinkedIn profile—by paying attention to the visual rhetoric of the avatar, limiting and shaping information, being moderately active, and being polite and positive—is an increasingly necessary step to creating the self-proficient digital literacy that potential employers and clients expect.

ACKNOWLEDGMENT

Dr. Moe Folk generously shared creative and editorial ideas which ended up making their way into this chapter. He is also the colleague who shared the classroom experience of having a businessman offer feedback to students on their resumés—a very timely example which I appreciate being able to share here.

REFERENCES

Banks, A. (2005). *Race, rhetoric, and technology: Searching for higher ground*. New York, NY: Routledge.

Cooper, M. (2010). Being linked to the matrix: Biology, technology, and writing. In Selber, S. (Ed.), *Rhetorics and technologies: New directions in writing and communication* (pp. 15–32). Columbia, SC: University of South Carolina Press.

Dye, D., & Bleckloff, M. (2003). *Amazing Gracie: A dog's tale*. New York, NY: Workman.

Eckstart, A., & Sterry, D. (2010). *The essential guide to getting your book published: How to write it, sell it, and market it...successfully.* New York, NY: Workman.

Estol, L. (n.d.). *Animal welfare group.* LinkedIn. Retrieved from http://www.LinkedIn.com/groupsDirectory?itemactio n=mclk&anetid=89180&impid=&pgkey =anet_search_results&actpref=anetsrch_ name&trk=anetsrch_name&goback=. gdr_1327510683784_1

Flemings, H. E. (2006). *The brand YU life: Rethinking who you are through personal brand management.* Canton, MI: Third Generation Publishing.

Folk, M. (2009). *Then a miracle occurs: Digital composition pedagogy, expertise, and style.* Unpublished doctoral dissertation, Michigan Technological University, Houghton, MI.

Garrett, B. (n.d.). *Dog lovers.* LinkedIn. Retrieved from http://www.LinkedIn.com/groups/Dog-Lovers77136?itemac tion=mclk&anetid=77136&impid=&pgke y=anet_search_results&actpref=anetsrch_ name&trk=anetsrch_name&goback=. gdr_1327510683790_1

Keller, D. (2007). Thinking rhetorically. In Selfe, C. (Ed.), *Multimodal composition: Resources for teachers* (pp. 49–63). Cresskill, NJ: Hampton Press.

LinkedIn. (2011). *About us.* LinkedIn press center. Retrieved May 15, 2012, from http://press.linkedin.com/about

Miller, C. (2010). Rhetoric, technology, and the pushmi-pullyu. In Selber, S. A. (Ed.), *Rhetorics and technologies: New direction in writing and communication* (pp. ix–xii). Columbia, SC: University of South Carolina Press.

Napolitano, F. (n.d.). *Pet friendly animal lovers.* LinkedIn. Retrieved from http://www.LinkedIn.com/groups/Pet-Friendly-Animal-Lovers-89012?itemacti on=mclk&anetid=89012&impid=&pgkey =anet_search_results&actpref=anetsrch_ name&trk=anetsrch_name&goback=. gdr_1327510683792_1

Pepsi-Cola® Company. (2005). *Over one hundred years of fun and refreshment: The Pepsi-Cola® story.* Retrieved from http://pepsiusa.com/downloads/PepsiLegacy_Book.pdf

Rohrer, C. (n.d.). *Pet industry international.* LinkedIn. Retrieved from http://www.LinkedIn.com/groupsDirectory?item action=mclk&anetid=86951&impid=&pgk ey=anet_search_results&actpref=anetsrch_ name&trk=anetsrch_name&goback=. gdr_1327511692593_1

Selber, S. (2004). *Mulitiliteracies for a digital age.* Carbondale, IL: Southern Illinois University Press.

Selber, S. (2010). Introduction. In Selber, S. (Ed.), *Rhetorics and technologies: New directions in writing and communication* (pp. 1–14). Columbia, SC: University of South Carolina Press.

Selfe, C. (Ed.). (2007). *Multimodal composition: Resources for teachers.* Cresskill, NJ: Hampton Press.

Selfe, C., & Hawisher, G. (2004). *Literate lives in the information age: Narratives of literacy from the United States.* New York, NY: Routledge.

Selfe, C. L. (2004). Toward new media texts: Taking up the challenges of visual literacy. In Wysocki, A., Johnson-Eilola, J., Selfe, C., & Sirc, G. (Eds.), *Writing new media: Theory and applications for expanding the teaching of composition* (pp. 67–110). Logan, UT: Utah State University Press.

Smith, C., & Kanalley, C. (2011, May 25). Fired over Facebook: 13 posts that got people CANNED. *The Huffington Post*. Retrieved May 10, 2012, from http://www.huffington-post.com/2010/07/26/fired-over-facebook-posts_n_659170.html#s115707&title=Swiss_Woman_Caught

Wronko, G. (2008). *Labrador retriever owners*. LinkedIn. Retrieved May 5, 2012 from http://www.LinkedIn.com/groupsDirectory?it emaction=mclk&anetid=1087707&impid=&p gkey=anet_search_results&actpref=anetsrch_name&trk=anetsrch_name&goback=. gdr_1327510683786_1

Wysocki, A. (2003). Seriously visible. In Hocks, M. E., & Kendrick, M. (Eds.), *Eloquent images: Word and image in the age of new media* (pp. 37–59). Cambridge, MA: The MIT Press.

Wysocki, A. (2004). Opening new media to writing: Openings and justifications. In Wysocki, A., Johnson-Eilola, J., Selfe, C., & Sirc, G. (Eds.), *Writing new media: Theory and applications for expanding the teaching of composition* (pp. 1–42). Logan, UT: Utah State University Press.

ADDITIONAL READING

Banks, A. (2011). *Digital griots: African American rhetoric in a multimedia age*. Carbondale, IL: Southern Illinois University Press.

Barscherer, T. (2011). *Switching codes: Thinking through digital technology in the humanities and the arts*. Chicago, IL: University of Chicago Press.

Berry, D. (2012). *Understanding digital humanities*. New York, NY: Palgrave Macmillan. doi:10.1057/9780230371934

Blackcoffe Design Inc. (2009). *1,000 icons, symbols, and pictograms: Visual communications for every language*. Minneapolis, MN: Rockport Publishers.

Brooke, C. G. (2009). *Lingua fracta: Towards a rhetoric of new media*. Creskill, NJ: Hampton Press Inc.

Brummett, B. (2008). *A rhetoric of style*. Carbondale, IL: Southern Illinois University Press.

Cody, C. (2012). *The art of writing and speaking the English language: Word-study and composition & rhetoric*. Oxford, NC: Oxford City Press.

Couture, B. (2004). *Private, the public, and the published: Reconciling private lives and public rhetoric*. Logan, UT: Utah State University Press.

Gee, J. (2011). *Social linguistics and literacies: Ideology in discourses*. New York, NY: Routledge.

Gold, M. (Ed.). (2012). *Debates in the digital humanities*. Minneapolis, MN: University of Minnesota Press.

Golombisky, K., & Hagen, R. (2010). *White space is not your enemy: A beginners guide to communicating visually through graphic, Web, and multimedia design*. New York, NY: Focal Press.

Hafez, K. (2007). *The myth of media globalization* (Skinner, A., Trans.). Malden, MA: Polity Press.

Herrington, A., Hodgson, K., & Moran, C. (Eds.). (2009). *Teaching the new writing: Technology, change, and assessment in the 21st-century classroom*. New York, NY: Teacher's College Press.

Lidwell, W., Holden, K., & Butler, J. (2010). *Universal principles of design, revised and updated: 125 ways to enhance usability, influence perception, increase appeal, make better design decisions, and teach through design*. Minneapolis, MN: Rockport Publishers.

Lynn, S. (2010). *Rhetoric and composition.* New York, NY: Cambridge University Press. doi:10.1017/CBO9780511780172

Malamed, C. (2011). *Visual language for designers: Principles for creating graphics that people understand.* Minneapolis, MN: Rockport Publishers.

McKee, H. (2007). *Digital writing research: Technologies, methodologies, and ethical issues.* New York, NY: Hampton Press.

Palmeri, J. (2012). *Remixing composition: A history of multimodal writing pedagogy.* Carbondale, IL: Southern Illinois University Press.

Peeples, T. (2002). *Professional writing and rhetoric: Readings from the field.* New York, NY: Longman.

Pratt, A. (2012). *Interactive design: An introduction to the theory and application of user-centered design.* Minneapolis, MN: Rockport Publishers.

Samara, T. (2007). *Design elements: A graphic style manual.* Minneapolis, MN: Rockport Publishers.

Selfe, C. (2007). *Resources in technical communication: Outcomes and approaches.* Amityville, NY: Baywood Publishing Co.

Sherwin, D. (2010). *Creative workshop: 80 challenges to sharpen your design skills.* Blue Ash, OH: HOW Books.

Shipka, J. (2011). *Towards a composition made whole.* Pittsburgh, PA: University of Pittsburgh.

Urbanski, H. (2010). *Writing and the digital generation: Essays on new media rhetoric.* Jefferson, NC: McFarland.

Weisser, C. (2002). *Moving beyond academic discourse: Composition studies and the public sphere.* Carbondale, IL: Southern Illinois University Press.

Welch, N. (2008). *Living room: Teaching public writing in a privatized world.* Portsmouth, NH: Boyton/Cook.

Wheeler, A. (2009). *Designing brand identity: An essential guide for the whole branding team.* New York, NY: Wiley.

Williams, R. (2008). *The non-designer's design book.* Berkeley, CA: Peachpit Press.

KEY TERMS AND DEFINITIONS

Avatar: The icon or image that visually represents a person in an online community.

Community: A LinkedIn community is begun by an individual or by people who agree to share the duties of moderating the community. Communities are centered around a shared, identifiable interest, e.g. Social Media Networking, Professional Pet Walkers etc. Communities are created for the purpose of connecting with others in the field through posting job openings, conversation threads, and announcements of new products.

Conversation Thread: Begun when an individual posts a comment, article, or directly asks for discussion of a topic. Each time an individual posts a comment to a thread, their name and avatar are shown next to the comment; those who post regularly are shown in a side bar on the community page as "Top Influencers This Week" in the group.

Login Identity: The name or identity a person uses to access a community through a computer. Unlike social networking sites, LinkedIn communities operate under the assumption that a person is using their legal name/real identity, as opposed to a identity created for the purpose of posting in a particular kind of community.

Moderate Use: To use LinkedIn moderately includes belonging to at least half a dozen discussion groups, contributing to a discussion thread once a week, and starting a discussion thread at least once every three or four months – or posting an update to a still popular thread that you began.

Moderator: An individual who is responsible for a specific community and is considered the "owner" of a community; the moderator decides if the community will be open to anyone who wishes to post, or if the group will be closed, so that people must apply to the moderator to join the group. Moderators can delete conversation threads and keep specific individuals from posting in the community by barring their login identity.

Popular Thread: A thread of conversation within a community that continues to be posted to. Threads can be popular for several days, weeks, or even months.

Top Influencers: The five people in a community whose comments and/or conversation threads are generating the most activity within a community. An individual's avatar and name appear in a sidebar when they are a Top Influencer for the week within a community.

Chapter 15
Online Identity Formation and Digital Ethos Building in the Chinese Blogosphere

Zixue Tai
University of Kentucky, USA

Yonghua Zhang
Shanghai University, China

ABSTRACT

Exponential growth in the past decade has turned the Chinese blogosphere into the largest blogging space in the world. Through studying some of the most popular blog sites and bloggers, this chapter critically examines a number of their key defining features such as rhetorical strategies and persuasive approaches in building popular ethos and unique online identities in order to attract a steady user base. It also discusses some of the personal, topical, social, cultural, and political factors of the emerging Chinese culture of blogs and blogging against the particular backdrop of China's state-controlled media and communication environment.

INTRODUCTION

The Internet sector has registered dazzling growth in the past two decades in China, and Internet-based technologies and applications have increasingly integrated into various aspects of Chinese people's everyday life. China now boasts the largest online population in the world, with 513 million people connected to the Internet as of December 2011 (China Internet Network Infor-

mation Center [CNNIC], 2012). As a result, such an ever-expanding online communicative space provides Chinese netizens with handy access to an exploding base of user-generated information and has created innovative ways of information production and dissemination in a highly controlled media environment. Similarly, a vigorous line of academic discussion has emerged to disentangle the dynamics and nature of the unprecedented path of socio-poli-cultural transformations triggered by the Internet in China (e.g., Tai, 2006; Yang, 2009).

DOI: 10.4018/978-1-4666-2663-8.ch015

Coinciding with the ongoing Internet revolution is the quick rise of the Chinese blogosphere and the mainstreaming of blogs as a socio-cultural phenomenon in China in the past decade (Tai, 2012). By December 2011, China's blogging population reached 319 million (CNNIC, 2012), making it by far the largest blog community in the world. Meanwhile, compared with their counterparts in cross-national settings, Chinese netizens display a much higher propensity to both contribute to, and rely on, user-generated content (UGC) on the Internet (Tai, 2006). It has also been noted that Internet use and social network size positively contribute to an individual's tendency to vent opinions in Chinese cyberspace (Shen, Wang, Guo, & Guo, 2009). The particular nature of the content expressed online among different user groups and in specific cyber territories, however, remains under-explored to a large extent, yet enlightenment in the particulars of blog content and interactive dynamics between bloggers and followers is critical in understanding China's evolving Internet landscape.

Goldhaber (2006) argued that the Internet epitomizes what he terms, "attention economy" —an "all-encompassing system" that, "revolves primarily around paying, receiving, and seeking … the attention of other human beings." It is an ecology in which attention is intrinsically limited and thus becomes the scarce resource, and new modalities of grabbing attention are the hallmarks of success. This premise certainly holds true for the ocean of Internet-based information in general and the vast blogosphere in particular. Through studying some of the most popular blogging sites and bloggers, this chapter offers a critical assessment of key features such as rhetorical strategies and persuasive approaches in building unique online ethos in communicating to a steady audience base. It also discusses some of the personal, topical, social, cultural, and political factors in the emerging Chinese culture of blogs and blogging against the particular backdrop of China's media and communication environment.

BACKGROUND

User-generated content, including material originating from a variety of user bases encompassing Bulletin Board Systems (BBS), online chat rooms, Internet forums, blogs and microblogs, has been afforded an unusually prominent place in Chinese cyber life. This claim finds undisputable support from the results of six waves of cross-national surveys of global Internet use from 2006 to 2012 by Universal McCann, a New York-based global media-marketing consultancy firm and a subsidiary of the Interpublic Group of Companies Inc. (IPG). As the world's largest and longest-running survey on the impact of social media on the global society, its findings consistently point to the pattern that Chinese netizens display a much higher propensity than their counterparts in other countries to consume (i.e., read blogs) and contribute to content (i.e., write blogs) in the blogosphere (Universal McCann, 2009; 2010; 2012). For example, in its Wave 5 study, Universal McCann (2010) found that 79.6% of Chinese netizens reported having used blogs in the past six months (compared to 46.7% in the United States and a global average of 64.5%), leading the rest of the countries surveyed. In its recently released Wave 6 report, Chinese Internet users are again leading the global trend in microblog use, with 71.5% reporting use in the past six months and 49% of Chinese individuals expressing a preference for using microblog services as a tool for self-expression (compared to a global mean of 32%) (Universal McCann, 2012).

Chinese netizens' unmistakable penchant for user-generated content on the Internet can be best contextualized in a highly controlled media and information environment, both on- and offline. Although economic liberalization and market reform in the past three decades has led to a fundamental shift from ideological indoctrination to mass appeal in the conventional media sector, the state still maintains a formidable presence and exercises tight control over what the media can and cannot publish. Likewise, in the online environment, the Chinese state has developed and implemented a multi-tiered and multifaceted technology-centered surveillance system in order to monitor and filter out "unhealthy" information in Chinese cyberspace (Tai, 2010). In particular, all Internet sites engaging in any type of content publishing must be licensed by relevant state authorities and must comply with demands from official censors, which in essence ensures that state-sanctioned information permeates major portal sites. This kind of state-controlled arrangement, not unexpectedly, produces the effect of channeling user interest to more unconventional, user-centered platforms such as blogs, chat rooms, and other social media applications where user-contributed content dominates.

Over the years, there has been no lack of proclamations of the Internet as a democratic, liberating and even utopian space of public communication. Blogs and other social media are at the forefront of user-generated content revolution. Therefore, it is natural that they have occupied the center stage of current debates on the formation and trends of "we the media" (Gillmor, 2006), "public's journalism," (Haas, 2005) and "participatory communication" (Russo, Watkins, Kelly, & Chan, 2008) that would otherwise be unfilled by the legacy media. Globally, the blogosphere has established itself as a viable discursive space for information sharing, opinion expression, issue debates, and position deliberations (Barlow, 2008). Similarly, Coleman (2005) equated blogs to "democratic listening posts" that create a "bridge between the private, subjective sphere of self-expression and the socially fragile civic sphere in which publics can form and act" and "lower the threshold of entry to the global debate for traditionally unheard or marginalised voices" (p. 277). In specific relation to the Chinese blogosphere, Yonghua Zhang (2008) described blogging as "a new form for public participation" that serves as a de facto online "discourse sphere" where diversified opinions can be vented. In a similar vein, Guobin Yang

(2009) discussed blogs as a popular platform for online activism in China.

A blog is, in essence, a Web space that serves as a personal journal comprising discrete entries organized in reverse chronological order. Yet at the same time, the attraction of this intimate, personally revelatory space lies in its public accessibility and open access. This reversal of individual privacy (personally initiated and controllable actions) to the public spotlight (in full public view) was termed "publicy" by Mark Federman (2009). This intermingling of private communication into the public domain, as MacDougall (2005) argued, leads to "a certain degree of spectacle and celebrity (and self-) worship associated with blogs" (p. 586). As is the case with worshipping in general, blogging space, by only opening itself to opinioned information along favored lines of arguments, "extends the hemophilic tendency in humans" (MacDougall, 2005, p. 579) and creates "social herding and group polarization" (Coleman, 2005, p. 278). On the other hand, there are opportunities for celebrication in the blogosphere. Among the millions of blog pages, only a small number can stand out and establish a palpable presence in the online terrains. As a result, it is critical that distinct identities and ethos be created and maintained in those blog sites for them to register as more than just a passing glance.

Aristotle noted in the 4th Century B.C. in his *Rhetoric* that a central element of persuasive communication is ethos, which is "established during discourse … when one portrays himself or herself as having practical wisdom, good moral character, and a concern for the audience" (Frobish, 2003, p. 19). In the words of Halloran (1984), ethos is constructed in "a characteristic manner of holding and expressing ideas" (p. 71). As a gigantic but widely dispersed discursive space, a systematic, panoramic, and critical examination of cherished identities and popular ethos across the Chinese blogosphere therefore sheds light on public sentiments, prevailing beliefs, and dominant moral value evident in Chinese cyberspace.

DIGITAL ETHOS AND ONLINE IDENTITY IN THE CHINESE BLOGOSPHERE

Compared with its inception in the United States and in Europe, blogging in China had a later start: its formal birth did not come until 2002 when the term *Boke* (meaning both blog sites and bloggers) was coined and formally introduced to netizens in China (Tai, 2009). Since then, it has registered phenomenal growth and has blossomed into a populist, grassroots cultural phenomenon—which Tai (2012) branded "the Chinese culture of blog-mongering" ——that has become the hallmark of Chinese cyberspace. In the rise of blogs from rarity to mainstream, a limited number of trendsetting blogs and bloggers whose popularity ranges from millions to hundreds of millions of hits have significantly shaped public ethos and defined the contours of the Chinese blogosphere.

Since its formal advent in 2002, the Chinese blogosphere has witnessed a number of spectacular events and eye-catching episodes that have triggered huge numbers of visits to the blog sites and turned the term *Boke* into a household word for netizens in China. *Bokes* mainly fall into three categories: (1) blogs in relation to bizarre, unconventional, anti-mainstream topics (typically off limits to the state-sanctioned legacy media and their online publications); (2) blogs with high appeal to youth-dominated constituencies; (3) and blogs in connection to hot-button issues of public concern (Tai, 2012).

There has been substantial discussion (e.g., Tai, 2006; Yang, 2009) among academics concerning the role of blogs in shaping public opinion, the nature of blogs as an empowering tool for Chinese civil society, and the potential of blogs in promoting democratic debates in China both online and offline. Unexpectedly, in the early years of blog development in China, Chinese netizens were introduced to the blog world through a series of incidents of grassroots celebrication and unorthodox entertainment, known as spoofing,

that otherwise would not be possible with the state-censored conventional media.

Spoofing Fun

A noticeable category of Chinese blog incidents catching waves of eye-balls were characterized by a populist appeal developed through fame-mongering stunts launched in both online platforms and offline media outlets or through *Egao* (spoofs or parodies/satirical imitations). Furong Jiejie (translated as "Sister Hibiscus") and Tianxian Meimei (translated as "Sister Fairy") are among the typical examples of fame-mongering. The case of Furong Jiejie,[1] which has been going viral on the Web in China for years, unfolded in 2005. She debuted by posting self-portraits and video clips of provocative poses on the BBS systems of Qinghua (Tsinghua) University and Beijing (Peking) University in early 2005, and she became known online through her pertinacious trials despite repeated deletions by network administrators.

As time went by, her flourishing fame and rising demand for interconnection led her to open dedicated blog spaces at Bokee.com and the blog section of Sina.com. Her blog caused such a continuous splash that her story found its way into headlines in a number of media outlets, including the *Washington Post* (see page A15 of the July 17, 2005 edition). Her online fame spilled over to the offline world, and she was invited to star in the 2010 comedy "The Double Life," which was released in theaters in Chinese cities in April 2010. If blogging was the channel through which she got into the public spotlight, then the audacity of countering traditional aesthetic tastes and keeping up with it in the face of popular ridicule and disdain appeared to accelerate her rising online fame.

Another case of celebrication is that of Tianxian Meimei,[2] the nickname of a young woman in a village inhabited by China's Qiang nationality in southwestern Sichuan Province. Her case was a similar story of cyber celebrication driven by the use of blogging in collaboration with offline media publicity. Her celebrity status was the result of the blogging of a Beijing-based traveler who spotted her during his tour of the mountainous areas in the region in 2005. He took a picture of this young beauty in her traditional Qiang costume and posted the picture online while discussing his travel encounters upon returning to Beijing. His marvels at her astounding beauty echoed online—with lots and lots of people. As the photo immediately captivated the attention of a large number of netizens, including tens of thousands of follow-up postings within a few days, he revisited the place, took more pictures, and posted them online. Within a few months, these pictures attracted the attention of two million visitors. The commercial potential of this episode resulted in a collaboration between the man and Tianxian Meimei, who had her own blog on the blog section of Sina.com.

With rising cyber fame came opportunities in conventional media: Tianxian Meimei was picked to star in a leading role in two award-winning movies in 2007—*A Postman of Paradise* and *Er Ma's Wedding*—and played leading characters in three TV dramas by June 2010 (Tai, 2012). Thanks to insatiable demand for more information on her, CCTV (China's only national television station) produced a full-length documentary based on her, including her childhood life and being forced to drop out of school because of the financial hardships within her family. In her case, the explosive online fame appears to be connected to her status as a member of an ethnic minority group and popular interest in the cultural values and life experience of ethnic minority groups.

Slightly different from these cases are incidents of *Egao*, or spoofs, a new online genre of satirical parody that uses blogging as their main vehicle. Such spoofs, which usually take place in the form of (re)production/(re)creations of audiovisual pieces of conventional media products in China's blogosphere, poke fun at current subjects in the mainstream media and express individual disfavor

with hot-button topics/issues in society. The first highly eye-catching *Egao* was *The Bloody Case of the Steamed Bun*, a 20-minute video clip by Hu Ge, a sound engineer from Shanghai. It was a light-hearted parody of the nationally advertised movie *The Promise* (2005), directed by the famous Chinese award-winning director Chen Kaige. Albeit a big-budget production, *The Promise* fared miserably at the box-office in China, receiving much ridicule and criticism from the public.

Using images from the movie, Hu created the light-hearted parody video clip with a wickedly funny storyline. By early 2006, this parody piece had been viewed online tens of millions of times, ironically surpassing the audience size of the movie by a wide margin. Involving such issues as what composes violation of intellectual property rights and the range of "fair use" of copyrighted material, the case of "The Bloody Case of the Steamed Bun" aroused heated controversies in both online and offline media. As spoofs generally subvert some serious works of culture and re-construct them into funny parodies of grassroots entertainment, some analysts have explored three major factors behind the new genre's subversive potential: First, the genre of spoofs problematizes the conventional copyright protection framework; second, it disrupts the status quo of the centralized media system in China; and, third, it offers an alternative means for individuals to engage with social and political issues inside a heavily censored and highly constrained discursive environment. (Gong & Yang, 2010; Meng, 2009; Tai, 2012).

It is worth highlighting that a common strategy of success connecting all of the above high-profile incidents is the anti-establishment and self-made nature of gaze-garnering in the blogosphere. For the rise of Sister Furong and Tianxian Meimei from unknowns to cyber icons, stunts in staging public events and in getting into the spotlight followed the well-trodden trajectories of cultural celebrities and social elites. In the case of *Egao*, the inevitable target of spoof is popular cultural products and names endorsed by mainstream media.

Blogkeeping Public Affairs

While the above instances are from the domain of entertainment, the next category of landmark events related to Chinese blogging emerges in the sphere of public discourse on hot social and political issues. In these cases, the blogosphere serves as a new public space for grassroots surveillance or for public deliberation as a process of public opinion formation in an era of unprecedented economic expansion and social transformation. Online democracy therefore naturally becomes the focus of debates among diversified bases of individuals and online groups. Blog postings on issues of broad social concerns come to the frontlines of cyber attention from time to time. This is especially true when controversial issues are not covered by state-monitored conventional media and blogs under these circumstances form viable alternative sources (and sometimes the only alternate source) of information. "The Chongqing Nail House Incident" and "The South China Tiger Photos Incident" are among the cases in which blogging played a vital role in creating a special type of grassroots journalism where relevant users publish first-hand eyewitness information amidst accolades and encouragement from fellow netizens.

The Chongqing Nail House Incident is a landmark event in which blogging served as a platform to focus public attention on the evolving controversy of individual property rights protection in China during a time of rapid real estate development and relocation. The story of "the nail house" dates back to 2004 and the commencement of a real estate development project in an old residential area of Chongqing Municipality. This project sought to build a shopping center on the site where Wu Ping and her husband Yang Wu owned a small two-story "nail" house. To make way for the construction of the shopping center, their house, just like other houses at the site, was targeted for demolition. This is a common scenario across China's booming urban areas where new high-rise commercial developments replace old residential districts. Things took an unusual turn, however, taking this incident into the spotlight of the local, national, and even international media.

In September of 2004, demolition work started, and 280 households were moved out soon afterward. But Wu and Yang refused to move because they were dissatisfied with the terms of the proposed compensation offered by the government-approved developer. As negotiations between the property developer and the house-owner couple were still ongoing, things turned ugly on another front. In October 2004, the water supply for their house was cut off, and on February 2, 2005, the electricity was disconnected. After the construction team came to work at the site, the way from the house to outside streets was blocked. But the couple did not yield. A protracted standoff pursued for months, as neither the property developer nor the house owners wanted to budge.

The couple's home became known as "the nail house" in reference to their lonely placement in an otherwise leveled field. The case caught the attention of netizens through constant updates on its latest developments in a blog run by a young man who called himself Zola Zhou in early 2007. Netizens threw their support behind the house owner on other blogs and forums, which amplified public attention on this dispute. Pictures of the standalone house were reposted here and there across the Internet, and the pictures were lauded as a symbol of resistance against real estate developer greed. Conventional media were quick to pick up the story, and reporters from many parts of the country flocked to this house to snap their own pictures and run their own stories. The couple showed up for TV interviews on numerous occasions, and public pressure built up on the developer to resolve the dispute. Eventually, the house owner and the developer reached a settlement on compensation terms on April 2, 2007. As this was a case involving property rights, it aroused much discussion on property rights protection in China and triggered heated debates about flaws in the existing legal framework in China in this area. Moreover, this incident becomes a source of inspiration for individuals trapped in similar situations for addressing their grievances against aggressive developers.

The South China Tiger Photos Incident is another case of online communication in which the blogosphere made it possible for dispersed individuals to collaborate and expose falsified information involving local government bureaucrats. The following report by China Daily on October 18, 2007, titled "Photo of 'Extinct' Tiger Sparks Controversy," offered some background of this incident:

A newly-released photo, which Chinese forestry authorities say proves the continuing existence of wild South China tigers which have been thought to be extinct, has sparked heated controversy from Internet citizens, questioning its authenticity. The digital picture, purporting to be a wild South China tiger crouching in the midst of green bushes, was released by the Forestry Department of northwest China's Shaanxi Province at a news conference on October 12. Zhou Zhenglong, 52, a farmer and former hunter in Chengguan Township of Shaanxi's Zhenping County, photographed the tiger with a digital camera and on film on the afternoon of October 3, a department spokesman said. Experts had confirmed the 40 digital pictures and 31 film photographs are genuine, the spokesman told reporters.[3]

Because the South China tiger is a rare species believed to be extinct in the wild, the story naturally made national headlines, including those on the major Chinese Web portals. Controversies immediately followed as netizens, in response to this story, fell into two groups: those who challenged it and those who defended it. Each group used blogging as the chief channel to argue their case. Individuals who challenged the authenticity of the picture, and thus the truthfulness of the picture, evolved from a few dozens to a national base of well-informed netizens; many of them provided detailed analysis of how the picture had been processed using Adobe Photoshop technolo-

gies. As more evidence accumulated on multiple fronts, especially after the discovery of a poster published years prior to the photos that featured an identical picture, it became clear that the photos were doctored digitally. This triggered a public uproar that led to a formal investigation by a team of national experts, including official agencies such as the State Forestry Bureau. Eventually, Zhou Zhenglong was charged with deception, and about a dozen of officials were disciplined for collusion and cover-ups. This case well illustrates the empowering potential of the blogosphere in bringing together like-minded individuals on self-appointed truth-seeking missions to confront official transgressions.

Writing Fame

The popularization of blogging in China has spawned two distinct types of bloggers. A-bloggers have a wide base of followers (often in the millions) and are trendsetters and opinion leaders in blogging; they have developed their unique style of blogging and are quintessential cyber celebrities. B-bloggers, on the other hand, only have a small base of readers, and, as opinion followers, they often contribute to online debates but do not influence the course of these debates. Due to the limited nature of the human attention span, only a very small segment of bloggers can become A-bloggers, so the vast majority of them remain in the B-blogger category. In the discussion that follows, we have chosen to focus on these A-bloggers: Han Han, Acosta, Xu Jinglei and Sha Minnong, whose fame pervades every corner of the Chinese blogosphere.

Han Han, arguably the most famous blogger in China, enjoys a nationwide fame both in the on- and offline world.[4] He writes with a sharp, often satirical style and is especially popular among the younger population, especially high school and university students. Since he opened his blog with Sina.com, he has remained one of the most-read bloggers as measured by the number of fans and visitor clicks. A scrutiny of his postings in recent years revealed that the content varied between narratives of his activities, episodes of his daily life experiences, and comments/remarks on a series of hot issues such as transportation facilities, high gas prices, problems of exam-oriented education in Chinese schools, and rising housing prices. Posts on exam-oriented education took up the greatest proportion. And his blog entries (posts) on rising housing prices display a similar sharp (and satirical at certain points) style as that shown in his writings published offline.

For example, in a popular posting titled "Composition (class) for students should be abolished" (June 15, 2007), Han Han wrote: "Composition (class) [in Chinese schools] does not only weaken your writing talent, but also tells you at the subconscious level that it is normal to say things with your tongue in the cheek, and that doing so is the key to survival… It is composition that kills many students' interest in literature." The sharp, critical tone is most typical of the style of his presentation. Among his more recent blog entries, in a widely read one (2,367,685 visits, followed by 22,918 comments as of January 2012), entitled "It will drop immediately, and drop to below 1,000 (RMB *yuan* per square meter)" posted on February 22, 2011, he talked about how he felt about the housing prices: when the housing price was RMB¥ 3,000 (approximately US$ 154) per square meter in downtown Shanghai in 2000, he predicted it would drop immediately and then drop to below RMB¥ 1,000 per square meter because he felt the 3,000-*yuan*-per-square-meter price was unreasonably high considering the level of income for the average citizens. "Of course, I was wrong," he wrote. The post also discussed housing prices in the context of stagnant income growth and soaring inflation for daily necessities. These sentences in this posting resonate well with his poignant style of light-hearted satire:

I cannot help wondering how those friends with a monthly income of RMB¥ 2,000 to 3,000 live their normal life in this city [Shanghai]… The high-rise

housing complexes in this city don't belong to them, and the only thing they could do to these facilities is to look at them in admiration. Fortunately, thanks to the benevolence of the government, one doesn't have to pay eye-maintaining tax for watching the city.[5] *(Han, 2007, February 22)*

At the end of 2011, Han Han posted three entries on democracy, revolution, and freedom, respectively—not coincidentally; these are all three highly sensitive topics in China's sociopolitical culture. In these blog posts, Han Han bluntly expressed his preference for reform over revolution, cited a lack of civil spirit as a major problem with the Chinese citizenry, said democracy was a long process to be pursued gradually, and commented on the significance of seeking the cooperation of the government rather than pressuring the government on political reform. These blog posts immediately triggered intense discussion and debates online, so much so that it got mainstream media websites like *People's Daily*'s website (people.com.cn) involved as well. International media such as the *New York Times* and the *Wall Street Journal* also published blog posts commenting on Han Han's blog.

Netizens' comments and readings on these three blog posts by Han Han basically fell into three categories: 1) those who objected to his views, which they saw as too conservative because the views aligned rather closely to official views; 2) those who saw in his posts a sophisticated understanding of the Chinese reality; and 3) those who embraced his remarks as telling the sad but unfortunate truth in the modern world they inhabited. Regardless of the divergent comments on these blog entries, the fact remains that they have been among the most-read entries, and as such have highlighted Han Han's influence as a prominent blogger regardless of where a reader stands on these issues.

In terms of rhetorical strategies Han Han has used in his blogs, two approaches are particularly worth mentioning here: the use of a question-

and-answer type of dialogic form to present his arguments, and the noticeable presence of a satirical touch and sharp tone. The question-and-answer format is fittingly employed in the above-mentioned blog postings on revolution and democracy, two serious topics in the realm of public discourse, especially within the Chinese context. And this dialogic form, which smacks of the genre of Socratic dialogues in ancient Greece, serves to articulate the blog writer's ideas forcefully and to add to the cogency of his arguments because the answers are expected to respond to the various points raised in his questions. The strategy of a satirical touch and sharp tone contributes to his identity as a "spokesperson" for young people's discontent on such social evils as corruption, abuse of power, the widening gap between the rich and poor, and the souring inflation that is beleaguering the everyday life of average folks.

Another shining star among the so-called grassroots bloggers is Acosta,[6] a young man whose real name/identity is unknown but who remains intriguing to many. In 2006, Acosta became the first "grassroots blogger attracting over 10 million hits" in China, and has remained as one of the most popular bloggers ever since ("Sina Blog," 2006). Our close examination of his blogs in recent years shows that Acosta's blog posts are mostly expressions of personal emotions and reflections/thoughts on life, while some blogs disclose his impressions on books he has read and on TV programs he has watched, while others provide comments on public events.

Among the blog entries that reveal his emotions are those mourning the victims of the massive earthquake in Wenchuan, Sichuan Province. As of May 2012, Acosta's blog site has registered over 224 million hits. This clearly attests to his celebrity status as a blogger. His overwhelming popularity online, as a matter of fact, has triggered such public intrigue online that the phrase "Who is Acosta?" has been listed as one of the most-inquired items on China's No. 1 search engine, Baidu (see Piao, 2006). The popular appeal for

Acosta's blogging rests with his ability to feel the pulse of China's youth in their 20s and 30s and his skills in empathizing with them on their emotions toward, and struggles with, everyday experiences, as can be seen from the testimonials by followers of his blogs.

One young netizen ascribed the tremendous attention Acosta has received to his ability to arouse people's memories of their "most self-reminiscent" moments; he/she admitted that Acosta reminds "you and me in the past, at present and in the future," and that "his blog is the most truthful monologue in our innermost," and that "our present situation and our past experiences are like the hazy yet real mirror images of Acosta, some joyful, some sad."[7]

This in a sense reveals the key to Acosta's success in capturing the fancies of the young netizens: he makes a great effort to feel the pulse of the young netizens and appeal to their emotions, thus making them feel that he shares their feelings—whether these feelings are positive ones or negative ones. The fact that he remains an unidentified author would normally be regarded as a disadvantage to the credibility of his blog. However, Acosta has used his "invisibility" to great advantage in a masterful way, as he has been very careful in not revealing any information in relation to his true identity. Acosta has been amazingly successful in keeping his identity a mystery in an era of the ubiquitous public gaze, and in doing so, he has managed to fixate public attention to the messages in his blogs, rather than his own celebrication.

Xu Jinglei is a Chinese movie actress, director, and producer with an established track record of success in the industry. She is referred to in the Chinese media as one of the Four Young *Dan* (meaning a specific female role in Chinese operas) actresses, along with Zhao Wei, Zhou Xun, and Zhang Ziyi. Starting in October 2005, Xu has been running one of the most popular blog sites in the country.[8] Her blog entries mainly focus on what is happening in her daily life. In less than four months after she started her blog, she set the record of registering over 10 million visits. In July 2007, she further broke the record by achieving 100 million hits on July 12 within that year. In November 2010, she decided to stop updating her blogs on Sina.com. On April 27, 2011, she migrated her writing to microblogs on QQ.com, and in about two months, her micro blog posts had already attracted a fan base of 10 million.[9] As of May 2012, her microblog followers exceeded 23 million.

Xu's blog entries in a sense are really diaries of daily life published on the Web. Millions of hits on these blog posts indicate that the casual-chat style of writing proves engaging to the audience, and her focus on familiar, everyday types of topical matters relates her affectionately with the commoners. Additionally, her choice of the title for her site—Lao Xu's Blog—is quite suggestive, as Lao is a popular way of addressing long-time pals, old friends, or intimate peers in the Chinese culture. The following two posts from her blog site offer a taste of her style and provide a glimpse of her online ethos:

(2010-11-12) [Chinese way of writing the date for November 12, 2010] The holiday is to end soon. It's said to be very cold in Beijing. Mom sent me photos of Little Friend Niu Niu, and my little cats. I, miss, you!

And I've now lost sense for various activities—for work, for writing, for finding words in my habitual kinds of talks... The long-awaited holiday, can it really change a person? Anyway, I've become lazy, relaxed, carried away by inertia, and fond of the having-nothing-to-do feeling. When I go back, I absolutely need a strong push.

My good nephew, you've grown bigger again![10]

With the use of "you" and a couple of photos foregrounded in familiar settings, the blog post created the impression as if she were chatting to old pals.

Here is another one, narrating her living experience in Los Angeles, in which she pokes fun at wearing high-heel shoes:

(2010-10-29) New secret for losing weight during holiday. The so-called "holiday" in Los Angeles has lasted for about a month. During this period, I've attended many meetings—English ones! I was dizzy listening... Couldn't understand more than half of what was said... Turned directly from an eloquent speaker to one who stammers... The little bit of self-confidence formed before has been beaten again. I'm determined to continue to fiercely feed myself in English! I cannot believe that I really don't have a talent for language. Or, is it because the Chinese language is so good that I cannot easily embrace another language... Hei, hei...

But I've found a new method for losing weight! Wear high-heel shoes! March forward in the day with a pair of two-inches-high-heel shoes! It looks very cool. Ha, Ha...[11]

While the three bloggers above are all young (in their late 20s to mid-30s), the next prominent blogger in our discussion features a retired senior citizen. Sha Minnong has homesteaded both on Sina and on Sohu in his blogging, having attracted an aggregate of more than 302 million hits on his Sina site and 372 million hits on his Sohu site as of May 2012.[12] He has varied credentials: He was dispatched to the countryside for Mao's youth reeducation campaigns and lived there away from his home for ten years; he served as the editor-in-chief for a few newspapers specializing in the stock market and worked as the deputy editor-in-chief of the *Modern Express* (*Xiandai Kuaibao*), a major newspaper affiliated with the Xinhua News Agency; and he is the author of several books.

Considered a top blogger in the area of commercial and financial affairs, he devoted more than half of his blog posts (895 out of 1611, as of February 2012) to content related to the stock market. The next big content category in his blog is "*Sha Niu Jiashu,*" with 481 posts. *(Niu* means "ox" in English; Sha was born in the Year of the Ox according to the Chinese lunar calendar, hence the pseudonym *Sha Niu. Jiashu* literally means "family letter.") His blog had attracted 295,808,542 visits as of February 2012. In 2007, his blog, called "Lao Sha's Blog," won the title of "Number One Chinese-Language Blog" from Sohu.com, and he was named "2007 Sina.com Man of the Year."

His blog posts on the stock exchange displayed an in-depth understanding of the subject matter and are therefore highly informative, structured into numbered sections, with each section reporting or quoting some information from various sources to support his points, often including famous international and domestic sources for information in this area, and then offering Sha's own comments. But interestingly, among the 14 blog posts he lists as "writings that I take pride in," almost none is focused on the stock exchange. Instead, they mostly tell stories of Sha's life experiences, yet Sha seems to be able to find some connection between these and his current major concerns with the economy, or more specifically with the stock market.

In one of these blog entries, titled "'Sweetness' Brought by My Elder Brother," Sha recalled an episode in his childhood in which his elder brother was the main character. It took place during the extremely hard period of time known as "the three (consecutive) years of natural disasters." His brother, famed writer Sha Yexin, was already studying in Shanghai East China Normal University (then called East China Normal College) while Sha Minnong, ten years younger, was a second-grade student in elementary school. During these years of extreme hardships, a scarcity of food and various products was the most marked feature of the country's economy. Rice, oil, meat, vegetables, sugar, and so on were all rationed. In Sha's recollection, in those years, "if a student in elementary school had a candy in his/her mouth,

other students surrounding him/her would all be watering in their mouths!"

As a university student, Yexin, Minnong's only elder brother, received a candy coupon each month. With six candy coupons accumulated from repeatedly exchanging one month's coupon for the following month with his classmates, his elder brother bought quite a few candies and poured them all out onto the table when he went home during summer vacation. All eight kids in the family grabbed a handful of candies immediately. "I remember putting a candy in my mouth without taking off the wrapping paper first," Sha wrote. After narrating this story, Sha concluded this blog entry with a paragraph that compared what his brother did with what individual stock investors do:

I feel that many individual securities investors are just like my brother (who saved candy coupons); they accumulate their savings bit by bit and exchange the accumulated savings for shares (like the candy coupons), and wish that these will bring some joy to their families, to their relatives and to their next generation![13]

Sha Minnong's popularity is tied to a few factors. First, his matter-of-fact style suits the major content area of his blog—stock exchange. This style is also conducive to trustworthiness because it makes the information seem objective. Second, his perceived knowledge in the stock market as a former editor of a few newspapers in the area also contributes to the credibility of his blog. Third, the way he interweaves information from external sources with his concise comments/ analyses serves to make his blog posts highly informative and to impress the netizens with his "insider's perspectives," a quality that, as a rule, adds to the credibility of the communicator (in this case, the blogger Sha Minnong). Fourth, as tens of millions of Chinese people have engaged in the stock exchange, and yet most of them are not experts in the field of financial and stock markets, easily readable writings focused in this area are welcome to them.

CONCLUSION

Since its formal birth in 2002, the Chinese blogosphere has experienced phenomenal growth. China now boasts by far the largest blogger population in the world. To put things into perspective, just its blogging population of 319 million people (i.e., those who reported owning at least one blog space/page, as of year-end 2011) (CNNIC, 2012) surpasses the *entire* Internet-using population of 313 million people in the United States, the No. 2 country in the world in terms of both Internet and blog use.[14] This explosive expansion is accompanied by the increasingly prominent role that blogging has played in shaping the landscape of Chinese cyber culture and the mainstreaming of blogging in the routine lifeworlds of Chinese netizens.

As Halloran (1982) noted, ethos has its most concrete reference as a "habitual gathering place" in the Greek lexicon (p. 60). The study of popular blog sites and bloggers, in that sense, is revealing for the collective ethos across the blogosphere in China. Through examining high-profile bloggers and popular blog sites in China, this chapter presented a critical analysis of common rhetorical strategies and persuasive approaches used by bloggers to construct online ethos in Chinese cyberspace. Several themes can be discerned throughout our discussions in regard to popular ethos and celebrated identities in the Chinese blogosphere.

First, the blogosphere has become the frontier for grassroots netizens to produce, disseminate and consume content. As pointed out earlier, Chinese Internet users have displayed an unparalleled propensity for user-generated content in their cyber life. Blogging continues to push this drive to new heights. At the same time, it has been noted that "even as the blogosphere continues to expand, only a few blogs are likely to emerge as focal

points" (Drezner & Farrell, 2004, p. 35). Thus it is not surprising that a relatively small number of bloggers and blogs can make an impact in the ever-expanding Chinese blog landscape.

There has been considerable deliberation among academia as to what constitutes influence in the blogosphere and how we can measure it. Gill (2004), for example, conceptualized measuring influence in the domains of "audience reach, media adoption and political necessity." In a similar vein, Azman, Millard and Weal (2010) likened the emergence of the blogosphere to the 17th-century phenomena of pamphleteering, and they proposed defining power and influence in the blogosphere as the "ability to produce effects among others within a decision-making process." With specific relevance to the various cases examined in this chapter, a few common threads stand out in differentiating high-profile bloggers from conventional blog writers: their blog sites typically attract high volumes of traffic as manifested in hits, links, and repostings; their blog posts frequently trigger heated discussions among fellow bloggers struggling for recognition; and their sphere of influence spills over to the conventional media from time to time to generate points of debate and coverage.

A second key theme to emerge is that the blogosphere has become a hotbed of mass collaboration and collective action under particular circumstances. In a way, notwithstanding the severe limitations and constraints emanating from state authorities, the online space has created certain platforms for mass participation and for public discourse on issues of public interest. The blogosphere, therefore, can serve to varying degrees as a viable barometer to gauge public sentiment and emotion on a wide range of topics and issues. In this process, certain bloggers are in a unique position to shape public discourse and set debate agendas.

Third, to borrow from the useful concept of "publicy" (Federman, 2009), the blogosphere intersects private conversations and public discourse by situating personally initiated actions in publicly viewable spaces. Celebrity bloggers are the ones who have found a niche way to construct their public ethos based on their private entity. Rapport with the online public, however, can be built and maintained in different ways such as through the display of care, empathy, knowledge, and trustworthiness, which was demonstrated in the numerous examples we have discussed in the chapter.

From its inception, the Internet has been created with a spirit of public sharing and grassroots participation. Blogging, however, has been a major breakthrough in empowering netizens to find innovative ways to build their own identities and construct individual ethos in an online space marked with their distinctive branding. What makes the blogosphere different from other popular technologies enabling public participation such as bulletin board systems and chat rooms is that the latter are more likely associated with a group/collective formation while the former are most typically linked to individual origination. The blogosphere, therefore, is inherently more conducive to individual identity building and ethos formulation because the labeled space most often features persuasive writings carved out of personal, and indeed often time personable, reflexive discourse and ruminations.

A fourth theme, as demonstrated by the numerous cases throughout the chapter, reveals that what resonates well with public sentiments in the Chinese blogosphere points mostly to the type of content and discourse typically residing outside of the domains of the state-sanctioned mainstream media and state-approved online space. Elsewhere, Tai (2006) has cogently argued that user-generated content is fondly treasured by Chinese Internet users due to the highly controlled nature of the (offline and online) information environment in China. Evidence from big-name bloggers analyzed here clearly reinforces that observation, as the attention-grabbing and the eye-catching unmistakably concerns the mundane, the unconventional, and the non-(and often anti-)official.

Unlike in Western democracies, where the power of blogs often demonstrates their abilities to bring politics to "focal points" (Farrell & Drezner, 2008), none of the celebrity bloggers we examined engaged in political discussions as their focus. There is no doubt that this is a smart strategy, not only because any political advocacy outside of official lines becomes an easy target for government crackdown and suppression, but toeing the official line is not likely to create any success in attracting accolades from fellow netizens.

Finally, the development of the Internet in China in general and the growth of the Chinese blogosphere in particular have benefited enormously from the vigorous support of the state on many fronts, and yet, at the same time, the state has maintained a formidable presence in the online world in China in terms of content regulation and censoring. This duality of state involvement will, in all likelihood, continue into the foreseeable future; however, specific twists and turns will continue to unfold in the years to come as the Chinese blogosphere pushes new boundaries and enters unchartered territories.

REFERENCES

Azman, N., Millard, D. D., & Weal, M. J. (2010). *Issues in measuring power and influence in the blogosphere*. Web Science Conference 2010. Retrieved May 20, 2012, from http://journal.webscience.org/344/

Barlow, A. (2008). *Blogging America: The new public sphere*. Westport, CT: Praeger.

China Internet Network Information Center (CNNIC). (2012). *The 29th statistical report on Internet development in China*. Retrieved February 20, 2012, from http://cnnic.com.cn/dtygg/dtgg/201201/W020120116337628870651.pdf

Coleman, S. (2005). Blogs and new politics of listening. *The Political Quarterly, 76*(2), 272–280. doi:10.1111/j.1467-923X.2005.00679.x

Drezner, D. W., & Farrell, H. (2004). Web of influence. *Foreign Policy, 145*, 32–40. doi:10.2307/4152942

Farrel, H., & Drezner, D. (2008). The power and politics of blogs. *Public Choice, 134*(1/2), 15–30. doi:doi:10.1007/s11127-007-9198-1

Federman, M. (2009). *McLuhan thinking: Integral awareness in the connected society.* Retrieved February 10, 2012, from http://individual.utoronto.ca/markfederman/IntegralAwarenessintheConnectedSociety.pdf

Frobish, T. S. (2003). An origin of a theory: A comparison of ethos in the Homeric Iliad with that found in Aristotle's Rhetoric. *Rhetoric Review, 22*(1), 16–30. doi:10.1207/S15327981RR2201_2

Gill, K. E. (2004). How can we measure the influence of the blogosphere? *WWW2004 Proceedings.*

Gillmor, D. (2006). *We the media: Grassroots journalism by the people, for the people.* Sebastopol, CA: O'Reilly.

Goldhaber, M. H. (2006). The value of openness in an attention economy. *First Monday, 11*(6). Retrieved from http://firstmonday.org/htbin/cgiwrap/bin/ojs/index.php/fm/article/view/1334/1254

Gong, H., & Yang, X. (2010). Digitized parody: The politics of *egao* in contemporary China. *China Information, 24*(1), 2–26. doi:doi:10.1177/0920203X09350249

Halloran, S. M. (1984). Aristotle's concept of ethos, or if not his somebody else's. *Rhetoric Review, 1*(1), 58–63. doi:10.1080/07350198209359037

Halloran, S. M. (1984). The birth of molecular biology: An essay in the rhetorical criticism of scientific discourse. *Rhetoric Review, 3*(1), 70–83. doi:10.1080/07350198409359083

MacDougall, R. (2005). Identity, electronic ethos, and blogs: A technologic analysis of symbolic exchange on the new news medium. *The American Behavioral Scientist, 49*(4), 575–599. doi:10.1177/0002764205280922

Meng, B. (2009). Regulating egao: Futile efforts of recentralization? In Zhang, X., & Zheng, Y. (Eds.), *China's information and communications technology revolution: Social changes and state responses* (pp. 53–67). New York, NY: Routledge.

Piao, Y. (2006, May 29). *After all, who is Acosta?* Retrieved July 16, 2012, from http://blog.sina.com.cn/s/blog_49191240010003my.html

Russo, A., Watkins, J., Kelly, L., & Chan, S. (2008). Participatory communication with social media. *Curator: The Museum Journal, 51*(1), 21–31. doi:10.1111/j.2151-6952.2008.tb00292.x

Sha, M. (2011, July 23). *The sweeties my brother brought back.* Retrieved July 16, 2012, from http://blog.sina.com.cn/s/blog_4c497d3a0102dqtj.html

Shen, F., Wang, N., Guo, G., & Guo, L. (2009). Online network size, efficacy, and opinion expression: Assessing the impacts of Internet use in China. *International Journal of Public Opinion Research, 21*(4), 451–476. doi:10.1093/ijpor/edp046

Sina Blog. (2006, June 2). *Acosta: The first grassroots blogger attracting over 10 million hits.* Retrieved July 16, 2012, from http://blog.sina.com.cn/lm/8/2006/0602/1642.html

Tai, Z. (2006). *The Internet in China: Cyberspace and civil society.* New York, NY: Routeldge.

Tai, Z. (2009). The rise of the Chinese blogosphere . In Dumova, T., & Fiordo, R. (Eds.), *Handbook of research on social interaction technologies and collaboration software: Concepts and trends* (pp. 67–79). Hershey, PA: IGI Global. doi:10.4018/978-1-60566-368-5.ch007

Tai, Z. (2010). Casting the ubiquitous net of information control: Internet surveillance in China from Golden Shield to Green Dam. *International Journal of Advanced Pervasive and Ubiquitous Computing, 2*(1), 53–70. doi:10.4018/japuc.2010010104

Tai, Z. (2012). Fame, fantasy, fanfare and fun: The blossoming of the Chinese culture of blogmongering . In Dumova, T., & Fiordo, R. (Eds.), *Blogging in the global society: Cultural, Political and geographical aspects* (pp. 37–54).

Universal McCann. (2009). *Power to the people: Social media tracker – Wave 4*. Retrieved June 4, 2011, from http://universalmccann.bitecp.com/wave4/Wave4.pdf

Universal McCann. (2010). *The socialisation of brands: Social media tracker – Wave 5*. Retrieved January 5, 2012, from http://www.umww.com/global/knowledge/view?id=128

Universal McCann. (2010). *The business of social: Social media tracker – Wave 6*. Retrieved March 5, 2012, from http://www.umww.com/global/knowledge/view?Id=226

Xu, J. (2010a, November 20). *State of hibernation during the holidays: More about Niu Niu*. Retrieved July 16, 2012, from http://blog.sina.com.cn/s/blog_46f37fb50100mv5i.html

Xu, J. (2010b, October 29). *New secret to losing weight during the holidays*. Retrieved July 16, 2012, from: http://blog.sina.com.cn/s/blog_46f37fb50100mekc.html

Yang, G. (2009). *The power of Internet in China: Citizen activism online*. New York, NY: Columbia University Press.

Zhang, Y. (2008). Impact of intentional social actions on applications of the Internet technology. [Zhongguo Wangluo Chuanbo Yanjiu]. *China Internet Communication Research, 2*(1), 98–105.

ADDITIONAL READING

Esarey, A., & Xiao, Q. (2008). Political expression in the Chinese blogosphere: Below the radar. *Asian Survey, 48*(5), 752-772. DOI: AS.2008.48.5.752

Hearn, K. (2009). The management of China's blogosphere boke (blog). *Continuum: Journal of Media & Cultural Studies, 23*(6), 887–901. doi:10.1080/10304310903294770

Herold, D. K. (2009). Editorial notes: Cultural politics and political culture of Web 2.0 in Asia. *Knowledge, Technology & Policy, 22*(2), 89–94. doi:10.1007/s12130-009-9076-x

Lai, C. H. (2011). A multifaceted perspective on blogs and society: Examples of blogospheres in Southeast and East Asia. *Journal of International Communication, 17*(1), 51–72. doi:10.1080/13216597.2011.559156

Li, S. (2010). The online public space and popular ethos in China. *Media Culture & Society, 32*(1), 63–83. doi:10.1177/0163443709350098

MacKinnon, R. (2008). Flatter world and thicker walls? Blogs, censorship and civic discourse in China. *Public Choice, 134*(1/2), 31–46. doi:doi:10.1007/s11127-007-9199-0

Morozov, E. (2011). *The net delusion: The dark side of Internet freedom*. New York, NY: PublicAffairs. doi:10.1017/S1537592711004026

Piao, Y. (2006, May 29). *After all, who is Acosta?* Sina Blog. Retrieved July 16, 2012, from http://blog.sina.com.cn/s/blog_49191240010003my.html

Wang, S., & Hong, J. (2010). Discourse behind the forbidden realm: Internet surveillance and its implications on China's blogosphere. *Telematics and Informatics, 27*(1), 67–78. doi:10.1016/j.tele.2009.03.004

Yuan, W. (2010). E-democracy@China: Does it work? *Chinese Journal of Communication, 3*(4), 488–503. doi:10.1080/17544750.2010.516581

Zhang, X., & Zheng, Y. (Eds.). (2009). *China's information and communications technology: Social changes and state responses*. New York, NY: Routledge.

Zhou, X. (2009). The political blogosphere in China: A content analysis of the blogs regarding the dismissal of Shanghai leader Chen Liangyu. *New Media & Society, 11*(6), 1003–1022. doi:10.1177/1461444809336552

KEY TERMS AND DEFINITIONS

A-bloggers: Bloggers who are successful in setting trends, leading debates, and attracting attention through their blog writing.

B-bloggers: Those blog writers who develop a small base of followers, and who may contribute to, but do not shape the course of, online discussions of major topics and debates.

Boke: This is the Chinese term for both blogs and bloggers.

Egao: A popular practice online in China to create parodies or satires of established celebrities or mainstream cultural products through artistic recreations and poaching.

Netizen: coined by merging two words "Internet" and "citizen," it refers to any person who participates in producing or consuming content on the Internet. It corresponds to the Chinese term *wangmin*.

Publicy: A term coined by Mark Federman to refer to the dual nature of blogs – it is considered private as a blog site is associated with a particular individual, and promotes private revelations and conversations; at the same time, it is public because blogs are publicly accessible on the Internet.

User-Generated Content (UGC): It refers to any type of content that is created by individual Internet users rather than commercial or institutional sponsors. Common examples of UGC include messages published on bulletin board systems, social media sites, Twitter accounts, and personal blogs.

ENDNOTES

[1] http://blog.sina.com.cn/frjj

[2] http://blog.sina.com.cn/txmm

[3] http://www.chinadaily.com.cn/china/2007-10/18/content_6188481.htm

[4] http://blog.sina.com.cn/twocold

[5] http://blog.sina.com.cn/s/blog_4701280b01017i4g.html

[6] http://blog.sina.com.cn/u/1456252804

[7] http://blog.sina.com.cn/s/blog_49191240010003my.html

[8] http://blog.sina.com.cn/xujinglei

[9] http://t.qq.com/xujinglei?pgv_ref=im.group.hot

[10] http://blog.sina.com.cn/s/blog_46f37fb50100mv5i.html

[11] http://blog.sina.com.cn/s/blog_46f37fb50100mekc.html

[12] http://blog.sina.com.cn/shaminnong and http://shaminnong.blog.sohu.com/

[13] http://blog.sina.com.cn/s/blog_4c497d3a0102dqtj.html

[14] Internet user statistics concerning the United States come from Internet World Stats, available at http://www.internetworldstats.com/stats.htm

Chapter 16
Theory and Application:
Using Social Networking to Build Online Credibility

Misty L. Knight
Shippensburg University of Pennsylvania, USA

Abigail Goben
University of Illinois at Chicago, USA

Richard A. Knight
Shippensburg University of Pennsylvania, USA

Aaron W. Dobbs
Shippensburg University of Pennsylvania, USA

ABSTRACT

Scholars are increasingly engaging with their peers in synchronous and asynchronous online forums. In order to adapt to this current trend, librarians and faculty must consider the nuances of computer-mediated communication and learn to understand the potential benefits and hazards of creating online identities that may round out others' perceptions. It can be overwhelming and confusing to determine how to best present oneself or to "create" a credible identity. Through the introduction and explanation of communication concepts and theories, this chapter discusses online credibility, or ethos, and examples of those who have successfully built online credibility.

INTRODUCTION

Your online ethos. Your credibility. Why should you care? Online ethos is how your audience sees you, or your organization, based upon their observation and interpretation of the way you present yourself in online media. This chapter will examine ethos as personal or institutional credibility in social media communication channels. Phrased in the terms of the medium, why should

your readers care about what you say and look forward to your next post? The ubiquity and accessibility of these channels has radically changed the communication patterns of individuals, groups, and businesses and will continue to influence scholarly and professional communication in the future (Kietzmann, Hermkens, McCarthy, & Silvestre, 2011). Beginning with a thorough examination of related research, this chapter will discuss what it means to have online credibility, provide specific examples, and discuss directions for users to build their own online ethos.

DOI: 10.4018/978-1-4666-2663-8.ch016

Social media is a pervasive tool for personal and professional networking. Interviewers, applicants, friends, colleagues, and others are using forums, e-mail discussion lists, and social networking sites such as Twitter, Flickr, and Facebook to communicate. Social media is the web of Internet- and mobile-based communication channels such as blogs, Twitter, and the like that allow the generation and exchange of user-created content (Kaplan & Haenlein, 2010). As faculty, students, and librarians develop and use the online communication channels, formally referred to as "computer-mediated communication" (CMC) in the literature, that are made available to them, the scope of their face-to-face (f2f) communication must change along with advances in technology (Berger, 2005; Hovick, Meyers, & Timmerman, 2003; Ishii, 2006; Ramirez, Zhang, McGrew, & Lin, 2007).

This chapter will address the importance of developing scholarly ethos with regard to social networking as it relies on communication theory to explore credibility through online impressions and identity creation. As social networking tools rise and fade, users need to understand the importance of, and methods for, establishing credibility across platforms. Examples of poor use of social media abound, from Facebook sites set up and then abandoned to lawsuits threatened for making factual statements about a website (Lawson, 2009).[1] By introducing communication concepts and theories, this chapter will explain strategies for avoiding these pitfalls as it discusses building one's online credibility, or ethos, while carefully considering the audience, the impact of a lessening of traditional non-verbal cues, and a consideration of

other communication concepts that must adapt from f2f to online format. The authors will also cite examples of librarians who currently have influential online presences to demonstrate the successful use of social media to build credibility. After analyzing these exemplars, we will offer suggestions and best practices for individuals,

libraries, and other institutions to build online credibility and discuss how this credibility can affect scholarly communication in the future.

Computer-Mediated Communication and Identity Formation

Previous predictions of poor development of individual relationships via CMC have been rejected by current research, which indicates that CMC has the "ability to convey rich, multidimensional messages through text and has potential for developing even 'hyperpersonal' relations (Ramirez, Zhang, McGrew, & Lin, 2007, p. 493). Librarians and others must learn to tap into that ability in order to build online followings for their personal professional development and to access their patrons in social media settings. Indeed, research indicates that active participation in the process is critical to successful CMC (Ramirez, Walther, Burgoon, & Sunnafrank, 2002). A discussion of CMC's impact on the development of personal and professional relationships cannot be conducted without an understanding of the pivotal theory that highlights the role of distinct communication factors which influence identity formation: Joseph Walther's Social Information Processing Theory (SIPT).

As mentioned above, early research and communication theories suggested that CMC would never allow users to communicate meaning fully. Theories such as the Media Richness Theory suggest that the lack of social context cues is likely to create a more hostile communication environment due to the possibility of multiple meanings in a message combined with the inability to read context cues (Daft, Lengel, & Trevino, 1987). Other theories such as Social Presence Theory suggest that participants may feel as if they are in a void or that they are, in a sense, alone in the communication process (Short, Williams, & Christie, 1976). While Social Presence Theory was developed in response to telecommunication, scholars regularly applied it to CMC in its early days. These remained the prevailing theories through the mid-1990s, when Walther began publishing his groundbreaking work in the realities of creating personal communication and, in fact, relationships via CMC in such a way as to shape future understanding of its potential. Considering that these were the prevailing theories that guided common perception of the practicality and utility of CMC, Walther's work in the realities of creating personal communication via CMC was groundbreaking.

Walther identified several communicative elements that are instrumental in creating various interpretations of messages exchanged using CMC. SIPT focuses on the role of impressions and influences on message and impression creation, particularly those that are identified as affiliative motives and temporal effects (Walther, 1992; Walther, 1994; Walther, 1995; Walther, 1996; Walther & Burgoon, 1992; Walther & Parks, 2002). Affiliative motives are shown through an attempt to identify or bond with others. Temporal effects deal with the amount of time participants have to respond or communicate with others. Communicators must consider these temporal effects as they select synchronous channels such as Instant Message systems or asynchronous channels such as email discussion lists or blogs. SIPT posits that during computer-mediated interactions, communicators are motivated to create impressions and formulate relationships despite obvious problems that are created by the medium (e.g., an absence of nonverbal communication cues), and that they will quickly adapt their behaviors to utilize any coding mechanisms that the medium will provide (Walther, 1992). Walther (1992) suggested that nonverbal cues are replaced by verbal cues of affinity. The temporal effects are primarily associated with the more transactional, rather than simultaneous, nature of CMC, which results in a slower social development than f2f communication (Walther, 1992).

While social media presents advantages that are obviously appreciated and utilized by the modern user, this preferred communication channel may also present delays or miscommunications of which all users should be aware. Additionally, people should be aware of cultural differences and the varying technological acumen of users. To prevent these problems, users must understand the studies that indicate that CMC can enhance instructional capabilities as it is used in an academic context when properly applied (Matthies, 2004; Rockman, 2003; Timmerman & Kruepke, 2006). Therefore, this chapter will explore the components and applications of developing a credible online identity.

As technology expands, the pervasive nature of social media typifies the function of CMC as a key dimension of one's social identity. Users continue to talk, share feelings, and exchange ideas, but they are now also able to develop overt or perhaps even alternative identities, "playing roles that may extend to changing gender and age, as well as ethnicity, social, or professional status" (Amaral & Monteiro, 2002, p. 577). Self-disclosure is often immediate when characteristics are displayed readily, which creates a hastening effect on the stages of interpersonal interaction. Thus, while SIPT cites potential temporal hindrance in social development, shared information about one's identity is often up for public display before significant interaction takes place. When considering online identity, many CMC users will make decisions about levels of identity revelation (real-identity, nicknames, or complete anonymity) and its impact on credibility and connections with other participants, but research has indicated that identity revelation has little to no impact on learners in terms of motivation or learning (Yu, 2012).

Questions about impression formation and CMC have unsurprisingly become commonplace as technology has evolved to a point where control of shared information has become tenuous. "With the advent of new social technologies, users no longer have to rely on an individual's self-composed emails, chat statements, or personal web pages to garner impressions about a subject" (Tong, Van Der Heide, Langwell, & Walther, 2008, p. 531).

While identity formation and social interaction have moved away from f2f for many interactions, the permanent, adapted impressions, as well as their ready availability, are no longer subject to a source's self-monitoring. While social judgments are continually adapted to each message that is received on a computer monitor, they are not always aligned with the impression that the sender desires, so misinterpretation may result. For example, Facebook users can easily search for any individual's personal information, past contributions, posts, or pictures posted on his or her Wall or Timeline. A further complication lies in the fact that other Facebook subscribers can tag other individuals in posts or photos without their prior consent. Therefore, it is critical that users understand the costs and the rewards to this option, as well as steps to strategically improve the CMC experience.

Ethos in Online Identity and Impression Creation

Those participating in building an online presence must be concerned with impression creation and management. Further, active participants are susceptible to the feedback offered by those with whom they interact (Schlosser, 2005). Ellison, Heino, and Gibbs (2006) explained that there are both "capacities" and "constraints" in a user's ability to connect with another participant in CMC. The capacities are the "aspects of technology that enhance our ability to connect with one another, enact change, and so forth," while the "constraints are those aspects of technology that hinder our ability to achieve these goals" (Ellison, Heino, & Gibbs, 2006, p. 417). They continue, stating, "Internet interactions allowed individuals to better express aspects of their true selves" (p. 418). The selves expressed in impression formation and identity management come through far more than just the context of the text typed on a listserv, website, or other venue for CMC. As an example, most social network-

ing sites, such as Facebook, Twitter, flickr, etc., encourage or require a personal profile wherein participants establish the role they wish to play within that network or website. This allows other users to have a context for the participant and to decide their desired level of engagement. Users allow participants to reduce their interpersonal uncertainty before choosing to communicate in a public forum (Antheunis, Valkenburg, & Peter, 2010). CMC users "form impressions based on textual and visual cues in concert" (Van Der Heide, D'Angelo, & Schumaker, 2012, p. 99).

One such cue consideration in impression formation is the selection of photographic images such as avatars and site adornment chosen for CMC by the author and creator of the site (Van Der Heide, D'Angelo, & Schumaker, 2012; Walther, Sovacek, & Tidwell, 2001; and Wang, Moon, Kwon, Evan, & Stefanone, 2010). These images may include actual photographs of the communicator used to create a closeness (immediacy) or images that reflect the identity she or he wishes to convey to users. Additionally, Ellison, Heino, and Gibbs (2006) cited a variety of sources listing additional considerations including a poster's e-mail address, links on an individual's homepage, timing of e-mail messages and posts, and expressions and use of language. Young, Kelsey, and Lancaster (2011) also addressed the timing of correspondence in addition to the use of emoticons to create a more personal level of communication. Such elements signal users of contextual cues which aid in the development of an online identity.

The availability of resources enabling communicators to more effectively monitor their self-presentation along with a lack of physical environment and nonverbal cues undoubtedly has a profound effect on the way that individuals establish ethos or credibility online. Despite the lack of physical environment, viewers are able to develop accurate impressions in a relatively quick manner from social networking and other types of websites (Van Der Heide, D'Angelo, & Schumaker, 2012). Of note are the concerns that instituting a credible persona while simultaneously evaluating that of others within the same environment may create norms not necessarily based in traditional forms of ethos. For example, a similar situation occurs when students observe presentations of classmates and model their own performance after what they have seen regardless of the "rules" taught in class or text. This explains what is known as a "recursive way in which participants developed rules for assessing others (e.g., avoid people in sitting poses) while also applying these rubrics to their own self-presentational messages (e.g., don't show self in sitting pose)" (Ellison, Heino, & Gibbs, 2006, p. 429). In their research, Ellison, Heino, and Gibbs (2006) concluded that communicators quickly become cognizant of the online environment and its associations with deceptive practices that forced them to "present themselves as credible" (p. 430). With a larger stress on visual and narrative cues, ethos-building changes from more traditional environments in other ways as well.

Social media users often establish a home base, page, or "wall" to display their written discourse, photographs, videos, and hyperlinks of interest, and in so doing form identity as well as ethos. One complication that arises from this newly established mode of self-presentation is that other individuals within a network also regularly participate in the ongoing discourse. In establishing ethos, "what complicates these sites from an impression formation perspective is that people other than the person about whom the site is focused also contribute information to the site. Such postings may or may not include secondhand descriptions about the target individual and his or her conduct" (Walther, Van Der Heide, Kim, Westerman, & Tong, 2008, p. 29). With the inclusion of this secondary noise, there is a much greater chance that evaluations of credibility will be influenced by information provided by other participants within a network in addition to that provided by the individual. When considering interactive online formats, the

statements, photographs, and responses by one's "friends" will undoubtedly have the potential to create enhanced credibility for the user. Unfortunately, the idea of guilt by association may also come into play because viewers make judgments about the user based on the types of friends' pictures, comments, and posts that are placed on the venue (Utz, 2010; Walther, VanDer Heide, Hamel, & Shulman, 2009). Even though users often have the opportunity to limit such interactions (e.g., by blocking their "friends" from posting anything on a wall or home page), the act of limitation in and of itself may affect an observer's judgment of an individual's ethos if that observer becomes curious about what the person has to hide. In this consideration of CMC, the user also needs to consider the community they are joining and what defines it. Examining groups in social networks, Backstrom, et al. (2006) found that individuals choose which social network to join based on some of the infrastructure of the site. From this we can assume that joining a particular kind of community, e.g. Facebook instead of Twitter, as a place to engage inherently impacts the definition of the user's identity. This is particularly affected by the changing nature of social media. If the primary point of engagement is seen as an outdated social media community (e.g., MySpace), this informs and influences how users outside the community perceive those within a community. While it is difficult for users to constantly migrate their identity to new platforms, they must recognize that new technologies and locations are important for community development.

The significance of the establishment of initial ethos should not be underestimated. The value of first impressions have long been noted and supported in research that stresses the role of initial impressions as a determining factor of how a relationship will develop in the future (Sunnafrank & Ramirez, 2004). Recent studies have shown, particularly when using CMC, that the valence of initial impressions may determine "how or even if relationships continue to develop beyond initial stages" (Ramirez, 2007, p. 54).

Establishing ethos in an online environment is filled with new challenges dealing with impression management and communicative norms. Rather than an overt or purposeful approach to building ethos, individuals creating online identities may passively allow others to build their ethos for them. In fact, these individuals are often unaware of the impressions that are formed about them because the impressions are based on the discourse of others. This discourse may be affected, as previously noted, by the choice of where the online identity is built. Because identification of types of users can be immediately presumed based on the location of the user profile, users must be aware of the connotations of different online communities and the potential benefits or detriments of that affiliation. This results in a playing field that establishes its own rules as it moves forward. Upon coalescence of these factors, when we consider that initial credibility becomes paramount in the establishment of future relationships (let alone how the communication in such relationships may develop), it becomes clear that an understanding of ethos in an online environment becomes crucial in social, work, and learning environments.

Let's Be Practical: Incentives and Examples

While we have considered the history and research supporting the use of CMC, librarians and the library community provide a case study that can be scaled upwards and related to faculty (and perhaps other professionals) as a larger community. Librarians have two different populations with whom they need to engage: other librarians and library patrons. As a community themselves, librarians fall into many types of sub-communities based on region, type of library (academic, public, medical, etc.), and professional interest. In representing their places of work, librarians also have the challenge of engaging their users not only for their personal ethos and credibility, but also to develop the credibility of their institution. Many libraries are engaging with their users through some form of social media, with official presences being managed by individuals, ad hoc, and formal/informal committees, policies, and participants.

As they contemplate this engagement, librarians and faculty may perceive social media engagement with the dilemma of adding extra work to an already heavy work-load. Will this effort with social media even count when looking for tenure and promotion? According to a number of researchers, it should (Gratch-Lindauer, 2002; Matthies, 2004; Rockman, 2003). As higher education assesses student learning, accrediting agencies have begun to consider the learning outcomes for students that can be directly attributed to information literacy, a direct goal of the online work librarians are conducting (Gratch-Lindauer, 2002). Further, research clearly indicates that "extra-class communication (ECC) allows for an expanded breadth of topics" that can build both student motivation and trust (Young, Kelsey, & Lancaster, 2011, p. 372). Much of Walther's groundwork research (1992; 1994; 1995; 1997) stressed the importance of relationship-building through CMC. Armed with this information, librarians can make their case by reminding administrators that there are a number of valid reasons for valuing such work. Most institutions of higher education promote long-term engagement in the community. An online library presence is obviously useful to students and faculty, but the public community utility should not be ignored. Additionally, this online presence can allow active sharing and educating with both the institution and the community of scholars at-large. When considering examples, one needs to look no further than the research in the hard sciences that has become social with collaborators who frequently work together while on a global scale. Finally, professional societies model the value of an online presence as they are using social networking as a way to engage their members, demonstrating their belief that there is value in sharing communication in this differently rigorous venue.

As librarians choose to engage with their communities on behalf of their institution, they must consider format, frequency, community, identity, and first impression. Because of the ever-changing nature of their patrons or students graduating, scientific communities' evolving librarians need to be aware of where their patrons are engaging online and need to determine what role they can take. This must be done keeping in mind the desired level of engagement of the community, other opportunities for outreach, and the workload of the staff managing online accounts. As mentioned previously, when users engage viewers, frequency of posts and updated communication is an important consideration. Research indicates that one of the best ways to build a persona of immediacy or closeness is to engage in frequent communication (Young et al., 2011).

Libraries engage with users across many social media sites. Seattle Public Library engages users through the daily blog Shelf Talk (http://shelftalk.spl.org/). In this venue, Seattle librarians talk about issues facing users such as digital accessibility, recommend books on user-identified topics, highlight community events, and alert patrons to library updates. The University of Wisconsin-Madison libraries maintain a Twitter account (http://twitter.com/UWMadLibraries) where they send out search tips to students, notify potential patrons of library events, and encourage students to send comments. Alternatively, the New York Public Library (http://twitter.com/nypl) focuses more on trivia and advocacy. Many academic libraries such as the Ezra Lehman Memorial Library (https://www.facebook.com/shiplibrary) are active on Facebook, where they may feature images of students and faculty using the building or highlighting digital special collections in addition to providing more traditional research-related information to their community.

Librarians, specifically, serve as excellent examples of people using social networking to evaluate and develop online credibility. Because their professional responsibilities include iden-

tifying and evaluating information, as well as assisting others in learning how to do the same, many librarians have determined trustworthy peers from within their online networks. By reviewing successful librarians who have developed their professional reputations through online social networks such as blogs, we can glean ideas about the best practices that help librarians engage their communities to develop their personal credibility. There are many online tools that librarians have used to help each other and establish their online identity. These include blogs, professional networks such as Linked In, social sites such as Twitter, Friendfeed, Facebook, and even topical email discussion lists. More recently, professional collaboration sites such as Mendeley and VIVO have further expanded scholarly communication channels. Each of these areas provides an opportunity to develop an online persona and credibility through those networks. This review focuses specifically on blogging to allow a clearer contrast between those highlighted.

Jessamyn West blogs at Librarian.net and is one of the pioneer library bloggers, with archives accessible to 2000, though she indicated her blogging began in 1999 (West, 2007). West's work was recognized early by the New York Times Magazine (West, 2003). She continues to write her blog, focusing on the digital divide, serving libraries in rural areas, and touching on major challenges facing most libraries in current times. As a result of her excellent writing and reputation, West wrote a book, Without a Net: Librarians Bridging the Digital Divide, and participated in a "Room for Debate" column in the New York Times (West, 2011a). West identifies more closely with public libraries but is also an independent scholar and demonstrates how she continues to be engaged with the library community, thus gaining both traditional and non-traditional recognition for her work.

Another positive example is Meredith Farkas, the Head of Instructional Services at Portland University and adjunct faculty for San Jose State

University. She blogs at Information wants to be free and has been regularly blogging about libraries and technology since 2004. Her posts are regularly lengthy, combining her personal experiences with information gathered from other library professionals, citing other blog posts, articles, presentations, and papers. Farkas's blog has led her to publish a regular column on use of technology in practice in American Libraries, the society journal for the American Library Association, as well as to publish a book titled Social Software in Libraries: Building Collaboration, Communication and Community Online (Farkas, 2012a).

A number of lessons can be learned by modeling these online librarian writers. Connections with others are critical. West and Farkas link to other blogs, and they are willing to evaluate other blogs as well. They provide comments, praise, and critiques. Additionally, they encourage the same in return. They encourage or recognize when others critique them. This engagement furthers the conversation, which regularly happens in the comments on the posts, and is live and visible on the web. A benefit of this openness is that it promotes professional conversation in the everyday experience. Both also model another crucial element: They have opinions and they are willing to share them. Simply sharing or mimicking another online opinion isn't enough. Blog posts can have both an internal and external focus for the professional. For example, Farkas (2012b) tackled the internal focus in her blog post "Classic blunder #1- Let's just try and see what happens." Responding to a fellow librarian's post, she provided an analysis of the topic presented by her colleague, pointing out the flaws that she noted and providing examples from her own work history. While she is critiquing, however, she also finds much to praise about the original topic and identifies that as well. Further, she engages with both her readers and the colleague whose work she critiqued in conversation in the comments of this blog post (2012b). West (2011b) provided a provocative exploration with an external focus

in her post "The Kindle lending experience from a patron's perspective 'a wolf in book's clothing.'" In the post, West described the processes of a vendor from the user perspective, which is followed by a spirited round of comments and opinions about the behavior users, vendors, and libraries (West, 2012). While West does not engage as frequently in the comments as Farkas does in these two examples, both librarians demonstrate attention and participation in the conversation after starting the discussion.

Another commonality of successful online librarians is that they provide enough personal information that the public knows they are not automatons. However, they refrain from getting too personal in their blogs so that they remain professional sources. Readers will also note that the best of these sites have excellent writing. Poor writing skills will not build online credibility, and they won't help the reader. Finally, longevity is essential. If you are willing to step out online, you must be prepared to stay with the course. An online presence isn't a presence if it merely lasts weeks or months.

We have highlighted two long-term, stable, strongly credible librarian bloggers; however, there are many other strong examples to refer to for potential best practices. Contemporaries of Farkas and West include academic librarians Jason Griffey, who blogs about technology in libraries at Pattern Recognition and co-authored a book on blogging for libraries (Griffey and Coombs, 2008); and Karen Schneider, who has blogged from the perspective of both a library vendor as well as a library director at Free Range Librarian, providing insight from two occasionally contentious roles within libraries. Public librarians David Lee King, who writes about libraries and outreach at DavidLeeKing.com, and Sarah Houghton (2011), who focuses on libraries, technology, and web services at the Librarian in Black, provide perspective outside the more targeted academic focus of librarians who usually center on higher education. Nor is decade-long longevity required for online credibility. An example of a more recent entrant to blogging who has developed strong online credibility is Andromeda Yelton, who blogs at Andromeda Yelton: Across Divided Networks. Yelton is leveraging her more recently established online credibility in the f2f and CMC opportunities of the American Library Association, where she established a continuing education community for librarians to collaborate while improving their computer coding skills through the Code Academy's Code Year (Landgraf, 2012).

Each blogger has a particular style that can be identified and potentially emulated by those seeking to further develop their online credibility, some of which include:

- **Regular Posting:** While daily posts are not necessary, a regular presence of posting demonstrates longevity and reliability.
- **Relation of Current Events:** This can demonstrate personal continuing education or awareness of the community and engagement, but must go beyond merely trumpeting the latest headlines. Providing an analysis and/or opinion allows others in the community to further develop an impression of authority.
- **Targeted Audience:** The featured bloggers here are all librarians writing for the library community and yet within that narrow community, clearer sub-audiences are identifiable. By speaking to a specific audience, the bloggers can focus their message and more easily find those with similar interests.
- **Engage with Others in the Community and Beyond:** Because no one works in a vacuum, it is important to engage with other users. For blogging, this can be referring to another blog where an idea was begun or responding publicly to comments made either on the blog site or in other venues. Being able and willing to engage with oth-

ers may open up the user for criticism and controversy, where others may form an opinion based on responses; however, this approach also allows the user to manage his or her online identity more clearly.

- **Participation in More than One Format of Social Media:** Though not highlighted specifically here, all of these bloggers can also be found participating in professional social media on other sites and they often include their posts as part of their participation there, or use the interactions on other sites as conversation points for their blogs.

These ideas, however, rely upon the interpretation by the reader of the social media channel being used, communities being engaged, and participation of the readers and author(s). While the bullet points listed are practical for most types of interactions, it is important that the authors consider these and other factors and their specific intentions when tailoring their messages and developing their personalities.

While we have highlighted blogging, there are numerous other opportunities in the various other social networks. Participation in a discussion list allows for regular and immediate interaction between colleagues. The advantage here is the dialog between peers who are at a variety of levels. In many library and information science programs, students are encouraged (occasionally required) to join a discussion list in the profession in an area that interests them. This allows these students to see the daily engagement between peers, learn about current topics that are issues within the field, and to begin to engage future colleagues while still undertaking their education. For the faculty, it is an opportunity to find others who are working on a specific problem or to provide insight of a solution at one's own institution. Discussion lists often surround professional societies, allowing the membership to develop a deeper connection so their participation at in-person events is only part of the conversation. Advantages to faculty with discussion lists may be a level of privacy and searchability, depending on the requirements to join the group and whether the list is publicly indexed.

Participation in a public-oriented or non-librarian social network such as Twitter or Facebook can be used in similar ways to a discussion list, with only a small insular network interacting in a semi-private online space, but it can also be used to develop credibility with a wider audience outside of one's field. Many social networking platforms require users to select individuals to follow or allow them to select their followers. Next, the user composes statements and crafts opinions which are presented through the social networking platform to a wide audience of professional friends and colleagues. This allows the conversation to expand and grow. For example, Twitter is a popular messaging service for libraries and librarians (Milstein, 2009). Considered a "micro blogging" medium for people to put information onto the web, librarians can send a short and simple message (Twitter allows a maximum of 140 characters per message) that contains useful information. Milstein (2009) suggested that, "a carefully crafted post can convey a good deal of information without taking a lot of time to read or write" (p. 17). One of the more influential librarian Twitter users is Sarah Houghton, who tweets under the handle @theLiB. Houghton's credibility in speaking about current topics, engaging in conversations with her peers, and providing interesting subject matter to her followers gave her the opportunity to participate as an official Twitterer for the 2012 State of the Union Address, where she and others live-tweeted the speech as well as had the opportunity to interact with senior officials in the Obama Administration and ask questions on behalf of themselves and their followers (Houghton, 2012).

THE CREDIBILITY ADVANTAGE

The importance of developing online credibility remains questioned in most fields of scholarship. While the value of traditionally published peer-reviewed or edited articles retains its luster, the ease of publication and participation has greatly lowered the bar to accessing the opportunity of making oneself heard. This ease of publication also means that readers and other professionals in the field are more skeptical about giving their trust. After that trust is earned, the profiled librarian bloggers have shown how their dedication, engagement, and good writing have translated this online to offline work in more traditional settings such as newspapers and scholarly journals. In addition to developing their own online credibility, it is critical for librarians and the campus-wide community to promote information literacy for students as well as the evaluation of faculty and administrators (Matthies, 2004) because these students are already emerging scholars whose reputations are being developed even before they take their place within various professions. Improving information literacy instruction and awareness of online credibility by librarians as educators or examples allows everyone in the scholarly community cycle to better evaluate future colleagues and be more informed of their own online reputation. The upcoming generations of faculty and researchers will have always had access to this type of social knowledge building and may better understand that people put a lot of effort into social media brand development in an effort to promote education, develop community, make money, and provide transparency into research.

To those who have been in educational institutions for a number of years, the opportunities that CMC and social media present may seem like recent technological developments with a steep learning curve. However, we must keep in mind that to the incoming student body, and particularly those entering graduate school, blogs may have always been a common medium for communicating news and educational material as well as social content. It is imperative that faculty take advantage of the tools that are available and to consider the positive changes that are waiting once online publishing and communication with students in particular can flourish. As Richardson (2010) asserted, the development and adoption of easier content creation and audience connection requires faculty to adopt new methods of communicating and interacting with students. In addition to creating new venues for learning, pedagogical effectiveness in general will be enhanced by learning to interact with students in media with which they are proficient and familiar.

Another advantage associated with social media is the availability of open and timely feedback. Historically, publication and dissemination tools were limited by the time costs of printing, distribution, and a formal review process. The rapid pace of blogging and social media allows for easier and more responsive development of writing and thinking skills which hitherto have been ephemeral or under-utilized (Solomon & Schrum, 2010). As students and faculty engage in online discussions and critiques, the experience changes contemporary coursework formatting as well as affects future curriculum, assessment, and teaching/learning styles. It creates opportunities for students to engage not only with others at their college or university but also opens a limitless network for co-authorship worldwide. Likewise, faculty can use blogs as an opportunity to develop skills for writing, as well as to define their specific disciplines by demonstrating capabilities early in their teaching careers while becoming more effective in the primary roles of teaching and service. Like students, faculty members have limitless prospects in terms of collaboration that can serve as a boon to their career potential far beyond the confines of a classroom. Through the accelerated utilization of blogging in particular, November (2009) asserted that educators have a greater network of collaborators and able to connect with both their community and peers to improve student experi-

ences and engagement. By communicating more effectively and sharing their productivity with professionals in an open network, faculty members greatly improve their chances of finding effective relationships with potential collaborators within their own fields, as well as for interdisciplinary research opportunities. In addition to the obvious benefits this presents, communication potential is also no longer hindered by physical distance. The frustrations that are caused by ineffective attempts at faculty collaboration have long been noted within institutions, let alone among faculty from different organizations. Faculty resistance to collaboration, due to the perceived limitations that have caused teaching to become what Wheelan (2004) referred to as the "second most private activity" (p. 2), has the potential to be greatly reduced when the spheres of influence involved are opened by CMC. Teamwork and partnership will undoubtedly flourish relationally and professionally when such practical factors as schedules, time zones, and actual financial costs are eliminated due to the proliferation of ideas that blogs, social networks and the like can provide.

As a concrete example of social media providing opportunities for collaboration across universities and various platforms, one could consider this particular book chapter. The call for publication for this book was shared through a professional development blog, Library Writer's Blog, which Goben highlighted on the social network Friendfeed in search of potential collaborators. Dobbs, a librarian with whom Goben had worked in both online and offline settings, suggested working together and bringing in colleagues from his university's Human Communication Studies department. The authors worked both through more traditional means of emailing documents and using Google Docs to allow simultaneous editing.

Members who are new to online communities are advised to refrain from interaction and "lurk" for a period of time instead. This enables new members to gain an understanding of the norms or the culture of the community (Preece, Nonnecke, & Andrews, 2004). This period of time allows the participants to develop more accurate perceptions of the active participants. That said, Preece, Nonnecke, and Andrews (2004) found that active participants were more satisfied with their online interactions than lurkers. So, efforts should be made to leave the lurker role behind and move into a more active role as soon as one understands the online community's norms.

CONCLUSION

This chapter has addressed the importance of scholarly ethos with regard to social networking using communication theory to explore credibility through online impressions and identity creation. The need for users to understand the importance and methods of establishing credibility carry through from legacy to cutting edge communication channels. Using communication concepts and theories, the authors have highlighted ways to strengthen one's online credibility, or ethos, by carefully considering their social networking audience, the impact of fewer traditional non-verbal cues, and being mindful of other communication concepts when making the transition from f2f interactions to online communication.

Discussion of effective behaviors and practices of currently credible librarians with influential online presences highlighted successful uses of social media to build both traditional scholarly and online credibility. While social media is new to many educators, it is often a ubiquitous communications channel in current students' reality. While it is ubiquitous, conscious choices about identity management should be made in the creation and maintenance of an online presence, personal or institutional. These considerations should include targeting communications to and for desired audiences, selecting and covering particular topics or subjects, understanding the norms of the communications channel(s) selected, and choosing appropriate platforms for engagement. Self-monitoring of one's online presence has

become de rigeur. Without active monitoring and management of reader impressions, others may come to define the author's ethos. This can work out well, so long as the ethos created is positive.

Based on the review of literature and the discussion within this chapter, we can summarize a few basic suggestions for those librarians attempting to create a credible online presence within social media. First, one should do some basic audience analysis to maintain a firm grasp on perceptions of selected social media outlets (e.g., is this a passing fad or a trend with staying power? Does this outlet reach the intended audience?). In building one's presence, attempt to use textual cues that support visual cues for a stronger and more accurate impression creation (e.g., users should describe themselves in much the same way their posted photos show them). While users may select pseudonyms or nicknames, there is no need to spend a great deal of time choosing these as research indicates that there is no marked difference in viewer impressions of the user based on naming. Next, users must be vigilant about communicating regularly and frequently through posts and other correspondence with their audiences. In said correspondence, users should personalize all communication by using strategies that build immediacy or a close relationship (e.g., users should address others by first name when possible and use collective pronouns such as "we" and "our" rather than "my" or "your"). Emoticons are also helpful in creating personality because they attempt to bridge the chasm created by the lack of nonverbal cues present in f2f interactions. In closing, remember to think conversation, not mountaintop sage, so write knowledgeably about personally interesting topics, analyze topics more than repeat announcements, cite and link to sources, and write casually but well.

A successful online presence can build or strengthen the case for libraries. When the library is perceived as active in the scholarly and social life of its community, it supports the traditional library information dissemination role and remains a visible part of its overarching institution. Conversely, if the online presence is haphazard and disorganized (or, worse yet, significantly out of date) this status reflects poorly on the organization and may rebound negatively on the library. Following the successful examples of the highlighted librarian bloggers, people can improve ethos by developing personal engagement with their readers and their intellectual community, remaining timely in their outbound communications, and linking to others in their professional circles.

Although social media is a much-used form of interaction across our society, its effects have just begun to show themselves in the academic arena. Currently the world of student Facebook use, for example, rarely intersects with the educator's use of the same medium. We feel that this will change, just as it has with previous computer-mediated channels of communication. Therefore, it is vital that forward-thinking academicians begin creating a presence in social media now, and also heed the advice recounted in this chapter to ensure that their credibility is strong. As the definition of scholarship expands to embrace social media, being on the cusp of this wave can only strengthen librarians' and scholars' influence in the academic arena.

REFERENCES

Amaral, M. J., & Monteiro, M. B. (2002). To be without being seen: Computer-mediated communication and social identity management. *Small Group Research, 33*, 575–589. doi:10.1177/104649602237171

Antheunis, M. L., Valkenburg, P. A., & Peter, J. (2010). Getting acquainted through social network sites: Testing a model of online uncertainty reduction and social attraction. *Computers in Human Behavior, 26*, 100–109. doi:10.1016/j.chb.2009.07.005

Backstrom, L., Huttenlocher, D., Kleinberg, J., & Lan, X. (2006). Group formation in large social networks: Membership, growth and evolution. *Proceedings of 12th International Conference on Knowledge Discovery in Data Mining* (pp. 44-54). New York, NY: ACM Press.

Berger, C. (2005). Interpersonal communication: Theoretical perspectives, future prospects. *The Journal of Communication, 55*, 415–447. doi:10.1111/j.1460-2466.2005.tb02680.x

Daft, R. L., Lengel, R. H., & Trevino, L. K. (1987, September). Message equivocality, media selection, and manager performance: Implications for information systems. *Management Information Systems Quarterly*, 355–366. doi:10.2307/248682

Ellison, N., Heino, R., & Gibbs, J. (2006). Managing impressions online: Self-presentation processes in the online dating environment. *Journal of Computer-Mediated Communication, 11*, 415–441. doi:10.1111/j.1083-6101.2006.00020.x

Farkas, M. (2012a). *Information wants to be free.* Retrieved from http://meredith.wolfwater.com/wordpress/

Farkas, M. (2012b, January 28). *Classic blunder #1 - Let's just try it and see what happens! Information wants to be free.* Retrieved from http://meredith.wolfwater.com/wordpress/2012/01/28/classic-blunder-1-lets-just-try-it-and-see-what-happens/

Gratch-Lindauer, B. (2002). Comparing the regional accreditation standards: Outcomes assessment and other trends. *Journal of Academic Librarianship, 28*(1), 14–25. doi:10.1016/S0099-1333(01)00280-4

Griffey, J., & Coombs, K. (2008). *Library blogging.* Columbus, OH: Linworth.

Houghton, S. (2011). Archives. *Librarian in Black.* Retrieved from http://librarianinblack.net/librarian-inblack/archives

Houghton, S. (2012). Miss librarian goes to Washington. *Librarian in Black*. Retrieved from http://librarianinblack.net/librarianinblack/2012/01/sotu2.html

Hovick, S. R., Meyers, R., & Timmerman, C. E. (2003). E-mail communication in workplace romantic relationships. *Communication Studies, 54,* 468–482. doi:10.1080/10510970309363304

Ishii, K. (2006). Implications of mobility: The uses of personal communication media in everyday life. *The Journal of Communication, 56,* 346–365. doi:10.1111/j.1460-2466.2006.00023.x

Kaplan, A. M., & Haenlein, M. (2010). Users of the world, unite! The challenges and opportunities of social media. *Business Horizons, 53,* 59–68. doi:10.1016/j.bushor.2009.09.003

Kietzmann, J. H., Hermkens, K., McCarthy, I. P., & Silvestre, B. S. (2011). Social media? Get serious! Understanding the functional building blocks of social media. *Business Horizons, 54,* 241–251. doi:10.1016/j.bushor.2011.01.005

Landgraf, G. (2012). Code year librarians geek out. *American Libraries*. Retrieved from http://americanlibrariesmagazine.org/inside-scoop/code-year-librarians-geek-out

Lawson, S. (2009, July 13). Clinical reader: From zero to negative sixty with one bogus threat. *See Also: a Weblog by Steve Lawson*. Retrieved from http://stevelawson.name/seealso/archives/2009/07/clinical_reader_from_zero_to_negative_sixty_with_one_bogus_threat.html

Matthies, B. (2004). The road to faculty-librarian collaboration. *Academic Exchange Quarterly, 8*(4), 135–141.

Milstein, S. (2009). Twitter for libraries (and librarians). *Computers in Libraries, 29,* 17–18.

November, A. C. (2009). *Empowering students with technology*. Thousand Oaks, CA: Corwin Press.

Preece, J., Nonnecke, B., & Andrews, D. (2004). The top five reasons for lurking: Improving community experiences for everyone. *Computers in Human Behavior, 20,* 201–223. doi:10.1016/j.chb.2003.10.015

Ramirez, A. Jr. (2007). The effect of anticipated future interaction and initial impression valence on relational communication in computer-mediated interaction. *Communication Studies, 58,* 53–70. doi:10.1080/10510970601168699

Ramirez, A. Jr, Walther, J. B., Burgoon, J. K., & Sunnafrank, M. (2002). Information seeking strategies, uncertainty, and computer-mediated communication: Towards a conceptual approach. *Human Communication Research, 28,* 213–228.

Ramirez, A. Jr, Zhang, S., McGrew, C., & Lin, S. (2007). Relational communication in computer-mediated interaction revisited: A comparison of participant-observer perspectives. *Communication Monographs, 74,* 492–516. doi:10.1080/03637750701716586

Richardson, W. (2010). *Blogs, wikis, podcasts, and other powerful web tools for classrooms*. Thousand Oaks, CA: Corwin Press.

Rockman, I. F. (2003). Integrating information literacy into the learning outcomes of academic disciplines: A critical 21st-century issue. *College & Research Libraries News, 64*(9), 612–615.

Schlosser, A. E. (2005). Posting versus lurking: Communicating in a multiple audience context. *The Journal of Consumer Research, 32,* 260–265. doi:10.1086/432235

Short, J. A., Williams, E., & Christie, B. (1976). *The social psychology of telecommunications*. London, UK: Wiley.

Solomon, G., & Schrum, L. (2010). *Web 2.0: How-to for educators*. Eugene, OR: International Society for Technology in Education.

Sunnafrank, M., & Ramirez, A. Jr. (2004). At first sight: Persistent relational effects of get-acquainted conversations. *Journal of Social and Personal Relationships, 21*, 361–379. doi:10.1177/0265407504042837

Timmerman, C. E., & Kruepke, K. A. (2006). Computer-assisted instruction, media richness, and college student performance. *Communication Education, 55*, 73–104.

Tong, S. T., Van Der Heide, B., & Langwell, L. (2008). Too much of a good thing? The relationship between number of friends and interpersonal impressions on Facebook. *Journal of Computer-Mediated Communication, 13*, 531–549. doi:10.1111/j.1083-6101.2008.00409.x

Utz, S. (2010). Show me your friends and I will tell you what type of person you are: How one's profile, number of friends, and type of friends influence impression formation on social network sites. *Journal of Computer-Mediated Communication, 15*, 314–335. doi:10.1111/j.1083-6101.2010.01522.x

Van Der Heide, B., D'Angelo, J. D., & Schumaker, E. M. (2012). The effects of verbal versus photographic self-presentation on impression formation in Facebook. *The Journal of Communication, 62*, 98–116. doi:10.1111/j.1460-2466.2011.01617.x

Walther, J. B. (1992). Interpersonal effects in computer-mediated interaction: A relational perspective. *Communication Research, 19*, 52–89. doi:10.1177/009365092019001003

Walther, J. B. (1994). Anticipated ongoing interaction versus channel effects on relational communication in computer-mediated interaction. *Human Communication Research, 20*, 473–501. doi:10.1111/j.1468-2958.1994.tb00332.x

Walther, J. B. (1995). Relational aspects of computer-mediated communication: Experimental observations over time. *Organization Science, 6*, 186–203. doi:10.1287/orsc.6.2.186

Walther, J. B. (1996). Computer-mediated communication: Impersonal, interpersonal, and hyperpersonal interaction. *Communication Research, 23*, 3–43. doi:10.1177/009365096023001001

Walther, J. B. (2007). Selective self-presentation in computer-mediated communication: Hyperpersonal dimensions of technology, language, and cognition. *Computers in Human Behavior, 23*, 2538–2557. doi:10.1016/j.chb.2006.05.002

Walther, J. B., & Burgoon, J. K. (1992). Relational communication in computer-mediated interaction. *Human Communication Research, 19*, 50–88. doi:10.1111/j.1468-2958.1992.tb00295.x

Walther, J. B., & Parks, M. R. (2002). Cues filtered out, cues filtered in: Computer-mediated communication and relationships . In Knapp, M. L., & Daly, J. A. (Eds.), *Handbook of interpersonal communication* (3rd ed., pp. 529–563). Thousand Oaks, CA: Sage.

Walther, J. B., Slovacek, C. L., & Tidwell, L. C. (2001). Is a picture worth a thousand words? Photographic images in long-term and short-term computer-mediated communication. *Communication Research, 28*, 105–134. doi:10.1177/009365001028001004

Walther, J. B., Van Der Heide, B., Hamel, L., & Shulman, H. (2009). Self-generated versus other-generated statements and impressions in computer-mediated communication: A test of warranting theory using Facebook. *Communication Research, 36*, 229–253. doi:10.1177/0093650208330251

Walther, J. B., Van Der Heide, B., Kim, S., Westerman, D., & Tong, S. T. (2008). The role of friends' appearance and behavior on evaluations of individuals on Facebook: Are we known by the company we keep? *Human Communication Research, 34*, 28–49. doi:10.1111/j.1468-2958.2007.00312.x

Wang, S. S., Moon, S. I., Kwon, K. H., Evans, C. A., & Stefanone, M. A. (2010). Face off: Implications of visual cues on initiating friendship in Facebook. *Computers in Human Behavior, 26*, 226–234. doi:10.1016/j.chb.2009.10.001

West, J. (2003, September 29). Meet the new site, same as the old site. *Librarian.net*. Retrieved from http://www.librarian.net/stax/19/meet-the-new-site-same-as-the-old-site/

West, J. (2007, January 1). About. *Librarian.net*. Retrieved from http://www.librarian.net/about/

West, J. (2011a, February 4). More about power than gender. *New York Times*. Retrieved from http://www.nytimes.com/roomfordebate/2011/02/02/where-are-the-women-in-wikipedia/more-about-power-than-gender

West, J. (2011b, November 15). The Kindle lending experience from the patron's perspective "a wolf in book's clothing." *Librarian.net*. Retrieved from http://www.librarian.net/stax/3725/the-kindle-lending-experience-from-a-patrons-perspective-a-wolf-in-books-clothing/

Wheelan, S. A. (2004). *Faculty groups: From frustration to collaboration*. Thousand Oaks, CA: Corwin Press.

Young, S., Kelsey, D., & Lancaster, A. (2011). Predicted outcome value of e-mail communication: Factors that foster professional relational development between students and teachers. *Communication Education, 60*(4), 371–388. doi:10.1080/03634523.2011.563388

Yu, F. Y. (2012). Any effects of different levels of online user identity revelation? *Journal of Educational Technology & Society, 15*(1), 64–77.

ADDITIONAL READING

Beason, L. (2001). Ethos and error: How business people react to errors. *College Composition and Communication, 53*(1), 33–64. doi:10.2307/359061

Commoncraft. (2011). *Protecting reputations online in plain English* [Video file]. Retrieved from http://www.commoncraft.com/video/protecting-reputations-online

Griffey, J. (2012). *Pattern recognition*. Retrieved from http://jasongriffey.net/wp/

Houghton, S. (2012). *Librarian in Black*. Retrieved from http://librarianinblack.net/librarianinblack/

King, D. L. (2012). *David Lee King*. Retrieved from http://www.davidleeking.com/

Milstein, S. (2008). *Twitter for business* [Video file]. Retrieved from http://www.youtube.com/watch?v=lUR2E8l3bi8

Schneider, K. (2012). *Free Range Librarian*. Retrieved from http://freerangelibrarian.com/

Walther, J. B., Deandrea, D. C., & Tong, S. T. (2010). Computer-mediated communication versus vocal communication and the attenuation of pre-interaction impressions. *Media Psychology, 13*, 364–386. doi:10.1080/15213269.2010.524913

Walther, J. B., Liang, Y., Deandrea, D. C., Tong, S. T., Carr, C., Spottswood, E. L., & Amichai-Hamburger, Y. (2011). The effect of feedback on identity-shift in computer-mediated communication. *Media Psychology, 14*, 1–26. doi:10.1080/15213269.2010.547832

West, J. (2012). *Librarian.net*. Retrieved from http://www.librarian.net/

Yelton, A. (2012). *Andromeda Yelton: Across divided networks*. Retrieved from http://andromedayelton.com/

ENDNOTES

[1.] In 2009, a website which described itself as an aggregator of medical information was launched. The site was reviewed by medical librarian Nicole Dettmar in a blog post where Dettmar highlighted areas of concern about the newly launched site. Specifically, Dettmar noted that there was a graphic on the website implying endorsement by the National Library of Medicine (NLM) and pointed her readers to the NLM policy that specifically stated that the organization does not provide endorsements. She also identified copyrighted images on the website, used with crediting lines but without official permission for use of the works on a commercial website. Dettmar asked that the website address these concerns or remove her blog from the list of recommended medical library blogs as she felt it would be damaging to her own online credibility to be affiliated with the website. The staff of the site contacted Dettmar via Twitter, noting that they would be taking down the copyrighted images but then threatening her with a lawsuit if she did not remove her initial blog post. Over the next four days, staff speaking for the website volleyed communications with Dettmar and other librarians (Lawson, 2009) and ultimately became a case study librarians use to demonstrate poor use of social media.

Chapter 17
Ethical Challenges for User-Generated Content Publishing:
Comparing Public Service Media and Commercial Media

Ceren Sözeri
Galatasaray University, Turkey

ABSTRACT

Mainstream online media is gradually encouraging user contributions to boost brand loyalty and to attract new users; however, former "passive" audience members who become users are not able to become true participants in the process of online content production. The adoption of user-generated content in media content results in new legal and ethical challenges within online media organizations. To deal with these challenges, media companies have restricted users through adhesion contracts and editorial strictures unlike anything encountered in the users' past media consumption experiences. However, these contractual precautions are targeted to protect the media organizations' editorial purposes or reputations rather than to engage ethical issues that can also ensure them credibility. It is expected that some public service media strive to play a vital role in deliberative culture; on the other hand, some commercial global media have noticed the importance of worthwhile user-generated content even though all of them are far from "read-write" media providers due to the lack of an established guiding ethos for publishing user-generated content.

INTRODUCTION

The mainstream online media gradually encourages user contributions including articles, comments, photos, and videos to boost brand loyalty and attract new users. However, the adaptation of user-generated content has introduced new

DOI: 10.4018/978-1-4666-2663-8.ch017

social, legal, and ethical challenges. In addition to the question of how user contributions are involved in the newsgathering process, content that includes abusive language, offensive comments, or trolling is considered a threat to the credibility of news organizations. Benkler (2006) defined credibility as quality that is measured by some objective criteria and relevance (p. 7). Moreover, general measures of media credibility, such as comparing perceptions of the "believability, reliability, fairness, lack of bias, balance, community affiliation, ease of use, completeness, composure, sociability, accuracy, or attractiveness of the media themselves" (Chung, Kim, & Kim, 2010, p. 673) inherently include user-generated contend as part of media content.

To deal with credibility concerns, the media organizations tend to have users adopt agreements in the form of "terms of use" or "privacy policy" to protect the organization's rights and conduct; additionally, the media organizations reduce their own risk of liability for the unlawful acts of others. Although it is expected that content providers take steps to help control the most extreme content (Williams, Calow, & Lee, 2011), these user agreements are treated as adhesion contracts and protect, in fact, only one side's rights without assuming any responsibility for the content, for which the users assume all the risk.

This chapter examines how media publishers handle the ethical and legal challenges of publishing user contributions by comparing public service media and commercial media. For this research, the user contribution patterns of three different types of public service broadcasters (SBS, the Special Broadcasting Service of Australia; the BBC, the Public Service Broadcaster of the UK; and PBS, the Public Broadcasting Service of the USA) were analyzed, and on the other side, CNN, NBC, the *New York Times*, Reuters, *Le Monde* and *The Guardian*'s user contribution strategies were examined in terms of their conditions, editorial guidelines and privacy policies. Toward that end, the next part of this research focuses on the theoretical background of audience evolution, audience engagement in content production, and the resulting consequences for the new media environment. In this section, unlike past media consumption experiences, the contractual relationship between the users and the media organization is examined in terms of the scope of the contracts, the moderation of user-generated content, and the ethos adopted by commercial and public service

media organizations. In the last part, the potential of user-generated content and unanswered ethical concerns are discussed from the perspective of the diversity of point of view, the credibility of media organizations, and the deliberative online media environment.

THEORETICAL BACKGROUND

Changing the relationship between news organizations and the audience is one of the significant indicators of digital culture (Deuze, 2006). "The people formerly known as the audience" (Rosen, 2006) are no longer passive recipients of media; rather, they participate, debate, create, and share. Axel Bruns (2005) described them as produsers because of their engagement in non-traditional forms of content production, and their involvement in produsage, which refers to user-led content creation environments. To some extent, this reconceptualization of the audience and their "newfound" production capabilities has replaced established media organizations. At the same time, these organizations intend to adopt this emerging audience factor into their business strategies (Napoli, 2010). They are providing new functions such as hosting and search functionality for massive aggregations of content produced by others (Napoli, 2009).

Although online media organizations are increasingly promoting bottom-up participatory culture (Jenkins, 2006) and creative activity (Svoen, 2007), the technical and content quality of user-generated content is still guaranteed through the choice of traditional media "gatekeepers" (Organisation for Economic Co-operation and Development, 2007). Moreover, this process reflects the conflicting expectations between media corporations and consumers. The content that comes from the users does not always meet the expectations of the media organization and other users in terms of quality. On one hand, Shirky (2008) wrote that these concerns sometimes arise from professional self-defense such as "professionals see the world through a lens created by other members of their profession" (p. 69). Some industry observers (e.g., Chadwick, 2008) also pointed out the threat from people equipped with new technologies such as cameras and mobile phones. On the other hand, Lessig (2006) tended to link credibility with reliability more than quality. From this standpoint, it can be said that transparency, accountability, and well-mentored user contributions raise trust, and accordingly, credibility. Nonetheless, as Jenkins (2006) contended, while corporations see the user participation mainly as a commodity, "consumers on the other side are asserting a right to participate in the culture, on their own terms, when and where they wish" (p.169).

In addition, the networked information economy already is pushing industrial producers (like the mass media) toward low-cost, low-quality productions (Benkler, 2006). It cannot be denied that user-generated content sometimes plays a vital role in covering some events. According to Wardle and Williams (2008), 9/11 was a very important wake-up call to media organisations about the potential of the user-generated content. Thus, Gillmor (2010) criticized the ethics and business models of big media companies that benefit from the majority of useful contributions without taking any risk or providing remuneration, writing, "This is not just unethical; it's also unsustainable in the long run, because the people who give freely of their time won't be satisfied to see mega-corporations rake in the financial value of what others have created" (p. 59).

Media organizations limit speech and their liability while controlling the use of information that they produce through online user agreements (Ekstrand, 2002). These agreements are widely considered adhesion contracts that are prepared by one party and are accepted by the party in a weaker position, and are also presented for signature on a "take-it-or-leave it" basis (Ekstrand 2002; Hartzog 2010). Unlike their previous media consumption

experiences, the contracts in the form of "terms of use," which are mostly located at the bottom of the home page, are usually unfavorable to the user in terms of intellectual property rights and privacy (Hartzog 2010).

Victoria Ekstrand's (2002) analysis of online user-agreements offered by the top fifty U.S. daily circulation newspapers in the United States showed that the organizations merely protect their own rights through a multitude of restrictions. As Ekstrand (2002) noted, half of the fifty online news user agreements stipulated that use of the site was an acceptance to the terms (p. 607). On the other hand, whether affirmative action was required or not (e.g., "clickwrap" or "browsewrap" agreements),[1] only a third of regular Internet users say they read the terms and conditions, disclaimers, and guidelines for posting comment (Williams, Calow, & Lee, 2011). Hartzog (2010) argued that the court should differentiate passive online media users from interactive users, transactional consumers, and businesses because when compared to passive online users, other iterations of online users can be charged more reliably with regard to the terms of use. As it stands, interactive media users or "produsers" are bound by these contracts regardless of the methods of affirmation of the agreements or their consciousness.

As Esktrand (2002) contended, between 40 and 50 of the user agreements studied contained copyright restrictions, including the redistribution and retransmission of content. Moreover, sixteen of them prevented the reuse of material on their sites exceeding the reach of copyright law, creating a new property right by virtue of the agreement. However, the proliferation of contractual restrictions may threaten diversity of content and deliberative democracy as well as fair use, which is described as the reproduction of another's material for non-commercial use, and is tolerated even in the Copyright Act (Abruzzi, 2010). In that context, Napoli (2009) suggested that contemporary media policymakers, policy advocates, and researchers need to recognize the right of "access to audience" in order to promote greater media diversity.

According to optimists, the public service media, which is no longer defined as merely public service broadcasting (Flew, Cunningham, & Bruns, 2008), could be a solution for the media to establish and maintain a participative online culture and the provision of media content of social, cultural and public value (Dwyer, 2010; Flew, Cunningham, & Bruns, 2008; Moe 2008). An easily accessible, publicly-funded, independently-managed online public service media (Coleman qtd. In Moe, 2008) guarantees the diversity of representation and maximization of participation and pluralism. Therefore, from being "read-only" broadcasting institutions, they must be transformed into "read-write" media providers that offer a place to the audience as co-creators who adopt the underlying principles and objectives for the new online media environment (Flew, Cunningham, & Bruns, 2008). Moe (2008) contended that the public service media function as a media policy tool that allocates resources, manages media sectors, and guides editorial policy.

Although not largely visible, the relationship between how interactivity and credibility are perceived within online news sources is significant (Chung, Nam, & Stefanone, 2012). A worthwhile debate is taking place on the different ethical and practical approaches of the public service media and commercial media in order to establish online credibility for the future of the media.

Challenges for User Contribution Publishing

This section examines the comparison of public service media and commercial media approaches to user-generated content through their user agreements and privacy policies. For this study, the user contribution strategies of three different types of public service broadcasters (SBS, the Special Broadcasting Service of Australia; BBC, the Public

Service Broadcaster of the UK; and PBS, the Public Broadcasting Service of the USA) were analyzed. SBS is a unique multicultural public service broadcasting service, and BBC has an educative mission with programs and services of high quality for strengthening nation building and cultural improvement. Even though it is in charge of a similar mission, however, PBS was established as a predominantly commercial broadcasting system. On the other side, the global well-known media organization that includes CNN, NBC, the *New York Times*, Reuters, *Le Monde* and *The Guardian* might reflect alternative commercial approaches over user-generated content because of their different corporate cultures, despite increasing the global media culture thanks to technological development and interaction between journalists (Hallin & Mancini, 2004).

Based on the synthesis of related work above, the ethical and legal challenges of publishing user-generated content were examined according to their potential for creating deliberative, trustworthy communication with their users, and their transparency and accountability in terms of editorial policy was assessed. As will be seen below, most of the agreements include similar statements about restrictions or submission rules; therefore, the tone of the agreements, user-friendly design factors, and supplementary documents such as discussion about ethical issues regarding moderation or quality within the comments will be evaluated to reveal their nuances.

Almost every media organization welcomes user-generated content, provided users adhere to the agreements and register with the site by giving accurate personal information. All the agreements cover access to content described as information, artwork, graphic materials, and software, among others. With the exception of the BBC, all of the agreements are made for personal use. For personal use, a valid e-mail address, a password, and some information about location are mostly enough to register. Location is important for some of the services. For instance, the BBC does not provide some content outside the UK, and Reuters refuses to provide any services to countries against which the UK, the US, and the UN have trade sanctions. For general usage, the minimum age is 13; however, few of the organizations request the year of birth at registration. The age requirement is an admonitory remark in CNN iReport's (2011) terms of use: "CNN iReport is available for individuals aged 13 years or older. If you are under 13, please exit this website immediately." Except for Le Monde, once registered, the user can access a large part of the content for free. On the other hand, quite different from their American counterparts, *Le Monde* and the BBC offer more guidance for less experienced contributors. From this standpoint, it may be more accurate to say that *Le Monde* especially also assumes an educative mission like public service media to increase its active and participatory users and enhance digital culture according to its founding objective *"une monde meilleur"* [a better world] and its influential position as a reference point for other French newspapers (Fottorino, 2009; Eveno, 2001).

The registered user can also choose to preserve his/her anonymity by choosing a fake user name to display on the site. Many users prefer to remain anonymous, especially on comments and forum sections, and therefore anonymity has become a highly controversial topic for web communities and user-generated content. Dan Wright (2010), the Ethics, Innovation and News Standards' Global Editor of Reuters, opened the question of anonymity up for discussion and invited users to sign comments with their real names to increase the quality of discussion, but many users did not welcome his invitation. Here is one negative response to Wright's call: "There is no freedom of speech unless there is anonymity. Forcing people to use their name will just force them to lie – as they have to do in the real world. And they will hate you for it, and be a little less angry at the world in general." Posted by AntonBerg (Reuters, October 8, 2010, 22:26 edt.).

While some editors believe that anonymity can encourage trolling, Jeff Jarvis (as cited in de Castella & Brown, 2011) argued that "to ban online anonymity in order to prevent trolling would be to remove the right of whistleblowers and dissidents to get their message across."

Although almost all contracts in the form of terms of use include similar (or even the same language) regarding restriction, the public service broadcasting contracts seem to promote personal usage, i.e., fair use, more than commercial online media. In particular, the BBC and the SBS have allowed users to retrieve content for personal or educational purposes. It may be more accurate to say that the BBC and the SBS seem aware of their vital role and have set strategies to create deliberative online media environment. The research on the BBC contends that audience contribution in news production during some important events like 9/11 or the London Bombings in 2005 drove the BBC to embrace user contribution in a number of ways (Wardle & Williams, 2008). The SBS also emphasizes their mission regarding user-generated to harness a range of views from Australia's diverse communities. PBS, on the other hand, follows a more inefficient strategy to incorporate the user contribution in their content.

Most commercial online media organizations encourage user-generated content as much as public service broadcasting. CNN iReport is one of the most successful examples of user-generated content management. In addition to the CNN web page's terms of use, iReport has its own particular terms of use and guidelines. Reuters and *The Guardian* are also working through their user contribution strategies. However, in addition to the generally restrictive structure of their agreements, commercial media's terms of use are less flexible. For example, NBC's terms of service begin with the words "Restriction on use." The general restrictions are usually aligned as defamatory, obscene, threatening, harassing, pornographic, abusive, libelous, deceptive, fraudulent language or material, impersonation, interference in other users' right to privacy, usage of the content for commercial purposes, or self-promotion. Only on *Le Monde* is there a restriction of historical revisionism and negationism (namely denying the genocide of the Jews in World War II).

Seeking the Best Moderation Method

In addition to contractual restriction, editors also invite the users to be polite and respectful to other users in a kind but firm tone: "Debate, but don't attack. In a community full of opinions and preferences, people always disagree. NYT encourages active discussions and welcomes heated debate on the Services, but personal attacks are a direct violation of these Terms of Service and are grounds for immediate and permanent suspension of access to all or part of the Service" (*New York Times*, Terms of Service, March 16, 2011).

"Be aware that you may be misunderstood, so try to be clear about what you are saying, and expect that people may understand your contribution differently than you intended. ... We want this to be a welcoming space for intelligent discussion, and we expect participants to help us achieve this by notifying us of potential problems and helping each other to keep conversations inviting and appropriate" (*The Guardian*, Community standards and participation guidelines, May 7, 2009).

Respecter l'esprit des discussions engagées, sans interférer par des messages de dérision ou hors sujet [Respect the spirit of the discussions, without interfering by derisive or off-topic messages]. (Le Monde, 2005, "Chats, mode d'emploi").

However, neither law nor the guidelines are enough to prevent trolling and offensive language, especially in comments on stories or on forum pages. Richard Baum, Global Editor for Consumer Media of Reuters (as cited in Wright 2010), admitted that racism and other hate language is not always caught by their software filters. All user-generated content is moderated by relevant

staff or division or outsourced to content providers who decide to publish, edit, or delete the content. Some of them use software filters simultaneously. In general, the media organisations are adopting pre-moderation, post-moderation, and reactive moderation[2] for different kinds of user-generated content. The online media services use all types of moderation for different kinds of user contributions. For example, sensitive areas (the pages designed for children) or comments on stories are usually prioritized, whereas post-moderation is used in limited areas that offer low risk for generating offensive or abusive content. However, sometimes the moderator prefers to use all types of moderation according to the nature of the content.

Last year, Reuters decided to elaborate their comment moderation system to create more civil and thoughtful conversation and reduce the delay of the publications of good comments. The new process grants a kind of VIP status to people who have had comments approved previously. When a user reaches a level of recognized user, his/her comments are published instantly, but the editors continue the review process after publication and may remove the comments if they do not meet the standards. In other words, Reuters allows users to rate one another's comments and feed those ratings into a global reputation system, which is also used by *The Guardian*.

However, this process is considered censorship and/or an elitist or biased attitude by some of their users: "Legal risk, such as what? We live in America, protected by the First Amendment. You especially are not liable for the content published for your users. I don't agree with this, it sounds more like an excuse" (Posted by cjohnweb, March 16, 2011, 19:23 EDT).

Some requested more initiatives from the editors and the moderators to ignore inappropriate content: "Letting users participate in flagging comments and thereby adjusting the commentator's own level; and implementing a level for people who choose to post under their own names, as verified by a $1 charge against their credit card.

The biggest single danger is that since it is Reuters employees doing the flagging, their own biases will affect their judgment" (Posted by Patrick Bowman, October 8, 2010, 17:02 EDT).

Some companies are stricter than the others about their methods: "Les décisions des modérateurs ne peuvent être contestées. En validant votre formulaire d'inscription, vous vous engagez à respecter leur autorité, en même temps que les règles ci-dessus" [The decisions of moderators cannot be challenged. By validating your registration form, you agree to respect their authority besides the above rules] (*Le Monde*, "Mode d'emploi des forums").

Finally, the BBC, the SBS, Reuters and *The Guardian* are more transparent about their moderation methods to achieve deliberative conversation, but it is hard to say that either commercial online media or public service broadcastings involve their users in their moderation process.

Copyright and Privacy Issues

It is generally accepted that the owner of user-generated content is the user, but he/she also grants a perpetual worldwide, royalty-free, nonexclusive, fully transferable and without limitation license by submitting the content. Online media services gain all the rights to change, edit and use the materials in user contribution. In comparison, all of the text, comments, and audiovisual content of the services are protected by copyright act, even for personal use. From this standpoint, it can be said that derivative and creative content is prevented by these adhesion contracts in the public domain even for fair use, as pointed out by Abruzzi (2010, p. 89). It seems that because of practical issues, both the commercial and public online media services have protected their copyrights and reduced their risk on almost the same grounds. However, unlike other commercial services, NBC has added a stipulation including the refusal of users' creative ideas on the grounds that "the possibility of future misunderstandings when projects developed by

NBC's employees and agents might seem to be similar to creative works submitted by users" (NBC, Terms of Service, August 14, 2007).

It should be noted that although there hasn't been any significant indication of such in the recent research on credibility perception for online news sources (Chung, Nam, & Stefanone, 2012), the privacy policy is one of the important components in the online sphere (Lauer & Deng, 2007). Today, some online media users are, at the same time, transactional consumers of the media companies using the paid service. Both the commercial online media and public service broadcasting gather personal and non-personal information from the users. While the personal information includes name, e-mail address, and credit card numbers submitted at registration, the IP addresses and the cookies constitute non-personal information that records the users' online usage patterns.

It is observed that although they publish their privacy policy in detail on their home pages, the commercial online media organizations have a more mercantile bent than their counterparts. In other words, most commercial services use personal and non-personal information to send newsletters or announce promotions, and sometimes share said information with third parties unless the user prevents this by checking a box while registering. Public service broadcasting seems more straightforward in using this collected information.

Transparency and Accountability

In an environment in which the media organizations promote user-generated content in almost every form providing the guidelines to produce quality content, there are fewer guidelines for the staff who handle this kind of content, excluding the public service broadcasting like the BBC and the SBS. For example, while CNN iReport provides excellent guidelines for keen users and technical support via its professional team, there is no mention of their editorial criteria for the evaluation process. The slogan of iReport Team roughly is "you can produce a worthwhile story if you listen to us" (CNN, Meet Team iReport, February 11, 2012).

In addition to the responsive and participatory approach of some commercial online news organizations like Reuters and *The Guardian*, the BBC and the SBS seem to follow more transparent and accountable strategies involving user contributions in their content although their educative, quality-oriented mission in society can prevent a courageous participatory strategy.

To Ensure Worthwhile Content through the Contracts

User-generated content has the potential to enhance good journalism and build trustworthy relationships between news organizations and their users. In addition to its importance during high profile news events such as 9/11, the London Bombings in 2005, or the Minnesota bridge collapse in 2007, some comments engendered a move "toward a more thoughtful conversation on stories" on Reuters (2010), showing that creative discussions between editors, bloggers, and users can create a more deliberative editorial policy. All online media organization in the sample of this analysis noticed the phenomenon of *audience evolution* (Napoli, 2010) and enforced the new participatory tendencies of audiences in different ways.

However, the myriad legal and ethical challenges usually orient the organizations to take restrictive measures to reduce legal risk and protect their credibility. At the same time, the similar restrictive adhesion contracts that reduce their liability offer only a "permitted" form of interactivity (Cover, 2006) rather than participatory culture.

Alternatively, some analysts have noticed the potential of public service broadcasting to create a deliberative media environment for the near future; accordingly, some public service organizations that can influence the media culture in their

country, like the BBC and SBS (and also ABC in Australia), respond to changes determining their organizational strategy and editorial policy. SBS in particular promotes multi-language user-generated content via virtual community centres (VCCs) on its website within the frame of well-defined standards. Nonetheless, it cannot be said that all public service media organizations intend to adopt a new strategy for user-generated content in their area of influence.

In more liberal media cultures like that of the USA, commercial media organizations are more enthusiastic about incorporating user contributions in their content. It is clear that the users' contributions regarding some sensational events make it impossible to ignore the users' contribution to the production of news. The news managers of the *New York Times* have, for example, admitted that texts and cell phone video were the only way they had to cover post-election protests in Iran (as cited in Leach, 2009).

However, one should not forget that said users are not a homogenous community, trolling and leaving unqualified or misleading commentary, termed "sock puppets" by Dan Gillmor (2010). Sock puppets are a serious problem for both media organizations and other users, but Gillmor claimed that these are a small percentage of total users and sock puppetry has never gone out of style in the traditional media. Moreover, in this research some discussion regarding the moderation of user contribution showed that the investigative and creative ideas of users can reveal more worthwhile methods to handle user-generated content.

As of yet, no one has found an ideal solution. Some analysts highlight some dilemmas for ethical challenges. At a conference on Online Journalism Ethics by Poynter, journalists agreed that clear standards for user-generated content should be adopted, and at the same time reconciled with the existing standards for journalists inside the organization (Martin, 2007).

Furthermore, as Kiesow (2011) wrote, usage standards and the privacy policies of media organizations need to be put forth as the unanswered questions for the publishers: "Do your website's terms of service provide appropriate protections to your organization as well as to your readers? Do anonymous commenters deserve protection similar to anonymous news sources?"

As a result, user-generated content publishing need clearer standards and more transparent editorial processes that should be linked with accountability. Hence, in addition to simply continuing to rely on "gatekeeper" adhesion contracts, the user-generated content guidelines for both users and staff must provide a responsible contact that gives a reason for whatever decisions are taken and can ensure a trustworthy relationship between users and the media organizations.

FUTURE RESEARCH DIRECTIONS

In addition to general measures, transparency, accountability and off-course reliability are salient components of media credibility in the abundance of content. Hernandez (2011) predicted that the new king will be credibility, which can be created by anyone: "Can I reliably trust you to tell me what is going on? If the answer is yes, then I don't care if you work out of a newsroom or out of your garage." In addition to the reliability to which Hernandez referred, it is expected that the ethos of the user-generated content of news organizations (or their approaches to the diversity of views) can also be components of credibility in our polarized and growing news area in the near future. Though the research of Chung, Kim, and Kim (2010) indicated a weak relationship between interactivity and credibility perception, it is clear that online media organizations gradually need more people to provide live footage, newsworthy sound clips and stories, or other contributions to their analysis. In other words, it can be claimed that credibility is not solely involved in the media organization anymore; user-generated content is also in competition with the institution for credibility. Thus, currently, the perception of the

credibility of active and passive online media users should be examined respectively for future research. Therefore, referring to the reconceptualization of Napoli (2010), the audience should also be scrutinized to understand its tendency to provide potential new perceptions of credibility.

Additionally, the legal perspective is also very important even thought it is conspicuously lacking in this study because it is always changing and is probably best handled by scholars with legal backgrounds.

CONCLUSION

The media users in the new online media environment have become participants in the process of content production. While media organizations invite more contributions from the users, at the same time they police them through adhesion contracts in the forms of "terms of use" and "privacy policies" to reduce their liability and protect their copyright and the rights of their affiliates. Therefore, it cannot be said that the user contribution strategies of the online media organizations promote a participatory digital culture providing a discussion platform or supporting new creative user-generated content ignoring their copyright in case of fair use. On the other hand, they have minimum editorial standards for user-generated content and most of them are not reconciled with the standards for the staff. Although there have been noteworthy endeavours to determine more efficient methods to increase quality content and ensure worthwhile conversation, it seems that most of the time, the focus is on their editorial purposes or reputation rather than ethical issues.

It is expected that the public service media can play an avant-garde role for establishing a deliberative democratic culture for the future of online media. The public service online media that can influence their media culture have a trailblazing mission more intent on harnessing the affordances of user-generated content, but their educative and culturally elitist missions can become an obstacle in fostering courageous participatory strategy.

However, considering the efforts of some commercial online media organizations to seek a more efficient editorial policy for user-generated content, no large difference currently exists between the public service and commercial online media in terms of contractual issues and editorial policy. Finally, it is not easy to transform "read-only" broadcasting institutions to "read-write" media providers through the restrictive, "permitted" user-generated content strategies that proliferate today.

REFERENCES

Abruzzi, B. E. (2010). Copyright, free expression, and the enforceability of "personal use- only" and other use restrictive online terms of use. *Santa Clara Computer and High-Technology Law Journal, 26*(1), 85–140.

AntonBerg. (2010, October 8). Comment posted to "Toward a more thoughtful conversation on stories." *Reuters.* Retrieved from http://blogs.reuters.com/fulldisclosure/2010/09/27/toward-a-more-thoughtful-conversation-on-stories/?cp=all#comments

BBC. (2010, March). *Privacy and cookies.* Retrieved from http://www.bbc.co.uk/privacy/

BBC. (2011, January 19). *Terms of use of BBC online services—Personal use.* Retrieved from http://www.bbc.co.uk/terms/personal.shtml#5

Benkler, Y. (2006). *The wealth of networks: How social production transforms markets and freedom.* New Haven, CT: Yale University Press.

Bruns, A. (2005). *Some exploratory notes on produsers and produsage.* Institute for Distributed Creativity. Retrieved from http://distributedcreativity.typepad.com/idc_texts/2005/11/some_explorator.html

Chadwick, P. (2008, May). *Adapting to digital technologies: Ethics and privacy.* Paper presented at Future of Journalism Summit, Sydney, NSW.

Chung, J. C., Kim, H., & Kim, J. H. (2010). An anatomy of the credibility of online newspapers. *Online Information Review, 34*(5), 669–685. doi:10.1108/14684521011084564

Chung, J. C., Nam, Y., & Stefanone, M. A. (2012). Exploring online news credibility: The relative influence of traditional and technological factors. *Journal of Computer-Mediated Communication, 17,* 171–186. doi:10.1111/j.1083-6101.2011.01565.x

Cjohnweb. (2010, October 8). *Comment posted to Toward a more thoughtful conversation on stories.* Reuters. Retrieved from http://blogs.reuters.com/fulldisclosure/2010/09/27/toward-a-more-thoughtful-conversation-on-stories/?cp=all#comments

CNN iReport (2011, May 5). *CNN iReport terms of use.* CNNi Report. Retrieved from http://ireport.cnn.com/terms.jspa

CNN iReport (2012, May 10). *Meet team iReport.* CNN iReport. Retrieved from http://ireport.cnn.com/blogs/ireport-blog/2009/11/18/meet-team-ireport

CNN iReport. (n.d.). *Community guidelines.* CNN iReport. Retrieved from http://ireport.cnn.com/guidelines.jspa

Cover, R. (2006). Audience inter/active: Interactive media, narrative control and reconceiving audience history. *New Media & Society, 8*(1), 139–158. doi:10.1177/1461444806059922

De Castella, T., & Brown, V. (2011). Trolling: Who does it and why? *BBC News Magazine.* Retrieved from http://www.bbc.co.uk/news/magazine-14898564

Deuze, M. (2006). Participation, remediation, bricolage: Considering principal components of a digital culture. *The Information Society, 22*, 63–75. doi:10.1080/01972240600567170

Dwyer, T. (2006). *Media convergence*. London, UK: Open University Press, McGraw-Hill Education, McGraw-Hill House.

Ekstrand, V. S. (2002). Online news: User agreements and implications for readers. *Journalism & Mass Media Quarterly, 79*(3), 602–618. doi:10.1177/107769900207900305

Eveno, P. (2001). *Le journal Le Monde une histoire d'Indépendance*. Paris, France: Editions Odile Jacob.

Flew, T., Cunningham, S., & Bruns, A. (2008). *Social innovation, user-created content and the future of the ABC and SBS as public service media*. ABC_SBS_Inquiry_Submission. Retrieved from http://eprints.qut.edu.au/16948/

Fottorino, E. (2009). Le Monde: Portrait d'un quotidien. *Le Monde*. Retrieved from http://medias.lemonde.fr/medias/pdf_obj/200912.pdf

Gillmor, D. (2010). *Mediactive*. Retrieved from http://mediactive.com/wp-content/uploads/2010/12/mediactive_gillmor.pdf

Guidance, B. B. C. (n.d.). *Moderation, hosting, escalation and user management, part 1: Hosting, moderation and escalation*. Retrieved from http://www.bbc.co.uk/guidelines/editorialguidelines/page/guidance-moderation-hosting

Hallin, D. C., & Mancini, P. (2004). *Comparing media systems three models of media and politics*. Cambridge, UK: Cambridge University Press. doi:10.1017/CBO9780511790867

Hartzog, W. (2010). The new price to play: Are passive online media users bound by terms of use. *Communication Law and Policy, 15*(4), 405–433. doi:10.1080/10811680.2010.512514

Hernandez, R. (2011, December 20). Robert Hernandez: For journalism's future, the killer app is credibility. *Nieman Reports*. Retrieved from http://www.niemanlab.org/2011/12/robert-hernandez-for-journalisms-future-the-killer-app-is-credibility/

Jenkins, H. (2006). *Convergence culture where old and new media collide*. New York, NY: New York University Press.

Kiesow, D. (2011, January 3). Plain Dealer settles comment lawsuit, limits of anonymity untested. *Poynter*. Retrieved from http://www.poynter.org/latest-news/media-lab/social-media/112641/plain-dealer-settles-comment-lawsuit-limits-of-anonymity-untested/

Lauer, T. W., & Deng, X. (2007). Building online trust through privacy practices. *International Journal of Information Security, 6*, 323–331. doi:10.1007/s10207-007-0028-8

Le Monde. (2005, March 21). Chats, mode d'emploi. *Le Monde*. Retrieved from http://www.lemonde.fr/services-aux-internautes/article/2003/02/19/chats-mode-d-emploi_309828_3388.html

Le Monde. (2011, May 5). La charte des blogs et les règles de conduite. *Le Monde*. Retrieved from http://www.lemonde.fr/services-aux-internautes/article/2004/12/03/la-charte-des-blogs-et-les-regles-de-conduite-sur-lemonde-fr_389436_3388.html

Le Monde. (2011, May 5). Blogs: Mode d'emploi. *Le Monde*. Retrieved from http://www.lemonde.fr/services-aux-internautes/article/2004/11/10/blogs-mode-d-emploi_386588_3388.html

Le Monde Forums. (n.d.). Mode d'emploi et règles de conduite. *Le Monde*. Retrieved from http://forums.lemonde.fr/perl/faq_french.pl?Cat=

Leach, J. (2009). Creating ethical bridges from journalism to digital news. *Nieman Reports*. Retrieved from http://www.nieman.harvard.edu/reportsitem.aspx?id=101899

Lessig, L. (2006). *Code: Version 2.0*. New York, NY: Basic Books.

Martin, M. (2007, 25 January). Online journalism ethics: Guidelines from the conference. *Poynter*. Retrieved from http://www.poynter.org/uncategorized/80445/online-journalism-ethics-guidelines-from-the-conference/#ugc

Moe, H. (2008). Dissemination and dialogue in the public sphere: A case for public service media online. *Media Culture & Society, 30*(3), 319–336. doi:10.1177/0163443708088790

Napoli, P. M. (2009, April). *Media policy in the era of user-generated and distributed content: Transitioning from access to the media to access to audiences*. Paper presented at the Media in Transition Conference Massachusetts Institute of Technology, Cambridge, MA.

Napoli, P. M. (2010). *Audience evolution: New technologies and the transformation of media audiences*. New York, NY: Columbia University Press.

NBC. (2007, August 14). *Terms of service*. NBC. Retrieved from http://www.nbc.com/privacy-policy/terms-of-service/

NBC. (2007, August 14). *Privacy policy*. NBC. Retrieved from http://www.nbc.com/privacy-policy/

New York Times. (2011, March 16). *Terms of service*. Retrieved from http://www.nytimes.com/content/help/rights/privacy/highlights/privacy-highlights.html

New York Times. (2011, April 18). *Privacy policy*. Retrieved from http://www.nytimes.com/content/help/rights/privacy/highlights/privacy-highlights.html

Organisation for Economic Co-operation and Development. (2007). *Participative web: User-created content*. Paris, France: OECD.

PatrickBowman. (2010, October 8). *Comment posted to Toward a more thoughtful conversation on stories*. Reuters. Retrieved from http://blogs.reuters.com/fulldisclosure/2010/09/27/toward-a-more-thoughtful-conversation-on-stories/?cp=all#comments

PBS. (2011). *PBS editorial standards and policies*. Retrieved from http://www.pbs.org/about/media/about/cms_page_media/35/PBS%20Editorial%20Standards%20and%20Policies.pdf

PBS. (2011). *PBS terms of use*. Retrieved from http://www.pbs.org/about/policies/terms-of-use/

PBS. (2011). *PBS privacy policy*. Retrieved from http://www.pbs.org/about/policies/privacy-policy/

Reuters. (2011). *Privacy policy*. Retrieved from http://www.reuters.com/privacy-policy

Reuters. (2011). *Terms of use*. Retrieved from http://www.reuters.com/terms-of-use

Rosen, J. (2006, June 30). The people formerly known as the audience. *Huffington Post*. Retrieved from http://www.huffingtonpost.com

SBS. (2011). *SBS policies & publications*. Retrieved from http://www.sbs.com.au/aboutus/corporate/view/id/114/h/Terms-and-Conditions

SBS. (2011). *SBS privacy statement*. Retrieved from http://www.sbs.com.au/aboutus/corporate/view/id/113/h/SBS-Privacy-Statement

Shirky, C. (2008). *Here comes everybody: The power of organizing without organizations*. New York, NY: Penguin Books.

Svoen, B. (2007). Consumers, participants, and creators: Young people's diverse use of television and new media. *Computers in Entertainment, 5*(2). doi:10.1145/1279540.1279545

The Guardian. (2009, May 7). *Community standards and participation guidelines*. The Guardian. Retrieved from http://www.guardian.co.uk/community-standards

The Guardian. (2009, May 7). *Frequently asked questions about community on the Guardian website: Everything you've ever wanted to know about community on the Guardian website*. The Guardian. Retrieved from http://www.guardian.co.uk/community-faqs

The Guardian. (2010, March 19). *Terms of service*. The Guardian, Retrieved from http://www.guardian.co.uk/help/terms-of-service

The Guardian. (2011, July 23). *Privacy policy*. The Guardian, Retrieved from http://www.guardian.co.uk/help/privacy-policy

Wardle, C., & Williams, D. (2008). *Understanding its impact upon contributors, non-contributors and BBC News*. Research Report of Cardiff School of Journalism, Media and Cultural Studies. Retrieved from http://www.cardiff.ac.uk/jomec/research/researchgroups/journalismstudies/fundedprojects/usergeneratedcontent.html

Williams, A., Calow, D., & Lee, A. (2011). *Digital media contracts*. New York, NY: Oxford University Press.

Wright, D. (2010, 27 September). Toward a more thoughtful conversation on stories. *Reuters*. Retrieved from http://blogs.reuters.com/fulldisclosure/2010/09/27/toward-a-more-thoughtful-conversation-on-stories/?cp=1#comments

ADDITIONAL READING

Bruns, A. (2008). *Beyond public service broadcasting: Produsage at the ABC*. Paper presented at ABC Digital Media Forum. Retrieved from http://produsage.org/node/23

Burns, M. E. (2008). Remembering public service broadcasting: Liberty and security in early ABC online interactive sites. *New Media & Society*, *11*, 115–132.

Burns, M. E. (2008). Public service broadcasting meets the Internet at the Australian Broadcasting Corporation (1995–2000). *Continuum: Journal of Media & Cultural Studies*, *22*(6), 867–881. doi:10.1080/10304310802419395

Carpentier, N. (2011). *Media and participation: A site of ideological–democratic struggle*. Bristol, UK: Intellect.

Deibert, R., Palfrey, J., Rohozinski, R., & Zittrain, R. (Eds.). (2010). *Access controlled: The shaping of power, rights, and rule in cyberspace*. Boston, MA: The MIT Press.

Donders, K., & Pauwels, C. (2010). The introduction of an ex ante evaluation for new media services: Is 'Europe' asking for it, or does public service broadcasting need it? *International Journal of Media and Cultural Politics*, *6*(2), 133–148. doi:10.1386/mcp.6.2.133_1

Dwyer, T., & Nightingale, V. (Eds.). (2007). *New media worlds: Challenge for convergence*. South Melbourne, Australia: Oxford University Press.

Keen, A. (2007). *The cult of amateur*. New York, NY: Random House.

Meikle, G., & Redden, G. (Eds.). (2010). *News online: Transformations and continuities*. London, UK: Palgrave MacMillan.

Paulussen, S., Heinonen, A., Domingo, D., & Quandt, T. (2007). Doing it together: Citizen participation in the professional news making process. *Observatorio (OBS*)*. *Journal*, *3*, 131–154.

Proulx, S. (2009, May). *Forms of user's contribution in online environments: Mechanisms of mutual recognition between contributors*. Paper presented at COST 298 Conference: The Good, The Bad and The Challenging: The user and the future of information and communication technologies, Copenhagen, Denmark.

Smith, M. A., & Kollock, P. (Eds.). (1999). *Communities in cyberspace*. London, UK: Routledge. doi:10.5117/9789056290818

Thurman, N. (2008). Forums for citizen journalists? Adoption of user-generated content initiatives by online news media. *New Media & Society*, *10*, 139–157. doi:10.1177/1461444807085325

KEY TERMS AND DEFINITIONS

Adhesion Contract: A kind of contract that is prepared by one side and is accepted by other sides (users) and is also presented for signature on a 'take it or leave it' basis.

Moderation of User Contribution: Review and elimination of user-generated content to create a more worthwhile conversation.

Participatory Media: Provides a platform for users to share, discuss and participate in news production and conversation.

Privacy Policy: A legal document that is prepared by publishers that discloses how personal information is used by publishers.

Public Service Media: As cited in Flew, Cunningham & Bruns (2008), the public service media is no longer defined as merely public service broadcasting.

Transparency: Responsibility of the publisher to inform the users about their user-generated content policy, their moderation methods, and the decisions that are taken, etc.

User-Generated Content: Content that is created outside of professional practice and by users or producers, in Bruns' terminology.

ENDNOTES

[1] A clickwrap agreement is requires an individual to click on a button to indicate assent but browsewrap agreements consider to have been agreed to before a user browses the website

[2] In pre-moderation, all posts are moderated before they appear online, in post-moderation, all posts are moderated after they appear online, but in reactive moderation, posts are only moderated after the moderator receives an alert from a user

Chapter 18
The Cyber–Propelled Egyptian Revolution and the De/Construction of Ethos

Samaa Gamie
Lincoln University, USA

ABSTRACT

This chapter examines two key Egyptian Facebook pages that became the voice and face of the youth movement that ignited the Egyptian revolution. The "Kolena Khaled Said" and "We are all Khaled Said" Facebook pages, respectively, represent the Arabic and international branch of the Egyptian human rights campaign against police violence. This chapter explores the complexity of ethos construction in activist digital discourses by analyzing the visual and textual elements of each page, the communal ethos, and the sense of shared identity that emerge in these digital social networks, as well as the internal and external challenges posed to their emergent ethos. The results of the analysis of both Facebook campaigns indicate the powerful role anonymity plays in activist digital discourses in creating a communal ethos that, combined with the visual and textual elements of the page, are able to achieve massive outreach that legitimates their calls for reform, activism, and revolutionary work. The chapter reflects upon the possibilities of integrating the critical study of ethos in composition teaching.

INTRODUCTION

The January 25th 2011 Egyptian revolution not only marked a popular uprising against a long-ruling dictator but also a technological revolution in which cyberspace and social networking sites were used as tools for political activism and as means of organizing and coordinating revolution-

DOI: 10.4018/978-1-4666-2663-8.ch018

ary work. The Egyptian cyber-propelled youth revolution organized around a few key movements that launched on Facebook: The January 25 Movement, 6th of April Youth Movement, and We Are all Khaled Said. These three movements initiated an Egyptian revolution by spreading their message to millions of Egyptians of all ages and socio-economic backgrounds, and they were a key factor in the success of the revolution that ended a thirty-year dictatorship and started a new phase of democratic reforms in Egypt. Hence, in this paper, I will examine We Are all Khaled Said, a cyber movement that ran two concurrent, yet separate, campaigns and Facebook pages: one in Arabic based in Egypt and another in English based in the UK. In my analysis, I will explore the construction of the communal digital ethos of these key youth cyber movements and the challenges posed to their communal digital ethos by pro-regime, counter-revolutionary pockets. The chapter aims to explore the role authorship and anonymity play in these activist digital discourses and how these digital users discursively and visually merge Western humanism, Islamism, and post-sectarian Egyptian nationalism in the formation of their communal digital ethos legitimating their calls for reform, activism, and revolutionary work.

BACKGROUND

It is undeniable that digital social networks have emerged as the fermenting ground for the propellers and proponents of democracy and reform. Theodore Roszak (1986) saw in information technology the potential to "concentrate political power, to create new forms of social obfuscation and domination" (p. xii), thus presenting opportunities for social control and suppression of freedoms and rights while also increasing the power of the individual. With the increased power of the individual comes the decentralization of power and knowledge, where a group member's take on national or international policies occupies center stage alongside the views of politicians and the elite, leading to the democratization of this digital sphere. Simultaneously entrenching chaos and confusion, while renewing the sense of community, and creating rich opportunities for social contact and growth. This is no less evident in the discourses that emerged pre and post-the Egyptian revolution.

This sense of community and shared identity, in both material and virtual social spaces, arises

from "an understandable dream expressing a desire for selves that are transparent to one another, (with) relationships of mutual identification, social closeness and comfort" (Young, 1990, p. 300). In essence, through identification or a series of identifications, cooperation, social cohesion and ultimately persuasion, the community identity or ethos is achieved. Herein, Burke's (1950) conception of identification can assist us in conceptualizing digital ethos as a discourse community whereupon the individuals and the cyber community collaborate in the formation of this communal ethos. In this sense, we can view ethos as a textual system with stated and unstated conventions for exercising power that extend beyond sociological lines. Such ethos that emerges from this discourse community is unstable and dynamic—possessing great complexity as the digital discourse community presents itself as "'an unstated assemblage of faults, fissures, and heterogeneous layers' and a network of intersecting systems, institutions, values and practices" (Porter, 1997, p. 107), stressing the intertextuality of such discourse communities and the ethos that emerges from, within and outside them. In viewing ethos as a "network of discursive practices" and conventions (Porter, 1997, p. 110), ethos can be seen as a discourse with an established set of conventions for knowledge production and dissemination.

This view of ethos as a discourse community lines up with Laura Gurak's (1999) discussion of ethos as a group quality expressing the cultural and moral tone of a community that allows group members, who share similar concerns and interests, to create a strong community ethos and to mobilize and spread relevant information more quickly and effectively. According to Gurak (1999), this communal sense of ethos promotes conformity and discourages dissent creating insular digital communities where speed takes precedence over accuracy and the beliefs of the community supersede the communal and individual responsibility of citizens to make informed decisions. Thus, the ethos of an online community is built on "a feeling of connectedness that confers a space of belonging" (Foster, 1997) and a desire for unity and cohesion whereby differences within and among individual communities are seen as threatening to that communal ethos and are, thus, bracketed, further alienating these communities from one another and from dissenting members within them. However, what seems apparent is that most online communities exercise social control and coercion against their members, silencing dissension, and replicating the same forms of ostracism, cyber shaming, and punishment exercised in their physical spaces, which will be more evident in the analysis of the above two Egyptian activist sites.

Understandably, ethos is a central component of any rhetorical situation, whether we view it as one's good or moral character, one's credibility, an element of style (as Quintilian would have us see it), a "dwelling place" in which we should consider the situation and context within which rhetoric is applied (Hyde, 2004), a group quality, or a network of communal discursive practices that is ideally "multi-voiced and authentic." One's publically constructed ethos presents itself as an important element in cyber discourses where, due to the extreme speed of news circulation and production and the laxity or absence of fact-checking by cyber users of posts and news, one post by the rhetor or another's counter-post(s) can put one's credibility and character in question. This is no less evident in the cyber-propelled Arab spring revolutions; the Egyptian revolution is a strong case in point.

In the digital domain, constructing an ethos of a listserv, a Facebook group, or a public persona has its challenges; however, the more visible a group or listserv becomes and the more public traction it gains, the more challenges are posed to its public ethos by detractors or opponents. In this light, maintaining the constructed perception of one's good character and trustworthiness in the eyes of the thousands or millions of Web followers becomes more difficult. The challenge is most evi-

dent when it comes to social and political activist sites that call for social, economic, and political reform and justice. These groups have sprouted in the cyber domain and have created a massive following. In countries of the Arab spring, such cyber pockets have been the fermenting grounds for massive demonstrations, political activism, and subsequent revolutions and regime change. The power that such groups have garnered has meant that they had to face not only challengers of their own members questioning the feasibility of massive action and continued public outcry for reforms and revolutionary action, but also the massive number of covert pro-regime members and groups that have used the power and the speed of news circulation and production on the web to counter the revolutionary and activist calls of these groups and present them as Western agents and opportunists. These pro-regime and anti-reform groups frame the calls for reform and action as destructive, unpatriotic, and borderline treasonous, thereby warranting calls for gunning down (in both literal and figurative terms) the most visible and public figures of these groups. Hence, the result is the discrediting of these activist groups and the public shaming of their visible members, which amounts to acts of cyber lynching against these activists and organizations and also creates conditions that make actual beatings, disappearances, and deaths more palatable to the general public. (The vilifying of the April 6th Youth Movement is a case in point that I will not be able to address in this chapter.)

The cyber discourses that emerged preceding, during, and following the Egyptian revolution carry within them the notions of classical rhetoric and the exploration and articulation of the truth of the physical realities of people living in Egypt under the pre and post-revolution governments. The need to expound the common good and advance the common group perception of the truth and the just to the general public by the activist groups becomes central to the discourses that have emerged in the digital domain, whether on a Facebook group or a blog that invites others to comment and respond to. These groups construct an intricate ethos—a public digital self that advocates morality, truth, justice, and democracy and educates the public about their rights and roles in the affirmation of these rights, claiming access to the truth and using this digital public venue to disperse their found truth and the means to its realization. Despite the public nature of their message and their advocacy, the identities of the founders of the groups mostly remain unknown to the public, as the groups struggle to maintain their anonymity for fear that public exposure would threaten the success of their causes. The "Kolena Khaled Said" Group will be a case in point which I will address later in the chapter.

In the digital discourses that emerged in the Egyptian cyber sphere, as in classical rhetoric, a great emphasis has been placed on truth and transparency (which, in this case, does not seem to mutually exclude anonymity) as central to the construction or perception of the digital orator's ethos and good character. As with Plato, the digital orators or groups should maintain a good moral character showcased in their posts and analyses of events, showcase that they have knowledge of the truth, the just, and the good and should advocate incessantly for the truth upon which the majority of the group members have agreed, i.e., that complies with their announced beliefs and values. Just as with Plato, other classical rhetoricians such as Aristotle and Isocrates gave no exceptions for advocating the truth because truth was seen as central to the construction and preservation of one's ethos. The contention between these classical views on ethos lies in whether the goal is "being" good or "seeming" good, whether the advocacy should be for truths unequivocally or non-truths employed for the common good, as in the case of Quintilian, and whether that truth is determined by the individual or the community. These are all questions that translate to digital discourses, where the circulation of half-truths

or blatant lies in public or group posts often casts doubt on the ethos of the producers and circulators of such discourses.

In modern rhetorical theory and digital discourses nonetheless, a state of flux and uncertainty governs the conception of truth; this gives rise to the questionable ethical reliability of relativism. In the realm of cyber rhetorics, truth, now substituted for personal opinion or perceived self-evident facts, has become evasive and what is determined as the common good is as relative as the common evil. This reflects the disjointed and blurred reality of cyber discourses, where fact-checking is sometimes nonexistent and "truth" is almost impossible to ascertain. (An example of which was the claim that the Egyptian military forces had desecrated the Abbasseya Mosque in Cairo by entering with their shoes on, committing a sacrilegious act that even the occupying British troops during the colonization of Egypt had not done, an analogy used in the same accusatory post, allegedly by the Imam of the mosque. Shortly after, another post circulated, allegedly by the same Imam, attesting to the contrary and showing a picture of tens of military shoes at the entrance of the mosque. The truth is somehow lost in both posts, and for digital users, the version of the story fitting one's worldview, whether vilifying the military or supporting it, will be the correct version.)

The different ideological, theological, political pockets and groups in cyberspace have transmitted the fragmentation of perceived realities and identities into the digital domain. The idea of having a digital rhetor, as envisioned by Isocrates, as being a model, an ideal persuasive character equipped to defend himself against his detractors and demonstrate the worthiness of his character, becomes impossible. For what do we know about the virtues of our digital rhetors? We most likely have never met them in person, tested their truthfulness or wisdom, or known whether their ethos is a product of moral development or training, or part of the surface construction of the text or (i.e., an element of style). This is where the dilemma of ethos construction and preservation erupts in the activist discourses circulating on the World Wide Web. Kolena Khaled Said is a case in point.

Textual and Visual Ethos Construction of the Kolena Khaled Said Facebook Page

The first Facebook page I will be examining is the Arabic Kolena Khaled Said (i.e., We are all Khaled Said) page representing an Egyptian human rights campaign against police brutality in Egypt. This Facebook page, which is administered out of Egypt, is named after Khaled Said, a 28-year-old Egyptian businessman from the coastal city of Alexandria who was tortured to death at the hands of two police officers. The page was founded on 6/10/2010 by Wael Ghonim, an Egyptian Google executive who used the pseudonym El-Shaheed (i.e. The Martyr). Though the use of "martyr" evokes a prominent Islamic symbol referencing the sanctified memory of martyrs who sacrifice themselves for the good of their Muslim community, it is used here to reference not only Khaled Said but also all the other victims of police brutality. Choosing to remain anonymous and use this pseudonym, Wael Ghoneim said he wanted "to increase the bond between the people and the group through my unknown personality. This way we create an army of volunteers" (Giglio, 2011, p. 3). Unlike within the classical conceptions of rhetoric that focus on the persona of the speaker or the traditional theorizations of online discourses that require the authenticity and identification of authorship as central to ethos construction, in this online community, anonymity of authorship became its greatest strength erecting a strong digital public ethos, directing its members and fans to focus on the message of the page rather than distract from it by focusing on the person behind the message. Ghonim added that his anonymity

was the reason people believed and trusted him and expressed his concern that publicity would corrupt the movement (Giglio, 2011). Such fears turned out to be accurate when Ghonim's identity was uncovered and the massive public backlash that ensued vilified him and his movement.

With the founder's anonymity intact, the page turned into one of Egypt's largest activist sites with a constant stream of photos, videos, and news, with 350,000 fans on January 14th, 2011, ten days before the start of the revolution. On January 14, 2011, after the eruption of protests in Tunisia, the page invited its fans to protest on January 25th. In three days, more than 50,000 people answered "Yes." The page called its fans to spread the word about the protests, marking the first spark of the revolution which was ignited on January 25th with mass demonstrations in all major Egyptian cities. On January 28th, 2011, Wael Ghonim was detained by Egyptian authorities and later released after two weeks. During this time rumors started circulating about his identity as El Shaheed, the identity which Ghonim later confirmed in a Newsweek interview after his release (Giglio, 2011).

The exposure of the page's founder's identity led to a massive tide of verbal attacks that vilified not only Ghonim, but also the movement itself casting their motives both as ethically questionable. Counter-revolutionary forces and many ordinary Egyptians accused him of being an American agent, an Israeli spy, a traitor, even an Islamist, who has profited from the fame he gained at the expense of the people. Rumors about an upcoming multi-million dollar book deal in the making did not help either (even though Ghonim announced that he will be donating all proceeds from the book to the families of the Egyptian revolution martyrs, families, with a portion going to buy books to public schools). Such negative wide exposure led to further animosity from other revolutionaries towards Ghonim who thought he took too much credit, while others slammed him for not doing enough to fight the Mubarak regime (Giglio, 2011). Following the toppling of Mubarak and the increased attacks on his character, Ghonim moved to the sidelines of the revolution and gave the site over to other administrators in an attempt to preserve the movement, which seems to have altogether succeeded since the site consists of 2.3 million fans. Ghonim's insistence that no leader is needed for the movement or any activist or revolutionary movement "keeps with the notion of the decentralized ethos of Internet organizing" (Giglio 2011, p. 2), and reflects an animosity toward the very idea of a leader, a public sentiment which, Giglio hypothesized, stems from Egypt's long history of corrupt autocrats. Despite Ghonim's disappearance from the public sphere and his severing all ties with "We Are All Khaled Said," this movement and other youth movements have been largely discredited and vilified in the eyes of some of the Egyptian public that, due to anti-revolutionary propaganda, have come to view them as traitors and Western and masonry agents recruited to destroy Egypt.

In the profile page of Kolena Khaled Said, an Arabic statement reads: "We dream of an Egypt that respects human rights, a government chosen by the people, an independent country that warrants the respect of the world." The statement articulates the mission of the Facebook campaign and page, which is to use this digital venue to realize the dreams of the community, dreams that most Egyptians can identify with and thereby help establish its noble activist ethos that seeks the common good of the people. The "Personal Information" heading in the About section of the page states that the page was started "to defend Khaled Said, May Allah have mercy on his soul, against all the injustice that he faced," adding that Allah willed that this page become a podium to defend the right of every Egyptian to an honorable life. The intermixing of the discourse of post-sectarian Egyptian nationalism, Western humanism, and Islamism is evident in the textual analysis of this digital community's statements.

The reference to human rights, democracy, and independence evokes the discourse of Western humanism that can appeal to all Egyptians and Arabs, both Muslim and Christian, whereas the Islamist references, as in Al-Shaheed, Allah, His mercy and will, appeal more closely to the Muslim members of that digital community, clearly identifying its admin as Muslim, and assisting in constructing an ethos that embraces Islamism and humanism nonetheless. In a country like Egypt, with more than 90% of the population Muslim and where one's religious identity is a central component of one's identity construction, it is crucial to appeal to this vast segment of society to ensure their inclusion in the constructed ethos of that activist community. Despite the Islamist references, the Western humanistic references implicitly evoke the dream of a post-sectarian Egypt where every Egyptian—regardless of religion or affiliation—will get their full rights and freedoms.

The page's "Personal Interests" section starts off with, Who are we and what do we want? In defining the movement's role and purpose, it is able to establish its digital ethos and invite others with similar beliefs and values to join in its causes. All the posts are written in Arabic and the responses are from Arabic-speaking members or fans. The movement states that this page is for all Egyptians regardless of their religion, age, gender, education, socio-economic status, or political affiliation who have united because they want their country to become a better one and they want Khaled Said to have his rights as a symbol for all Egyptians' rights. The page identifies the primary audience of its human rights' campaign as Egyptian, while the secondary audience is most likely Arabs or fluent Arabic speakers. The page adds that it aims to defend the rights of all Egyptians, inform the public about the human rights' issues and violations in Egypt, and educate them of their constitutional rights, adding that the realization of these rights will not be accomplished unless "we" (all Egyptians) are united as one and are joined with the largest number of Egyptians on the Internet and elsewhere. Secondly, the admin lays out the rules for contributing to the page, which include: The contributors' comments should not stray from the admin's post (in case one of the web users wishes to start another discussion topic, then posting it on the wall is more appropriate) and should not use inappropriate language or be disrespectful. The admin prohibits comments or posts that publicize other sites or Facebook pages and any arguments between supporters or opponents of any political/ public figure, party, or group, unequivocally stating that the repetition of any such violation can result in blocking the person from the page. The admin ends by calling for a respectful public conversation, stressing that the main goal is to understand and love each other despite our differences. The call for unity constructs a digital ethos and aims at creating alliances and bridging differences between the members of the Egyptian populace through that digital space; the hope is that this unity will be translated to the physical spaces of their real communities.

In the new timeline, the background picture of the Facebook page shows hordes of Egyptians (of all ages) gathered in Tahrir Square during the protests in the revolution. The Muslims are seen praying while the Christians are surrounding them creating a human shield against possible attacks by the regime thugs or security forces. Banners and Egyptian flags appear in the background. On the right-hand side of the page, an Egyptian flag bearing the sign of the cross and the crescent, is emblematic of the unity among Muslims and Christians as partners in one nation. The profile picture is that of the Egyptian flag with its red, white, and black and the eagle in the center. The flag symbolizes the pride of Egyptians and represents the sacrifice of those who died in the revolution and the blood they shed in love and honor of their country. Both pictures remind the viewers of the spirit of unity that spread throughout Egypt during the revolution, as well as the revolution's

promise to end the sectarianism, prejudice, and injustice and to convey the message that Egypt will rise and progress only with unity among its Muslims and Christians. The Egyptian nationalist message evoked by both pictures aids in visually constructing the page's patriotic ethos—a unifying patriotism that seeks the good of all Egyptians, regardless of age or religion.

The early visual posts, included shortly after the inception of the page, feature a picture of the young Khaled Said, full of life and promise with a background showing the Egyptian flag, and a black ribbon on the side with the words: "Egypt's martyr" written in Arabic in white font. The post accompanying the picture presents Khaled as the symbol of oppression and injustice, and it asks all the members to change their profile picture to Khaled's picture in support of his rights, stressing that unifying over a common goal will be the means by which "we" can get his rights back. This picture of Khaled contrasts sharply with another posted picture of Khaled's mutilated face, broken jaw, and shattered teeth. The viewers are implicitly invited to reflect on this young man's life that was cut short in the most horrific manner. Whether young or old, most viewers cannot help but feel a great sense of sorrow for this young man and a sense of outrage over the brutal methods which the Egyptian police use. Most mothers or fathers might see their own children in his youth, while the young might see themselves in him and how they could, with a twist of fate, have been in his shoes. Another picture shows Khaled's mother, in black, mourning her dead son, while others show Egyptians holding Khaled's picture in the silent stand that took place in different parts of Egypt against the injustice he faced to tell the story of other victims of police brutality and violence. All the pictures support the page's activist message: Defend the human rights of all Egyptians and educate the people about their constitutional rights and the violation of these rights, thus inviting the members to reflect on the horrific ending of

Khaled and all similar victims, while also providing analysis of the violation of citizens' rights and humanity by the police.

One of the conversation threads started after a July 24, 2012 post by the admin about Mina Danielle, one of the young political activists and martyrs of the Egyptian revolution. The accompanying picture shows him running away from a cloud of tear gas that engulfed the background of the picture, almost completely concealing it and the dangers behind it. The intensity of the tear gas in the background gives a glimpse of the dangers these young protestors faced when they stormed the streets of Egypt demanding the rights and freedoms for all Egyptians. The post reads: The martyr of freedom Mina Daniel, On the Friday of Rage in the vicinity of the Interior Ministry (identifying where the picture was taken). The comments made on the picture, by both Muslims and Christians, mostly identify Mina Daniel as a martyr and hero of the revolution. In line with the ethos of the group, the posts generally reflected the belief in a post-sectarian Egypt that does not discriminate between its Muslim and Christian citizens and iterated the discourse of universal humanism and respect for all citizens and all the martyrs who died in the revolution regardless of their religion. However, a few comments took insult to these comments and cynically questioned what one called the "cheapened" use of the word martyr (an Islamist reference) to refer to a Christian whom they called an infidel, flaming the admin with derogatory comments, accusing him of being hypocritical, and reminding him that one day everyone will stand before God to be judged on everything one says, stressing that not everyone should be called a martyr. The conversation that emerged from that one post about an Egyptian hero who died in the revolution defending the rights and freedoms of others gets diluted to the mere notion of infidel or martyr, thus reducing his sacrifice to a simplistic binary that reinstates factionism and sectarianism, all of which are contrary to the page's ethos and discourse on unity, love, respect

of difference, and understanding of otherness. The deviation of these posts from the communal ethos of this page prompted others to come to the defense of the memory of Mina Daniel and to affirm his status as a hero within the Islamic and Christian traditions and defend the discourse of universal humanism and post-sectarianism.

The other pictures, cartoons, videos, and textual posts zoom in on national issues and news and document the page's and campaign's expanded focus from police violence to the different events that transpired in Egypt during and after the revolution. The activist nationalist message of the page, the covering of locally-related events in Egypt, and the use of Arabic as the only language of communication align well with this campaign's primarily Egyptian audience (possible also Arab) and local outreach.

Textual and Visual Ethos Construction of We are all Khaled Said Facebook Page

The We are all Khaled Said Facebook page represents the international arm of the Kolena Khaled Said campaign which was founded on 7/19/2010. Initially administered anonymously out of the United Kingdom until a few months after the revolution, the administrator used the pseudonym El-Shaheed (i.e. The Martyr), evoking a prominent Islamist discourse about the sanctified memory of martyrs, who sacrifice themselves for the good of their Muslim community. The page, which uses English as its primary language of communication, commemorates the memory of Khaled Said as the symbol and face of all the other victims of police brutality in Egypt and provides a venue to educate Egyptians and other international web users, who are more comfortable using English, about the human rights abuses at the hands of the Egyptian police and create international exposure for its activist campaign. This Egyptian activist site currently has more than 239,000 fans from different nationalities, which is evident from the

Western and Arabic names of the posting members, who also often opt to write their nationality at the end of their posts. Unlike the Arabic Kolena Khalid Said page, this page does not allow its members to make new posts on its wall, restricting that right solely to the admin of the page, while putting no restrictions allowing on the members' comments on the admin's posts. (Though this seems quite restrictive, none of the members seems to have taken it as issue or commented on it.)

In the "About" section of the page, the emotionally subdued account of the murder of Khaled Said presents a discursive digital ethos of a community that wishes to be viewed as more logos- than pathos-oriented, stating that Khaled Said was "tortured to death at the hands of two police officers," recounting how "several eye witnesses described how Khaled was taken by the two policemen into the entrance of a residential building where he was brutally punched and kicked" and his head was banged into the wall, the staircase, and entrance steps. The absence of an impassioned account of the murder and of any graphic description of the battered condition of Khaled's body and face, whose brutally disfigured picture circulated widely on the net, shows the page's interest in appealing more directly to the minds of the audience and not explicitly playing on their emotions, which, due to the intensity of the public debate on whether Khaled Said deserved to die in this manner or not, would distract from the message of the page. The drug-using claims made against Khaled Said by the police prompted some Egyptians to profess openly that he deserved what transpired. The page stresses the need for change, the need for putting "an end to all violence by any Egyptian policeman" and martial laws, and calls for "freedom and dignity back," as well as a true representative democracy. The fine line that the page tried to walk between the reality of police brutality in Egypt and the story of Khaled Said aims at drawing the web users' attention to the goal behind the page, which is to see Egyptians unified around establishing a new era of human

rights in Egypt while preventing further factionism from flaring up as people take sides in the Khaled Said debate. The page advocates a general call for dignity that many Egyptians can identify with, not only for themselves, but also for their children who may one day face the same fate as Khaled Said as well if the police continue to operate with brutal methods. This logical call for dignity and human rights for all Egyptians is one that few could contest.

The page's call for an end to police brutality and the violation of people's rights and humanity presents the movement and its members as activists trying to realize real change in real communities through their digital activism by reaching out to members of their community, educating them about the human rights abuses in Egypt, and starting a conversation about the routes to prevent these abuses. Such an emerging communal digital ethos aligns closely with classical rhetoric and the importance of advocating the common good and educating the masses about it, not to mention the unequivocal commitment to the noble ideals of a rhetor and his community, all of which are in line with Aristotelian and Isocratic conceptions of ethos. The investment in a fair and ideal society is one that also aligns well with Plato's theory of ethos and the belief that the noble rhetor will be the venue through which the epistemic truth can be conveyed to the public. This message has resounded well with users of the page and garnered support across Egypt and around the world, as evidenced by the way Egypt erupted in massive protests in all its major cities, calling for human rights, justice, freedom, and democracy for all Egyptians.

There are multiple pictures and videos posted on the page that support the discursive message of the page and this digital community. The most prominent image is the gray scale background poster-like picture of Khaled Said with the Arabic phrase (one of the very few uses of Arabic on the page): "May God have mercy on him" partially appearing at the bottom. Viewing the picture in full scale, one can see the background is filled with scattered Arabic letters and statements in red and black colors, which contrast with the grayscale of Khaled's picture. One of the enlarged black letters includes Arabic statements asking God to bless Khaled's soul and avenge his death from those who would commit the most heinous crimes against the people. The same statements are repeated in black font at the bottom of the picture. These references effectively merge the discourses of Islamism and Western humanism, which are evoked in the references to "Allah" and the calls for His vengeance, as well as the references to the violation of human rights of Egyptians through torture, heinous crimes, and murder, which are no better illustrated than with the disfigured face of Khaled all page users are more than familiar with. The intermixed use of Islamism and Western humanism is the means by which this page is able to construct its ethos. This Islamist discourse, which accounts for the only Arabic posts by the admin, can easily garner the Arab and Muslim digital users' approval and support because it aligns with their religious beliefs, while the references to human rights, which constitute the majority of the admin's posts, appeal to the non-Muslim users.

In the Facebook timeline, the sepia tone profile picture of the young Khaled Said is separated with a straight black line running down the middle of his face and picture, indicating the two sides of his story, the one promoted by the police that he died of a drug overdose and the one put forth by the pictures of his disfigured face that relates police violence. The red Arabic scribbling on both sides of the slightly-parted picture says "Egypt's right" on the left-hand side and "Khaled's right" on the right-hand side of the picture, giving both individual and national rights equal weight, evoking nationalist sentiments in its Egyptian audience and expressing the indivisibility of human rights from one's conception or articulation of nationhood. The bright red color contrasts sharply with the sepia tone of the rest of the picture since it evokes the color of blood on the page as if visually splashing the blood Khaled and many others have shed at the brutal hands of the police. The pictures of Khaled Said draw in various audiences across generational lines: The young can identify with his youth and sympathize for his sad ending, while the old can see their own children in him and what they would never like to see happen to their own. The digital ethos of the page and movement grounds itself in what is right and morally just despite the consequences, thereby aligning itself more closely with the classical conceptions of ethos.

The early pictures posted by the administrator on the page shortly after its inception in 2011 show Egyptian and international supporters who sent pictures documenting their silent stand in support of Khaled Said's rights for justice and against torture in Egypt. Other pictures and posts expose the systematic torture of Egyptians at the hands of the police, visually and textually telling the story of other victims of police brutality. The pictures of Khaled Said and other victims of the police invite the community members to reflect upon, and provide analysis of, the systematic violation of citizens' human rights by the police. The absence of the graphic picture of Khaled's disfigured face aligns with the logos-oriented ethos the page is constructing, which does not aim at inflaming the emotions of the masses or intensifying the debate on both sides of those proponents—or opponents—of police violence in alleged drug cases.

One of the expanded conversation threads on the page effectively articulates the dilemma of the administrator as to whether exposing his identity will be a judicious act. The first post by the administrator was on May 10, 2011 and read: "I wanted you, members of this page, to be the first to know. I am about to make an important decision that could have many consequences on my life. I am now seriously considering disclosing my identity. I have always kept my identity secret but now I hope I can help Egypt's transition to freedom and democracy… I hope it's the right decision." The hesitation and almost palpable fear of the "many" unknown consequences of such a disclosure is evident in the emphasis on the fact that s/he has "always" protected the secrecy of his/her identity. The admin's reasoning behind considering ending the state of anonymity about his/her identity is framed within the classical parameters of the common good, which would be accomplished by aiding Egypt in its transition to freedom and democracy through public talks and engaging the page's primary international audience. The responses of the community members were partly positive, with some members wishing the admin good luck. However, the majority of the comments discouraged the admin from such a move, talking about the importance of the admin's safety. Some of the members' comments revolved around this space being a public venue that is more about the people contributing than about someone's personal space or ownership, openly stating that the admin shouldn't do it because "It's really not about your name." Other comments stressed the power and reassurance that this digital community finds in this anonymity without which a collective voice of resistance that "isn't about one individual" would be lost. From this comment thread, the preservation of the communal ethos of that digital space and its cohesion through the maintenance of the admin's anonymity became the core of the debate. Despite the absence of any later posts about this topic, the admin disclosed his identity shortly afterwards, identifying himself as Mohamed Ibrahim Mostafa—an Egyptian living in the UK. The disclosure of the admin's identity did not appear to negatively affect the community's members because the members continued to comment, although the average number of comments to posts decreased from hundreds of replies to tens. However, there is no evidence to substantiate a direct correlation between the reduced number of comments and the disclosure of the admin's identity.

The other pictures, cartoons, videos, or textual posts on the page shed light on the national and international topics of interest, such as that of the Syrian uprising and other protests taking place around the world; however, the focus is mostly directed to current events in Egypt relating pre and post-revolution. All the visual and textual elements document central events that have become carved in the public's memory of the pre- and post-Arab Spring and the Egyptian revolution. The communal visual digital ethos supports the movement's discursive ethos of an activist campaign that advocates the rights of all citizens to dignity and justice in Egypt and around the world and believes that societies can uphold and gain these universal rights through unified action and communal support and education. The activist message of the page, the covering of events beyond Egypt, and the use of English as the only language of communication (despite occasional Arabic posts by members) align well with this campaign's international audience and outreach.

Synthesis of the Textual Analysis of Both Facebook Pages

In constructing the communal digital ethos, the digital users' comments and responses to the admin's posts, in both pages, employ a mix of Western humanism, Islamism, and post-sectarian nationalism. The posts call for human rights, democracy, and freedom for all. The employment of Islamist discourse revolves around making references to Al-Shaheed, God or Allah, His blessing, and His mercy and justice, which embody the faith in a supreme and all-encompassing being who possesses the power to bolster and assist noble and just rhetors in accomplishing their just causes. The post-sectarian nationalism is evident in the call for unity among all Egyptians, both Christian and Muslim (evident in the Kolena Khaled Said page), and for unity among all people of the world—every religion, race, and nationality (evident in We are all Khaled Said page), and aids in the realization of the dreams of justice and freedom for every oppressed person on the face of this earth. All these discursive elements assist in constructing and preserving a communal ethos that believes in, and advocates, universal humanism and respect for otherness while simultaneously instituting its discourse within the parameters of social control and regulation instated by the admin and complied with by the page's web users.

Issues, Controversies, Problems

Much has been spoken about the potential of the worldwide web as a liberatory space in which multiple subjectivities and realities can be negotiated and altered and in which multiple communication opportunities can be created that do not exist in "'real' space" (Gruber, 2003; Hawisher and Sullivan, 1999). In computer-mediated communication, the premise upon which technological empowerment is based is the reduction, absence or elimination of the technology user's entire corporeal body from the discourse environment (De Pew, 2003). Though, in this space, the absent body of the user becomes the source of power and appeal, anonymity often puts into question the user's digital ethos because one's authority and motives appear undetected or unconvincing to other online users. Traditionally, digital ethos has been firmly grounded in the identification of the speaker's identity and authority, as seen in Jim Kapoun's 1998 article in which he established the general criteria for evaluating websites that became popular in many contexts. These five elements include: Accuracy, Authority, Objectivity, Currency, and Coverage, where authority here can only be established with the articulation of the identity and qualification of the writer, which are the same criteria he used to assess print (qtd in Apostel and Folk, 2004). Despite the use of these criteria in evaluating digital discourses, the limitations of such criteria becomes no less evident in activist discourses in which activists choose (or have to, out of necessity) for reasons of expedience or safety to remain anonymous. As evident from my analysis of the two activist Facebook pages,

revolutionaries and activists are able to function more effectively and safely anonymously. Such anonymity and facelessness preserve their digital ethos, attract users to their pages, and garner more support for their causes, inciting more people's trust in the activists' selfless efforts. However, as the analysis illustrates, once the activists' identities are exposed, they become targeted by overt and covert regime operatives, thus posing risks to their constructed ethos in both their digital and physical communities and possibly affecting the sustainability and relevance of their digital communities and activist agendas. In these cases, anonymity in such digital communities is not only called for but it is the most powerful and effective ethos to be used in this rhetorical context to ensure the success of one's activist goals. The decision of the admin of We are all Khaled Said, Mohamed Ibrahim, to expose his identity did not receive as much negative publicity as the blowing of the cover of anonymity from Wael Ghonim. Whereas the latter's cover was exposed forcibly during the height of the revolution, the former chose to expose his identity after what appeared to be the "success" of the revolution. This brings up questions of expedience and kairos and what factors determine the appropriateness of the use of, or the unveiling of, anonymity without jeopardizing one's ethos or message. Thus, the classical notions of authorship are contested widely in these digital discourses, thereby highlighting the need for new theorizations about ethos and postcolonial/ post-print constructions of authorship.

Also evident from the analysis of both pages is the increased emphasis on ethos construction and preservation through the discursive and visual elements on the page. However, the preservation (which poses the most challenge) of the pages' digital ethos and activist persona required replicating forms of social control and regulation in these digital spaces. While one page's admin allowed fans or members to post their topics of interest on the page, another restricted this right only allowing these digital users to comment on the page's posts. In the interest of self, or more so, community-preservation, limiting posts by members, yet allowing comments, focuses the members on the topics of discussion and allows for increased accuracy, the cross-checking of information, and for a more engaged readership for those closely following the page or those committed to its message. Furthermore, despite the freedom of fans/members to post their own topics on Kolena Khaled Said, the list of prohibitions supplied by the admin attempts to regulate that digital space and restrict the page's users. The analysis illustrates that as activists and revolutionaries broadcast their causes, they attempt to maintain control (at least partially) over their message and the method of its dispersal. Through exercising a subversive form of dictatorship, activists are able to shield their digital spaces and preserve the authenticity of their causes from the chaotic nature of online communications. However, as evident from the analysis, despite the exercise of power and domination, these spaces have been successful in creating communal pockets for advocating democracy and propelling political protests and regime change in Egypt and other countries of the Arab spring in addition to allowing the fermentation of anti-revolutionary, democracy-opposing pockets. The research further shows that these communal pockets exercise coercion against their dissenting members and replicate the same forms of ostracism, shaming, and punishment exercised in *real* physical spaces, all in an attempt to preserve the communal ethos that bolsters the success and outreach of their digital agenda.

Solutions and Recommendations

Evident from the above analysis, the central role ethos occupies in legitimizing identity, experience, and digital discourses can affect rhetors in both their digital and real spaces. Thus possibilities of integrating the critical study of ethos in composition teaching becomes of great importance as our students can study and articulate how perceptions of ethos shape knowledge production, dissemination, and achieve or hinder persuasion in both the digital and physical public domain. The need to study the construction of a reputable ethos and the challenges posed to it will allow our students, especially those occupying the margins of our society and academic culture, to uncover the means through which their own, as well as their community's ethos, can be reshaped and reconstructed, creating a discursive territory within which our students can better examine the relationship between digital texts and their and others' lived experiences.

FUTURE RESEARCH DIRECTIONS

The importance of ethos in identity formation in both our physical and virtual spaces amplifies the need for more research on ethos construction, re/de/construction, and preservation in digital discourses. Furthermore, the need arises for new theorizations on postcolonial/post-print constructions of ethos and authorship stemming from the need to create a space within our scholarship for the study of anonymity as an expediently legitimate form of authorship. In addition, the examination of the various constructions of ethos in different social, cultural, and political contexts in online discourses from different parts of the world will provide a richer understanding of the complex, culturally and linguistically-situated, and diverse applications of ethos in the digital domain.

CONCLUSION

This chapter examined two Egyptian facebook pages: "Kolena Khaled Said" and "We are all Khaled Said." Both pages advocated an activist message and played a key role in the propelling of the Egyptian revolution, utilizing their digital spaces as means for organizing and coordinating revolutionary work. The analysis explored the complexity of ethos construction in these activist digital discourses and the power of anonymity in ethos construction and preservation in activist discourses, which, contrary to long-held biases against the credibility of anonymous sources in academia, as seen in part by the demonization of Wikipedia, here resulted in the emergence of a positive, powerful communal ethos and a palpable sense of shared identity in these digital social networks. This sense of shared identity and fate assisted these digital communities in successfully achieving massive outreach for their activist calls. In our scholarship, we attempt to map out these virtual spaces with their "multiple heterogeneous borders where different histories, languages, experiences, and voices intermingle amidst diverse relations of power and privilege" (Giroux, 1992, p. 169), with the knowledge that we will never fully exhaust such spaces.

REFERENCES

Burke, K. (1950). *A rhetoric of motives*. London, UK: Prentice Hall.

Facebook. (2012). *Kolena Khaled Said*. Retrieved July 26, 2012, from www.facebook.com/ElShaheed

Facebook. (2012). *We are all Khaled Said*. Retrieved July 26, 2012, from www.facebook.com/elshaheeed.co.uk

Foster, D. (1997). Community and identity in the electronic village. In Porter, D. (Ed.), *Internet culture* (pp. 23–37). New York, NY: Routledge.

Giglio, M. (2011a, Oct 31). The reluctant revolutionary. *Newsweek, 158* (18), 45-45.

Giglio, M. (2011b, Feb. 21). The Facebook freedom fighter. *Newsweek, 157* (18), 14-17.

Giroux, H. A. (1992). *Border crossings: Cultural workers and the politics of education*. New York, NY: Routledge.

Gurak, L. J. (1999). The promise and the peril of social action in cyberspace: Ethos, delivery, and the protests over MarketPlace and the Clipper Chip. In Smith, M. A., & Kollock, P. (Eds.), *Communities in cyberspace* (pp. 244–263). London, UK: Routledge.

Hawisher, G. E., & Sullivan, P. (Eds.). (1999). *Passions, pedagogies and 21st century technologies* (pp. 268–291). Logan, UT: Utah State University Press.

Hyde, M. J. (Ed.). (2004). *The ethos of rhetoric*. Columbia, SC: University of South Carolina Press.

Porter, J. E. (1992). *Audience and rhetoric*. NJ: Prentice Hall.

Roszak, T. (1986). *The cult of information: The folklore of computers and the true art of thinking*. New York, NY: Pantheon Books.

Young, I. M. (1990). The ideal community and the politics of difference. In Nicholson, L. (Ed.), *Feminism/postmodernism* (pp. 300–323). New York, NY: Routledge.

ADDITIONAL READING

Burke, K. (1973). *A philosophy of literary form* (3rd ed.). CA: U of California Press.

Gruber, S. (2001). The rhetorics of three women activist groups on the Web: Building and transforming communities. In Gray-Rosendale, L., & Gruber, S. (Eds.), *Alternative rhetorics: Challenges to the rhetorical tradition* (pp. 77–92). Albany, NY: State U of New York P.

Gurak, L. (1997). *Persuasion and privacy in cyberspace*. New Haven, CT: Yale UP.

Hawk, B. (2004). Toward a rhetoric of network (media) culture: Notes on polarities and potentiality. *JAC: Special Issue on Mark C. Taylor and Emerging Network Culture, 24*(4), 831–850.

Hocks, M. (2003). Understanding visual rhetoric in digital writing environments. *College Composition and Communication, 54*(4), 629–656. doi:10.2307/3594188

Kress, G., & van Leeuwen, T. (2006). *Reading images: The grammar of visual design* (2nd ed.). New York, NY: Routledge. doi:10.1016/S8755-4615(01)00042-1

Lanham, R. A. (1993). *The electronic word: Democracy, technology, and the arts.* Chicago, IL: U of Chicago Press.

Lankes, R. D. (2007). Credibility on the Internet: Shifting from authority to reliability. *The Journal of Documentation, 64*(5), 667–686. doi:10.1108/00220410810899709

Lockard, J. (1996). Progressive politics, electronic individualism and the myth of virtual community. In Porter, D. (Ed.), *Internet culture* (pp. 219–231). New York, NY: Routledge.

Plato,. (1952). *Gorgias* (Hembold, W. C., Trans.). New York, NY: Macmillian.

Porter, J. (2009). Recovering delivery for digital rhetoric. *Computers and Composition, 26*(4), 207–224. doi:10.1016/j.compcom.2009.09.004

Smith, M., & Kollock, P. (Eds.). (1999). *Communities in cyberspace.* London, UK: Routledge. doi:10.5117/9789056290818

Ulmer, G. (1994). Preface and part 1. In Ulmer, G. (Ed.), *Heuritics: The logic of invention* (p. xi-42). Baltimore, MD: Johns Hopkins UP.

Warnick, B. (2004). Online ethos: Source credibility in an "authorless" environment. *The American Behavioral Scientist, 48*(2), 256–265. doi:10.1177/0002764204267273

KEY TERMS AND DEFINITIONS

Communal Ethos: It refers to a network of discursive practices and conventions for knowledge production and dissemination to which a certain community has consented.

Discourse Community: It refers to a localized and temporal restraining system, defined by a common body of texts or conventions.

Epistemic Truth: It is the arrival at truth whether through divine inspiration (as with Plato), analysis, perception, or contemplation.

Ethos: It usually refers to one's good or moral character or credibility.

Identification: It means to identify with one's audience's needs and norms in a more empathetic and dialogic approach, ultimately achieving persuasion.

Section 5
Games

Chapter 19

"I Rolled the Dice with Trade Chat and This is What I Got":
Demonstrating Context–Dependent Credibility in Virtual Worlds

Wendi Sierra
North Carolina State University, USA

Doug Eyman
George Mason University, USA

ABSTRACT

In this chapter, the authors extend Warnick's (2007) appropriation of Toulmin's (1958) "field-dependency" as applied through an ecological lens to examine credibility and ethos in the virtual world of a massive multiplayer online game. The authors theorize that ethos in such virtual environments is context-dependent—that it is in the interaction between designed game and user action/communication that ethos is engineered in a process that is fundamentally different from both websites (which are static) and other social media (where the environment is not nearly as much of an actor in the development of ethos/credibility). To better understand how players (as inhabitants of the game ecology) view the establishment of ethos, the authors collected in-game chat and near-game forum posts that included responses to requests for assistance or invitations to join a guild, and we asked our participants to evaluate these texts. The chapter uses the data collected about the perception of ethos to identify three key elements for successful demonstration of credibility in multiplayer games: specificity, demonstrated expertise, and experience.

INTRODUCTION

Massive multiplayer online games such as *World of Warcraft*, *League of Legends*, *Rift*, and *City of Heroes* take place in virtual worlds; these worlds are both programmed (by the game designers) and socially constructed (by the game players), and it is this intersection of digitally-mediated environment and action that provides a unique rhetorical ecology for each multi-player game. Because each game provides its own virtual world, the study of rhetorical activities takes place in communication networks that are both bounded and programmed.

DOI: 10.4018/978-1-4666-2663-8.ch019

In our review of the literature, we noted that a direct consideration of ethos has not yet been addressed by game studies scholars or by those in rhetoric and writing: even important works specifically interested in the rhetorics and literacies of games, such as Bogost's (2007) *Persuasive Games* and Selfe and Hawisher's (2007) *Gaming Lives in the Twenty-First Century*, for instance, address questions of agency, identification, and ethics, but they do not look at ethos as constructed by either the player or the game (or any combination thereof). We believe that digital games—and multiplayer games in particular—are ideally situated to allow scholars to consider the role of context in the development of ethos. Halloran (1982) pointed out that "most concrete meaning given for the term [ethos] in the Greek lexicon is 'a habitual gathering place,'" (p. 60), and in this chapter we examine not only the ways in which ethos is located in the person of the rhetor and the construction of the argument, but also within the designed ecology of the "habitual gathering place" of *World of Warcraft*. In the following sections, we examine the differences between analyzing ethos in websites/relatively static digital texts and analyzing ethos in interactive digital environments, present a theoretical framework based on an ecological view of game activity, and use in-game and near-game texts as prompts for player interviews to help us see the interrelationships of player, text, and game design as they impact the invention and deployment of ethos in game-based communications.

Analyzing written texts for the use of ethos as one of the three artistic proofs of classical rhetoric (as identified by Aristotle) has been a staple of rhetoric and composition courses, although it has shifted from the classical approach of proof inherent in the text itself to a form that is both inhabited by the rhetor and bolstered by drawing on networks of expertise (see Warnick, 2007). And over the past decade, computers and writing scholars have applied and refined the analysis of ethos as embodied in digital texts (Brent, 1997;

Heba, 1997; Fogg, 2002; Grover, 2002; LaGrandeur, 2003). For websites, we might be able to use the same basic analytic approach as printed text, but we also need to address a number of elements that arise from the position of the websites within networks: "[W]hat other sites link to the site in question, whether its content is supported by other content in the knowledge system ... how well the site functions, and whether it compares favorably with other sites in the same genre" (Warnick, 2007, p. 49). The digital nature of the text also allows the use of tools that orient the analysis to the role of the site's infrastructure and its place within the network itself—see, for instance, Jim Ridolfo's (2006) "Comprehensive Online Document Evaluation," which provides instruction on using three network analysis tools to uncover both geographies and owners of digitally networked sites, along with two additional web-based tools for examining the changes over time that a given website experiences.

Multiplayer online games, however, have not received as much attention in terms of the development of rhetorical approaches to assessing ethos, whether exercised by the design of the game itself or by the activities and interactions of the players. One of the distinctions between the rhetorical situations of websites and games is the level of interaction. While a website exists within a network of links and associations, and the interface of any given site may require more or less user interaction, games require a much higher degree of engagement and interaction in order to function; this interactivity takes place between and among the players, the non-player characters (software agents), the programmed game mechanics (the "rules") and the designed virtual environment that contains the primary interactions of play. We would suggest that understanding the role of ethos in games calls for a distinct analytic approach that considers the ways that environment, expertise, and rhetorical moves interact.

Both contemporary and classical approaches to the development of ethos focus on the interplay of the reputation and credibility of the rhetor and the effectiveness of the construction of the argument itself. Crowley and Hawhee (2009), drawing on Aristotle, suggested that ethical proof can be either situated or invented: situated ethos is based on the reputation of the rhetor within his or her community while invented ethos is created by constructing a persona that inhabits and demonstrates authority solely within the text (p. 198).

Situated ethos can be enhanced over time by building up a reputation that is tied to a particular discourse community; as Halloran (1982) explained its use in the classical tradition, "to have *ethos* is to manifest the virtues most valued by the culture to and for which one speaks" (p. 60). In the game environments we have studied, situated ethos can be deployed when the audience is relatively small, such as when communicating with one's guild; however, much of the in-game (and near-game) communication that we examined was delivered to a much broader audience. Crowley and Hawhee (2009) noted that composing for broad audiences requires the rhetor to rely on invented ethos rather than situated ethos (pp. 199-200), but even this approach is constrained in the game because textual communication is limited to short messages and real-time discussion (which is not recorded in the game itself). Given the constraints of the game—the difficulty of building an individual reputation in a world that generally does not record the histories, arguments, and actions of the players, and the inability to construct texts beyond very brief elocutions within the in-game chat feature—we wondered whether players could effectively develop and use ethos, and if so, what would those approaches teach us about the rhetorical contexts of gameplay?

In *Rhetoric Online*, Warnick (2007) proposed an adaptation of Stephen Toulmin's (1958) model of field-dependence as a framework for examining ethos in online texts. Using this approach, "the credibility of an argument is evaluated according to the standards indigenous to the field in which the argument is made;" thus, "users may judge sites according to the procedures, content quality and usefulness, functionality, and values and norms important in the field in which the online site operates" (p. 49). Toulmin's use of field focuses on broad disciplinary categories such as law or medicine, and as Warnick explained, the concept of field serves as "an epistemological context in which arguments [take] shape" and provides criteria for judging these arguments (p. 49). In this chapter, we extend the concept of field-dependency, shifting from an epistemological view of the field to an ecological view of field-as-context. Rather than identifying an epistemological foundation that constrains available modes of argument (and therefore dictates the form of ethos), we focus on both the programmed constraints of the virtual world and the interactions of the players within the designed ecosystem of the game.

A scientific term originally applied to research on interactions in specific natural environments[1], "ecology" as a metaphor for complex, interconnected relationships has a rich history of use in writing studies (Cooper, 1986; Syverson, 1999; Nardi & O'Day, 1999; Spinuzzi & Zachry, 2000; Spinuzzi, 2003; Blythe, 2007). The basic scientific definition of ecology is "the study of the relationships of organisms to their environment and to one another. ... Ecology is a study of interactions" (Brewer, 1988, p. 1). For this study, we draw particularly on the work of Nardi and O'Day (1999) and Syverson (1999) to develop an ecological lens that we then train on the specific ecology of a single multiplayer online game.

Nardi and O'Day's (1999) approach to an ecological view was defined as an "information ecology": "a system of people, practices, values, and technologies in a particular local environment" (p. 49). They argued that "the ecology metaphor provides a distinctive, powerful set of organizing properties around which to have conversations. The ecological metaphor suggests several key properties of many environments in which technology is used. An information ecology is a complex system of parts and relationships" (p.

50). While Nardi and O'Day located "ecologies" in specific material locations (such as libraries, schools, and hospitals), we see the framework as equally applicable to specific virtual locations as well. And even when those locations are intended to represent worlds, the scale of these worlds is designed to work with individual players, which resonates with Nardi and O'Day's articulation of an ecology as "a place that is scaled to individuals" (p. 50). For our purposes, the two most important elements of the ecological metaphor are that "an information ecology is marked by strong interrelationships and dependencies among its different parts," (p. 51) and that "locality is a particularly important attribute of information ecologies" (p. 55). An additional appeal of drawing on the information ecology framework is that it allows a very focused and bounded view of a specific ecological instance (in our case, the game's virtual world along with the outside-the-game structures that support player interaction); although the situation of an ecology within a larger technological system is acknowledged, the focus of the lens is on the local interaction—in other words, we can focus on the game *as* the world rather than how the game functions *in* the larger world of technological, social, political, and economic networks.

In *The Wealth of Reality: An Ecology of Composition*, Syverson (1999) took a broader view of the scale of an ecological framework. She applied the ecology metaphor to writing studies by arguing that:

[W]riters, readers, and texts form ... a complex system of self-organizing, adaptive, and dynamic interactions. But even beyond this level of complexity, they are actually situated in an ecology, a larger system that includes environmental structures, such as pens, paper, computers, books, telephones, fax machines, photocopiers, printing presses, and other natural and human-constructed features, as well as other complex systems operating at various levels of scale, such as families, global

economies, publishing systems, theoretical frames, academic disciplines, and language itself. (p. 5)

Thus, if we see a game environment as a fully realized ecology, we need to be sure to investigate and delineate both environmental structures (such as the space of play and the objects that can be manipulated within the game) and these other complex systems, which include the rules of the game, the social and economic systems that are available to players, and the additional locations outside of the virtual environment of the game that support in-game activities as well as player development and interaction.

We begin, then with an overview of the ecology of one of the most popular (and most researched) games, *World of Warcraft*.

World of Warcraft: Game Ecology and Infrastructure

We selected *World of Warcraft* as the focus of our study primarily due to our own level of engagement with the game. In order to use an ecological frame in our research, we need to be able to articulate the elements of the environment with a high degree of confidence. We also believe that it is important to have experienced the game in order to position ourselves as qualified to undertake this kind of research. We have each been playing *World of Warcraft* (*WoW*) for a number of years (Doug began playing in 2006, Wendi in 2007) and have experienced the changes in the game as major expansions have been added and refined. We have both played with a number of different player communities (called "guilds") and we have each played several character classes in both Alliance and Horde game factions. We also have very different play styles, so our views of the interactions within the game come from different perspectives. We believe that our commitment to both understanding and playing the game is a requirement of the kind of study we are undertaking; it speaks to our own ethos as researchers,[2] but we also cannot fully understand the game-as-ecology without a deep understanding of the game's underlying principles and the way that players interact with those principles.

World of Warcraft: A Brief Overview

World of Warcraft is a multiplayer game: the initial game software is purchased once but subsequent play requires a monthly subscription fee (currently around fifteen dollars per month). The game software for each expansion is also purchased separately and is required for players to access to expanded gameplay options once they have logged in to the game. Much of the designed world and underlying rule-based infrastructure resides on the player's computer, but the interaction between player and world and among players (and a record of activities and possessions) takes place in real time on the game servers.

Economically, players can be viewed as subscribers—and in 2010, *WoW* boasted over 12 million subscribers. The number of players has declined since, and Blizzard has reported a current subscriber base of 10.2 million players in December 2011 (Holisky, 2012). Because there is such a large number of players, Blizzard operates a number of distinct servers that players can choose to log into. These servers are geographically distributed, and they are also designated as serving three different play styles: Player-versus-Environment (PvE), Player-versus-Player (PvP), or Role-Playing (RP). Each of these different forms of play take place on every server; however, the server designation indicates differences in rules and supported activities that foreground one type of play over another.

Types of Play and Player Actions

One of the reasons that *WoW* is particularly appropriate for an ecological approach is that it is a very rich and complex game that provides a wide range of different activities for players to engage. While we won't cover all possible activities here, a review of the main player actions is necessary to understand the responses of our study participants because they approach the game from different preferred play types. Although different servers support and privilege different play types, each server provides opportunities to engage all of the available activities. Players can work individually (called "soloing") as they work to level up their characters in general PvE (player-versus-environment) play, they can work collaboratively by joining a guild and participating in collaborative play (which may involve assembling a group of players to defeat difficult monsters in designer-created content, either in "dungeons" limited to five players or "raids" designed for up to 40 players), or they can play competitively against other players in PvP (player-versus-player). PvP gaming involves team matches against other players in either battlegrounds or arenas.

The different approaches to the main forms of play color the way that players interact with and experience the ecology of the game world. For instance, while Doug (as a player) tends to follow quests, one of his colleagues skips all of the quests (and the narrative or "lore" of the game embedded in those quests) in favor of leveling solely through completing dungeons—and this leads to a difference in perspective about what kinds of gear (weapons and armor) should be acquired and what gaming skills should be practiced and perfected through many hours of play. Because we are taking an ecological approach, it is necessary to work with participants who can represent a range of these different views of the environment and their preferred interactions with it (and with other players).

In addition to these primary modes of play, *WoW* offers a number of additional elements that add complexity to the overall gaming experience while simultaneously supporting the different motivations of players. These elements map fairly neatly onto Yee's (2005) taxonomy of player motivations within multi-user domains. The three primary motivations, identified through large-scale surveys of players, include self-directed achievement (through advancement, the development of expertise in manipulating game mechanics, or competition with other players), social interaction (including socializing, developing relationships with other players, and performing in groups or guilds), and immersion in the world of the game (via exploration, role playing, and character customization). Raiding and PvP play clearly map onto both the social interaction and achievement motivations, but *WoW* also provides an achievement system that awards points for meeting specific goals (such as defeating a particularly difficult boss in an instance or raid, acquiring special objects, accumulating a large number of pets or mounts, or developing high levels of skill in a profession). Finally, the game world is quite extensive, which provides many opportunities for players to explore as they map the game space and encounter its objects and inhabitants.

Environment: The Designed World of WoW

The name of the virtual world in which the game activity of *World of Warcraft* takes place is Azeroth (essentially the name of the primary planet in the game). Some areas of the world are unavailable to players until they reach certain milestones (such as reaching a higher level, acquiring a flying mount, or gaining sufficient reputation with the inhabitants of certain locales).

The design of this world uses a physics not unlike our own (although considerable license is taken with the aid of magic); the design of the world, moreover, is finite and the rules of its topographies do circumscribe the actions and range of motions available to the player. This is important to note because an ecological analysis is facilitated by the bounded nature of the game world.

In addition to the game world's environment, the designers have also placed objects with which the player is either required to interact (in order to further a storyline or manipulate the environment, such as key to open a lock or a package delivered from one place to another) or may optionally examine (such as the many books scattered throughout the game world that chronicle its history, as told from a number of different perspectives).

Finally, the game also contains a vast array of non-player characters (NPCs); software agents whose task it is to give and receive quests, to further the player's experience of the game's narrative or history, to purchase and sell player-usable objects, or to provide assistance to the player (e.g. the city guards who help players find specific vendors).

Kurt Squire (2006) has argued that digital games "offer *designed experiences,* in which participants learn through a grammar of *doing* and *being*" (p. 19, emphasis in original), recognizing the game world as a designed ecology that serves a specific purpose and that facilitates both doing and being for the game player highlights the importance of accounting for and evaluating the environment (and not just the player action/interaction) as a critical element of the game ecology.

Rules and Procedures: Mechanics and Strategies

We define game *mechanics* as the underlying algorithms and procedures that determine how player actions are carried out in the game.[3] Mechanics affect the result of each action as modified by player choices. For instance, a player can choose among different types of weapons that may do more or less damage depending on a number of complex factors that include class and racial bonuses for using particular kinds of weapons, special abilities that add to or multiply weapon damage, and the other gear the player has acquired. For each class, players make choices about the optimal deployment of special abilities and attacks, and these choices (called "rotations") are also governed by the mathematical rules for action within the game.

The following example (from the WoWWiki page of advice on building a PvE-oriented Death Knight class character) shows just one of the many factors that are in play during any given encounter (in this case, the chance to hit an opponent in combat):

A death knight's hit cap is 8%. With two-handed weapons at level 80, this is a hit rating of 262. For Death Coil, Icy Touch and Howling Blast, the cap is 17% hit or 445.91 hit rating. Death knights choosing to dual wield have an additional penalty of 19% (16% with talent) to hit with both weapons, but this only affects auto-attack damage. Special attacks such as Blood Strike have the normal 8% cap. Raising hit over 8% is expensive, so ignoring the 19% miss and investing in other stats is the correct strategy. (WoWWiki, Death Knight PvE Guide, 2012)

Expert players invest a considerable amount of time learning about how all of the game mechanics interact and using that knowledge to make informed choices about what kinds of gear to acquire, how to modify that gear to improve the character's abilities, and what order of attacks and special abilities to use in a given situation. These mechanics have changed with each major expansion, so a constant commitment of time and effort is required to maintain knowledge and expertise about game mechanics. With enough time in the game, even non-expert players develop an intuitive sense of the game mechanics, but without a full understanding of how they work, a player will not become an expert or elite player in the game.

Similarly, expert players strive to understand game *strategies* as well as game mechanics. Strategies are the tactics used by the player to respond to a specific task or encounter—ranging from how to defeat a particular boss in an instance or raid (typically a collaborative strategy), to how to accomplish a difficult quest task, to how to most effectively battle a player of another class in a PvP situation. An example of this latter approach may be described as in the scenario below,

this time from the WoWWiki page of advice on playing a PvP-oriented Death Knight (2012):

Death knights are well equipped to handle any threat that a mage has to offer, but the opening sequence is key. The trump cards of this fight will be Anti-Magic Shell, Anti-Magic Zone, as well as Death Grip.

How the opening sequence is handled will determine the rest of the fight. The mage will generally run at you and Frost Nova you in place. If they get the frost nova off, Deep Freeze is sure to follow. To counter this, you can build runic power before fight starts by placing Death and Decay on the ground and using your horn of winter. ... If the mage tries to Polymorph you, immediately use Lichborne. It turns you undead, which makes you immune to being polymorphed. Do NOT use Anti-Magic Shell to absorb Polymorph, as you will need it to absorb damaging spells. ... Do not try to use Scourge Strike against a mage because when you get close enough to a mage for melee, you will need the runes for Death Strike. Apply spell interrupts like Mind Freeze and Strangulate as necessary. Keep enough runic power for Anti-Magic Shell, should he/she use the Shatter combo. There is no bad time to use it since all their attacks are magic based. The rest of the fight should solve itself in your favor so long as you continue to deal damage since mages have no healing skills.

The interplay between mechanics (rules) and strategies (player decisions) reflects the ecological relationship between the environment and its effects on the inhabitants of the environment. Additionally, in our study, we found that a player's ability to convey a sense of expertise (specifically in terms of knowledge about both mechanics and strategies) was an important consideration for our participants' evaluation of their credibility and authority.

Guilds as Social Networks

An important social aspect of the game is forming and participating in guilds. Guilds are loosely structured organizations that bring players together to accomplish shared goals, assist each other with individual goals, and share resources. The game provides incentives for cooperation (such as providing game content that cannot be completed individually), and guild formation takes advantage of these incentives while at the same time providing social and organizational structure. Guilds provide an additional layer of identity and community for players and they serve as in-game social networks. Guilds have their own communication channels within the game, and players tend to chat with each other regardless of whether they are working toward a common goal or playing individually. There are also opportunities to participate in guild chat function even when not playing the game (via smartphone and web apps), thus maintaining social ties even during non-play times. Guilds gain advantages as they progress through particular game challenges, so many guilds recruit new members who can strengthen the guild's roster and help the guild proceed to accrue more benefits. Players in guilds refer to each other as "guildies" or "guildmates" and one component of our research asked our participants to respond to guild recruitment ads.

In-Game and Near-Game Communication Channels

In addition to the guild channels, players can communicate with each other privately (via "whisper"), or they can speak aloud in text that can be read by anyone in the immediate vicinity ("say"). Players can also join chat channels, such as the General channel, the Trade channel (often very active, and usually not focused on trade per se), and the Looking for Group channel. Each of these primary channels operates within a specific zone (a specific ecosystem such as a swamp or desert, or within the cities). City-based channels tend to be the most active as there are always a large number of players carrying out tasks within the cities at any given time. More focused channels, similar to guild chat, allow players who are

in parties or raid groups to communicate directly just with those other players in the party or raid.

These in-game communication channels are supplemented by a wide range of out-of-game channels, such as official *WoW*-sponsored forums, unofficial forums (such as the *elitist jerks* website), and game wikis (such as wowhead.com). We describe these external sites as "near-game" channels because players will often refer to these sites (or post queries to them) while playing the game.

METHODS

Understanding the game as ecology and identifying how ethos and credibility are constructed within the designed space of the game requires both familiarity with the game system and access to members of the game ecology. Our primary research question—"How do players identify credibility and ethos in other players?"—required us to both identify game spaces in which players might be trying to demonstrate authority or ethos, and to be able to ask players to interpret these demonstrations.

Because little research has been done into how players identify and define credibility and ethos, we selected semi-structured interviews with a limited number of players as the most appropriate means of data collection. With only six interview participants, this survey is clearly exploratory. Instead of making definitive or statistically significant claims, we hope to show how researchers might begin the process of identifying the unique ways that authority and credibility are represented in massively multiplayer online games. The results of an exploratory qualitative study like this one can provide a solid foundation for larger scale quantitative projects.

Answering our question necessitated two stages of data collection. First, we needed to identify several texts for our participants to respond to. Thus, we took several sample texts from in-game and near-game sources, texts that displayed either players in the process of actively seeking help and offering information to each other or players persuading others to join their groups. This collection of texts provided us with examples of players displaying, or attempting to display, authority and credibility. However, for our purposes, we were not interested in how players (as speakers) attempt to establish authority, but rather how other players (as inhabitants of the game ecology) interpret these acts as successful or not. To examine how the community identifies authority and evaluates information in these spaces, we interviewed six players, providing them with the sample texts we collected and asking them to respond.

Selecting Text: Trade Chat

The first part of our study design, as mentioned above, involved acquiring texts for our participant players to evaluate. Based on our knowledge of the game, we drew texts from three locations, two in-game modes of communication and one near-game. Our first grouping of texts were drawn from an in-game chat channel. *World of Warcraft* features a "Trade Chat" channel that allows all players of the same faction on the same server to communicate in-game.[4] Players use this chat for a variety of needs, from looking for help and advice to selling items they have crafted or bragging about their prowess. Over a period of three weeks in November and December, we recorded conversations from trade chat, looking specifically for discussions in which players were asking for advice and receiving responses.

Trade chat is often the most immediate option players have while in game to connect with the larger community, but it is also limited. The in-game chat window only occupies a very small amount of the player's screen space. This keeps the screen clear for play, but in turn means that players can usually only post very short messages to via the chat application. A long message will take up the chat window and disappear immediately as the

next player posts to the channel. As the example below demonstrates, players typically respond to requests for advice or assistance in short phrases, with the longest responses rarely being more than a full sentence. The three sample texts were chosen based on the number of responses the initial question received, length of those responses, and the diversity of responses. The sample shown here was chosen as one example of a player requesting assistance and the responses that request received:

User 1: Which healer is best for arenas?
User 2: Depends on comp.
User 3: ^
User 4: Druids.
User 2: lol.
User 5: Shaman for shammy lock mage.
User 3: Shaman for spell cleaves, pally for melee cleaves, priests only really work for rmp/fmp.
User 6: Pally = target acquired.
User 7: Priest is for jungle cleave too.
User 3: Forgot about jungle cleave.
User 3: Druids work in a few comps, but with how poor they are right now you're better off with a shaman.

In this sample text, User 1 asks a question about a specific aspect of the game (healing in a player vs. player arena setting) and receives several responses. While we recognize that some of the responses in this sample are too short for our interviewees to respond to (User 4's simple answer: "druids"), many users respond in detail, making this discussion useful for our study.

Finally, it should be noted that the texts players post in these channels are also not saved by any in-game mechanic (although, as mentioned below, there are extra-game ways to record them). This means that, once a chat has gone off a player's screen, it is no longer visible and cannot be retrieved. Advice that players receive through in-game chat, as well as comments that they make, must be heeded immediately or they will be lost. While our study did not look at the kinds of questions that were asked in trade chat, we can imagine that this aspect of the game ecology impacts not only how players respond, but also the kinds of questions they ask in this chat system.

Selecting Text: Guild Recruitment Ads

Our second set of in-game texts was drawn from guild recruitment ads. Participation in a guild often shapes one's experience of the game, as many quests and encounters require players to collaborate in a group to succeed. Furthermore, guilds often have different goals, requirements, and atmospheres. For players, choosing a guild that matches their play style is crucially important. Guilds may advertise themselves through an in-game system that allows them to post a short description of themselves and invite others to join. These ads are slightly less ephemeral than comments posted in trade chat. Guilds can post them when they like and an ad will remain in the guild recruitment system until the guild decides to delete it. However, guilds are given a 250-character limit for their ad, forcing them to be extremely concise when persuading others to join their group. As shown in the following sample, this enforced brevity requires guilds to make decisions about which aspects they wish to highlight for others:

We teach fights, help with rotations/class. We will also help with gear. We're a friendly guild that strives to have fun. Fun > Gear type of guild. Through having all friendly/nice people we make it the best envirmment [sic] to play.

In this example, the guild has chosen to emphasize their amiability, repeatedly mentioning how fun and friendly they are. In doing so, they are forced to leave out other details players might consider important when selecting a guild. From these short and focused texts we hoped to

discover if certain attitudes or play styles would make players seem more or less knowledgeable to our group of interviewees.

Selecting Text: Forum Texts

Finally, we selected one set of near-game texts. *WoW*'s official forums provide an extra-game site where players can appeal to a much broader audience than they would find in-game (because in-game chat channels are limited by server and faction, whereas forums are open not only to all players but also to the public). As with the in-game chat client, players use the official forums for a variety of purposes; technical support, bug reports, guild ads, advice and strategy guides, and general complaining are all a part of the lively forum community. Forums chats are generally longer than the in-game chats we have discussed, since forum posts will stay on the forums page for much longer than chats will stay in the in-game window, and thus players have the opportunity to be more thorough in their responses. However, forums are also much less immediate source of help for players; they must go out of the game to post on the forums and generally must wait longer for a response. And, of course, not all of the players in the game will check the forums. Similar to our trade chat selections, we focused primarily on requests for help and their responses, looking to see how our group of players would evaluate the advice being given. Below is an abbreviated sample of a text shown to participants:

Player 1: I'm getting tired of being vote kicked for asking for 5 seconds to stop for mana in <insert any 45-50 dungeon here> in the past 48 hours, after a large pull or consecutive pulls where a lot of damage went out and could not be mitigated (which is fine, it happens) healers sometimes need mana. I ask the group to stop for a sec while I drink. When they don't, I am forced to run so far back wards to exit combat, or even at times run through the portal to port back to SW/IF so I can drink. Every time I did this, I was vote kicked. What am I supposed to do? Stand there with no mana and very little means to heal watching the group slowly die only to get vote kicked a few min later for being a "bad healer?"

Player 2: You should'nt be out of mana so close to the portal that you can still run out. Heck at that level with that gear I am surprised you need to drink at all. Pay attention to your spell usage, pay attention to your mana and dont ever run out the portal, if you lose the the tank because of oom you can always shadowmeld but i would probably boot a healer that ran out regardless of how good they are.

Player 3: Switch to resto when que pops, and go into the dungeon drinking

Player 4: Communicate with your group. As soon as you get out of combat, drink. If the tank runs to pull and dies it's their own fault not yours, not exactly rocket science.

Player 5: Set up a macro: "I am changing specs, then I am drinking water, run at the mobs at your own risk". Usually when I put that in when the dungeon pops, people wait.

Player 6: OR better yet, just be in your resto spec and drink to full mana BEFORE you queue, since the chance of your queue popping up as a healer is infinitely higher than dps. I'm not saying your kicking is necessarily justified, but there are little things you can do to mitigate that unpleasant experience.

These responses, and even the question posed, are decidedly different from those posted in trade chat. The player asking the question provides a very detailed description of the situation she finds herself in. The majority of responses to this player are two or three sentences long, and players give more detailed responses than those responding via in-game chat.

Selecting Participants and Conducting Interviews

We feel that the most valuable means of understanding how ethos operates within a specific ecology requires the selection of participants who are themselves integral actors within that ecology. When studying gaming communities, researchers must have some level of familiarity with the game and varieties of game and game play styles to effectively identify participants. In this study, we interviewed six longtime players (each with at least four years of play experience) who have all been guildmates with one or both of the researchers. However, despite the similarities in their length of play, these players engage the game (and by extension the community) in dramatically different ways. Two of our participants, Vyeris and Wepuenka, have been very casual players throughout their gaming experience and have participated primarily in more relaxed guilds. The second pair, Raszun and Pickles, have participated in a variety of different guilds but were at the time of the interviews in a hardcore PvP guild.[5] The final group, Zizzle and Tarja, have been focused on top-tier end game PvE content; that is, they are playing against the most recently released, and therefore most difficult, encounters.

We began each interview with a question about the player's own experience: Can you tell us about a time when you have made a request about game tactics or mechanics? How did other players respond to your request? In this way, we hoped to find some information about our group of players and how they searched for information or sought assistance with the game. Each player was then presented with the same three texts from trade chat, three texts from the forums and five guild ads. For chat texts and forum postings text our player-participants were asked the following questions:

Q1: Based on your experience in WoW, what do you imagine is the context of this exchange?

That is, how would you characterize the players in terms of their experience and motivations?

Q2: Which player do you think is most sincere and why? Which is least sincere?

Q3: Would you listen to the advice that is being provided by these players? Why or why not?

Q4: Which players seem most authoritative and why?

Q5: How would you respond to a similar request?

Each of these questions seeks to get at some aspect of our larger question: What makes another player seem credible in the eyes of the community? The guild ad texts, which feature players actively trying to recruit other players rather than players asking for and providing advice, warranted their own set of questions. Nonetheless, these questions are similar in their aim of helping to construct a user-created definition of credibility:

Q1: Based on this ad, what is your impression of what this guild is like?

Q2: Based on the ad text, what are the overall goals of the guild?

Q3: Do you think they can achieve them?

Q4: Based on the ad text, do they sound knowledgeable?

Q5: Would you consider joining if you met the requirements?

Technology Notes

This study was made possible through the use of several programs that made our data collection and interview process proceed smoothly. In order to collect our in-game chat data, we used a player-created game modification called WoWScribe. This mod allows players to record in-game chats to a file that is then saved to their computer. Because *WoW* is a lively game with a community of active players, it is often difficult to capture chat manually, and once a chat has gone off a player's screen she cannot retrieve it (there

is no built-in log). The players we interviewed are spread throughout the country, and so to conduct interviews we contacted them through Skype and recorded our discussion using a Skype add-on called MP3 Skype Recorder.

ANALYSIS AND DISCUSSION

After conducting our interviews we examined the players' responses and looked for common themes. While we asked players to evaluate which of the responses in the texts were most authoritative, to establish a definition of credibility in WoW, we were not as concerned with *which* texts players found to be more authoritative, but rather their means of identifying authority. Across the six interviews, three attributes were repeatedly mentioned as signs of authority: specificity, extra-game research, and experience.

Specificity

Specificity was mentioned more often than any other attribute, and mentioned by all players in response to the various texts. Tarja's response to one of the forum threads is very typical of responses players gave when asked which responder seems most authoritative:

Player 5 gives a really good answer because they're very specific, about what to do and how to do it. They even tell you buttons to push instead of just strafe to the side while casting. Player 5 is pretty authoritative, I would listen to player 5.

Our participants responded positively to answers that mention specifics, viewing specificity as a sign of greater knowledge about the game. In this instance, Tarja identified a player's response as the most authoritative because the respondent gave a very exact description of how a problem should be solved and what buttons one would need to push. In other words, this player's response covered both mechanics (how the game architecture privileges one way of acting over another) and tactics (how players should act in game space based on those mechanics). She did not consider the other players' responses (players who provided the same general discussion of mechanics but did not address the issue of tactics) to be as authoritative.

However, while the responses our interviewed players gave seem to suggest that specificity is important, across players there was some disagreement both on the level of specificity and on how to appropriately respond. In one instance two players disagreed on which response was most authoritative, but they both mentioned specificity as the reason the response they choose was better. Players were shown the following dialogue:

User 1: What spec is good for dk in pvp?
User 2: Whatever you feel like, Frost is easiest, Blood is difficult to kill, and Unholy has most damage output.
User 3: No hell no Unholy is insane at that lvl ill own someone the same level or higher lvl the time. It dont matter just DoT and run then DG and hit Scourge strike and constantly cast Death coil. then repeat.

Vyeris, one of our casual players, selected User 3's response as the most authoritative. He said:

User 3's more specific, he has a definite opinion on a particular spec. He's giving a specific answer whereas User 2 is kinda a generalized, generic answer. Not really a specific answer to what the question was. User 3 is more authoritative because he gives specific skills and a specific rotation for doing damage.

In this segment Vyeris responded similarly to Tarja, highlighting the identification of a certain play style (mentioning the order that spells should be hit in this case, as opposed to buttons in the last example). In this case, Vyeris is responding

to a discussion of tactics. User 3 gives a specific rotation, a set of abilities to play and the order to play them in. Vyeris felt that the second answer is far too broad to really be meaningful or helpful. However, while Vyeris dismissed the answer provided by User 2, Raszun identified this answer as the more useful and authoritative one.

User 2 is giving a smaller but still detailed answer. User 2's advice is a little more helpful. It seems like 3 said unholy is the spec that he would do but he doesn't give details, he's like you do this, but that might not be the way that you want to play... I would go with 2 over 3 because 2 is giving information on the other specs.

Raszun, like Vyeris, expressed the value he places in answers that provide specific information. However, unlike Vyeris, he rejected User 3's answer, suggesting that the tactics might not be the best strategy, or even the way a particular player wants to play. In this case a discussion of tactics without any mention of mechanics, essentially a discussion of how to play with no suggestion of why to play, appeared to Raszun as lacking ethos. Thus, he selected the answer that might at first seem broader as the one with the most useful detail.

Research

While specificity was mentioned most often and clearly seems to be the best indicator players have for identifying credible sources, research was also mentioned as a positive sign of authority and knowledge. When asked about their personal experiences in asking for help, many players responded that they typically did their own research first, and then asked questions. Players did not seem to see outside research as the only means of developing knowledge, nor were outside sources granted automatic authority. Instead, research seemed to be one means of establishing a consensus, with one's personal experience and feedback from other players as equally important steps. This seems to suggest that players valued Lankes's (2008) suggestion that credibility is shifting from an authority model, where users automatically trust third party authorities, to a reliability model, where users look for consensus among many sources. Vyeris responded that before asking other players "I'll try to go and research it, using tools like wowhead.com and *elitist jerks*." While he values the information found at these extra-game sites, Vyeris still returns to the game to balance this information against both his own experience and the opinions of other players.

One of the websites Vyeris mentioned to us, wowhead.com, was also mentioned in one of our trade chat texts:

User 1: Is there a cheese vendor in orgrimmar?
User 2: I'm not even gonna go there.
User 3: Punch a tauren and maybe he'll tell u.
User 1: Seriously I need Fine Aged Cheddar for a quest.
User 4: Wowhead.com bro.

All of our players responded that in this set of responses User 4 gave the best answer, but several players had some reservations about calling this response authoritative. Pickles went so far as to say that none of the responses seemed authoritative to him: "User 4 just gives him a link so he can go find it himself and gain some independence. I wouldn't say any of them are authoritative, but the best is probably User 4 because he gives the guy the link so he can learn to use the system." When pressed, Pickles explained that the reason he felt User 4 lacked authority, despite providing a good solution, was the lack of details User 4 gave. Simply mentioning the outside site (relying on an authority model of ethos), though a good reference tool, is not enough for other players to perceive a sense of ethos in User 4.

Vyeris also selected User 4 as the most authoritative of the group, but noted that overall he would not call this an authoritative response. When asked how he would respond, he replied, "If I knew the answer I would give them the answer," but added that he would also tell the player

asking the question where she could go to find the answer for herself, citing the old adage about giving a person a fish or teaching the person to fish. In this case, he would do both, once more suggesting that the players in this survey value reliability over multiple sources instead of the authority of a single source.

Finally, it should be mentioned that one of our players, Wepuenka, questioned the motives of the original poster for even asking for such easily accessed information. "Anybody who's played *WoW* for any amount of time knows that there's resources you can go to," she says. "You don't have to ask in trade chat." Zizzle was also suspicious of the whole exchange, characterizing all players in this instance (including the asker of the question) as trolling, saying ridiculous or insulting things just to get a reaction out of other players. Whether legitimate or not, the venue User 1 chose to ask the question (trade chat, sometimes known for trolling) may cause other players to doubt the sincerity. This is an example of the way the designed communication avenue, in concert with the interaction of environment and inhabitant, functions ecologically to shape the expectations and responses of users.

Experience

The last indicator of authority that was mentioned at some point by all participants was personal experience, although the value of experience was more ambiguous than either the importance of specificity or extra-game research. Most players responded quite positively when they assumed that a speaker had been playing the game for a number of years, but some also wrote off answers as being simply one's experience. This echoes Chen's (2009) observations of a raiding group, which culminated in his realization that becoming an expert player required more than simply playing the game well. Chen (2009) argued that expert players must also "move in various social circles and communicate effectively...

use third-party tools and other resources that have been taken up by expert players."

The following text, taken from a forum post, had several responses that suggested the players had been part of the game for a number of years:

Player 1: Why warriors? Why are we the only melee class getting nerf?

Player 2: Because you can live without being the unquestioned #1 DPS at the end of this expansion. You've already been so with TBC and Wrath, so be happy.

Player 3: Legend has it that at the end of Vanilla, well geared warriors were so OP people brought them to raids to DPS! As arms! That alone is OP as holy hell.

Player 4: After the buffs, Ret is going to more than likely be the #1 melee DPS spec outside of rogues with Legendaries. Nothing wrong with Arms being competitive, but Fury would have been over Legendary rogues in T13 without the slight nerf.

Player 5: Our damage is being lowered because we're doing nearly the same amount of damage as other melee classes. Also, because warriors have shined in past expansions (even though most of the players who were around to see this probably don't even play anymore), the class needs to be punished for the game developers mistakes apparently

Player 6: To be fair, mechanics were very, very different at the end of Vanilla than they are now.

For this forum text, Tarja indicated it was difficult to determine a most authoritative answer, since the question is really more of a complaint than a request for advice. However, when pressed, she selected Player 5, because that player "seems like they've probably played for a while, they mentioned past expansions and that kinda helps." Vyeris also selected Player 5 as the best answer: "I would say 5 is giving a fairly comprehensive answer, he's an experienced player as well. He

talks about how the warrior class has fared in other expansions." Vyeris and Tarja identified this player's experience as a sense of game history. Because Player 5 can discuss how the game has been played in the past, they both viewed his answer on the current state of the game as the best response to a question about current game mechanics. However, when experience is viewed more as subjective play style and not as historical knowledge of mechanics, it is not valued nearly as much.

In a separate discussion, relying on experience was seen to be ineffective by our players. Both Pickles and Raszun, our two PvP players, were critical of User 3's response in the chat discussion of death knights (quoted above). Pickles explained "I would say that's just his experience, he's just answering in the knowledge he has with unholy. He's just explaining his own play style." Pickles repeated the word "just" three times in as many sentences, emphasizing that because User 3 seemed to only have experience with one type of play, the answer he gave cannot be considered authoritative. Raszun's response explained further why one might not want to only listen to another player's experience: "that might not be the way that you want to play." Both players are hesitant to listen to advice that seems grounded in a narrow experience. In this case, User 3's advice, though grounded in experience, appears to our players to be divorced from a thorough understanding of game mechanics or tactics.

CONCLUSION

In this project, we have theorized that the development and demonstration of ethos within a multiplayer game is contingent upon the specific ecology in which it is deployed and that understanding ethos in such a context requires a deep investigation of the interactions between the designed spaces and narratives of the game and the actions of the users, who see themselves as inhabitants of the game's ecosystem. If we extend the notion of field-dependency for the evaluation of ethos to include a full ecological profile, we need to also identify the key interactions and perceptions of the game's inhabitants; thus, our pilot study here focused on how players (as inhabitants of the specific ecosystem under study) responded to ethos as represented in in-game and near-game communications. We can see from our analysis that our participants used both the original Aristotelian approach to determining ethos from the quality of the argument itself (in terms of specificity), but they also clearly value assessing the ethos of the speakers based on their demonstrated knowledge and expertise (which comes from a commitment to playing the game and to understanding both its mechanics and strategies).

In this study, we have begun to sketch out a theory of ethos that resides not just in the rhetor or in the text, but in the complex interactions of rhetor, text, and designed (and inhabited) environment. There is a concern that our findings may be too specific to the environment or ecology we have been studying—will the results be different for different gaming environments? But we believe that our overall theory of interaction-based ethos should work in a number of digital spaces—particularly in cases where players or users cannot rely on knowledge of the reputation of the speaker to make judgments about reputation (such as massive open online courses).

Our next steps will extend this project by drawing in more participants both from *World of Warcraft* and other multiplayer games (from a variety of genres) and examining a wider range of in-game and near-game communications. We believe that both our theoretical framework and methodology represent effective approaches to the rhetorical study of games (and other virtual environments). While the overviews of the game environment, mechanics, strategies, narratives, and design presented here are necessarily limited, a full treatment of the ecology of several multiplayer games would be a logical next step

in the development of our theories and methods. There are certainly research questions for which this approach may be valuable, including investigations of game culture and game ecology (which is especially important as more and more communication venues adopt game-like elements), and we hope that we have provided a model that other researchers may draw upon for the exploration and investigation of games as rich contexts for rhetorical communication and activity.

REFERENCES

Blythe, S. (2007). Agencies, ecologies, and the mundane artifacts in our midst. In Takayoshi, P., & Sullivan, P. (Eds.), *Labor, writing technologies, and the shaping of composition in the academy* (pp. 167–186). Cresskill, NJ: Hampton Press.

Bogost, I. (2007). *Persuasive games: The expressive power of videogames*. Cambridge, MA: MIT Press.

Bramwell, A. (1989). *Ecology in the 20th century: A history*. New Haven, CT: Yale University Press.

Brent, D. (1997). Rhetorics of the web: Implications for teachers of literacy. *Kairos: A Journal of Rhetoric, Technology, and Pedagogy, 2*(1). Retrieved from http://kairos.technorhetoric.net/2.1/binder.html?features/brent/wayin.htm

Brewer, R. *The science of ecology*. Philadelphia, PA: Saunders College Publishing.

Chen, M. (2009). *Social dimensions of expertise in World of Warcraft players. Transformative Works and Cultures, 2*. Retrieved from http://dx.doi.org/10.3983/twc.2009.0072.

Cooper, M. (1986). The ecology of writing. *College English, 48*(4), 364–375. doi:10.2307/377264

Crowley, S., & Hawhee, D. (2009). *Ancient rhetorics for contemporary students* (4th ed.). New York, NY: Pearson/Longman.

Fogg, B. J. (2002). *Prominence-interpretation theory: Explaining how people assess credibility*. A Research Report by the Stanford Persuasive Technology Lab. Retrieved from http://www.captology.stanford.edu/PIT.html

Grover, M. (2002). Analyzing the rhetoric of websites. *Kairos: A Journal of Rhetoric, Technology, and Pedagogy, 7*(2). Retrieved from: http://kairos.technorhetoric.net/7.2/binder.html?sectiontwo/grover/

Halloran, S. M. (1982). Aristotle's concept of ethos, or if not his, somebody else's. *Rhetoric Review, 1*(1), 58–63. doi:10.1080/07350198209359037

Heba, G. (1997). HyperRhetoric: Multimedia, literacy, and the future of composition. *Computers and Composition, 14*(1), 19–44. doi:10.1016/S8755-4615(97)90036-0

Holisky, A. (19 Feb 2012). World of Warcraft subscriber numbers dip 100,000 to 10.2 million. *WOW Insider*. Retrieved from http://wow.joystiq.com/2012/02/09/world-of-warcraft-subscriber-numbers/

LaGrandeur, K. (2003). Digital images and classical persuasion. In Hocks, M. E., & Kendrick, M. R. (Eds.), *Eloquent images: Word and image in the age of new media* (pp. 117–136). Cambridge, MA: MIT Press.

Lankes, D. R. (2008). Credibility on the Internet: Shifting from authority to reliability. *The Journal of Documentation, 64*(5), 667–686. doi:10.1108/00220410810899709

Nardi, B. (2010). *My life as a Night Elf priest*. Ann Arbor, MI: University of Michigan Press.

Nardi, B. A., & O'Day, V. L. (1999). *Information ecologies: Using technology with heart*. Cambridge, MA: MIT Press.

Ridolfo, J. (2006). (C)omprehensive (O)nline (D)ocument (E)valuation. *Kairos: A Journal of Rhetoric, Technology, and Pedagogy 10*(2). Retrieved from http://kairos.technorhetoric.net/10.2/binder.html?praxis/ridolfo/index.html

Selfe, C. L., & Hawisher, G. (2007). *Gaming lives in the twenty-first century: Literate connections*. New York, NY: Palgrave Macmillan. doi:10.1057/9780230601765

Sicart, M. (2008). Defining game mechanics. *Game Studies, 8*(2). Retrieved from http://gamestudies.org/0802/articles/sicart

Spinuzzi, C. (2003). *Tracing genres through organizations: A sociocultural approach to information design*. Cambridge, MA: MIT Press.

Spinuzzi, C., & Zachry, M. (2000). Genre ecologies: an open system approach to understanding and constructing documentation. *Journal of Computer Documentation, 24*(3), 169–181. doi:10.1145/344599.344646

Squire, K. (2006). From content to context: Videogames as designed experiences. *Educational Researcher, 35*(8), 19–29. doi:10.3102/0013189X035008019

Syverson, M. (1999). *The wealth of reality: An ecology of composition*. Carbondale, IL: Southern Illinois University Press.

Toulmin, S. (1958). *The uses of argument*. Cambridge, UK: Cambridge UP.

Warnick, B. (2007). *Rhetoric online: Persuasion and politics on the World Wide Web*. New York, NY: Peter Lang.

WoW Wiki. (2012). *Death knight PvE guide*. Retrieved from http://www.wowwiki.com/Death_knight_PvE_guide

WoWWiki. (2012). *Death knight PvP guide.* Retrieved from http://www.wowwiki.com/Death_knight_PvP_guide

Yee, N. (2005). A model of player motivations. *The Daedalus Project.* Retrieved from http://www.nickyee.com/daedalus/archives/001298.php

ADDITIONAL READING

Anthropy, A. (2012). *Rise of the video game zinesters: How freaks, normals, amateurs, artists, dreamers, drop-outs, queers, housewives, and people like you are taking back an art form.* New York, NY: Seven Stories Press.

Bainbridge, W. S. (2010). *The Warcraft civilization: Social science in a virtual world.* Cambridge, MA: MIT Press.

Bogost, I. (2007). *Persuasive games: The expressive power of video games.* Cambridge, MA: MIT Press.

Bogost, I. (2011). *How to do things with video games.* Minneapolis, MN: U of Minnesota Press.

Buckingham, D., & Burn, A. (2007). Game literacy in theory and practice. *Journal of Educational Multimedia and Hypermedia, 16*(3), 323–349.

Carr, D., & Oliver, M. (2009). Tanks, chauffeurs, and backseat drivers: Competence in MMORPGs. *Eludamos, 3*(1), 43-53. Retrieved from http://www.eludamos.org/index.php/eludamos/article/view/vol3no1-6/107

Castronova, E. (2006). *Synthetic worlds: The business and culture of online games.* University of Chicago Press.

Chen, C., Sun, C., & Hsieah, J. (2008). Player guild dynamics and evolution in massively multiplayer online games. *Cyberpsychology & Behavior, 11*(3), 293–301. doi:10.1089/cpb.2007.0066

Chen, M. (2009). Communication, coordination, and camaraderie in World of Warcraft. *Games and Culture, 4,* 331–339.

Consalvo, M. (2006). Game analysis: Developing a methodological toolkit for the qualitative study of games. *Game Studies, 6*(1). Retrieved from http://gamestudies.org/0601/articles/consalvo_dutton

Consalvo, M. (2012). Gaining advantage: How video game players define and negotiate cheating. In Zagal, J. P. (Ed.), *The video game ethics reader* (pp. 85–102). San Diego, CA: Cognella.

Corneliussen, H., & Rettberg, J. W. (2008). *Digital culture, play, and identity, a World of Warcraft reader.* Cambridge, MA: MIT Press.

Cuddy, L., & Nordlinger, J. (2009). *World of Warcraft and philosophy: Wrath of the philosopher king.* Chicago, IL: Open Court.

Downing, S. (2010). Online gaming and the social construction of virtual victimization. *Eludamos, 4*(2), 287-301. Retrieved from http://www.eludamos.org/index.php/eludamos/article/view/vol4no2-11/190

Ee, A., & Cho, H. (2012). What makes an MMORPG leader? A social cognitive theory-based approach to understanding the formation of leadership capabilities in massively multiplayer online role-playing games. *Eludamos, 6*(1), 15-24. Retrieved from http://www.eludamos.org/index.php/eludamos/article/view/vol6no1-4/6-1-4-html

Elverdam, C., & Aarseth, E. (2007). Game classification and game design: Construction through critical analysis. *Games and Culture, 2*(1), 3–22. doi:10.1177/1555412006286892

Gee, J. P. (2003/2007). *What video games have to teach us about learning and literacy* (rev. ed.). New York, NY: Palgrave Macmillan. doi:10.1145/950566.950595

Genette, G. (1997). *Paratexts: Thresholds of inter-pretation.* Cambridge, MA: Cambridge University Press. doi:10.1017/CBO9780511549373

Holmevik, J. (2012). *Inter/vention: Free play in the age of electracy.* Cambridge, MA: MIT Press.

Huizinga, J. (1950). *Homo ludens: A study of the play element in culture.* Boston, MA: Beacon Press.

Jenkins, H. (2006). *Convergence culture: Where old and new media collide.* New York, NY: New York University Press.

Jørgensen, K. (2008). Audio and gameplay: An analysis of PvP battlegrounds in World of Warcraft. *Game Studies, 8*(2). Retrieved from http://gamestudies.org/0802/articles/jorgensen

Juul, J. (2011). *Half-real: Video games between real rules and fictional worlds.* Cambridge, MA: MIT Press.

Kallio, K. P., Mayra, F., & Kaipainen, K. (2011). At least nine ways to play: Approaching gamer mentalities. *Games and Culture, 6*(4), 327–353. doi:10.1177/1555412010391089

Kendall, A., & McDougall, J. (2009). Just gaming: On being differently literate. *Eludamos, 3*(2), 245-260. Retrieved from http://www.eludamos.org/index.php/eludamos/article/view/vol3no2-8/135

Klastrup, L., & Tosca, S. (2009). "Because it just looks cool!" Fashion as character performance: The case of WoW. *Journal of Virtual Worlds Research, 1*(3), 4–17.

Lankoski, P. (2011). Player character engagement in computer games. *Games and Culture, 6*(4), 291–311. doi:10.1177/1555412010391088

Lehdonvirta, M., Nagashima, Y., Lehdonvirta, V., & Baba, A. (2012). The stoic male: How avatar gender affects help-seeking behavior in an online game. *Games and Culture, 7*(1), 29–47. doi:10.1177/1555412012440307

Maaninen, T. (2003). Interaction forms and communicative actions in games. *Game Studies, 3*(1). Retrieved from http://www.gamestudies.org/0301/manninen/

Meadows, M. S. (2007). *I avatar: The culture and consequences of having a second life.* New Riders.

Moberly, K. (2010). Commodifying scarcity: Society, struggle, and spectacle in World of Warcraft. *Eludamos, 4*(2), 197-213. Retrieved from http://www.eludamos.org/index.php/eludamos/article/view/vol4no2-7/179

Moeller, R. M., Esplin, B., & Conway, S. (2009). Cheesers, pullers, and glitters: The rhetoric of sportsmanship and the discourse of online sports gamers. *Game Studies, 9*(2). Retrieved from http://gamestudies.org/0902/articles/moeller_esplin_conway

Nakamura, L. (2009). Don't hate the player, hate the game: The racialization of labor in World of Warcraft. *Critical Studies in Media Communication, 26*(2), 128–144. doi:10.1080/15295030902860252

Nardi, B. A. (2010). *My life as a night elf priest, an anthropological account of World of Warcraft.* Ann Arbor, MI: University of Michigan Press.

Paul, C. A. (2011). Optimizing play: How theorycraft changes gameplay and design. *Game Studies, 11*(2). Retrieved from http://gamestudies.org/1102/articles/paul

Pearce, C., Boellstorff, T., & Nardi, B. (2011). *Communities of play: Emergent cultures in multiplayer games and virtual worlds*. Cambridge, MA: MIT Press.

Salen, K., & Zimmerman, E. (2004). *Rules of play: Game design fundamentals*. Cambridge, MA: MIT Press.

Salen, K., & Zimmerman, E. (2006). *The game design reader: A rules of play anthology*. Cambridge, MA: MIT Press.

Schmieder, C. (2009). World of maskcraft vs. World of queercraft? Communication, sex and gender in the online role-playing game World of Warcraft. *Journal of Gaming & Virtual Worlds*, *1*(1), 5–21. doi:10.1386/jgvw.1.1.5_1

Schulzke, M. (2011). How games support associational life: Using Tocqueville to understand the connection. *Games and Culture*, *6*(4), 354–372. doi:10.1177/1555412010391090

Smith, J. H. (2007). Tragedies of the ludic commons: Understanding cooperation in multiplayer games. *Game Studies*, *7*(1). Retrieved from http://gamestudies.org/0701/articles/smith

Steinkuehler, C. (2006). Massively multiplayer online video gaming as participation in a discourse. *Mind, Culture, and Activity*, *13*(1), 38–52. doi:10.1207/s15327884mca1301_4

Taylor, T. L. (2009). *Play between worlds: Exploring online game culture*. Cambridge, MA: MIT Press.

Voorhees, G. (2009). The character of difference: procedurally, rhetoric, and roleplaying games. *Game Studies*, *9*(2). Retrieved from http://gamestudies.org/0902/articles/voorhees

Waggoner, Z. (2009). *My avatar, my self: Identity in video role-playing games*. Jefferson, NC: McFarland.

Williams, D., Duecheneaut, N., Xiong, L., Zhang, Y., Yee, N., & Nickell, E. (2006). From tree house to barracks: The social life of guilds in World of Warcraft. *Games and Culture*, *1*(4), 338–361. doi:10.1177/1555412006292616

Yee, N. (2009). Befriending ogres and wood-elves: relationship formation and the social architecture of Norrath. *Game Studies*, *9*(1). Retrieved from http://gamestudies.org/0901/articles/yee

Zagal, J. P. (2010). *Ludoliteracy: Defining, understanding, and supporting games education*. Pittsburgh, PA: ETC Press.

Zagal, J. P. (Ed.). (2012). *The video game ethics reader*. San Diego, CA: Cognella.

KEYWORDS AND DEFINITIONS

Game Mechanics: Essentially, game "rules" – the underlying algorithms and procedures that determine how player actions are carried out in the game. Game mechanics describe how the game works (as opposed to how the player works within the game environment, about which see "game strategies").

Game Strategies: Tactics used by the player to respond to a specific task or encounter – these strategies may be learned through experience, advice from other players, online information sources, or developed through a close study of game mechanics.

Guilds and Guildies: Guilds are in-game, player-run communities. Players may only join one guild, and they must be invited by a guild member to join. The character of a guild varies greatly based its members and their goals in the game.

Information Ecology: "…a system of people, practices, values, and technologies in a particular local environment. In information ecologies, the spotlight is not on technology, but on human activities that are served by technology" (Nardi & O'Day, 1999).

In-Game, Near-Game, and Extra-Game Texts: A variety of texts contribute to understanding communication and games. In-game texts refer to communications that take place within the gameworld. Near-game texts refer to communications that do not take place within the game world, but are still part of the officially sanctioned game environment. Finally, extra-game texts refers to player-created resources that are not officially connected with the game, but nonetheless provide important information and resources to players.

Massive Multiplayer Online Role-Playing Game (MMORPG): A genre of videogame that requires players to take on the role of characters within a fictional setting or narrative that takes place in a persistent virtual world. The number of players who participate in an MMORPG ranges from several thousand to several million.

Non-Player Character (NPC): An in-game character that has been programmed into the game; NPCs usually serve to provide information or goods to players (also referred to as "player characters").

Rhetor: Any agent engaging in communicative action that is intended to be persuasive.

World of Warcraft: An MMORPG released in 2006 by Blizzard Entertainment. *WoW*, as it is affectionately called by players, has a monthly subscription fee, and as of December 2011 reported player subscriptions at 10.2 million worldwide.

ENDNOTES

[1] In 1866, zoologist Enrst Haeckel coined the term "ecology" to define an area of biology that aimed to study of interrelationships between organisms and the environment (Bramwell, 1989).

[2] Individually, we have encountered "studies" of *World of Warcraft* that were deeply flawed – and in most cases, it was clear that the researcher spent little time actually playing the game, yet professed to understand it well enough to analyze and critique. Our approach is aligned with (and inspired by) the more ethnographic methods used by Bonnie Nardi's (2010) *My Life as a Night Elf Priest*.

[3] We are aware of Sicart's (2008) proposed definition of mechanics "as methods invoked by agents for interacting with the game state" (para. 6) but find that it is too broad for our purposes, which require a distinction between the underlying algorithm and the action performed; however, we do agree with his stance that the mechanic provides the opportunity for agency for both players and programmed non-player characters.

[4] There is also a "General Chat" channel, but most players use Trade as their default channel.

[5] The four players in this survey that we have labeled hardcore were hesitant to label themselves as such. There were some indications that hardcore players no longer play the game to have fun, or focus on the game to the exclusion to other activities. For the purposes of this study, we are using hardcore to suggest that these players play the game at a high skill level.

Chapter 20
Documentary at Play

Inge Ejbye Sørensen
University of Copenhagen, Denmark

Anne Mette Thorhauge
University of Copenhagen, Denmark

ABSTRACT

Docu-games designate a versatile group of games that have in common an attempt to depict and reflect on aspects of reality such as military conflicts, historical periods, or contemporary political and socio-cultural issues. As such, docu-games have become a new communication tool for individuals or organizations. This chapter explores different perspectives on games as documentaries, going beyond the mere subject matter and visualization of docu-games to approach questions about simulations as statements about reality and gameplay as a tool for communicating statements about reality. Combining cognitive documentary and games theory with content analysis, the chapter offers a theoretical framework for understanding how docu-games reference the relationship between reality and game, as well as how they establish credibility in relation to these representations.

INTRODUCTION

Videogames are increasingly used as strategic communication tools that offer new ways of representing subject matter and depicting real-life situations. However, they differ in profound ways from more linear types of communication by integrating audiences in interactive modes of experience. This calls for a new understanding of the communication that takes place and new concepts for assessing the credibility of videogames as statements about reality.

In this context, docu-games are particularly interesting. Docu-games are a diverse group of games, but what they have in common is an attempt to depict and reflect on aspects of reality such as military conflict, historical periods, or contemporary political and socio-cultural issues. As such, docu-games have become a new tool for individuals or organizations to communicate their agendas, issues, and interests. However, in order to understand docu-games, it is important to explore the ways in which games can be understood as documentary and how they can make statements about reality.

This chapter will explore different perspectives on games as documentaries, going beyond the mere subject matter and visualisation of docu-games toward questions about simulations as conveyors of statements about reality (Bogost, 2007; Nieborg,

DOI: 10.4018/978-1-4666-2663-8.ch020

2004, September; Sisler, 2009). By applying current documentary theory and game studies to docu-games, we hope this chapter transcends and adds to current thinking about how games relate to notions of credibility, reality, and real life, and how they reference the relationship between the real world and games.

BETWEEN REPRESENTATION AND REALITY

The emergence of new hybrid and interactive forms of documentary such as newsgames, persuasive games, crowd-sourced documentaries, database documentaries and docu-games has rekindled the debates about the relationship between fact and fiction and representation and reality, debates which have raged since the birth of documentary film. Recently, and in following this tradition, critics have questioned the ways, if at all, docu-games can be seen as being able to refer to—and make credible statements about—reality.

In "Reality Play: Documentary Computer Games Beyond Fact and Fiction," Joost Raessens held the term "documentary game" up to critical review. Based on analyzing gaming experiences from playing these games and the statements they make about reality, Raessens (2006) concluded that documentary games occupy a space in between objectivity and subjectivity and therefore neither represent reality objectively nor exist as mere subjective renderings of the game designers. Inspired by Michael Renov's (2004, 2009) writings about the autobiographical documentary, Sanchez-Laws (2010) explored whether the immersive potential of digital storytelling might lead to a new documentary form, albeit one that depicts a first-person perspective on reality and therefore an ultimately subjective version of the truth. Similarly, comparing docu-games to other kinds of real-life simulations (e.g., the forensic modelling that is now accepted in courts in the

United States), Fullerton (2008) conceded that documentary as simulation holds the promise of accurate depictions of reality, but is sceptical toward docu-games as sources of information about the real world at this point in time. Again drawing on Renov, Fullerton argued that the involvement of the viewers leads to a subjective point of view, which undermines objectivity and creates an uncertain reference to reality. For these authors, docu-games' reference to reality is either indeterminate or, at best, subjective. However, Renov (2004, 2009) focused on documentary as the expressive strategies of a subject and to a lesser extent on the ways in which documentary texts position themselves vis-à-vis the reality they describe. Therefore, this chapter will instead combine game studies with the rhetoric and cognitive documentary theories of Carl Plantinga (1997, 2005), Ib Bondebjerg (2002, 2008, forthcoming 2012) and Paul Ward (2005, 2009) and make the case that docu-games provide a credible, useful and effective way of representing and reflecting on reality. From this perspective, docu-games and their relationship to reality can be understood in the same way as the documentary subgenres (the authoritative, the dramatized documentary, observational documentary and poetic reflective documentary) reference reality.

Following a brief overview of the relationship between documentary films and reality through the history of this genre, this chapter will outline current cognitive documentary film theory, as well as chart the relationship between audiovisual representation, narrative contextualization, and procedural rhetorics crucial to the understanding of the communicative function of computer games in contemporary game studies. These theoretical approaches will then be combined to inform the analysis of docu-games, most prominently KUMA Wars: Afghan Air Strikes and Global Conflicts: Afghanistan and in this way argue that docu-games can make credible and valid statements about reality.

THE ELEMENT OF DOCUMENTARY IN DOCUMENTARIES

Being programed systems and simulations, games are at the outset artificialities, constructions, and re-enactments, quite the opposite of the direct "imprint" of reality that has been the common-sense conception of the documentary. However, documentaries cannot be reduced to this narrow definition. Although, as we will explore below, the privileged position that reality occupies in documentaries has been explained in many different ways, documentary critics generally agree that documentary films have reality as their core subject matter. Furthermore, most documentary theorists have moved away from the perception that documentary has an indexical relationship to reality and most understand the relationship to reality in documentary films as contextual, complex and—some argue, contractual.

The commonplace understanding of documentary films is that they depict real life and that there is a one-to-one relationship between the content of the film and reality in the same way as there is in André Bazin's indexical relationship between the photographic image and reality (Gray, 1960). Indexicality in this context implies a direct link between image and reality similar to that between fire and smoke or foot and footprint. However, most documentary critics, theorists and, not least, documentary makers themselves understand this relationship as a more complex negotiation. Documentary scholar Stella Bruzzi (2000), for example, pointed out that indexicality renders virtually every documentary project an impossibility (p. 4), while Patricia Aufderheide argued that "there is no way of making a film without manipulating information" (p. 2).

From the very beginning of documentary history, freely interpreted enactments and re-constructions loosely based on real events have routinely been used in documentary films as aesthetic and narrative tools. Describing the selection of scenes, thematic, and narrative structure of

one of the very earliest documentaries, Auguste Lumière's Le Dejourner de Bébé (1895, France), Kevin Macdonald and Mark Cousins observed (1998) "[E]ven in the simplest of non-fiction films, the relationship between film and reality is not a straightforward or literal one, but one of metaphor" (p. 5). The earliest recorded incident of reconstruction or fakery took place as early as 1897 when Albert E Smith and J Stuart Blackton convincingly reconstructed the battle of Santiago Bay in a water-filled tub with cardboard ships and cigar smoke to provide footage of this key battle in the Spanish-American war (Smith & Blackton, 1922), and the famous ethnographic documentary Nanook of the North (1922, UK) was scripted by its maker, Robert Flaherty, who had Inuits enact activities that no longer took place in order to tell the story of their history, lives, and survival. In the middle of the 20th century, the Scottish documentary maker John Grierson pioneered and defined the documentary genre and is now often seen as the originator of, in particular, the authoritative documentary. However, for Grierson the relationship between reality and documentary films was always complex and creative. A prolific documenter of his own film-making practice both in writing and on film, Grierson saw his films precisely not as documents but as documentaries, which he described as "the creative interpretation of actuality" (Grierson, 1933; Ward, 2005). In the 1990s, documentary theory gained ground as an academic research area. Scholars like John Corner identified certain modalities that characterized different types of documentary films, such as the authoritative journalistic voice-over and interview that dominate the current affairs documentary (Corner, 1995, 1996, 1997, 2009), or the observational fly-on-the-wall footage in the observational documentary. In defining documentary film itself as well as its genres, Bill Nichols asserted (1994, 2009) that documentary is always ideologically inflected and thus breaks with the more innocent notions of documentary's relationship between representations and reality. The works of Corner and Nichols have been hugely influential in defining documentary and non-fiction film, as well their subgenres, positioning this area as one worthy of scholarly interest and establishing its principle traits, aesthetics and characteristics.

CREDIBILITY IN DOCU-GAMES

It is especially fruitful, however, to explore docu-games through the optics of the documentary and non-fiction film theories of Carl Plantinga (1997, 2005), Ib Bondebjerg (2002, 2008, forthcoming 2013) and Paul Ward (2005, 2009). Building on Corner and Nichols and rooting themselves in David Bordwell and Kirstin Thompson's neoformalist film theories, these theorists use a cognitive approach to documentary films. This cognitive approach is based on two innate interpretational frameworks, either the referential/assertive or the fictitious/playful, that the audience brings to the film—a framework brought forth by the stylistic and aesthetic elements, as well as the contractual relationship established in the meeting with the audience. Plantinga, with reference to speech act theory, anchors the relationship between reality and representation in the rhetorical and communicative function and the context in which this takes place. The bond between film and reality is established by the film's intention and rests as a contractual relationship between the documentary producer and its audience or receivers. Ib Bondebjerg expanded on Plantinga's theory and links rhetorical functions and forms of address to various epistemological approaches to reality. Key to the documentary theories of both Plantinga and Bondebjerg is the idea that the relationship between representation and reality is not a priori or an inherent property of the aesthetics, form, and style of documentary films; instead, the relationship between representation and reality stems from the context, the shared perception of the reality described, and the interaction between the documentary maker, its receiver, and the film.

Understanding documentary as constituted by its contextual and contractual relationships allowed these critics to free documentary film from being defined by its forms. Based on Nichols's typology, Plantinga (1997, 2005) operated with three documentary genres1. Each genre is defined by what Plantinga called "voice": the formal, the open and the poetic-reflexive. Voice is an expression of the underlying point of view of the filmmaker and is constituted by the use, organization, and orchestration of different stylistic, aesthetic, narrative, and rhetorical forms and tropes. For Plantinga, aesthetic forms and tropes are thus the building blocks that the overall genre or "voice" rests on. These forms do not constitute any reference to reality in themselves, but they are the elements which enable the rhetorical function that links documentary representation to reality. Building on Plantinga's documentary typology, Bondebjerg constitutes the dramatized documentary as a fourth documentary genre with its own aesthetics and reference to reality. Working in this same area, in his book Documentary: the Margins of Reality (2005), as well as his chapter "Drama-documentary: "The 'Flight 93' Films" (2009), Paul Ward explored the relationship between drama and documentary. Like Plantinga, neither Bondebjerg nor Ward disputed that certain forms, styles, aesthetics, and tropes are part of established conventions and the building blocks of documentary film, nor do they refute that the reference to reality is more central to the argument of certain documentary sub-genres such as the authoritative current affairs documentary, which is assertive in its statements about actuality more so than other types of documentary films such as the poetic-reflexive documentary. However, this reference to reality is precisely established in the voice of the documentary and is not inherent in its physical properties as such. Crucially, by including the dramatized documentary as a documentary genre, both Bondebjerg and Ward did away with any claim that reality can be found in the form of a film. For both authors, dramatizations, reconstruction, fictive scenarios and animations can be documentary. Paul Ward (2005) wrote:

Rather than seeing documentaries […] as an inevitably failed attempt to render experiences or certain situations directly, we should therefore recognize that the aesthetic choices made are merely the formal dimension and have no necessary say in whether or not something is a "documentary." What makes a documentary resides somewhere else, in the complex interaction between text, context, producer and spectator. (p. 11)

The use of archival material and documents, the mise-en-scène and aesthetic and stylistic features and tropes are simply formal features of the documentary film and cannot in themselves explain the difference between fiction and non-fiction films, as Ward (2005) hammered home:

It needs emphasising that the use of certain conventions and techniques is not the status of a text's status vis-à-vis the real world. If it were, then The Office (BBC, series one 2001; series two 2003, UK) would be a documentary; The Thin Blue Line (Errol Morris, 1988, US) would be a fiction film. (pp.11-12)

Thus, in a similar way that documentary films say something about reality without relying on indexical references, docu-games can reference reality and be documentary in their expression. The question, however, is in which ways "voice" is expressed in computer games in relation to the specific modes of expression in this form.

COMPUTER GAMES AS EXPRESSIVE MEDIA

Computer games combine traditional play—and game—phenomena with mediated communication. Thus, on one hand, they are games in roughly the same way as chess or football are games, and, on the other hand, they are audio-visual representations of elaborate game-worlds replete with meaning. This has transformed the question about videogames as expressive media

into a contested issue. On a basic level, the very idea of videogames as media has been questioned with reference to its kinship with traditional games. They have sometimes been seen as games rather than media and thus not as carriers of communication. Certain regulatory bodies still deal with videogames this way, but the technological and aesthetic development of videogames from the 1970s to the present has made the expressive aspects of videogames hard to deny. On the other hand, even when videogames are indeed seen as expressive media, their specific expressive aspects have remained a contested issue. In particular, the idea of videogames as narrative media has been an object of heated debate during the first years of the millennium between narratologists (Jenkins, 2004; Murray, 1997) and ludologists (Frasca, 2003; Juul, 2001, 2005) arguing about whether or not videogames are able to convey narrative meaning. The narrative characteristics of videogames have sometimes been seen in opposition to its characteristics as a simulation (Frasca, 2003) representing another principle of meaning and representation in the game. For instance, an increasing focus on the "procedural rhetorics" of videogames (Bogost, 2007) has drawn attention to the way games as programed systems and simulations may in themselves convey meaning independently of their narrative contextualisation and visual representation. The concept of simulation here represents a particular way of explaining the expressive characteristics of games, although it is not the only possible one, and in our analysis we will include several perspectives on the way meaning is conveyed in games.

Thus, videogames are extremely complex texts sharing semiotic characteristics with a range of other media and integrating them into programed systems that convey meaning in their own right. In order to understand how videogames may relate to reality, it is necessary to understand how they convey meaning in this way.

First of all, videogames are audiovisual representations. Indeed, during game production, considerable resources are invested in creating and fine-tuning the audiovisual appearance of games, not least in the so-called triple-A or large-budget productions, thus making the representational aspect a relevant object of analysis. However, seen in relation to the documentary issue, the audiovisual representations of videogames consist of computer graphics, i.e., visual reconstructions. The audiovisual representation of videogames do not possess the "indexical" nature that has traditionally been ascribed to the documentary photograph or film, granting them a privileged status with regard to the representation of reality (Gray, 1960). The visual representations within videogames do not "document" particular actions or situations as in the dramatized documentary. Instead, they are a re-enactment of particular actions or situations and should be analyzed with emphasis on the specific choices and interpretations underlying this enactment.

Secondly, videogames often include narratives, but in a way that differs considerably from other narrative media such as films and novels. As previously mentioned, this issue has caused considerable debate in the field of game studies, and it is not the aim of this chapter to present the entire discussion. However, in relation to the documentary issue it is worth mentioning two important characteristics of videogame narratives: their non-linear structure and their focus on player actions. Regarding the former, the focus of narratology has traditionally been on plotting, that is, on the intricate orchestration of action in time and the deliberate distribution of plot-relevant knowledge to the recipient (Bordwell, 1985). This close relationship between the concept of narrative and the sequential ordering of action is challenged in videogames in which the action and outcome are entirely dependent on players' choices and performance. Indeed, much videogame "action" is highly repetitive and would make no sense within a traditional notion of narrative. Narratives in videogames instead exist as potential embedded in the game world or emerging from its structure

(Jenkins, 2004) and serve to contextualize action in different ways. Obviously, this challenges one of the traditional documentary's most important functions—the "authoritative narrative" portraying a true state of affairs. With regard to the second characteristic of videogame narratives, their "raw material" is player action. Traditional narratology distinguishes between diegesis and mimesis as two basic narrative modes (Fulton, 2005). According to this distinction, stories can either be told as they typically are in novels or shown as they typically are in films, and this difference indicates very different narrative strategies. In comparison, videogames make player actions possible (Jenkins, 2004). Of course, videogames may retell or show story events as they often do in so-called cutscenes, that is, non-playable sequence where the plot is represented in a more traditional linear manner, but the actions made available to the player represent a new type of narrative building blocks and allow for a new set of narrative strategies. Thus, instead of bringing forth "authoritative narratives," videogames allow players to enact story events in different ways.

Finally, videogames are programed systems, and this turns the programed behaviour of the gameworld and its objects into an alternative source of meaning in the game. This characteristic has sometimes been described as "procedural rhetorics" (Bogost, 2007), that is, the expression of meaning through processes. According to this point of view, programed systems and simulations define particular behaviours and causal relationships that work in themselves as statements about the world. For instance, the well-known strategy game Civilization not only represents the encounters between its competing countries audiovisually and places them within a historical context, it also makes certain actions possible within the game and grants them different impact with regard to success or failure, thus making a particular statement regarding history and the forces that drive it. From a documentary point of view, this implies that "meaning in videogames is not through a re-creation of the world but through selectively modelling appropriate elements of that world" (Bogost, 2007, p. 46).

Bogost argued that "verbal, written and visual rhetorics inadequately account for the unique properties of procedural expression" (p. 29) making this a primary expressive feature of videogames. However, with regard to the documentary issues, it makes little sense to exclude the audiovisual or narrative aspects of videogames as important sources of meaning because videogames may make statements about reality by combining these sources of meaning. Indeed, many docu-games derive their meaning from the coupling of certain programed behaviour with certain representations and narrative contextualization. Whereas the programed behaviour of falling blocks may impregnate a certain potential for meaning, the visual representation of the blocks as bombs falling over a contemporary Afghan landscape would turn that programmed behavior into a particular statement. Furthermore, Bogost has more recently been criticized for his overly author-centred (or, in this case, designer-centred) perspective on meaning in videogames. Miguel Sicart (2011), for example, rightly argued that procedural rhetorics ascribe too much meaning to the designed aspects of the videogames instead of the dynamic aspects of playing the game (2011)2. This is a relevant criticism that has also been put forward elsewhere (Thorhauge, 2012), and, in accordance with this, our analysis below should be seen as dealing with the formal characteristics in games as a potential for meaning, encoded messages (Hall, 1980) that have to be activated or decoded by the players, who may, depending on their preferences, take the action in totally different directions.

In the following section we will discuss more thoroughly how the coupling between audiovisual representation, narrative contextualization, and procedural rhetorics may support documentary "voice" and genres and their epistemic relationship as proposed by Plantinga and Bondebjerg.

EPISTEMIC POSITIONS

As discussed in the previous section, we do not believe the relationship between docu-games and the reality they describe should be seen as more artificial or fragile than is the case in other media or genres. Rather, what is interesting with regard to games and their description of reality is that they represent new ways of conveying meaning and reality.

In this regard and as mentioned previously, Bondebjerg (2002, 2008, forthcoming 2013) distinguished between four documentary genres with four different ways of relating to the reality they describe: the authoritative; observational, dramatized, and poetic-reflective documentary. These distinctions include a set of genre characteristics regarding the purpose, structure, and aesthetics of these types of films and, importantly in this context, the different kinds of references to reality embedded in these documentary genres. The four genres proposed by Bondebjerg open up four different ways of relating to reality: epistemic authority, epistemic openness, epistemic hypothetical, and epistemic-aesthetic. Epistemic authority and epistemic openness have a strong anchorage in the presentation of facts in common. However, they differ with regard to their way of dealing with the facts. The authoritative documentary sets forth an assertive argument about the world, whereas the observational documentary sets out to show a piece of lived reality for the audience for the audience to interpret.3 On the other hand, the dramatized and poetic-reflexive are less anchored in the presentation of facts. The dramatized documentary takes various forms, whether as the dramatization of a real event or a fictitious event presented in a documentary form (i.e., a contra-factual "what if doc"). In both cases the dramatization is a hypothetical statement exploring how things might (have) be(en) or how people in communities might react should a certain situation arise. The focus of poetic-reflective documentary is often on the representation of reality in itself, and thus this documentary genre often takes a highly aesthetized form, drawing attention to its medium, the production process, and the very act of representing reality (Bondebjerg, 2008, forthcoming 2013).

In the following section we will analyze a number of cases in order to explore how the interplay between audiovisual representation, narrative contextualization, and procedural rhetorics may establish different—but equally valid and credible—types of references to reality, and we will discuss how this corresponds to the positions described above.

INTERACTIVE DATABASE DOCUMENTARIES: CREDIBILITY AS EPISTEMIC OPENNESS

As described in the previous sections, games differ from traditional types of narratives in being non-linear and by allowing for player actions. However, having said this, non-linearity and player/viewer interaction is precisely what characterizes some new forms of documentary. Online database documentaries allow the viewers to interact with sections of footage directly and decide their own narrative journey through this material based on, for example, following a specific character, viewing the events of a specific time or place, or simply being a flâneur moving through various clips. As such, they too are docu-games. For example Gaza/Sderot (Arte, 2009, France) explored the lives of a number of villagers in two neighbouring villages on opposite sides of the Israeli-Palestinian border. Here, villagers are confined to their own territory, but the viewer can cross the border and explore the similarities of the lived realities and common human reactions on both sides of the border. The Model Agency (Channel 4, 2009, UK) followed the lives and work at one of the UK's busiest model agencies. Here, the user can watch pre-made documentaries, which were also televised, or explore individual characters, situations, and

events more deeply by accessing and navigating all the footage from the series online. As is the case with these examples, database documentaries often take the form of observational documentaries and, as such, their reference to reality or epistemic credibility rests precisely with the ability of the viewers to watch and access the material in an as unobstructed, unedited, and unmediated way as possible. The Direct Cinema school of observational documentary, in particular, argued that a film's reference to reality depends on its footage being presented as unedited and observational. Following this line of argument, the procedural rhetorics of the database enable the viewers/gamers to decide their narrative path and access all the raw footage filmed. This creates an openness that does not undermine the credibility of the material, but, on the contrary, makes its reference to reality stronger by allowing an extended and expanded epistemic openness. But what about productions involving the more direct gameplay features we recognize from videogames? How do they work as statements about reality and what sort of epistemic positions do they point towards?

KUMA WARS AND GLOBAL CONFLICTS: TWO TAKES ON PRESENT-DAY AFGHANISTAN

KUMA Wars: Afghan Air Strikes and Global Conflicts: Afghanistan both deal with the issue of conflict in present-day Afghanistan, and both have a more or less declared ambition of making the player familiar with various "facts" about this conflict. However, they do so from very different angles involving different narrative contexts, audiovisual scenarios, and gameplay objectives. Global Conflicts (http://www.globalconflicts.eu/) is an educational game series dealing with conflict all over the world. It is deliberately designed for classroom teaching and offers different sorts of teaching resources such as teacher manuals and student assignments on its website. This specific episode is set in a small Afghan village where the school is attacked and the player has to figure out how and why this has taken place by talking to the locals. In comparison, KUMA Wars primarily deals with combat, recreating real life combat situations through a series of "missions" or episodes based on historical events. Depending on the availability of material, each mission comes with an array of supplementary "documentary" sources such as satellite imagery and news coverage. This specific mission is called Afghan Air Strikes and involves an encounter between the Afghan national army and Taliban warriors, introducing the possibility of ordering air strikes as a new and interesting gameplay feature. As indicated by this description, the two cases are situated within very different moral frameworks emphasizing, in the first case, citizenship, democracy and intercultural understanding and, in the second case, armed conflict and military operations. The aim of this analysis is not to evaluate these frameworks normatively but to describe how they are reflected in the audiovisual representation, narrative contextualization, and procedural rhetoric of the two games, as well as how different types of reality claims are established on this basis.

First of all, both games establish a certain "contract" with their audience by way of the specific context. As mentioned previously, Global Conflicts presents itself as a learning resource stating that the episodes are "developed with close attention to curriculum requirements and ease of use in classroom teaching." Furthermore, Global Conflicts: Afghanistan claims to represent as neutrally as possible various perspectives on the triggering event, stating that it "does not attempt to show what's right or wrong, but instead focuses on presenting the various perspectives that are present in the country." In this way, the game invokes the credibility and neutrality that is usually associated with classroom texts. Thus, Global Conflict places itself among a group of educational or "persuasive" games that also have the stated purpose of informing students/players

about contemporary political and global issues. On the other hand, Kuma Wars (www.kumawar. com) presents itself as "a free, high-end series of playable recreations of real events in modern combat" emphasizing the accuracy of its portrayal of real events as an important aspect of its gameplay quality; this is further underlined in the presentation of this episode in which information is split between "game info" and "real world event." Game info includes production notes, tips and tricks, screen shots, game maps, and level objectives, while information regarding the real world event includes mission details, chronology, satellite imagery, multimedia, news coverage, and global headlines as well as an introduction to tactics, forces, and weapons. In this way, the credibility of the game is based on the assumed integration between certain aspects of the real world event and the game design. In this way Kuma Wars belongs to a group of games that combines a well-known gameplay genre with real historical events and settings, pointing towards another type of referentiality and credibility.

The audiovisual representation in both games displays a rather stereotypical vision of a stony and sandy landscape with low buildings and sparse vegetation, locating the conflict in underdeveloped rural areas of Afghanistan. However, in Global Conflicts this landscape is inhabited by civilians, including children, going about their everyday routines among the stony walls, whereas Kuma Wars depicts a rather deserted area with empty buildings and seemingly idle civilians running away as the player approaches. Of course, both games suffer from the monotonous movement patterns and lack of detail typical for low-budget 3D productions, but the differences are nevertheless significant. Regarding narrative contextualization, the two games take very different perspectives on the conflict. As mentioned previously, videogame narratives have little of the intricate plotting and careful distribution of plot-information associated with traditional novels and fiction films. Rather, the game narrative should be seen as a way of contextualizing player actions by endowing them with particular meaning and implications in the game world. In this light, narrative contextualization has to do with the way the game is "staged" by way of roles, types of action, and the overall conflict in the game. In Global Conflicts, for instance, the player takes the role of a character called Michael who travels to Afghanistan to help his friend Alan, who is seemingly in trouble after establishing a school in a small village. The characters he meets along the way include an Afghani civilian, a mullah, a police officer, a Taliban warrior, and an ISAF soldier. The main conflict has to do with the hostility toward the school as articulated differently by the different characters. In stark contrast to this approach, Kuma Wars: Afghan Air Strike mainly approaches the conflict as a military operation. The player controls a squad of four soldiers from the Afghan Army engaging in armed conflict with Taliban warriors who have just ambushed a vehicle and fled into the mountains. The player has to navigate the characters through a rather hostile landscape and kill the insurgents before they kill him, but the deeper reasons for the conflict are not questioned or discussed in any way.

However, as argued above, the audiovisual representation and narrative contextualization cannot be seen as independent from the overall procedural rhetorics of the game, that is, the way they are integrated into a programed system with particular behaviours and chains of causation that do in themselves serve as statements about the reality in question. In the Global Conflicts episode, for instance, the main action available to the player is to engage in dialogue with the people involved in the conflict. More specifically, this involves choosing different directions in the dialogue as it is not possible to make up new questions or answers4. However, even within this rather simple construction, behaviours and casual relations are significant. For instance, in the ongoing exchange of statements, the player has to balance between politeness and insistence in order to get

the necessary information. In this way the design of the dialogue makes itself a point regarding the possible solutions to the conflict in the game: It is a continual balancing act of several conflicting considerations. This is further emphasized in the final part of the game, in which the player has to combine the different statements that have been collected into a set of arguments regarding the future of the school. These involve security, education, and cultural understanding, each of which, obviously, calls for very different solutions. The most important exception to this flexibility is the part of the game in which the player is taken hostage by Taliban warriors and interrogated. At this point all attempts at negotiation lead to the player's death implying that in such a situation the player can do nothing but comply. In this way, the procedural rhetoric of Global Conflicts concerns the challenge of communication and the pay-off between conflicting considerations in this war-torn area of the world.

While player actions in Global Conflicts are reduced to choosing a path through a predefined dialogue-structure, player actions in Kuma Wars are reduced to navigating the terrain and shooting. Civilians run and members of the Taliban shoot when approached, giving the player no choice but to flee or shoot back, making other types of interaction with non-player characters impossible. Instead, the range of possibilities made available to the player is limited to how to cross the terrain, how to alternate between the four characters at hand, and which weapons to use. With regard to the latter, an important consideration concerns when to call for an air strike, which can only be done a limited number of times, but which is rather effective when carried out. In this way, the "procedural rhetorics" of Kuma Wars concern the skills of the soldier and strategic considerations relating specifically to the implementation of military operations—alternative ways of dealing with the conflict are not integrated into the gameplay.

On the basis of this description, how are the two games effectuating credibility in their reference to reality? As mentioned previously, Global Conflicts presents itself as a teaching resource that makes the player aware of different perspectives on the conflict while avoiding taking a stance on its rights or wrongs. In practice, this is mainly implemented by way of a dialogue structure in which the player exchanges statements with particular characters representing parts in the conflict. Accordingly, evaluating the credibility of this game as a documentary entails evaluating whether the chosen perspectives are indeed relevant to the conflict and whether they are truthfully and neutrally described. Due to the dialogue structure, for instance, all perspectives—civilians, mullahs and members of the Taliban—are presented as equally significant, which is not necessarily the case, just as the focus on different cultural perspectives rules out alternative explanatory models such as territorial issues or international power struggles. In contrast, Kuma Wars (http://www.kumawar.com/) presents itself as "recreations of real events in modern combat" emphasizing instead the congruity between the real event and the specific game episode. In this situation, questions regarding the credibility concern the accuracy of detail: Is the game map comparable to the satellite imagery presented on the website, and does the gameplay involve those weapons and vehicles described as part of the real world event? With regard to this issue, the episode in question has certain shortcomings. The availability of health packs, for instance, represents a typical First Person Shooter (FPS) feature that bears little relation to reality, and, similarly, the player's ability to carry three large weapons around in an inventory lacks credibility. Furthermore, the Taliban warriors turn out to be in possession of a sniper rifle that is not described in the "real world event"5 . In this way, the contract and claims to credibility established on the website are not met by the actual game design. This is probably due to the wider goals of the two games. While Kuma War represents itself as an opportunity to "play the news," the wider context of the website also offers

pure entertainment games such as Dinohunters, which suggests that entertainment, rather than real life, is the focus. That is, the news element is perhaps just a pretext for the gameplay rather than the gameplay being a tool for presenting the news. In comparison, Global Conflicts is first and foremost a teaching resource and would work poorly as a game in its own right.

CREDIBILITY IN DOCUMENTARY GAMES

As mentioned in the introductory sections, much writing on docugames has asked whether games can be documentary texts at all and questioned whether they—qua games—can establish credible references to reality. We have argued that this implies a reductive and media-centric understanding that makes little sense in the broader context of documentary theory. Instead, we have tried to ask in what ways games can make credible references to reality, focusing on two cases belonging to very different genres of docu-games. By way of the context, these two games establish very different contracts and thus criteria of credibility, focusing respectively on balanced and fair representation of perspectives or accuracy of simulation. Credibility in this view is closely linked to the particular epistemic position or voice established by the contract and aesthetic means of the particular game in question. As shown in the analysis, the games enforce new stylistic and aesthetic ways of establishing "voice" in Plantinga's (2005) terms, including procedural rhetorics as a game-characteristic feature.

But how do the two games relate to the documentary genres and epistemic positions introduced above? Seen in relation to Bondebjerg's categories, both games and, indeed, videogames in general, might at a first glance be defined as epistemic-hypothetical because they deal with simulations of actions and processes rather that presentation of facts. However, this would once again put too much focus on general characteristics of the medium and less focus on particular meaning-making processes. In the context of docu-games, it is worth exploring both "facts" and "voice" further and in more detail. Games can indeed be designed and programed in correspondence with a chosen set of facts, and they can, in this way, be seen as presenting facts in the same manner as the authoritative and observational documentary. If Kuma Wars had actually kept its own promise and built up its simulation in correspondence with its analysis of the real world event, it could indeed have been seen as presenting facts. Furthermore, voice, as argued above, is a combination of aesthetic and stylistic choices including gameplay and cannot be reduced to one single aspect of a medium such as the programed and simulated nature of videogames. Looking at the two games in question, they do, at least in their stated purpose, share important characteristics with other epistemic positions such as the epistemic authority that Bondebjerg associates with the authoritative documentary genre.

The procedural rhetorics of Global Conflicts guide the player to specific learning outcomes and underline the importance of neutrality and equal perspectives. In doing so, a clear selection and simplification has taken place; certain perspectives, including those of the religious (the Mullah), the civic and cultural (the civilian) and the military (the Taliban and the ISAF soldier), are negotiated and are considered crucial to an understanding of the conflict, whereas other perspectives (e.g., gendered, generational and ethnic) are excluded. In this way the game selects and simplifies for the sake of its educational purpose while still seeking a representative version of events. The combination of the contract established in the game context and the particular design and stylistic choices involving a rather predetermined course of action in the gameplay puts forth an assertive argument about the world characteristic of the authoritative documentary. In this way Global Conflicts shares characteristics with the epistemic hypothetical as

well as the epistemic authoritative position. As stated in the analysis, Kuma Wars does not really meet its own criteria of credibility because the game design does not reflect those aspects of the real world events that are put forth, and for this reason it is difficult to define its possible affiliation with any of Bondebjerg's epistemic positions.

To conclude, the credibility of docu-games cannot be defined by the innate characteristics of the medium of videogames in isolation. Rather, it must be seen as a combination of the contractual relationship established by the context and the aesthetic and stylistic means being employed. In this regard, videogames do not differ from other media. However, their programed characteristics or procedural rhetorics represent a new source of (documentary) meaning and must be seriously considered as the discursive power of games, and not to mention the sheer number of games, continues to grow and grow.

REFERENCES

Aufderheide, P. (2007). *Documentary films—A very short introduction*. Oxford, UK: Oxford University Press.

Bogost, I. (2007). *Persuasive games: The expressive power of videogames*. Cambridge, MA: MIT Press.

Bondebjerg, I. (2002). The mediation of everyday life. In Jerslev, A. (Ed.), *Realism and "reality" in film and media* (pp. 159–193). Copenhagen, Denmark: Museum Tusculanum University of Copenhagen.

Bondebjerg, I. (2008). *Virkelighedens fortællinger den danske tv-dokumentarismes historie* (1. udgave ed.). Frederiksberg, Denmark: Samfundslitteratur.

Bondebjerg, I. (2013). (forthcoming). Engaging with reality: Documentary, politics and globalisation. *Publisher TBC.*

Bordwell, D. (1985). *Narration in the fiction film*. Madison, WI: University of Wisconsin Press.

Bruzzi, S. (2000). *New documentary: A critical introduction*. London, UK: Routledge.

Corner, J. (1995). *Television form and public address*. London, UK: Edward Arnold.

Corner, J. (1996). *The art of record: A critical introduction to documentary*. Manchester, UK: Manchester University Press.

Corner, J. (1997). Restyling the real: British television documentary in the 1990s. *Continuum, 11*(1), 11–21. doi:10.1080/10304319709359415

Corner, J. (2009). Documentary studies: Dimensions of transition and continuity. In Austin, T., & de Jong, W. (Eds.), *Rethinking documentary: New perspectives, new practices* (pp. 13–28). Maidenhead, UK: Open University Press.

Frasca, G. (2003). Simulation versus narrative: Introduction to ludology. In Wolf, M. J. P., & Perron, B. (Eds.), *The video game theory reader*. New York, NY: Routledge.

Fullerton, T. (2008). *Documentary games: Putting the player in the path of history*. Nashville, TN: Vanderbilt University Press.

Fulton, H. (2005). *Narrative and media*. Cambridge, UK: Cambridge University Press. doi:10.1017/CBO9780511811760

Gray, A. B. H. (1960). The ontology of the photographic image. *Film Quarterly, 13*(4), 4–9. doi:10.2307/1210183

Grierson, J. (1933). The documentary producer. *Cinema Quarterly, 2*(1), 7–9.

Hall, S. (1980). Encoding/decoding. In Hall, S., Hobson, D., Lowe, A., & Willis, P. (Eds.), *Culture, media, language: Working papers in cultural studies, 1972-79* (pp. 128–138). London, UK: Hutchinson.

Jenkins, H. (2004). Game design as narrative architecture. In Harrigan, P., & Wardrip-Fruin, N. (Eds.), *First person: New media as story, performance and game*. Cambridge, MA: MIT Press.

Juul, J. (2001). Games telling stories?—A brief note on games and narratives. *Game Studies, 1*(1). Retrieved from http://www.gamestudies. org/0101/juul-gts/

Juul, J. (2005). *Half-real: Video games between real rules and fictional worlds*. Cambridge, MA: MIT Press.

Macdonald, K., & Cousins, M. (1998). *Imagining reality: the Faber book of the documentary*. London, UK: Faber.

Murray, J. H. (1997). *Hamlet on the holodeck: The future of narrative in cyberspace*. New York, NY: Free Press.

Nichols, B. (1994). *Blurred boundaries: questions of meaning in contemporary culture*. Bloomington, IN: Indiana University Press.

Nichols, B. (2009). The question of evidence, the power of rhetoric and documentary film. In Austin, T., & de Jong, W. (Eds.), *Rethinking documentary: New perspectives, new practices* (pp. 29–37). Milton Keynes, UK: Open University Press.

Nieborg, D. B. (2004, September). *America's army: More than a game*. Paper presented at the meeting of the International Simulation and Gaming Association Conference, Munich, Germany.

Plantinga, C. R. (1997). *Rhetoric and representation in nonfiction film*. Cambridge, UK: Cambridge University Press.

Plantinga, C. R. (2005). What a documentary is, after all. *The Journal of Aesthetics and Art Criticism, 63*(2), 105–117. doi:10.1111/j.0021-8529.2005.00188.x

Raessens, J. (2006). Reality play: Documentary computer games beyond fact and fiction. *Popular Communication, 4*(3), 213–224. doi:10.1207/s15405710pc0403_5

Renov, M. (2004). *The subject of documentary.* Minneapolis, MN: University of Minnesota Press.

Renov, M. (2009). First-person films: Some theses of self-inscription. In Austin, T., & de Jong, W. (Eds.), *Rethinking documentary: New perspectives, new practices* (pp. 29–39). Maidenhead, UK: Open University Press.

Sanchez-Laws, A. L. (2010). Digital storytelling as an emerging documentary form. *Seminar.net, 6*(3). Retrieved from http://seminar.net/index.php/volume-6-issue-3-2010/161-digital-storytelling-as-an-emerging-documentary-form

Sicart, M. (2011). Against procedurality. *Game Studies, 11*(3). Retrieved from http://gamestudies.org/1103/articles/sicart_ap

Sisler, V. (2009). Palestine in pixels: The Holy Land, Arab-Israeli conflict, and reality construction in video games. *Middle East Journal of Culture and Communication, 2*(2), 275–292. doi:10.1163/187398509X12476683126509

Smith, A. E., & Blackton, J. S. (Eds.). (1922). *Two wheels and a crank.* London, UK: Faber.

Thorhauge, A. M. (2012). (in press). The rules of the game—The rules of the player? *Games and Culture.*

Ward, P. (2005). *Documentary: The margins of reality.* London, UK: Wallflower.

Ward, P. (2009). Drama-documentary: The "Flight 93" films. In Austin, T., & de Jong, W. (Eds.), *Rethinking documentary: New perspectives, new practices* (pp. 191–203). Maidenhead, UK: Open University Press.

ADDITIONAL READING

Bogost, I. (2011). *How to do things with video games.* Minneapolis: U of Minnesota Press.

Buckingham, D., & Burn, A. (2007). Game literacy in theory and practice. *Journal of Educational Multimedia and Hypermedia, 16*(3), 323–349.

Carr, D., & Oliver, M. (2009). Tanks, chauffeurs, and backseat drivers: Competence in MMORPGs. *Eludamos, 3*(1), 43-53. Retrieved from http://www.eludamos.org/index.php/eludamos/article/view/vol3no1-6/107

Holmevik, J. (2012). *Inter/vention: Free play in the age of electracy.* Cambridge, MA: MIT Press.

Huizinga, J. (1950). *Homo ludens: A study of the play element in culture.* Boston, MA: Beacon Press.

Jenkins, H. (2006). *Convergence culture: Where old and new media collide.* New York, NY: New York University Press.

Kendall, A., & McDougall, J. (2009). 'Just gaming': On being differently literate. *Eludamos, 3*(2), 245-260. Retrieved from http://www.eludamos.org/index.php/eludamos/article/view/vol3no2-8/135

Paul, C. A. (2011). Optimizing play: How theorycraft changes gameplay and design. *Game Studies, 11*(2), http://gamestudies.org/1102/articles/paul.

Pearce, C., Boellstorff, T., & Nardi, B. (2011). *Communities of play: Emergent cultures in multiplayer games and virtual worlds.* Cambridge, MA: MIT Press.

Salen, K., & Zimmerman, E. (2004). *Rules of play: Game design fundamentals.* Cambridge, MA: MIT Press.

Salen, K., & Zimmerman, E. (2006). *The game design reader: A rules of play anthology.* Cambridge, MA: MIT Press.

Selfe, C. L., & Hawisher, G. (2007). *Gaming lives in the twenty-first century: Literate connections*. New York, NY: Palgrave Macmillan. doi:10.1057/9780230601765

Taylor, T. L. (2009). *Play between worlds: Exploring online game culture*. Cambridge, MA: MIT Press.

Voorhees, G. (2009). The character of difference: procedurally, rhetoric, and roleplaying games. *Game Studies, 9*(2). Retrieved from http://gamestudies.org/0902/articles/voorhees

ENDNOTES

[1]   Plantinga also - albeit briefly - mentions the drama-documentary as a documentary genre, but he does not elaborate on it in detail, neither in *Rhetoric and Representation in Nonfiction Films* (1997) nor his 2005 essay What a Documentary Is, After All (2005).

[2]   In this way, procedural rhetoric can be seen as a game studies pendant to auteur theory in films studies focusing on the text as a product of the designer's personality and motives of persuasion. And just like auteur theory this perspective has its benefits and its shortcomings revealing, on the one hand, important aspects of the text's form and origins, but ignoring, on the other hand, the production process as a predominantly collective endeavour and the reception context as a crucial aspect of the meaning making process.

[3]   Obviously, this genre also involves a considerable amount of organisation, selection and direction from the production team and the "authenticity" of the observational form is also a construction, as explored by proponents of, for example, cinema verité and Direct Cinema.

[4]   The game also involves two minigames which involve slightly more interaction with the visual interface, but these minigames have little significance to the outcome of the game.

[5]   This may be due to a flaw in the description of the real world event rather than the game since the members of the Taliban could indeed be in possession of a sniper rifle. However, it breaks that congruity between the game episode and the documents about the real world event that constitutes the truth claim of the game.

Chapter 21
Press C→ to Play the Ocarina:
Rhetoric and Game Music

Dan W. Lawrence
Michigan Technological University, USA

ABSTRACT

The purpose of this chapter is to investigate the intersection where digital media studies meet rhetoric and rhetoric is re-introduced to musicology. In the recent academic excitement surrounding game studies, the music of games has been overshadowed. The author would like to call attention to the significance of game music and to consider a rhetorical method to approaching it that calls upon a rekindling of the history of coupling rhetoric with music. The author builds on this history by suggesting the foundation of a rhetorical framework for understanding the argumentative power of video game songs. He then moves to offer an approach for evaluating the ethos of game music that consists of assessing worlds and how they are carried through, and by, music. While 17th century baroque composers thought music to be fundamentally an issue of affections—and especially played off of emotional binaries such as joy/ sadness as a rhetorical approach—the author hope to here revive this lost art of applying rhetoric to music through broadening the discussion beyond the matter of human emotion. This rhetorical approach allows the individual a framework with which to evaluate the ethos of game music as it now appears through numerous mobile operating systems, online environments, and as remediated forms manifesting in/as cultural artifacts. As games become ubiquitous, so do their songs.

INTRODUCTION

The ethos of game music is found in its particular capacity to carry a virtual world. But game songs also carry experiential memories and mythical narratives, subtleties and inspirations, human sparks. The inviting ethos of game music has inspired a remix culture to recreate and reimagine

DOI: 10.4018/978-1-4666-2663-8.ch021

sounds across genres, across realities. This remix culture has developed an ethos of its own: one that welcomes musical boundaries to be crossed and encourages experimentation through the sharing and replaying of retro video game scores. Though there has been a great amount of heated debating surrounding the status of games as art—and thus its music—there seems to be a growing understanding that the affective, self-speaking power of game music will hold an important place in a global, musical culture. Yet, we must not be all-praising, and approach video game music with a critical ear and a rhetorical lens to seek for its meanings, interpret its purposes, peek at its origins and assess its value. Music in itself is an immersive, mysterious human activity. When coupled with the immersive nature of games, we must be doubly conscious. I offer in this chapter a few conjectures at how we might think about the credibility of video game music and how we can analyze it rhetorically. The approach I use is partially musical, partly theoretical, and wholly exploratory, as these are new ambiences in which we all wade.

Though much has been said about the significance of games in society, education, and in human expression and argumentative purpose—McGonigal (2011), Gee (2003), and Bogost (2007, 2011) respectively—it may be noted that the aspect of game music has been overshadowed by other concerns within the burgeoning web of scholarship surrounding digital media and game studies. The relationship between music and perception in cultural studies has been raised to a point of pivotal concern—with work that has explored the cultural significance of popular music and music videos by Frith, Goodman, and Grossberg (1993)—such that it no doubt warrants our attention and *has* warranted much attention. Yet there is room for alternative musics in this discussion, and there is room in the realm of rhetoric for the discussion of music. The music of games, so prevalent, and so fundamentally different in their composition, function, and form, has gone largely unassessed. In her careful work *Game Sound*, Karen Collins (2008) noted that "removing audio from games… can significantly affect gameplay" (p. 136). I would like to posit a turn to this statement: removing gameplay from the audio may significantly affect our conceptualization of the game. For example, what is it about game music that especially lends itself to remix culture, to fan-made music, to fervent re-imaginations and often, even, re-orchestration? So often, our exposure to video

game music is outside of traditional gameplay. This playing-beyond-the-game into the physical world, into the hands of amateur or experimenting musicians and experienced arrangers, is a significant element of game music. Game music inspires not just re-creation, but re-invention. Imagine a pedagogy of music or composition that promoted critical originality over recitation and memorization: improvising outside of boundaries, *play* with malleable rules.

I will begin below with an outline of the work that has attempted to connect rhetoric and music. I hope this will illuminate the difficulties that scholars have faced at this task. It will also serve to dig the academic shoebox out from underneath the bed where it has been kicked in frustration and irritation. Have we abandoned the connection between art and argument, song and meaning, rhetoric and music, not to mention how these connections affect ethos? What can we resurrect, and what should we? My purpose will be, as much as possible, to extract, examine, abstract, and re-situate with a nod to the roots of music and game studies. Of course, these positions are always in flux. I will offer an analogy here, knowing full well it may be trite. Ancient Greek observers noted the presence of strangely moving figures in the night sky. They were left unsure about their origin and named these celestial anomalies, in their language, something approximating *wanderering stars*. Yet, with time, not only were the patterns and movements of these planets—*wanderers*—charted, but the entire galactic system in which they spin and whirl has been observed, and the universe in which our galaxy moves. Yet, we wander still, and so do they. So, too, I play with an early theory of the significance of a collision between rhetoric and music and speculate what it is we might be able to do with it today in the light of game music, because of its peculiarities, idiosyncrasies, and the ever-changing influence of the technology with which we connect to each other and the influence we put on it.

MUSICA POETICA AND DIGITAL MUSICALITY

The relationship between rhetoric and music has been opened up, cast aside, disregarded, and altogether tossed around since its early works set out the tradition of *musica poetica*. One of the first to publish on the connection between rhetoric and music was Nikolaus Listenius (1533) in *Rudimenta Musicae Planae*. Musical theories with rhetorical characteristics—mostly discussions of pathos and emotional affect for the audience—predominated in 16th and 17th century German thought and praxis—though this connection did not continue. Indeed, contemporary musical theory is decidedly unconcerned with rhetorical considerations. Yet, new interest in this tenuous tie between rhetoric and music surfaced again in the (primarily) 1970s work of American scholar George J. Buelow when he sought to investigate the rhetorical groundwork set out by the German baroque composers Burmeister, Dressler, Mattheson. It may have been the near-concurrent explosion of cultural studies and its engulfment of music—and so much else—that drew attention away from this new research. Or, perhaps, like those German composers, the time had simply not yet come for a meaningful assessment of music as argument.

It is plausible, too, that the rhetoric of music has been overlooked because of the historical situatedness of what is known widely to the West as classical music, an umbrella term that is used to encompass periods that are distinguished usually as Baroque, Classical, and Romantic. Classical music can also refer to any art music that derives from these European traditions or precedes them, making it dangerous in the ease with which it can misguidedly group unrelated material together; this will play readily into the topic at hand as I unravel connections to prejudice, credibility, and issues with "retro" media. On another side of the matter, it is also plausible that the area of scholarship that promenades with both the study of rhetoric and the study of music has been dismissed because of its extraordinary ambivalence. Early baroque theories that developed from the tradition of *musica poetica* such as the doctrine of the affections sought to apply the rhetoric of oratory to musical composition with the pepper of emotive binaries (joy/sorrow). Yet no direct correlation between rhetorical tropes and those of music were cemented, nor was a serious attempt made to formulate a new conclusive and universally applicable theory of the rhetoric of music. Looking back with our vantage point, it seems that a rather confused commingling occurred considering the passionate fire of rhetoric and bubbling ebullience of that time's talented songsmiths. It's often the conditions under which ideas meet that are more important than what some call chemistry.

Joachim Burmeister (1606/1993) described what a musical poetics would entail in the opening to his treatise *Musica Poetica*:

[H]ow to put together a musical piece by combining melodic lines into a harmony adored with various affections of periods, in order to incline men's minds and hearts to various emotions. (p. 13)

Burmeister was restricted in his analysis by relegating the rhetorical potential of music to only have affective properties. He drew from Plato and Aristotle, who, although advocates of musical education for youths, thought of music as a method only to stir the passions. Burmeister (1993) wrote, too, that to experience these emotional powers of music, one must be "not altogether unmusical" (p. 57). This privileges those with a formal musical education and background. Music in contemporary 21st century digital society, however, is a sort of *lingua franca*. Instruments are mass produced and inexpensive. Digital composition and recording tools present themselves as freeware for public proliferation. Music is independently produced, published, and shared. We live very much in a musical age, and an age where music does more than move our moods, it affects our individuality and thus our ethos.

I do not believe we must endanger the emotional power of music in order to assess its rhetorical value. We must simply move beyond this designation. At the advent of scholastic controversies surrounding digital media studies, to re-open the case of rhetoric and music is to load the individual with a formidable arsenal with which to address the backlash against unfamiliar forms of communication. The issue, here, is one of credibility. I prefer the construction *digital* media studies rather than *new* media studies. Eventually there will come *newer* media, and the pre-existing nominal assignation will be misleading and then difficult to rid from our palate. Digital media as an academic construction is equally loaded with difficulties, but will work for the sake of this argument to refer to an interest in the manner through which meaning travels, how its transmission effects its qualities and reception, and how it is construed in and around data technology. The digital/analog binary is perhaps a false one, as lines are often blurred and jumped, sometimes with aesthetic intent and purpose as we come to understand the way that technology, meaning, and media shape reality. In other (more) words, we must arm ourselves with these rhetorical tools to protect against the infiltration of ideology into media, as it webs from mass produced content to the independent and self-produced. Often the question of credibility is not *who* produced an artifact but what ideology motivates its coming-of-being, overt or otherwise.

From digital media studies there has opened up a space for assessing the ever-growing history and significance of games. While the first pinpoint in the history of video games dates, perhaps, to the American 1940s and 1950s with *Tennis for Two!*, it is not until the late 1980s that formidable gaming companies such as Nintendo seized the home video game market (although systems like Atari and IntelliVision were certainly widespread before Nintendo). This home-gaming period was preceded by what is known to video game historians as the golden age of arcade gaming, which began in the late 1970s and ended after the North

American release of the Nintendo Entertainment System in 1985. At that point, the video game market shifted from the public arcade to the home console, to be connected to a television and enjoyed in the domestic sphere rather than the arcade. Though music and sound have been inextricably bound for decades, it is not until the home console market of gaming emerges that significant consideration for composition seems to arise. In 1986, Nintendo released *The Legend of Zelda*, a pioneering game featuring a now-standard element of gameplay: an overworld map that allows freeform navigation by the player. When the player is wandering in the overworld, a distinctive—and now iconic—melody by Koji Kondo plays. This "Overworld Theme" has been re-imagined and re-hashed for consecutive Nintendo Zelda titles up until the most recent release at the time of writing, *The Skyward Sword* (2011). The sense of the importance of in-game myth, legend, and history are transmitted from title to title in the Zelda series not only through the reappearance of similar characters, themes, and narratives, but perhaps most importantly through its music. Nobuo Uematsu, for example, composed the music for the continuity of the *Final Fantasy* series from 1987 to 2010, spanning fourteen games. Thus, as games shifted from the arcade to the home, so too the sounds of games shifted from attempting to grab a potential player's attention—arcade machines would entice walker-bys with loud noises and exciting sound effects—to becoming an integral part of a very different experience, a private experience. And it is an experience shaped largely by the venerable video game composers who have gone largely unnoticed.

GAME MUSIC

The Cultural Role of Game Music

The boundaries that are placed on digital media are constantly shifting. I wish to only set forth a definition of game music that may here apply to the argument rather than attempt to establish a universal guideline. The descending square tones that comprise the audio of *Space Invaders* (1978) is something altogether different than the orchestrations of Hans Zimmer for *Call of Duty: Modern Warfare 2* (2010). For this argument, I am placing three parameters: on games, music, and the amalgam that results from their sometimes precarious, often monumental collision. I mean to say that what constitutes video game music is not the same as what constitutes American folk music or 1960s British rock; game music takes the form of countless genres and is constantly re-inventing itself. This poses a problem, then, for a rhetorical approach. We might examine form and function rather than surface characteristics. But we can think, too, about how other genres influence game music and what it means to hear a "classical" piece in a contemporary game.

To wrangle in the many strands of possibilities for this discussion, we might hone in on those games of the 8 through 64-bit home console era. Radical developments occurred within this period—the move from primarily 2D to the capacity for 3D environments, for example—which is why I am interested especially in these compositions. These home console games shifted from compositions defined by their restriction to little memory (the ceaseless, memorable looping of Nintendo's 1985 *Super Mario Bros.*, for example) to orchestrated arias (Nobuo Uematsu's musical scoring of *Final Fantasy VII* with the increased memory capacity of the game disc). Yet, for the most part, there was little physical instrumentation—or analog recording—of music. This contentious space between digital and analog is particularly engaging when viewed with the lens of virtual reality. We might ask, what happens when a virtual character picks up an instrument and it is controlled by the player? Who is playing the instrument? How does this effect in-game composition? Rhetorically, who is the speaker? Where does ethos reside? Can the self, the player, be both speaker and audience? Is this not a crucial skill of both orators and musi-

cians, *to listen to the self*? And how might we use video games in the future as tools for musical composition (or is this dangerous)?

While many claims I make herein are applicable to the analysis of non-digital compositions, whether linked to gaming or not, the focus of my piece is to not only provide a vocabulary for discussing what is often dismissed as merely "retro" but also to hint at the necessity for addressing the myths and prejudice surrounding these digital game songs: that the work of talented and insightful composers has been reduced and expelled from the canon of worthwhile human music. And, too, these attempts at expunging game songs have seen considerable grassroots recoil. Few songs initiate such a common bond, nostalgia, and fervor as the songs of games. Yet, by giving them the status of "retro," we endanger their vitality as well as our own livelihood: we make our pasts retro, and somehow less than a real history. By retro-izing our own lives, we replace a human history with a technological one. Yet, the music of games allows us to relive worlds outside our own, worlds of the past, and to carry those worlds into the future.

I do not wish to suggest that any sort of nostalgic reversion take place nor that we should drown ourselves in the sometimes-stale halls of MIDI. However, countless composers—such as the musical collective Big Giant Circles with their *Imposter Nostalgia* (2011)—have taken the music of 8-64 bit console era games and recreated the aesthetic, repurposed the meanings therein. It seems that as important as the game songs themselves are the numerous remixes and recreations that stem from them. What is evident is that even within limitations there is art. And the 88 spring steel strings of a piano are as much a technological limitation as the 128 sound types of MIDI. Igor Stravinsky went further in his *Poetics of Music* and wrote that there is freedom only in these limitations (1993). Some have worked to warn of the digitalization of music, notably Jaron Lanier (2010) with his *You Are Not a Gadget: A Manifesto*, in which he painted a picture of a dire future where MIDI ringtones are still the norm, thousands of years later, "locked-in," because this is how they were first composed and delivered, and this is how they shall always be, unquestioned. Of course, with the upgraded memory capacity of smartphones, the user can already create or form unique call tones, even draft original compositions for a physical instrument, record with analog equipment, and transmit through the digital to be stored on the device. Wagnerian epics, in their entirety, can play from the no-longer-so-tiny speakers of mobile phones when that significant other wants to get in touch to send a reminder about the spinach that needs to be picked up. Though this is not the common experience, it exists in the realm of current technological and human possibility. Too, we cannot yet perfectly reproduce analog signals that have been digitally mediated. The question has been: What is lost when music is digitized? I ask: What can be gained?

Lanier's argument is much more nuanced than presented here. Yet, my point remains the same. Game music, at least that contextualized in this specific historic period with all its limitations, is not only worthy of examination because of its ubiquity and prominence in popular culture, but because of its artistic and human value that has been discredited or left unobserved because of the proliferation of opinions associating games with low art or no art at all. Roger Ebert (2010), the esteemed film critic, wrote in his *Chicago Sun-Times* blog:

No one in or out of the field has ever been able to cite a game worthy of comparison with the great poets, filmmakers, novelists and poets.

Yet Ebert provided us with his own definition of art:

Does art grow better the more it imitates nature? My notion is that it grows better the more it improves or alters nature through an passage

through what we might call the artist's soul, or vision [sic]. Countless artists have drawn countless nudes. They are all working from nature. Some of their paintings are masterpieces, most are very bad indeed [sic]. How do we tell the difference? We know. It is a matter, yes, of taste.

I argue that tastes have changed. The very nature of early game music—its brevity, memetic gravity, and reproducibility—have urged listeners toward recreation, to new art: art that inspires art. Art that inspires invention. Game music works, often, as an impetus to create anew, to reinvent a style, to playfully reproduce a melody in a new context, in a physical world rather than a virtual one. And here is where the element of Luddism is important. It is not that nascent composers distrust or find distasteful the music of their games and thus must recreate them. We have seen this often done when a pop song is covered and transferred from what the artist perceives as an archaic, perhaps disagreeable genre to one contemporary to his or her own time, such as Guns 'n Roses's successful hair-metal rendition of Bob Dylan's "Knockin' on Heaven's Door." Game music re-renditions seek to transfer between realities, from the virtual to a perceived, shared, communal cultural reality. A music of a separate time and place, not of human history, is granted new human life when it is reborn in our world. Are the original works not art, all their recreations, their re-orchestrations and symphony productions, their performance and recording and ability to carry not only emotion but meaning, to carry entire worlds?

In *How to Do Things With Video Games* Ian Bogost (2011), challenged Ebert's assertion. He wrote:

How then, can we understand the role of games in art? Satisfying Ebert's challenge that games simply need to get up off their proverbial couches and rise up to the authorial status of literature or film is not the way forward. (p. 11)

Bogost suggested that if there is art in games, or if people are now making "*artgames*," then their art is derived from their procedurality. Bogost suggested a "procedural rhetoric" which works less like an argument and more like an experience of a situation (2007). I do not believe that the traditional notion of rhetoric as argument must be entirely abandoned—and neither does Bogost—but I believe the ethos of game music can still be analyzed meaningfully by blending both classical and contemporary theories of rhetoric. Still, questions remain. Are video games art? I might suggest here that game music has the potential to be called art without straying too far from my purpose, though I will not now argue for its crowning up to the status of high art or Art which surely such critics as Roger Ebert would contest (and have done so). It is a question that need not concern us. We now make our own art. Simply, though importantly, my argument here is that we must recognize this medium and its many forms and a new framework may be proposed with which to aptly discuss it and deploy it. Having no vocabulary with which to discuss the sounds around us, we are left to be buoys in the sonic waves of our time. Having no vocabulary with which to evaluate the music in our eyes and minds, we are left with no entry to ethos, no ability to distinguish and discern and analyze. We must lay a groundwork.

Toward a Rekindling of Musical Rhetoric

I want to propose a start toward developing a framework for the rhetoric of game music. Some of these elements may be applied outside of game music, but I have developed these ideas with the connectivity of sound and world in mind. That is, there is a bond between the music of games and the world they present that is uncommon to music played and performed in our own world. And, unlike cinematic music, there is not necessarily a specific mood or human emotion in the mind of the composer, director, or editor that corresponds to a specific event. For example, a player often has the choice to engage a villain, opponent, or obstacle in a game, where in film a battle sequence is pre-planned as part of a plot build-up and greater schema. Some games are, of course, linear in build, but the music relays the whole of the ups-and-downs, often not shifting in a single "level" (e.g., the *Mario* main theme), though it might change with virtual locale (e.g., the *Mario* underwater theme). Still, these games are not always played linearly, and it is the user who determines his or her experience with the game: It can be shut off, resumed, shared ("here, it's your turn!"), and consumed in pieces (save states). Games, especially those of the 16-bit era, began to see larger virtual landscapes and greater playtimes: Narratives and experiences that covered more than 50 individual songs and approximately 40 hours of play time for experienced users were not uncommon, as seen in Square's *Chrono Trigger* (1995). And games continue to grow in scale, as memory becomes less of an issue and game companies grow to acquire larger staffs of writers, developers, and designers. Bethesda's 2011 *The Elder Scrolls V: Skyrim* can be played with new, randomly generating quests and content forever. Communities surrounding games modify the virtual worlds they love. Will we start to see gaming communities composing new music and producing remixes to accompany their mods?

Where the thinkers of the tradition of musica poetica left off, we can pick up the controller again. While rhetoric and music was a matter of linking only emotion and figure, we might now envision music—and the world it carries with it, and its cultural manifestations—as arguments themselves. I offer a few similarities between rhetorical terminology and musical elements. We might begin to think of melody in terms of central argument or, in compositional terms, thesis. The melody of a song is its most often remembered and recognized characteristic. Though musical compositions of games are often complex, with digital percussion, rhythmic bass, arpeggiated accompaniments, harmonies, and counterpoints, the melody is what is picked up by the listener and often the only surviving element in a remix or recreation of a game song. Certain features emphasize game melodies that might be mixed more loudly against other tracks, panned hard to one side in a stereophonic mix, or appear at a higher octave as to be distinguished from the mix. Similarly, writers and speakers applying rhetorical strategies to their work will be sure to emphasize their thesis, which is often the primary element that is carried away by the reader or listener. The argument is to be remembered, and even to be shared, reborn.

Further, the looped nature of game music plays with the rhetorical feature of repetition. Repetition in rhetorical style takes many forms. Anaphora is the repetition of the same word at the beginning of several lines and is a common move even in contemporary speeches, just as it was in classical oratory. We see it in countless political speeches advocating for an endless number of positions, beliefs, and ideas, from John F. Kennedy's delivery to the Texas Democratic State Committee on November 22nd, 1963 to George W. Bush's First Inaugural Address on January 20th, 2001:

We will defend our allies and our interests.

We will show purpose without arrogance.

We will meet aggression and bad faith with resolve and strength. (n.p.)

Game music employs anaphora in a similar sense in that it repeats a melodic phrase, or riff, in a consecutive pattern, where the first section of a musical sequence might be the same, followed by a variation, such as the snippet from :14 to :20 in the 16-bit version of the "Hyrule Overworld Theme" from Nintendo's *The Legend of Zelda: A Link to the Past*. Though anaphoric phrases do not necessarily constitute main themes or melodies in themselves, they offer a consistency of repetition

that may appeal to the listener. As in rhetorical oratory, rhetorical music must find a careful balance between repetition and variation, so as to not confuse the listener but to maintain emphasis.

I suggest, too, that we may think about tonal qualities in terms of mood. We might consider the example of David Wise's "Aquatic Ambience" from Nintendo's 16-bit *Donkey Kong Country* in 1994. We may note immediately that the rhythm/backing has considerable digital delay applied to the track. This rippling, wavering tonality creates a bond with the underwater aesthetic of the game world. Similarly, the physics of the underwater gameplay allow the user to bounce freely, much like the smooth, lazy, nearly syncopated rhythmic quality of the piece, as if to create the sense that the user were actually swimming. We must understand, though, that swimming, as a human experience, has no soundtrack. We might hear a sort of gushing, heavy static when we are underwater. But the composer has attempted to create an extension of the experience of swimming: the slow but forceful gliding of the limbs, the sense of open exploration and freedom, the tranquil and relaxing qualities of submergence. The melody of the song does not appear until 1:30, perhaps to give new breath to the otherwise ambient quality of the track (as the name suggests). Thus, the delay of the track's opening sequence from :00 to :45 is a tonal adjustment, a compositional choice, an effect that has been added to a primary recording of a digitally synthesized piano. And it is an effect that causes a sort of mood. Similarly, great oratorical rhetoricians will set a mood for their audience, a tone, to increase immersion, to acquire ethos, to play with pathos, to create logical consistency between message and feeling. Here, we have a valency, a meaningful connection between game song and world: The ripples of delay mirror—and even add to—the virtual ripples of the water.

Early game music also relied on a double-riff structure as a building block, exemplified by the 1986 *Legend of Zelda's* "Overworld Theme" which features the simultaneity of two lines. Whether these lines are mutually exclusive melodies or not, they define the arrangement. My intent is not to evaluate the relative merits of contrapuntal or polyphonic arrangement. It seems that a switch to a prevalent homophonic aesthetic occurs by the 64-bit console era. The homophonic structure, or what we might think of as the riff-over-complement trope (rather than the contrapuntal riff-and-complement), is the backbone of traditional American folk singer-songwriter compositions (voice-over-chords), German romantic art songs (e.g., Schubert's "Heidenröslein" from 1815), and contemporary vocal-over-pianist or "karaoke" combinations. It follows from the earlier analysis that we emphasize our melodies. However, vocal music has become the dominant form in contemporary popular composition. This is the arrangement with which we are familiar: melody and backdrop. Whether this is dangerously ideological or just a coincidence of artistic trends or development, I am not sure.

We might consider other practical elements of game music as well. Pitch shifts, or modulation of a track, could be thought of in terms of parallelism. Or we might consider the beats per minute (BPM) of a track as a similar quality to pacing and flow. Or we might consider polyphony (layering) in a track in terms of the depth or complexity of an argument; however, it would be wrong to privilege complex musical compositions just as it would be to privilege densely written arguments because brevity is so often key to rhetorical success. And of course we can think about time—the looped nature of game music is much different than the beginning/middle/end of a rhetorical speech or a five-paragraph-essay—in many ways as it applies to music. My point here is that the possibilities are inexhaustible. I only hope these building blocks here are those of a free music. I do not wish to chart a direct correspondence between oral and written rhetoric to a new rhetoric of music. But to place a musical rhetoric in these familiar terms is a place to start a conversation toward understanding how music works argumentatively. We

might wonder what music can help us learn about rhetoric, as well as what rhetoric can help us learn about music, and what both relationships can help us understand about human existence.

What this entails, then, is a few considerations. Recent research in new/digital media studies has focused on the externalized mind and its relation to media (e.g., Manovich 2002). Lisa Gitelman (2008), in her *Always Already New: Media, History, and the Data of Culture* brought a call of attention to the "logic of repetition." Similarly, Jaron Lanier (2010) discussed the "loop" in his *You Are Not a Gadget: A Manifesto*. The repetitious, looped riff was a necessary solution to saving space when memory was lacking in the early years of arcade games and later console game programming. While I will avoid designations of determinism and their relations to hardware limitations, it is important to note that there is often a symbiotic give and take between a medium and its standardized boundaries. While technological progression is not perfectly linear, modules must at one point be deemed "complete" in order for production to take place. These finished systems become the composite playground for developer, composer, and consumer.

We might move on, then, to the rhetorical heuristic of considering speaker, text, and audience, which may be helpful to answer a few questions and to begin to at least grapple with the various issues and problems surrounding a rhetorical analysis of game music. Regarding matters of the speaker, one might ask several questions about a game song. Who was its composer and what is their affiliation with the development team: independent or incorporated? We can borrow terminology from film theory: Is the song in question diegetic, extra-diegetic, or does it transverse these boundaries? The fundamental issue of speaker rests here: What is the origin of the song, both in the physical world of composing as well as the virtual reality of the game world? One might follow this line of investigation further. In *Ocarina of Time*, the enchanting "Zelda's Lullaby" is both an extra-diegetic element that plays as background during a dialogue scene in the castle garden as well as a song that one can "learn to play" on the virtual ocarina in order to cause effect in the game, such as opening a secret passage behind a waterfall. Thus, the player of the game becomes a player of music, even a composer. Even further, as the individual moves along in the narrative of the game, it is discovered that this lullaby is a song known only to the in-game royal family, thus playing the song at pivotal moments allows the player special permissions and noble associations. Here, conceptualizing the song as speaker leads to fascinatingly multi-faceted and rich rhetorical experimentations. Granting a song speakership, and thus narrative and argumentative power, is to investigate, too, its ethos. Does the melody have an in-game meaning? Does it have a history? How much information can music carry with it? And how do we de-code these messages?

It is evident that an analysis of the text itself is also necessary. This point might consider issues surrounding the medium of delivery. How does the song appear to us? Those games of special interest here, those in the boom of the home console era of gaming, would most likely be played—and thus heard—through mid-to-high end home televisions with attached tweeters. The transmission of the game sound is likely to have been relatively low fidelity. Further, the score for *Ocarina of Time*, again as an example, was composed in MIDI. These could be considered textual elements. Textual elements might also include actual tonal qualities of the music. We might consider in this dimension the historical situation of the song and if it might be "classified," though these sorts of nominations are often reductive, problematic, or specific to advanced knowledge of music history and theory: Is it more like a bossa nova or a samba, and what would either distinction mean for its relation to the game and the user's experience? This sort of privileged and specialized knowledge does not seem like a necessary factor for investigating the rhetorical situation of a game song. However,

it may be helpful in understanding its textuality and thus allowing the analyst a particular ethos in determining the metrical, rhythmic configurations at play.

Audience must also be considered. This does not simply mean "Who is the intended audience for this piece" but might even extend to evaluating the success of the piece in maintaining universality. Of course there are core markets for which games are designed, but with the increasing computational power of mobile devices, games have seen a widening field. Audience, here, is two-fold. There is the potential for the in-game character to hear diegetic music and there is the player, the user who listens. How much of the user's consciousness is embedded in an avatar? How immersed are we in our games? Each experience is different, surely. Considerations of audience might also garner manifold concerns regarding race, class, and gender. Has a particular composition somehow excluded a marginalized group from understanding or even progressing in-game? And there are problems relating to the physical nature of sound as well. How much is ingrained and how much is enculturated? We might seek explorations, somewhere between the studies of neuroscience and physics. How do sound waves relate to human emotion? Biofeedback on gamers hearing audio and those without would reveal some mysteries.

Since the aforementioned questions are beyond the scope of this chapter, I will continue to open a dialogue between rhetoric and music and now turn to three more distinctions and possible lines of thought. I will suggest two in-game properties to analyze: first, of univalence to discuss what we can think of as the logos of a game song, then of diegesis, or meaning in narrative context. Then, I will suggest a meta-level of rhetorical analysis. This is the study of the ethos of game music. Determining the ethos of game music is, altogether, a separate question and why it has warranted special attention in the context of the work here, at large. It is helpful to think of game song and game environment in terms of valency. The question is, to what degree is there symbiosis or connectivity between the rhetorical moves of a game song and the virtual reality that is being displayed? We might note that many game songs attempt something like direct valency, or a symmetry between the visual world represented and the song itself, such as various Nintendo composers' recurrent attempts at "desert" music, which always seems to be some kind of harmonic minor progression, somewhere between a cartoonish snake charmer ditty and the neo-classical, seen from *Mario Party 3's* "Spiny Desert Song" (2000) or in *Super Mario 3D Land* (2011). These direct attributions are inherently dangerous. Examples such as the common Western association of the flute with "dainty" femininity or the trumpet with "bold" masculinity are clear examples (Richards 2009).

This outlines the fundamental problem in associating rhetoric to music. It requires one to imagine that there is something resembling a one-to-one connection between the physical characteristics of a musical figure and either a human emotion, idea, argument, or meaning. Rather than assume that there are absolute connections, I take an equivocal ground. Of course, rapid tempos and well-timed cymbal crashes might jolt the player and thus result in physiological reactions such as an increased heart rate and maybe excitement. The Russian folk song "Korobeiniki," adapted as the main theme for *Tetris*, stands testament to this very simple reality about the relationship between cognition, body, and game song.

Where the baroque composers and theorists found too much ambiguity, we can find solid ground. Game music differs from art music in that game music has an experience directly connected to it. Game music differs from art music in that it does not assume a universal human experience or emotion, but necessitates that a programmatic representation of a reality be coupled with it. Thus, where there is univalence in much art music, there is predetermined valence in game music. However, the game music can be easily extracted,

reperformed, orchestrated, analogized. This does not separate it from its necessary history and meaning. Game music is narrative. Game music carries a particular universe and world with it. Thus, to evaluate the history of game music is to pay mind to the world that it carries. It will pay to differentiate this from other forms of media music experience. The necessary bond between linear, chronological cinematographic progression, for example, and the linear, chronological auditory progression formulate a direct bond, pushing for a necessary reading. The minor key progression coupled with the scene of two lovers in their final embrace signals an expectation. However, the more open-ended, exploratory nature of games coupled with the cyclical trope of looping allows for individual assignation of meaning to song in the larger picture of the game world that is carried.

Ethos might be best applied to the music-out-of-game, as it is analogized, re-digitized, interpreted, and tossed about the globe. The ethos of game music is, altogether, a separate question and why it has warranted special attention in the context of the work here, at large. While I have attempted, perhaps much like the early theorists of *musica poetica*, to break down a medium into manageable figures, pieces, and structures so that we might begin to talk about the rhetorical effectiveness of tropes and themes, the question of ethos is one that involves communication, interaction, and a greater cultural interest in the place of bit art and game music as a credible transmission of an argument. Game music should not be reduced to merely "background" or "mood" music. While it may have purposeful pathetic elements, its function is not limited to inciting a particular feeling or response in the player or user of a game technology. Rather, the ethos of game music is the question of the potential for survival of its kernel, its melody, its driving force. Can it exist outside of the valent bond linking game world and musical score? Will it be able to carry the meaning of the game outside its self? Will it transverse the digital to the analog? Ethos, here, is perhaps even a question of memetic value and longevity. I argue that game music constitutes a new digital folk modality rather than a genre or a medium. Much like American folk music carries a history, a legacy, a struggle, and a story, so, too, does the music of games. The evocative power of a single phrase—say, those first eight bars of the melody from David Wise's "Aquatic Ambience"—has the ability to evoke in the player, even outside of gameplay, the aesthetic world of that game, its tonal characteristics, its physics, its narrative, and its in-game history and geography. I know, though, that we cannot disregard that mysterious ability of music to move us emotionally, for some even spiritually, or otherwise. The ability for game music to evoke the recollection of entire worlds, an experiential memory, contributes to the ethos of game music, particularly when it is re-deployed in other contexts, whether in an audio remix or simply mentioned in a written text.

We must, of course, understand that game music is plural. But I have intentionally concentrated on original works by composers in the home console era I have defined because game music sometimes plays from classical pieces or adopts popular music. Yet, it is the original compositions of game music that have been the fodder for remix in our digital culture on sites like YouTube and on the physical stage. The strength of game music—and its core ethos—is that it has an experiential memory already tied to it. There can be secondary attributions—"who was I with and what was the weather like while I was playing this game," for example—but the shared connection is its first element of credibility. For example, a friend's cell phone rings and a brief snippet of a "Battle Theme" from *Final Fantasy* plays. When recognized, an entire world has been transmitted. It gives ethos to the individual who appreciates the music and has the cultural capital to speak freely about the game, and the game music itself carries a credibility. It is a music that quickly unites and engages individuals, spurs discussion, and locates them in a similar historical trajectory, even if virtual.

Game Music: Rhetoric and Remix

I have suggested that an effective way of approaching the issues I have raised is through an analysis of game music with a rhetorical approach. But we must move even further to understand how game music works in our world. Much like definitions of digital media, contemporary understandings of rhetoric are constantly shifting due, perhaps, to its expanding applications. So, too, the social construct of "music" shifts with time. And often the limitations of technology determine our apprehension of media. The American blogger Andy Baio has published the following analysis from the Whitburn Project on his blog, Waxy.org. The Whitburn Project is a group of individuals who have collected an incredible amount of data from Billboard top songs from the past 120 years and have covertly published the findings. Baio (2008) published this chart after encountering the documents; here are findings for the average length of a song:

- **1950s:** 2:30 (95 songs)
- **1960s:** 2:30 (250 songs)
- **1970s:** 3:30 (153 songs)
- **1980s:** 3:59 (142 songs)
- **1990s:** 4:00 (132 songs)
- **2000s:** 3:50 (58 songs)

We might wonder, of course, why in the 1970s the average song length jumped a minute beyond the 2:30 mark. Or why the capacity of the original 78rpm record, at 3:30, which debuted in the 1890s, seems to still have left its mark on music, considering the advent of the LP record in 1948 which increased recording time capacity. Yet, to this day, the average pop song length stays roughly the same as it was in 1890 (when it was around 2:45) (Baio 2008). The connection here to game music is evident. In the compositional days of 8-bit games, composers had very little time (memory) with which to work. Game songs often continue in this aesthetic tradition with short, melodic phrases as a defining element.

We can go back further to assess the importance of brevity and memory. The early starts of troubadours with their heroic *chanson de gestes* of the 12th-15th century and, of course, the Grecian epics remind us that the capacity of the mind to recall and remember more than one hundred lines of lyric per musical unit (song) is not impossible. Yet, with something like *La Chanson de Roland*, though the stanza lengths are variable, only one stanza in I-LXXXVII breaks a thirty-line cap. Though we can only speculate, one might wonder if the modern chorus started as a trope of repetition: a rhetorical and mnemonic technique. A defining characteristic of digital media is its ability to be broken down into units (Manovich 2002). It might be worth drawing, here, a connection to the early German theorists who attempted to break music down into units of sizable figures and manageable bits as well. Like new theories of digital media, these composers understood the significance of the sample size, the rhetorical reproducibility of the stand-alone fragment.

Beyond memory, we also are drawn to the uncanny of the digital, the human extracted. For example, synthetic reverb can be applied to the ocarina in *Ocarina of Time* by jiggling the control stick during a musical session, but it is not an analog reverb. The ocarina is one of the oldest human instruments, and it emits a tone that is reverberatory and haunting, often designed in a pitch range akin to the soprano spectrum of the human voice. Thus, some human element has been lost by using the standardized MIDI ocarina tone in the game, and the end result is neither analog nor natural. But, and as Karen Collins insisted in her *Game Sounds*, the restricted memory of cartridge-based games of the time lent themselves to MIDI. While we now have fully orchestrated scores for games such as film composer Hans Zimmer's recent work with Activision in the Call of Duty series (2010, 2011), 1998 was a different time for video game composers. Still, is there something in the characteristic beeps of bits? There is even a call to go to music that is lower in digital fidelity

than MIDI, with the recent interest in 8-bit music, which is a very square-sounding tonal mode that has become popular in some subcultures.

In *You Are Not a Gadget: A Manifesto* (2010), Lanier viewed the emergence of MIDI—which was standardized in 1982—as detrimental to the human condition. Worse, still, he wrote, is that it has become a concept that is "locked in." That is, he worries that, dependent on human survival, hundreds or thousands of years from now, we will still be stuck with MIDI tones and tunes, all around us. Yet, the limitation of definition has been around for a while. Lanier (2010) wrote:

People have played musical notes for a very long time. One of the oldest human-hewn extant artifacts is a flute that appears to have been made by Neanderthals about 75,000 years ago. The flute plays approximately in tune. Therefore it is likely that whoever played that old flute had a notion of discrete toots. So the idea of the note goes back very far indeed.

But as I pointed out earlier, no single, precise idea of a note was ever a mandatory part of the process of making music until the early 1980s, when MIDI appeared. Certainly, various ideas about notes were used to notate music before then, as well as to teach and to analyze, but the phenomenon of music was bigger than the concept of a note. (p. 134)

Yet, it seems that we are always playing with boundaries of notes, of meaning, and of music. Though Lanier—like Sherry Turkle (2011), in her *Alone Together*—might suggest that computer technologies are de-humanizing our lives, the MIDI music of *Ocarina of Time* has inspired creative re-interpretations, re-orchestrations, live performances (most notably *Play! A Video Game Symphony* and *The Hyrule Symphony*, which features the live performance of an actual ocarina that substitutes the MIDI ocarina melodies from the game). Clearly Kondo's work has inspired a generation, in some cases even working as the driving desire to pick up and learn an instrument in the first place: and often it is a self-education.

In 1999, Sharon King noted in the *New York Times* that since the 1998 release of *Ocarina of Time,* sales of actual sweet potato ocarinas had nearly doubled. This might be one corner of the intersection of virtual reality, music, and digital technology that would make Lanier smile.

While most musical albums are often recorded to a metronome, or nowadays, a digital click-track, with musicians in isolated booths or in the isolation of over-dubbing tracks to reduce microphone bleed, folk music is usually recorded "live" in the studio, with members together in the same room, to capture as much of the dynamicism and energy as might be possible. This allows for spontaneity in the form of improvisation, something like kairos, perhaps, as well as the room for strange (but sometimes "happy" mistakes) or inspired occurrences of chance, called aleatoricism. Unlike improvisation, aleatoricism allows for the player, the musician, to break free from pre-determined styles, modes, structures, and genres. Though game music is in some ways aleatoric, in that the player often can use music to alter reality, or the player can trigger certain sounds or tracks to play from their location, movement, or action in the game, there is much room for aleatoric experience in the recreation of game music. To follow the example of David Wise's "Aquatic Ambience," one can find hundreds of covers and remixes of just this one particular song on YouTube: some use a heavily distorted electric guitar in the metal style, some very moving acoustic arrangements, piano interpretations, drum and bass, techno. There are even theremins and beer bottles in the mix. Some are faithful to the original score, but many challenge the notion of a stable song. Some are flawless recreations, recorded cleanly and mixed pleasingly. Some are hasty and energetic improvisations, some shakily put together with bad lighting and peaking audio. But we look past this.

Figure 1. Koji Kondo's "Gerudo Valley Theme" re-interpreted for the marimba 11 years after the release of Ocarina. Screenshot from YouTube.com, user Mart0zz. Public image. 2008.

For in these videos there are the tracks of a world absorbed and recommunicated, an ethos related to the game that trumps the ethos of musicianship. For, musicians must transfer from one world to another: from digital to analog, from fictional worlds with different laws to a new world of our own. And here is where the ethos of game music is truly evident. It has inspired a cultural wave of recreation, of experimentation, of play. It has become a game outside the game itself, to mold and to bend and to make anew.

The looped nature of game music opens it up for criticism, then, but game music is not simply looped endlessly, relegated to the non-diegetic, background position. In Nintendo's *Ocarina of Time*, the ability for the player to use the virtual ocarina in the game allowed for diegetic control of sound. The in-game use of the ocarina alters the virtual reality, paralleling a very human use of music in everyday life. Similarly, and as Karen Collins (Collins) noted, some attempts at randomizing, or including what we might think of as "natural" musical elements such as improvisation and aleatoricism, are included in *Ocarina*'s program, quite explicitly:

Open form has been used successfully in a number of games. The Legend of Zelda: Ocarina of Time, for instance, used a variable sequencing structure in the Hyrule section of the gameplay. Since the player must spend a considerable amount of time running across Hyrule Field to gain access to other important gameplay areas, the sequences were played randomly to maintain interest and diversity. Every time the game was played, the song would play different. (p. 158)

While such an attempt would not satisfy Lanier—nor does it comfort me, entirely—it is a place to start. The sequences are still pre-determined, merely shuffled, in a sense. Such as it is, *Ocarina* is full of music, from the twelve trigger songs the player must learn in various areas of the game to advance and eventually reach the game's end to

the hundreds of non-diegetic compositions and snippets, including the "Hyrule Theme" (which is a re-interpretation of Kondo's own *The Legend of Zelda* main theme from 1988). It is no doubt that Kondo's work in *Ocarina* redefined the potential cultural impact of music in gaming and stretched the limits of MIDI, so much that these songs are still being played, re-imagined, and hummed happily around the world. Amateur and professional musicians alike have participated in the dismantling of their digitalness, their MIDI tonality, and have performed and recorded and even fully orchestrated these works with physical instruments, thereby lifting them from their technological embeddedness and breathing new, "human" life into them. But were they never human to begin with, and does re-playing them outside of the virtual reality ever truly de-situate them from their origin?

The simple answer of analogization might help us deal with these issues. A melody composed in MIDI is not forever digitized. Even still, digital musical production must not be de-humanizing. I mean that it is our responsibility to work with it, give and take, without succumbing to it. Anne R. Richards (2009), in her "Music, Transtextuality, and the World Wide Web," argued for regarding "aurality" as a significant factor in hypermedia. Games, of course, whether they appear as applications on mobile devices such as iPhones, high-end console devices, or are hosted on Internet sites, make use of sound in ways that other new digital media do not. Sounds in games are triggered by player interaction. Game music is changing, expanding, breaking boundaries, situated distinctly in a market niche but trying to be art, at times, and sometimes succeeding.

Jim Guthrie's recent work for independent studio Capybara Games Inc.'s *Superbrothers: Sword and Sorcery EP* game (2011), available as an application for devices with Macintosh's mobile operating system, is one such title that breaks boundaries between traditional game music and contemporary composition. The game's

Figure 2. Sword and Sorcery EP screenshot, Copyright Capybara Games Inc., 2011. The game's visuals incorporate a blend of pixelated "retro" "bit"-style with contemporary, high-contrast design. This blend of old and new closely parallels the musical style. Note the diffused light beam (hi-fidelity) against the "blocky" (lo-fidelity) hillside. Used with permission.

title reveals its close ties with music: It takes the musical "EP" (extended play), as if to suggest, perhaps, that here the game and album have been blended. Guthrie concurrently released *Sword and Sorcery LP* as a musical album, which includes 27 tracks from the "*EP.*" The game is a mix-genre homage/parody of 16 and 32-bit era side-scrolling and adventure style games, with music that blends the driving melodic phraseology born of the square sound of digitality customary to that period (early to mid 1990s) with organic tonalities, such as what might be "actual" glissandos recorded with a physical—or simulated—bass (Guthrie's "Lone Star"). Shinji Hosoe, composer for FromSoftware's 2010 release *3D Dot Game Heroes*, used a similar hybrid aesthetic to blend the lines between bit and physical, analog and digital. Will this hybridity, though, eventually lead to assimilation between the physical and digital, to where we can no longer distinguish, or must

the square digital 8-bit sound be forever doomed to seem merely digital, having become a culture norm? Or, will the digital and the physical emerge converged, a post-MIDI digital musical reality that can somehow have all the nuanced realities of the physical bow and string?

Bouënard et. al (2010) discussed their research in creating virtual instrument interfaces—and their drawbacks—in "Virtual Gesture Control and Synthesis of Music Performances: Qualitative Evaluation of Synthesized Timpani Exercises," explaining their attempts toward creating a virtual timpani player, linking realistic fluid images to sound. This work demonstrates, especially, the linkage between the physicality of musical performance and the analog quality of musicality. This apparent realism is still a subjective evaluation, however. Here, the physical activity of instrumenting the timpani is also part of the virtual representation of playing, much like the player adopting the ocarina in *Ocarina of Time*. But do we want this "realism"? And is it "realism" such that it could replace the physical timpani, or are we much further away? How much of music must be felt with the arms, the fingertips? Is it sufficient to deceive some, and what would that mean? With Guthrie's work, the blend of the retro with the "real" creates a hybrid style with pathos intended to be drenched in both nostalgia and a push against, or for, something new. That something new, perhaps, is the problematic reality of those born surrounded by integrated digital technology.

Alan Lomax (1968), collector of folk music and renowned ethnomusicologist, discussed in his seminal *Folk Song Style and Culture* the "redundancy" of vocal music, in contrast with the changing, spontaneous nature of speech acts: "Because of its heightened redundancy, singing attracts and holds the attention of groups; indeed, as in most primitive societies, it invites group participation" (p. 3). This conception seems to be at loggerheads with Lanier's (2010) vexation toward digital music production. Game music, until the turn to disc-based console platforms, was largely repetitive and looped endlessly, as Collins (2008) noted above. The break begins, perhaps, with *Ocarina*, where the twelve ocarina specific melodies, through the MIDI embodiment of the ocarina, attempt a sort of human vocal phraseology. That is, the melodies are ocarina-driven, and the ocarina resembles, in many ways, the tonality of the human voice.

I'll ask again: Is this what we want? Lomax (1968) wrote:

At first glance it would seem impossible that any group of individuals could delicately calibrate the manifold levels of phonation essential to produce a unified choral sound—the syllables, intervals, glides, attacks, releases, levels of emphasis, many voice qualities, etc. Nevertheless, singing with blended voices, like marching in step, is found on all six continents, some individuals and societies being more given to it than others. (p. 71)

These are human nuances, subtle, masterful techniques. Yet they are universally present. Does this mean, like the timpani, we will start seeing the emergence of attempted virtual representations of the voice, that is, if we can emulate each other to join the group, will the digital enter the chorus or will we kick it away? And where is its place? Can we start to think about the relationship between folk culture and digital culture? Is Eric Whitacre's virtual choir a call to McLuhan's global village? Can we start to think about the possibility for universality from digital music? These digital compositions from Kondo in *Ocarina of Time* have been the impetus for collaboration and artistic experimentation in the form of re-orchestration among non-professional associations as well as philharmonic orchestras, as well as through social media and electronic commerce ("Koji Kondo"), and through the live performance of these songs. There may not be an immediate solution to mediate the gap between the extremism of Glendinnings anti-digital "Neo-Luddism" or Lanier's watch-guarding of the organic human experience with the emergence of new digital technologies and the music which is produced with them (MIDI, multi-track audio editing software, solid state amplification, inexpensive computer microphones, and on and on) or produced for them (short web films, video games, mobile device applications). Still, the cultural effects of the work of Koji Kondo cannot be disregarded; now shines the vigilance of a new generation of musicians playing about in the gulf between low and high fidelity, digital and analog, plugged and unplugged.

Yet, Lanier's warning of the lock-in is a necessary reminder that we must retain the ability to distance ourselves from the digital so as to not become so absorbed that the self-reflection and awareness necessary to "flip the switch" becomes lost. Yes, perhaps there is another Luddism that co-exists with technology, a Luddism that instead of seeking to destroy would rather reconfigure the systems, the artifacts to something sustainable, something human again. And this is why in part there is still music in the streets, because MIDI melodies can awake at the touch of trumpet valves. These are the prevailing problems, this divide we place between ourselves and our technology, for better or for worse (or for neither).

Acknowledging Limitations

As with all writing and ideas, I expect the reader to find points of contestation, irritation, and disbelief with what I have set out to do here. The drawback of examining a "retro" medium has been stated already, despite its cultural ubiquity. I believe firmly that the prominence of the remediation of these forms displays their wide appeal and inherent success as argumentative pieces. Game music has gained popularity because it carries with it memories of entire worlds, each world distinct to the listener, but each world shared. The impetus to recreate this music is a demonstration of fascination with the game world as represented by the game song, and it also demonstrates the ability to harness a virtual ethos for use in the real world.

I can imagine several counter-arguments in advance and points of repudiation. I cannot address all of them, but will discuss several, here, to suggest some ways of working around them. The problems are (at least) as follows:

1. Game music is nostalgic, banal, and of no cultural value.
2. These ideas lose value when applied to contemporary game music that is instrumented physically.
3. The egregious amount of "bad" video game compositions outweigh the worthy.
4. Attention is being distracted from broader theories of music, technology, rhetoric, and being.
5. The greatest problems of *musica poetica* have been here repeated: ambiguity, structurelessness.

Though all of these issues deserve an extraordinary amount of attention, I will attempt here instead to dispel some myths and reassert some points to clarify my argument. There are, of course, limitless other concerns, many having to do with the changing nature of technology. Some posit that 24-bit depth audio is now of such a quality that analog and digital signals cannot be distinguished even by the trained ear and mind of an audio expert. This is of much concern to audiophiles, and rightly so. However, to concentrate on such technical differences might be to distract from greater concerns over purpose, meaning, and context. It should be noted, too, that memetic theory is often disregarded by scholars or seen as outdated, but as an ethotic element of rhetorical survival, and in the context of remediation and digital media, it stands as a meaningful way to assess cultural success and mass rhetorical appeal.

Game music is neither (entirely) nostalgic nor banal, but rather an important factor of, at least, American culture. The video game *Angry Birds* has been downloaded 500 million times across the world (Hodapp). This is a game most often played on a mobile device. While debates will continue over the artistic value of games and the status of high art and digital media, it is clear that digital games are increasingly important to life. The principles I have outlined and suggested are intended to open a new dialogue between digital media studies, game studies, rhetoric, and music. Though I have focused on a particular time period in the history of video games to elucidate my argument, I have not done so to purposefully exclude other types of game music, but only to solidify and hone in on particular phenomena. I do not see why considerations of rhetoric, valence, ethos, and analogization could not be analyzed in games that have fully orchestrated scores. These games whose music begins as analog then becomes digital are, perhaps, of equal interest, for varied reasons. I have examined here music that primarily began as bits, moved to someone's fingertips on a piano, back out into the world as analog waves, into a low-end computer microphone, and then uploaded to YouTube. Other potential patterns of digitization, analogization, instrumentation, and composition are of course of value and interest to both the rhetorician and the media scholar.

Though questions of virtual reality open up new issues in thinking about being, this is not the aim, here. I have, instead, attempted to rekindle interest in the dusty trail that begins where rhetoric and music converge. Especially of interest has been the issue of ethos, and game music clearly exemplifies the problematic of credibility. It is a medium that carries weight and influence, and it is a medium that is often subject to severe criticism and global misunderstanding. The incredible complexity of game music asserts its place in a world where games are no longer mere toys or objects of mere play but are worthy and influential tools, channels of engrossment into human understanding and meaning. I have attempted to provide some approaches for thinking about the rhetoric of game music and thus assess its credibility in all its forms. In doing so, I hope I have not reduced the masterful works of so many composers but instead shed new light upon them.

EXPLORING NEW WORLDS

Further areas of research interest abound. I have, in casting definitions of game music and its relationship to rhetoric, attempted to open a discussion, though I have had to set some parameters to qualify the scope of my argument. I encourage further research into the creation of musical compositions for games—exploring invention, especially—and how technological factors influence aesthetic choices. I can envision a prosperous future where applications of rhetoric

find their way into digital media studies and into considerations of aurality. I hope, too, to find new meaning beyond the binary of digital/analog to what exists under these categorizations. We must rethink rhetoric, code, language, and music to reunderstand them in new contexts: not just in the context of the games themselves but how those games influence our realities, effect our existing and emerging cultures, and lead us to live.

The bit aesthetic has staked a claim in cultural awareness perhaps because of its obtuse digitality. It is obviously representative of the units that compose digital media objects. But this has come to be both an embrace of the aesthetic and a critique of the retro. Nostalgia and prejudice have not had much a place in media studies, nor media histories, yet they are powerful human persuasions. Bit art carries a rhetorical importance somewhere between reality and representation, a new ground to investigate. As Slovenian philosopher Slavoj Žižek (2007) noted: "Virtual reality is a miserable idea" (qtd. In Wright, n.p.). When we game, are we just submerging ourselves even further in ideology? Or can game music help us think about universality, again? Is there room for continental philosophy in digital media studies? I hope so.

Evaluating game music rhetorically allows for the establishment of credibility. I have provided here only a brief introduction to the possibilities of assessing game music reproductions of game music as cultural objects. There is much more to be said about the relationship between music and rhetoric, as well as the ethos of game music. The omnipresence of digital bit-style music does not give it distinction. The individual must look at how the situatedness of a piece of game music works: to what does it harken? What is its story and inspiration and why does that matter? How was it composed and how was it digitalized? What instrumented musical style does it most resemble and does that have valency with the game world? How does the game song stand on its own? Beginning with these questions, it is possible to establish a place in the world for the meaningful assessment of the music of games. And perhaps we can eventually draw conclusions between the ties of music and the physical universe.

CONCLUSION

Operatic aria performance no longer belongs solely to the domain of Aida. Approximations of the requiem mass of Mozart are fitted into the metadiegetic reverberatory melodies of Koji Kondo in *The Legend of Zelda: Ocarina of Time*. Summon the digital hero from sleep, give her a virtual instrument and she appears in the narrative. Some questions still remain—such as "Is this art?"—and have largely acted as distractions from important considerations at hand. We can forego these controversial matters and examine instead the rhetorical qualities of game music. Should contemporary composers such as Kondo, Uematsu, and Masuda stand in the halls of Vivaldi, Schubert, and Borodin? The answer is undoubtedly yes. And, while the German baroque composers applied a knowledge of classical rhetoric to their compositions and theorized about these connections in the tradition of *musica poetica*, contemporary game composers apply new types of rhetoric to new rhetorical situations. It is a matter of not just attempting to elicit an emotion but to elicit a new reality.

The ethos, the *character* of these early game songs, is one that invites. Their very tonality may date them, yet they have not died. Unlike so much of our technology, which has been discarded, replaced, or "upgraded," game songs live on. They are remixed and remade. Performed on stage. Recorded with webcams in basements and bedrooms and garages. And it seems to be an authentic appreciation, a genuine fascination with a not-too distant past. Whether it is the beeping of a thirty

second 8-Bit Nintendo loop or a three-minute 64-bit bossa nova with an unforgettable melody, gamers seem to feel a strong sense of nostalgia, of history, of connection. Real, physical, human connections are made instantly over the shared recognition of ten, fifteen simple notes, and the connections stem from identification and identity centered around ethos. Seldom can so much be conveyed in a paragraph. Seldom can so much be relayed in a novel.

I have suggested here that rhetorical considerations of video game music should be raised. I proposed a rhetorical framework for the digital game auditory environment that can be considered, and should be considered, credible. I have suggested that the rhetoric of music might look beyond the one-to-one correlations suggested by the theorists of *musica poetica*—though we can now draw some connections with familiar language, such as riffs as repetition, arpeggios against melodies as contrast—but instead call upon its inherent ethos, the ethos of this form of game music with all its power to carry meaning. In expanding on the work done in the relationship between games and rhetoric—especially the procedural rhetoric of Bogost (2007) and ever-dated applications of literary theory and analysis—I put forth the notion of an atmospheric and existential rhetoric that addresses issues of reality, transmission, awareness, digitization, and analogization. I suggest, though, that there is still much room to explore this intersection at which the elusive but present tonalities of games meet the formidable scholarship of rhetorical studies and nascent interest in digital media. Here we can sing anew our songs anew. Here we can sing all new our songs all new.

REFERENCES

Baio, A. (2008). The Whitburn Project: 120 years of music chart history. *Waxy.org.* Retrieved from http://waxy.org/2008/05/the_whitburn_project/

Bogost, I. (2007). *Persuasive games: The expressive power of videogames.* Cambridge, MA: MIT Press.

Bogost, I. (2011). *How to do things with video games.* Minneapolis, MN: University of Minnesota Press.

Bouënard, A., Wanderly, M. M., & Gibet, S. (2010). Gesture control of sound synthesis: Analysis and classification of percussion gestures. *Acta Acustica united with Acustica, Special Issue on Natural and Virtual Instruments: Control. Gesture and Player Interaction, 96,* 668–677.

Burmeister, J. (1993). *Musical poetics* (Rivera, B. V., Trans.). Cambridge, MA: Harvard University Press.

Bush, G. W. (2001). First inaugural address. Retrieved from http://www.bartleby.com/124/pres66.html

Collins, K. (2008). *Game sound: An introduction to the history, theory, and practice of video game music and sound design.* Cambridge, MA: MIT Press.

Ebert, R. (2010). Video games can never be art. *Roger Ebert's Blog (Chicago-Sun Times).* Retrieved from http://blogs.suntimes.com/ebert/2010/04/video_games_can_never_be_art.html

Frith, S., Goodwin, L., & Grossberg, L. (Eds.). (1993). *Sound and vision: The music video reader.* New York, NY: Routledge.

Gitelman, L. (2008). *Always already new: Media history and the data of culture.* Cambridge, MA: MIT Press.

Glendinning, C. (1990). Notes toward a neo-Luddite manifesto. *The Utne Reader, 38,* 50–53.

Hodapp, E. (2012). *Angry Birds* hits half a billion downloads. *Touch Arcade.* Retrieved from http://toucharcade.com/2011/11/02/angry-birds-hits-half-a-billion-downloads/

Kondo, K. (n.d.). *Bandcamp.* Retrieved from http://bandcamp.com/tag/koji-kondo

Lanier, J. (2010). *You are not a gadget: A manifesto.* New York, NY: Knopf.

Lomax, A. (Ed.). (1968). *Folk song style and culture.* Washington, DC: American Association for the Advancement of Science.

Manovich, L. (2002). *The language of new media.* Cambridge, MA: MIT Press.

McGonigal, J. (2011). *Reality is broken: Why games make us better and how they can change the world.* New York, NY: The Penguin Press.

Richards, A. R. (2009). Music, transtextuality, and the world wide web. *Technical Communication Quarterly, 18,* 188–209. doi:10.1080/10572250802708337

Stravinsky, I. (1993). *Poetics of music in the form of six lessons.* Cambridge, MA: Harvard University Press.

Wright, B. (2007). *The reality of the virtual (Film lecture by Slavoj Žižek).* Saint Charles, IL: Olive Films.

ADDITIONAL READING

Amerika, M. (2004). Expanding the concept of writing: Notes on net art, digital narrative and viral ethics. *Leonardo, 37*, 9–13. doi:10.1162/002409404772827987

Bartel, D. (1997). *Musica poetica: Musical-rhetorical figures in German baroque music*. Lincoln, NE: University of Nebraska Press.

Bissell, T. (2011). *Extra lives: Why video games matter*. New York, NY: Vintage Books.

Buelow, G. J. (1973). Music, rhetoric, and the concept of the affections: A selective bibliography. *Notes, 30*, 250–259. doi:10.2307/895972

Butler, G. G. (1977). Fugue and rhetoric. *Journal of Music Therapy, 21*(1), 49–109. doi:10.2307/843479

Carr, D., Buckingham, D., Burn, A., & Schott, G. (2006). *Computer games: Text, narrative and play*. Cambridge, UK: Polity Press.

Ciccoricco, D. (2007). "Play, memory": *Shadow of the Colossus* and cognitive workouts. *Dichtung-Digital, 37*. Retrieved from http://dichtung-digital.mewi.unibas.ch/index.htm

Collins, K. (2008). *Game sound: An introduction to the history, theory, and practice of video game music and sound design*. Cambridge, MA: MIT Press.

Dressler, G. (2007). *Praecepta musicae poeticae* (Forgacs, R., Ed.). Champaign, IL: University of Illinois Press.

Floridi, L. (2009). Against digital ontology. *Synthese, 168*, 151–178. doi:10.1007/s11229-008-9334-6

Gee, J. P. (2003). *What video games have to teach us about learning and literacy*. New York, NY: Palgrave Macmillan. doi:10.1145/950566.950595

Geminiani, F. (1969). *A treatise of good taste in the art of musick*. New York, NY: Da Capo Press.

Gibson, J. (2008). A kind of eloquence even in music: Embracing different rhetorics in late seventeenth-century France. *The Journal of Musicology, 25*, 394–433. doi:10.1525/jm.2008.25.4.394

Harrison, D. (1990). Rhetoric and fugue: An analytical application. *Music Theory Spectrum, 12*(1), 1–42. doi:10.2307/746145

Katz, S. (1996). *The epistemic music of rhetoric: Toward the temporal dimension of affect in reader response and writing*. Carbondale, IL: Southern Illinois University Press.

Kim, J. (2001). Phenomenology of digital-being. *Human Studies, 24*, 87–111. doi:10.1023/A:1010763028785

Lindholdt, P. (1997). Luddism and its discontents. *American Quarterly, 49*, 866–873. doi:10.1353/aq.1997.0033

Means, L. (1984). Digitization as transformation: Some implications for the arts. *Leonardo, 17*, 195–199. doi:10.2307/1575190

Morley, D. (2006). *Media, modernity, technology: The geography of the new*. New York, NY: Routledge.

Ohmann, R. (1985). Literacy, technology, and monopoly capital. *College English, 47*, 675–689. doi:10.2307/376973

Selfe, C. L. (2009). The movement of air, the breath of meaning: Aurality and multimodal composing. *College Composition and Communication, 60*, 616–663.

Théberge, P. (2004). The network studio: Historical and technological paths to a new ideal in music making. *Social Studies of Science, 34,* 759–781. doi:10.1177/0306312704047173

Turley, A. C. (2001). Max Weber and the sociology of music. *Sociological Forum, 16,* 633–653. doi:10.1023/A:1012833928688

Well, B. (2002). Art in digital times: From technology to instrument. *Leonardo, 35,* 523–526, 528–537. doi:10.1162/002409402320774349

Zuberi, N. (2007). Is this the future? Black music and technology discourse. *Science Fiction Studies, 34,* 283–300.

KEY TERMS AND DEFINITIONS

Aleatoricism: The occurrence as well as the space (potential) for chance in music.

Analogization: The turning of a digital medium to an analog state. This does not require a reversion to a previous state. Media crafted through digital tools can be made analog.

Bit: In this chapter, bit is most often used as an adjective to describe a square tonality of music or the bit style of visual aesthetics. This is derived from the naming of processors of early console games and their associated musical aesthetic that resulted from limited memory capacity. The Nintendo Entertainment System, first released as the Family Computer (Famicom) in 1983, uses an 8-bit processor. Note that the limitation of the processor is not necessarily related to the lack of memory that resulted in the characteristic sonicality.

Diegesis: Narrative and its carried elements.

Digitization: The turning of an analog medium to a digital state.

Kairos: A conception of time. Contrasted with chronological, it depicts a reality governed by specific moments: opportunism rather than linear determinism.

Loop: A rhetorical trope formulated from the notion of repetition. Entails cyclicality. Game music of the early home console era was looped, usually in order to save memory. Thus, the same "song" is heard over, and over. Compositions were often written, at this time, with these hardware limitation in mind, often by programmers themselves.

MIDI: Musical instrument digital interface. Exemplifies the digitization and standardization of 128 instruments.

Valence: Apparent connectivity between a game song and game environment.

Compilation of References

Aberdour, M. (2007). Achieving quality in open source software. *IEEE Software, 24*(1), 58–64. doi:10.1109/MS.2007.2

Abruzzi, B. E. (2010). Copyright, free expression, and the enforceability of "personal use- only" and other use restrictive online terms of use. *Santa Clara Computer and High-Technology Law Journal, 26*(1), 85–140.

Acharya, A., Cutts, M., Dean, J., Haahr, P., Henzinger, M., Hoelzle, U., et al. (2005). *Information retrieval based on historical data.* (US Patent US 7,346,839 B2).

Adobe. (2011). *Adobe Photoshop CS5, our latest picture and image editor software.* Retrieved November 6, 2011, from http://www.adobe.com/products/photoshop/photoshop/

Agichtein, E., Brill, E., & Dumais, S. (2006). Improving web search ranking by incorporating user behavior information. *Proceedings of the 29th Annual International ACM SIGIR Conference on Research and Development in Information Retrieval* (pp. 19–26). Seattle, WA: ACM. DOI:10.1145/1148170.1148177

Agichtein, E., Brill, E., Dumais, S., & Ragno, R. (2006). Learning user interaction models for predicting web search result preferences. *Proceedings of the 29th Annual International ACM SIGIR Conference on Research and Development in Information Retrieval* (pp. 3–10).

Ahn, T., & Esarey, J. (2008). A dynamic model of generalized social trust. *Journal of Theoretical Politics, 20*(2), 151–180. doi:10.1177/0951629807085816

Alberts, A., Elkind, D., & Ginsberg, S. (2006). The personal fable and risk-taking in early adolescence. *Journal of Youth and Adolescence, 36*, 71–76. doi:10.1007/s10964-006-9144-4

Alberts, W., & van der Geest, T. (2011). Color matters: Color as trustworthiness cue in web sites. *Technical Communication, 58*(2), 149–160.

Alesina, A., & La Ferrara, E. (2002). Who trusts others? *Journal of Public Economics, 85*, 207–234. doi:10.1016/S0047-2727(01)00084-6

Allan, S., & Thorsen, E. (2011). Journalism, public service and BBC News online. In Meikle, G., & Redden, G. (Eds.), *News online: Transformations and continuities* (pp. 20–37). UK: Palgrave MacMillan.

Allard-Huver, F. (2011). *Transformation and circulation of the notion of "risk" in the European Commission.* Unpublished Master Thesis, University of Paris-Sorbonne, Paris.

Alvarez-Torres, M., Mishra, P., & Zhao, Y. (2001). Judging a book by its cover: Cultural stereotyping of interactive media and its effect on the recall of text information. *Journal of Educational Multimedia and Hypermedia, 10*(2), 161–183.

Amaral, M. J., & Monteiro, M. B. (2002). To be without being seen: Computer-mediated communication and social identity management. *Small Group Research, 33*, 575–589. doi:10.1177/104649602237171

American Council of Learned Societies Commission on Cyberinfrastructure for the Humanities and Social Sciences. (2006). *Our cultural commonwealth: The report of the American Council of Learned Societies Commission on Cyberinfrastructure for the Humanities and Social Sciences.* New York, NY: American Council of Learned Societies Commission. Retrieved from http://www.acls.org/uploadedfiles/publications/programs/our_cultural_commonwealth.pdf

Anderson, R., Chan, H., & Perrig, A. (2004). *Key infection: Smart trust for smart dust.* Paper presented at the 12th IEEE International Conference on Network Protocols, Santa Barbara, CA.

Andrews, S. (2007, July 12). Wikipedia vs. the old guard. *PC Pro, 154.* Retrieved November 8, 2010, from http://www.pcpro.co.uk/features/119640/wikipedia-vs-the-old-guard

Antheunis, M. L., Valkenburg, P. A., & Peter, J. (2010). Getting acquainted through social network sites: Testing a model of online uncertainty reduction and social attraction. *Computers in Human Behavior, 26,* 100–109. doi:10.1016/j.chb.2009.07.005

AntonBerg. (2010, October 8). Comment posted to "Toward a more thoughtful conversation on stories." *Reuters.* Retrieved from http://blogs.reuters.com/fulldisclosure/2010/09/27/toward-a-more-thoughtful-conversation-on-stories/?cp=all#comments

Apostel, S., & Folk, M. (2005). First phase information literacy on a fourth generation website: An argument for a new approach to website evaluation criteria. *Computers and Composition Online,* Spring. Retrieved February 5, 2012, from http://www.bgsu.edu/cconline/apostelfolk/c_and_c_online_apostel_folk/apostel_folk.htm

Apostel, S., & Folk, M. (2008). Shifting trends in evaluating the credibility of CMC. In Kelsey, S., & St. Amant, K. (Eds.), *Handbook of research on computer mediated communication* (pp. 185–195). Hershey, PA: Idea Group Reference. doi:10.4018/978-1-59904-863-5.ch014

Apple. (2011). *Mac OS X Lion – The world's most advanced OS.* Retrieved November 6, 2011, from http://www.apple.com/macosx/

Apple. (2012). *Apple in education.* Retrieved May 15, 2012, from http://www.apple.com/education/ibooks-textbooks/

Arditi, L. (2012). Cranston West prayer banner was removed Saturday, school chief says. *Providence Journal Breaking News.* Retrieved from http://news.providencejournal.com/breaking-news/2012/03/cranston-west-p-1.html

Aristotle. (1934). *Nichomachean ethics* (Rackham, H., Trans.). Cambridge, MA: Harvard University Press.

Aristotle. (1991). *On rhetoric: A theory of civic discourse* (Kennedy, G., Trans.). New York, NY: Oxford University Press.

Armstrong, C., & McAdams, M. (2009, April 16). Blogs of information: How gender cues and individual motivations influence perceptions of credibility. *Journal of Computer-Mediated Communication, 14,* 435–456. doi:10.1111/j.1083-6101.2009.01448.x

Associated Press. (2011). *Social media guidelines for AP employees.* Retrieved December 3, 2011, from http://www.ap.org/pages/about/pressreleases/documents/SocialMedia-GuidelinesNov.2011.pdf

Association for College and Research Libraries. (2000). *Information literacy competency standards for higher education.* Retrieved from http://www.ala.org/acrl/standards/informationliteracycompetency

Asur, S., Huberman, B., Szabo, G., & Wang, C. (2011). *Trends in social media: Persistence and decay.* Retrieved January 8, 2012, from http://www.hpl.hp.com/research/scl/papers/trends/trends_web.pdf

Aufderheide, P. (2007). *Documentary films—A very short introduction.* Oxford, UK: Oxford University Press.

Aula, A., Majaranta, P., & Räihä, K. J. (2005). *Eye-tracking reveals the personal styles for search result evaluation* (pp. 1058–1061). Human-Computer Interaction-INTERACT. doi:10.1007/11555261_104

Austria, J. L. (2007). Developing evaluation criteria for podcasts. *Libri, 57*(4), 179-207. Retrieved from http://www.librijournal.org/pdf/2007-4pp179-207.pdf

Axelrod, R. (1984). *The evolution of cooperation.* New York, NY: Basic Books.

Azman, N., Millard, D. D., & Weal, M. J. (2010). *Issues in measuring power and influence in the blogosphere.* Web Science Conference 2010. Retrieved May 20, 2012, from http://journal.webscience.org/344/

Bacharach, M., & Gambetta, D. (2000). Trust in signs. In Cook, K. (Ed.), *Trust in society* (pp. 148–184). New York, NY: Russell Sage.

Backstrom, L., Huttenlocher, D., Kleinberg, J., & Lan, X. (2006). Group formation in large social networks: Membership, growth and evolution. *Proceedings of 12th International Conference on Knowledge Discovery in Data Mining* (pp. 44-54). New York, NY: ACM Press.

Baio, A. (2008). The Whitburn Project: 120 years of music chart history. *Waxy.org*. Retrieved from http://waxy.org/2008/05/the_whitburn_project/

Ball, C. (2012). Logging on: Review of *Profession 2011* section on "Evaluating Digital Scholarship." *Kairos: A Journal of Rhetoric, Technology, and Pedagogy, 16*(2). Retrieved February 4, 2012 from http://www.technorhetoric.net/16.2/loggingon/lo-profession.html

Ball, C. (2004). Show, not tell: The value of new media scholarship. *Computers and Composition, 21,* 403–425.

Ball, C. (2012). Assessing scholarly multimedia: A rhetorical genre studies approach. *Technical Communication Quarterly, 21*(1), 61–77. doi:10.1080/10572252.2012.626390

Banks, A. (2005). *Race, rhetoric, and technology: Searching for higher ground*. New York, NY: Routledge.

Banning, S. A., & Sweetser, K. D. (2007). How much do they think it affects them and whom do they believe? Comparing the third-person effect and credibility of blogs and traditional media. *Communication Quarterly, 55*(4), 451–466. doi:10.1080/01463370701665114

Barlow, A. (2008). *Blogging America: The new public sphere*. Westport, CT: Praeger.

Barry, C. L., & Schamber, L. (1998). Users' criteria for relevance evaluation: A cross-situational comparison. *Information Processing & Management, 34*(2-3), 219–236. doi:10.1016/S0306-4573(97)00078-2

Barthes, R. (1968). *Elements of semiology*. New York, NY: Hill and Wang.

Barthes, R. (1977). *Image, music, text*. New York, NY: Hill and Wang.

Barthes, R. (1978). *Image, music, text* (Heath, S., Trans.). New York, NY: Hill and Wang.

Barzun, J., & Graff, H. (1992). *The modern researcher*. Boston, MA: Houghton Mifflin Co.

Basu, A., Marsh, S., & Dwyer, N. (2012). *Rendering unto Caesar the things that are Caesar's: Complex trust models and human understanding*. Paper presented at the 6th IFIP WG 11.11 International Conference on Trust Management, Surat, India.

Bates, M. J. (1989). The design of browsing and berrypicking techniques for the online search interface. *Online Information Review, 13*(5), 407–424. doi:10.1108/eb024320

Batt, R., Christopherson, S., Rightor, N., Van Jaarsveld, D., & Economic Policy Institute. (2001). *Networking: Work patterns and workforce policies for the new media industry*. Washington, DC: Economic Policy Institute.

Baudrillard, J. (1988). Simulacra and simulation. In Poster, M. (Ed.), *Jean Baudrillard: Selected writings* (pp. 166–184). Cambridge, MA: Polity Press.

Bayer, A. E., & Smart, J. C. (1991). Career publication patterns and collaborative "styles" in American academic science. *The Journal of Higher Education, 62*(6), 613–636. doi:10.2307/1982193

BBC. (2010, March). *Privacy and cookies*. Retrieved from http://www.bbc.co.uk/privacy/

BBC. (2011, January 19). *Terms of use of BBC online services—Personal use*. Retrieved from http://www.bbc.co.uk/terms/personal.shtml#5

Beason, L. (1991). Strategies for establishing an effective persona: An analysis of appeals to ethos in business speeches. *Journal of Business Communication, 28,* 326–347. doi:10.1177/002194369102800403

Beckett, C., & Mansell, R. (2008). Crossing boundaries: new media and networked journalism. *Communication, Culture & Critique, 1*(1), 92–104. doi:10.1111/j.1753-9137.2007.00010.x

Beel, J., Gipp, B., & Wilde, E. (2010). Academic search engine optimization (ASEO): Optimizing scholarly literature for Google Scholar & Co. *Journal of Scholarly Publishing, 41*(2), 176–190. doi:10.3138/jsp.41.2.176

Benkler, Y. (2006). *The wealth of networks*. New Haven, CT: Yale University Press.

Benkler, Y. (2006). *The wealth of networks: How social production transforms markets and freedom*. New Haven, CT: Yale University Press.

Bennett, S., Maton, K., & Kervin, L. (2008). The 'digital natives' debate: A critical review of the evidence. *British Journal of Educational Technology, 39,* 775–786. doi:10.1111/j.1467-8535.2007.00793.x

Berendt, B. (2011). Spam, opinions, and other relationships: Towards a comprehensive view of the Web. In Melucci, M., & Baeza-Yates, R. (Eds.), *Advanced topics in information retrieval* (pp. 51–82). Berlin, Germany: Springer. doi:10.1007/978-3-642-20946-8_3

Berger, C. (2005). Interpersonal communication: Theoretical perspectives, future prospects. *The Journal of Communication, 55*, 415–447. doi:10.1111/j.1460-2466.2005.tb02680.x

Berkman, R. (2004). *The skeptical business searcher: The information advisor's guide to evaluating web data, sites, and sources.* Medford, NJ: Information Today.

Bernal, V. (2006). Diaspora, cyberspace and political imagination: The Eritrean diaspora online. *Global Networks, 6*(2), 161–179. doi:10.1111/j.1471-0374.2006.00139.x

Bizzell, P., & Herzberg, B. (Eds.). (1990). *The rhetorical tradition: Readings from classical times to the present.* Boston, MA: Bedford Books.

Blakeslee, S. (2004). The CRAAP test. *LOEX Quarterly, 31*(3). Retrieved from http://commons.emich.edu/loex-quarterly/vol31/iss3/4

Bloch, A. (2005). The development potential of Zimbabweans in the diaspora: A survey of Zimbabweans living in the UK and South Africa. *IOM Migration Research Series, 17*.

Blythe, S. (2007). Agencies, ecologies, and the mundane artifacts in our midst. In Takayoshi, P., & Sullivan, P. (Eds.), *Labor, writing technologies, and the shaping of composition in the academy* (pp. 167–186). Cresskill, NJ: Hampton Press.

Bogost, I. (2007). *Persuasive games: The expressive power of videogames.* Cambridge, MA: MIT Press.

Bogost, I. (2011). *How to do things with video games.* Minneapolis, MN: University of Minnesota Press.

Bolter, J. D., & Grusin, R. (2000). *Remediation: Understanding new media.* Cambridge, MA: MIT Press.

Bondebjerg, I. (2008). *Virkelighedens fortællinger den danske tv-dokumentarismes historie* (1. udgave ed.). Frederiksberg, Denmark: Samfundslitteratur.

Bondebjerg, I. (2002). The mediation of everyday life. In Jerslev, A. (Ed.), *Realism and "reality" in film and media* (pp. 159–193). Copenhagen, Denmark: Museum Tusculanum University of Copenhagen.

Bondebjerg, I. (2013). (forthcoming). Engaging with reality: Documentary, politics and globalisation. *Publisher TBC.*

Booth, M. (2007, May 01). Grading Wikipedia. *Denver Post.* Retrieved November 8, 2010, from http://www.denverpost.com/search/ci_5786064

Bordwell, D. (1985). *Narration in the fiction film.* Madison, WI: University of Wisconsin Press.

Borlund, P. (2003). The concept of relevance in IR. *Journal of the American Society for Information, 54*(10), 913–925. doi:10.1002/asi.10286

Bouënard, A., Wanderly, M. M., & Gibet, S. (2010). Gesture control of sound synthesis: Analysis and classification of percussion gestures. *Acta Acustica united with Acustica, Special Issue on Natural and Virtual Instruments: Control. Gesture and Player Interaction, 96*, 668–677.

Bourdieu, P. (1986). The forms of capital. In Richardson, J. (Ed.), *Handbook of theory and research for the sociology of education* (pp. 241–258). New York, NY: Greenwood.

Bourdieu, P. (2005). The political field, the social science field, and the journalistic field. In Benson, R., & Neveu, E. (Eds.), *Bourdieu and the journalistic field* (pp. 29–47). Cambridge, UK: Polity Press.

Bourges-Waldegg, P., & Scrivener, S. (1998). Meaning, the central issue in cross-cultural HCI design. *Interacting with Computers, 9*(3), 287–309. doi:10.1016/S0953-5438(97)00032-5

Bowie, J. (2012, Spring). Rhetorical roots and media future: how podcasting fits into the computers and writing classroom. *Kairos: A Journal of Rhetoric, Technology, and Pedagogy, 16*. Retrieved from http://kairos.technorhetoric.net/16.2/topoi/bowie/index.html

Bowker, G. C., Baker, K., Millerand, F., & Ribes, D. (2007). Towards information infrastructure studies: Ways of knowing in a networked environment. In J. Hunsinger, L. Klastrup, & M. Allen (Eds.), *International handbook of Internet research.* New York, NY: Springer Verlag. Retrieved from http://interoperability.ucsd.edu/docs/07BowkerBaker_InfraStudies.pdf

Bowker, G. C., & Star, S. L. (1999). *Sorting things out: Classification and its consequences.* Cambridge, MA: MIT Press.

Bowman, S., & Willis, C. (2003). *We media: How audiences are shaping the future of news and information.* The Media Center at The American Press Institute. Retrieved October 11, 2011, from http://www.hypergene.net/wemedia/download/we_media.pdf

boyd, d. (2011). Social network sites as networked publics: Affordances, dynamics, and implications. In Z. Papacharissi (Ed.), *Networked self: Identity, community, and culture on social network sites* (pp. 39-58). New York, NY: Routledge.

Boyer, E. (1990). *Scholarship reconsidered: Priorities of the professoriate.* Washington, DC: Carnegie Foundation for the Advancement of Teaching.

Brabazon, T. (2006). The Google effect: Googling, blogging, wikis and the flattening of expertise. *Libri, 56*(3), 157–167. doi:10.1515/LIBR.2006.157

Bradshaw, P. (2011). *Mapping digital media: Social media and news.* Open Society Foundations. Retrieved January 3, 2012, from http://www.soros.org/initiatives/media/articles_publications/ publications/mapping-digital-media-social-media-and-news-20120117/ mapping-digital-media-social-media-20120119.pdf

Bradshaw, P. (2008). When journalists blog: How it changes what they do. *Nieman Reports, 62*(4), 50–52.

Brahnam, S. (2009). Building character for artificial conversational agents: Ethos, ethics, believability, and credibility. *PsychNology Journal, 7*(1), 9–47.

Bramwell, A. (1989). *Ecology in the 20th century: A history.* New Haven, CT: Yale University Press.

Brent, D. (1997). Rhetorics of the web: Implications for teachers of literacy. *Kairos: A Journal of Rhetoric, Technology, and Pedagogy, 2*(1). Retrieved from http://kairos.technorhetoric.net/2.1/binder.html?features/brent/wayin.htm

Brewer, R. *The science of ecology.* Philadelphia, PA: Saunders College Publishing.

Brodkey, L. (1987). Modernism and the scene(s) of writing. *National Council of Teachers of English, 49*(4), 396-418.

Brubaker, R. (2005). The "diaspora" diaspora. *Ethnic and Racial Studies, 28*(1), 1–19. doi:10.1080/0141987042000289997

Brumm, M., Colbert, S., Dahm, R., Drysdale, E., Dubbin, R., & Gwinn, P. … Hoskinson, J. (Director). (2006, July 31). Ned Lamont [Television series episode]. In S. Colbert et al. (Producers), *The Colbert Report.* New York, NY: Comedy Central.

Bruns, A. (2005). *Some exploratory notes on produsers and produsage.* Institute for Distributed Creativity. Retrieved from http://distributedcreativity.typepad.com/idc_texts/2005/11/some_explorator.html

Bruzzi, S. (2000). *New documentary: A critical introduction.* London, UK: Routledge.

Bryant, S. L., Forte, A., & Bruckman, A. (2005). Becoming Wikipedian: Transformation of participation in a collaborative online encyclopedia. *Proceedings of the 2005 International ACM SIGGROUP Conference on Supporting Group Work,* USA, (pp. 1-10).

Buchanan, R. (2004). Rhetoric, humanism, and design. In Handa, C. (Ed.), *Visual rhetoric in a digital world: A critical sourcebook* (pp. 228–259). Boston, MA: Bedford/St. Martin's.

Budd, A., & Clarke, A. (2006, March 7). *How to be a web design superhero.* Presented at the South by Southwest Interactive Festival 2006, Austin, TX.

Budd, A., Moll, C., & Collison, S. (2006). *CSS mastery: Advanced web standards solutions.* Berkeley, CA: Springer-Verlag.

Burke, K. (1950). *A rhetoric of motives.* London, UK: Prentice Hall.

Burkhardt, J. M., MacDonald, M. C., & Rathemacher, A. J. (2010). *Teaching information literacy: 50 standards-based exercises for college students.* Chicago, IL: American Library Association. doi:10.1080/10875301.2011.551069

Burkholder, J. (2010) Redefining sources as social acts: Genre theory in information literacy instruction. *Library Philosophy and Practice (e-journal),* paper 413. Retrieved from http://digitalcommons.unl.edu/libphilprac/413

Burmeister, J. (1993). *Musical poetics* (Rivera, B. V., Trans.). Cambridge, MA: Harvard University Press.

Bush, G. W. (2001). First inaugural address. Retrieved from http://www.bartleby.com/124/pres66.html

Campbell, J., Greenauer, N., Macaluso, K., & End, C. (2007). Unrealistic optimism in Internet events. *Computers in Human Behavior*, *23*, 1273–1284. doi:10.1016/j.chb.2004.12.005

Canavilhas, J. (2004). *Os jornalistas portugueses e a Internet*. Retrieved March3, 2012, from http://www.bocc.ubi.pt/pag/canavilhas-joao-jornalistas-portugueses-internet.pdf

Canavilhas, J., & Ivars-Nicolás, B. (2012). Uso y credibilidad de fuentes periodísticas 2.0 en Portugal y España. *El Profesional de la Información*, *21*(1), 63–69. doi:10.3145/epi.2012.ene.08

Canini, K., Suh, B., & Pirolli, P. L. (2011). *Finding credible information sources in social networks based on content and social structure*. Third IEEE International Conference on Social Computing (SocialCom), October 9-11, Boston, MA. Retrieved January 4, 2012, from http://www.parc.com/content/attachments/finding-credible-information-preprint.pdf

Carlson, E. (2009, July/August). What to look for when evaluating web sites. *Orthopedic Nursing*, *28*, 199–202. doi:10.1097/NOR.0b013e3181ada7a0

Carlson, M. (2011). Whither anonymity? Journalism and unnamed sources in a changing media environment. In Franklin, B., & Carlson, M. (Eds.), *Journalists, sources and credibility: New perspectives* (pp. 37–48). New York, NY: Routledge.

Carnegie Foundation for the Advancement of Teaching. (n.d.). *About Carnegie classification*. Retrieved May 15, 2012 from http://classifications.carnegiefoundation.org/index.php

Carnegie, T. A. M. (2009). Interface as exordium: The rhetoric of interactivity. *Computers and Composition*, *26*(3), 164–173. doi:10.1016/j.compcom.2009.05.005

Carterette, B., Kanoulas, E., & Yilmaz, E. (2012). Evaluating Web retrieval effectiveness. In Lewandowski, D. (Ed.), *Web search engine research* (pp. 105–137). Bingley, UK: Emerald. doi:10.1108/S1876-0562(2012)002012a007

Carvin, A. (2012). The gay girl in Damascus that wasn't. *Storify*. Retrieved from http://storify.com/acarvin/the-gay-girl-in-damascus-that-wasnt

Casaló, L. V., Cisneros, J., Flavián, C., & Guinalíu, M. (2009). Determinants of success in open source software networks. *Industrial Management & Data Systems*, *109*(4), 532–549. doi:10.1108/02635570910948650

Cassidy, W. P. (2007). Online news credibility: An examination of the perceptions of newspaper journalists. *Journal of Computer-Mediated Communication*, *7*, 478–498. doi:10.1111/j.1083-6101.2007.00334.x

Ceccarelli, L. (2001). *Shaping science with rhetoric: The cases of Dobzhansky, Schrodiedinger, and Wilson*. Chicago, IL: University of Chicago Press.

Chadwick, P. (2008, May). *Adapting to digital technologies: Ethics and privacy*. Paper presented at Future of Journalism Summit, Sydney, NSW.

Chan, D. (2011). Activist perspective: The social cost hidden in the Apple products. *Journal of Workplace Rights*, *15*(3/4), 363–365.

Charumbira, S. (2012, May 12). US-based Chimhina fundraises for arts. *The Standard*. Retrieved from http://www.thestandard.co.zw

Chen, S.-Y., & Rieh, S. Y. (2009). Take your time first, time your search later: How college students perceive time in web searching. *Proceedings of the 72nd Annual Meeting of the American Society for Information Science and Technology*.

Cheng, Q., Chen, F., & Yip, P. S. F. (2011). The Foxconn suicides and their media prominence: Is the Werther effect applicable in China? *BMC Public Health*, *11*, 841–851. doi:10.1186/1471-2458-11-841

Chen, M. (2009). *Social dimensions of expertise in World of Warcraft players. Transformative Works and Cultures, 2*. Retrieved from http://dx.doi.org/10.3983/twc.2009.0072

Cheseboro, J., & Bonsall, D. (1989). *Computer-mediated communication*. Tuscaloosa, AL: University of Alabama Press.

Chesney, T. (2006). An empirical examination of Wikipedia's credibility. *First Monday, 11*(11). Retrieved from http://firstmonday.org/htbin/cgiwrap/bin/ojs/index.php/fm/article/view/1413/1331

Cheverie, J., Boetcher, J., & Buschman, J. (2009, April). Digital scholarship in the university tenure and promotion process: A report on the Sixth Scholarly Communication Symposium at Georgetown University Library. *Journal of Scholarly Publishing, 40*(3), 219–230. doi:10.3138/jsp.40.3.219

Chiagouris, L., Long, M. M., & Plank, R. E. (2008). The consumption of online news: The relationship of attitudes toward the site and credibility. *Journal of Internet Commerce, 7*(4), 528–549. doi:10.1080/15332860802507396

China Internet Network Information Center (CNNIC). (2012). *The 29th statistical report on Internet development in China.* Retrieved February 20, 2012, from http://cnnic.com.cn/dtygg/dtgg/201201/W020120116337628870651.pdf

Christians, C. G., Glasser, T. L., McQuail, D., Nordenstreng, K., & White, R. A. (2009). *Normative theories of the media: Journalism in democratic societies.* Urbana, IL: University of Illinois Press.

Christley, S., & Madey, G. (2007). Analysis of activity in the open source software development community. *Proceedings of the 40th Hawaii International Conference on System Sciences.* Waikoloa, HI.

Chu, H. (2011). Factors affecting relevance judgment: A report from TREC legal track. *The Journal of Documentation, 67*(2), 264–278. doi:10.1108/00220411111109467

Chung, C. J., Nam, Y., & Stefanone, M. A. (2012). Exploring online news credibility: The relative influence of traditional and technological factors. *Journal of Computer-Mediated Communication, 17*(2), 171–186. doi:10.1111/j.1083-6101.2011.01565.x

Chung, C., Nam, Y., & Stefanone, M. (2012, January 13). Traditional and technological factors. *Journal of Computer-Mediated Communication, 17*, 171–186. doi:10.1111/j.1083-6101.2011.01565.x

Chung, J. C., Kim, H., & Kim, J. H. (2010). An anatomy of the credibility of online newspapers. *Online Information Review, 34*(5), 669–685. doi:10.1108/14684521011084564

Chung, J. C., Nam, Y., & Stefanone, M. A. (2012). Exploring online news credibility: The relative influence of traditional and technological factors. *Journal of Computer-Mediated Communication, 17*, 171–186. doi:10.1111/j.1083-6101.2011.01565.x

Cicero. (1959). *De oratore* (Sutton, E., & Rackham, H., Trans.). Cambridge, MA: Harvard University Press.

Cision & Bates. D. (2009). *Social media & online usage study.* George Washington University. Retrieved from www.gwu.edu/~newsctr/10/pdfs/gw_cision_sm_study_09.PDF

Cjohnweb. (2010, October 8). *Comment posted to Toward a more thoughtful conversation on stories.* Reuters. Retrieved from http://blogs.reuters.com/fulldisclosure/2010/09/27/toward-a-more-thoughtful-conversation-on-stories/?cp=all#comments

Clarke, K., Hardstone, G., Hartswood, M., Procter, R., & Rouncefield, M. (2006). Trust and organisational work. In Clarke, K., Hardstone, G., Rouncefield, M., & Sommerville, I. (Eds.), *Trust in technology: A socio-technical perspective* (pp. 1–20). Dordrecht, The Netherlands: Springer. doi:10.1007/1-4020-4258-2_1

Clarke, V. A., Lovegrove, H., Williams, A., & Macpherson, M. (2000). Unrealistic optimism and the health belief model. *Journal of Behavioral Medicine, 23*(4), 367–376. doi:10.1023/A:1005500917875

Clifford, J. (1994). Diasporas. *Cultural Anthropology, 9*(3), 302–338. doi:10.1525/can.1994.9.3.02a00040

CNN iReport (2011, May 5). *CNN iReport terms of use.* CNN iReport. Retrieved from http://ireport.cnn.com/terms.jspa

CNN iReport (2012, May 10). *Meet team iReport.* CNN iReport. Retrieved from http://ireport.cnn.com/blogs/ireport-blog/2009/11/18/meet-team-ireport

CNN iReport. (n.d.). *Community guidelines.* CNN iReport. Retrieved from http://ireport.cnn.com/guidelines.jspa

Cofta, P. (2006). *Distrust.* Paper presented at Eighth International Conference on Electronic Commerce, Fredericton, Canada.

Cofta, P. (2007). *Trust, complexity and control: Confidence in a convergent world.* Hoboken, NJ: John Wiley and Sons.

Cofta, P. (2011). The trustworthy and trusted Web. *Foundations and Trends in Web Science, 2*(4), 243–381. doi:10.1561/1800000016

Coleman, S. (2005). Blogs and new politics of listening. *The Political Quarterly, 76*(2), 272–280. doi:10.1111/j.1467-923X.2005.00679.x

Collins, K. (2008). *Game sound: An introduction to the history, theory, and practice of video game music and sound design.* Cambridge, MA: MIT Press.

Comments policy. (2012). Tampabay.com Retrieved January 22, 2012, from http://www.tampabay.com/universal/comment_guidelines.shtml

ComScore. (2010). *comScore reports global search market growth of 46 percent in 2009.* Retrieved from http://comscore.com/Press_Events/Press_Releases/2010/1/Global_Search_Market_Grows_46_Percent_in_2009

Conseil supérieur de l'audiovisuel. (2011, October 18). *Convention de la chaîne M6.* Retrieved February 12, 2012, from http://www.csa.fr/Espace-juridique/Conventions-des-editeurs/Convention-de-la-chaine-M6

Consumerist.com. (2006). *Sony's PSP blog flog revealed.* Retrieved from http://consumerist.com/2006/12/sonys-psp-blog-flog-revealed.html

Cooke, M. (2007). Baghdad burning: Women write war in Iraq. *World Literature Today, 81*(6), 23–26.

Cooper, M. (1986). The ecology of writing. *College English, 48*(4), 364–375. doi:10.2307/377264

Cooper, M. (2010). Being linked to the matrix: Biology, technology, and writing. In Selber, S. (Ed.), *Rhetorics and technologies: New directions in writing and communication* (pp. 15–32). Columbia, SC: University of South Carolina Press.

Cooper, W. S. (1971). A definition of relevance for information retrieval. *Information Storage and Retrieval, 7*(1), 19–37. doi:10.1016/0020-0271(71)90024-6

Corner, J. (1995). *Television form and public address.* London, UK: Edward Arnold.

Corner, J. (1996). *The art of record: A critical introduction to documentary.* Manchester, UK: Manchester University Press.

Corner, J. (1997). Restyling the real: British television documentary in the 1990s. *Continuum, 11*(1), 11–21. doi:10.1080/10304319709359415

Corner, J. (2009). Documentary studies: Dimensions of transition and continuity. In Austin, T., & de Jong, W. (Eds.), *Rethinking documentary: New perspectives, new practices* (pp. 13–28). Maidenhead, UK: Open University Press.

Couturat, L. (2002, March 19). *The logic of Leibniz in accordance with unpublished documents* (D. Rutherford & T. R. Monroe, Trans.). Retrieved December 7, 2011, from http://philosophyfaculty.ucsd.edu/faculty/rutherford/leibniz/contents.htm

Cover, R. (2006). Audience inter/active: Interactive media, narrative control and reconceiving audience history. *New Media & Society, 8*(1), 139–158. doi:10.1177/1461444806059922

Croft, W. B., Metzler, D., & Strohman, T. (2010). *Search engines: Information retrieval in practice.* Boston, MA: Pearson.

Crowley, S., & Hawhee, D. (2009). *Ancient rhetorics for contemporary students* (4th ed.). New York, NY: Pearson/Longman.

Crowston, K., Heckman, R., Annabi, H., & Massango, C. (2005). A structurational perspective on leadership in free/libre open source software development teams. *Proceedings of the 1st Conference on Open Source Systems (OSS),* Genova, Italy. Retrieved from http://floss.syr.edu

Crowston, K., & Howison, J. (2005). The social structure of free and open source software development. [from http://floss.syr.edu]. *First Monday, 10,* 1–27. Retrieved November 6, 2011

Cuadra, C. A., & Katter, R. V. (1967). Opening the black box of "relevance." *The Journal of Documentation,* 291–303. doi:10.1108/eb026436

Culliss, G. A. (2003). *Personalized search methods.* (US Patent US 6,539,377 B1).

Currie, L., Devlin, F., Emde, J., & Graves, K. (2010). Undergraduate search strategies and evaluation criteria. *New Library World, 111*(3), 113–124. doi:10.1108/03074801011027628

Cutrell, E., & Guan, Z. (2007). What are you looking for? An eye-tracking study of information usage in web search. *Proceedings of the SIGCHI Conference on Human Factors in Computing Systems* (pp. 407–416).

Daft, R. L., Lengel, R. H., & Trevino, L. K. (1987, September). Message equivocality, media selection, and manager performance: Implications for information systems. *Management Information Systems Quarterly*, 355–366. doi:10.2307/248682

Dahl, C. (2009). Undergraduate research in the public domain: the evaluation of non-academic sources online. *RSR. Reference Services Review*, *37*(2), 155–163. Retrieved from http://ecommons. usask.ca/xmlui/bitstream/handle/10388/281/Undergraduate%20research%20in%20the%20public%20domain-final-Feb.%207.pdf?sequence=1d oi:10.1108/00907320910957198

Dahlström, E. (2008, January 22). *Getting to the core of the Web*. Retrieved March 27, 2011, from http:// my.opera.com/MacDev_ed/blog/2008/01/22/core-web

Davallon, J. (2004). Objet concret, objet scientifique, objet de recherche. *Hermes*, *38*, 30–37. doi:10.4267/2042/9421

De Castella, T., & Brown, V. (2011). Trolling: Who does it and why? *BBC News Magazine*. Retrieved from http://www.bbc.co.uk/news/magazine-14898564

De Laat, P. (2005). Trusting virtual trust. *Ethics and Information Technology*, *7*(3), 167–180. doi:10.1007/s10676-006-0002-6

de Saussure, F. (1977). *Course in general linguistics*. Glasgow, UK: Fontana/Collins.

Dean, J. A., Gomes, B., Bharat, K., Harik, G., & Henzinger, M. H. (2002, March 2). *Methods and apparatus for employing usage statistics in document retrieval*. (US Patent App. US 2002/0123988 A1).

DeCamp, D. (2011a, November 17). Pinellas county commissioner Norm Roche has alter ego for online comments. *St. Petersburg Times*. Retrieved January 22, 2012, from http://www.tampabay.com/news/local-government/article1202065.ece

DeCamp, D. (2011b, November 18). Norm Roche's anonymous online snark strains city, county relations. *St. Petersburg Times*. Retrieved January 22, 2012, from http://www.tampabay.com/news/politics/local/norm-roches-anonymous-online-snark-strains-city-county-relations/1202287

December, J. (1996, August). Living in hypertext. *Ejournal, 6*. Retrieved from http://www.ucalgary.ca/ejournal/archive/v6n3/december/december.html

Del Bianco, V., Lavazza, L., Morasca, S., & Taibi, D. (2011). A survey on open source software trustworthiness. *IEEE Software*, *28*(5), 67–75. doi:10.1109/MS.2011.93

Denning, P., Horning, J., Parnas, D., & Weinstein, L. (2005). Wikipedia risks. *Communications of the ACM*, *48*(12), 152. doi:10.1145/1101779.1101804

Deuze, M. (2006). Participation, remediation, bricolage: Considering principal components of a digital culture. *The Information Society*, *22*, 63–75. doi:10.1080/01972240600567170

DeVoss, D., & Platt, J. (2011). Image manipulation and ethics in a digital–visual world. *Computers and Composition Online*. Retrieved May 12, 2012, from http://www.bgsu.edu/cconline/ethics_special_issue/DEVOSS_PLATT/

Diakopoulos, N., De Choudhury, M., & Naaman, M. (2012). *Finding and assessing social media information sources in the context of journalism*. CHI'12, May 5–10, 2012, Austin, Texas, USA. Retrieved February, 2, 2012, from http://research.microsoft.com/en-us/um/people/munmund/pubs/chi_2012.pdf

Dihydrogen Monoxide. (n.d.). *DHMO homepage*. Retrieved from http://www.dhmo.org

Dilger, B., & Rice, J. (Eds.). (2010). *From A to <A>: Keywords of markup*. Minneapolis, MN: U of M Press.

Dinkelman, A. L. (2010). Using course syllabi to assess research expectations of biology majors: Implications for further development of information literacy skills in the curriculum. *Issues in Science and Technology Librarianship, 60*. Retrieved from http://www.istl.org/10-winter/refereed3.html

Djamasbi, S., Siegel, M., Tullis, T., & Dai, R. (2010). *Efficiency, trust, and visual appeal: Usability testing through eye tracking*. Paper presented at the 43rd Hawaii International Conference on System Sciences (HICSS), Kauai, Hawaii.

Dochterman, M. A., & Stamp, G. H. (2010). Part one: The determination of web credibility: A thematic analysis of web user's judgments. *Qualitative Research Reports in Communication, 11*(1), 37–43. doi:10.1080/17459430903514791

Dochterman, M. A., & Stamp, G. H. (2010). Part two: The determination of web credibility: A theoretical model derived from qualitative data. *Qualitative Research Reports in Communication, 11*(1), 44–50. doi:10.1080/17459430903514809

Donath, J. (2006). *Signals, truth and design* [Video Recording]. Retrieved from http://www.ischool.berkeley.edu/about/events/dls09272006

Downey, D., Dumais, S., Liebling, D., & Horvitz, E. (2008). Understanding the relationship between searchers' queries and information goals. *Proceeding of the 17th ACM Conference on Information and Knowledge Management* (pp. 449–458).

Doyle, T., & Hammond, J. L. (2006). Net cred: Evaluating the Internet as a research source. *RSR. Reference Services Review, 34*(1), 56–70. Retrieved from http://search.proquest.com/docview/200503715?accountid=36823doi:10.1108/00907320610648761

Drezner, D. W., & Farrell, H. (2004). Web of influence. *Foreign Policy, 145*, 32–40. doi:10.2307/4152942

Dube, M. (2012, May 18). Where will the stars party? *The Sunday Mail*. Retrieved from http://www.sundaymail.co.zw

DuBois, P. (1991). *Torture and truth*. New York, NY: Routledge.

Dumais, S., Buscher, G., & Cutrell, E. (2010). *Individual differences in gaze patterns for web search*. Presented at the IIiX, New Brunswick, New Jersey.

Dwyer, N. (2011). *Traces of digital trust: An interactive design perspective*. Unpublished doctoral dissertation, Victoria University, Australia.

Dwyer, T. (2006). *Media convergence*. London, UK: Open University Press, McGraw-Hill Education, McGraw-Hill House.

Dye, D., & Bleckloff, M. (2003). *Amazing Gracie: A dog's tale*. New York, NY: Workman.

Dynamic backend generator (DBG): A scholarly middleware tool. (2006). *Vectors*. Retrieved May 7, 2012 from http://vectorsjournal.org/journal/blog/dbg-overview/

Eamon, M. K. (2004). Digital divide in computer access and use between poor and non-poor youth. *Journal of Sociology and Social Welfare, 31*(2), 91–113.

Eastin, M. (2008). Toward a cognitive developmental approach to youth perceptions of credibility. In M. J. Metzger, & A. J. Flanagin (Eds.), *Digital media, youth, and credibility*, (pp. 28-46). MacArthur Foundation Series on Digital Media and Learning. Cambridge, MA: MIT Press.

Ebert, R. (2010). Video games can never be art. *Roger Ebert's Blog (Chicago-Sun Times)*. Retrieved from http://blogs.suntimes.com/ebert/2010/04/video_games_can_never_be_art.html

Eckstart, A., & Sterry, D. (2010). *The essential guide to getting your book published: How to write it, sell it, and market it…successfully*. New York, NY: Workman.

Ecma International. (n.d.). *What is Ecma International?* Retrieved April 4, 2011, from http://www.ecma-international.org/memento/index.html

Eco, U. (1976). *A theory of semiotics*. Bloomington, IN: Indiania University Press.

Ede, L., & Lunsford, A. (1990). *Single texts/plural authors: Perspectives on collaborative writing*. Carbondale, IL: Southern Illinois University Press.

Edelman, B. (2010). *Hard-coding bias in Google "algorithmic" search results*. Retrieved April 11, 2011, from http://www.benedelman.org/hardcoding/

Edwards, P. N., Jackson, S. J., Bowker, G. C., & Knobel, C. P. (2007). *Understanding infrastructure: Dynamics, tensions, and design*. NSF Grant 0630263.

Edwards, P. N. (2003). Infrastructure and modernity: Force, time, and social organization in the history of sociotechnical systems. In Misa, T. J., Brey, P., & Feenberg, A. (Eds.), *Modernity and technology* (pp. 185–225). Cambridge, MA: MIT Press.

Egger, F. (2003). *From interactions to transactions: designing the trust experience for business-to-consumer electronic commerce*. Unpublished doctoral dissertation, Eindhoven University of Technology, The Netherlands.

Ekstrand, V. S. (2002). Online news: User agreements and implications for readers. *Journalism & Mass Media Quarterly, 79*(3), 602–618. doi:10.1177/107769900207900305

Elkind, D. (1967). Egocentrism in adolescence. *Child Development, 38*, 1025–1034. doi:10.2307/1127100

Elliott, M., & Scacchi, W. (2008). Mobilization of software developers: The free software movement. *Information Technology & People, 21*(1), 4–33. doi:10.1108/09593840810860315

Ellison, N., Heino, R., & Gibbs, J. (2006). Managing impressions online: Self-presentation processes in the online dating environment. *Journal of Computer-Mediated Communication, 11*, 415–441. doi:10.1111/j.1083-6101.2006.00020.x

English, K., Sweetser, K. D., & Ancu, M. (2011). YouTube-ification of political talk: An examination of persuasion appeals in viral video. *The American Behavioral Scientist, 55*(6), 733–748. doi:10.1177/0002764211398090

Enos, T. (Ed.). (1996). *Encyclopedia of rhetoric and composition: Communication from ancient times to the information age*. New York, NY: Garland Publishing.

Enos, T., & Borrowman, S. (2001). Authority and credibility: Classical rhetoric, the internet, and the teaching of techno-ethos. In Gray-Rosendale, L., & Gruber, S. (Eds.), *Alternative rhetorics: Challenges to the rhetorical tradition* (pp. 93–110). Albany, NY: State U of New York P.

Ensmenger, N. L. (2001). The "question of professionalism" in the computer fields. *IEEE Annals of the History of Computing, 23*(4), 56. doi:10.1109/85.969964

Ensmenger, N. L. (2003). Letting the "computer boys" take over: Technology and the politics of organizational transformation. *International Review of Social History, 48*(S11), 153. doi:10.1017/S0020859003001305

Epstein, S., Pacini, R., Denes-Raj, V., & Heier, H. (1996). Individual differences in intuitive-experiential and analytical-rational thinking styles. *Journal of Personality and Social Psychology, 71*, 390–405. doi:10.1037/0022-3514.71.2.390

Essjay controversy. (2010, November 08). *Wikipedia*. Retrieved November 8, 2010 from http://en.wikipedia.org/wiki/Essjay

Estol, L. (n.d.). *Animal welfare group*. LinkedIn. Retrieved from http://www.LinkedIn.com/groupsDirectory?itemaction=mclk&anetid=89180&impid=&pgkey=anet_search_results&actpref=anetsrch_name&trk=anetsrch_name&goback=.gdr_1327510683784_1

European Commission. (2009). *Towards a thematic strategy on the sustainable use of pesticides*. European Commission. *Environment*.

Eveno, P. (2001). *Le journal Le Monde une histoire d'Indépendance*. Paris, France: Editions Odile Jacob.

Facebook. (2012). *Kolena Khaled Said*. Retrieved July 26, 2012, from www.facebook.com/ElShaheed

Facebook. (2012). *We are all Khaled Said*. Retrieved July 26, 2012, from www.facebook.com/elshaheeed.co.uk

Farkas, M. (2012). *Information wants to be free*. Retrieved from http://meredith.wolfwater.com/wordpress/

Farkas, M. (2012b, January 28). *Classic blunder #1 - Let's just try it and see what happens! Information wants to be free*. Retrieved from http://meredith.wolfwater.com/wordpress/2012/01/28/classic-blunder-1-lets-just-try-it-and-see-what-happens/

Farrel, H., & Drezner, D. (2008). The power and politics of blogs. *Public Choice, 134*(1/2), 15–30. doi:doi:10.1007/s11127-007-9198-1

Federman, M. (2009). *McLuhan thinking: Integral awareness in the connected society*. Retrieved February 10, 2012, from http://individual.utoronto.ca/markfederman/IntegralAwarenessintheConnectedSociety.pdf

Feng, P. (2003). Studying standardization: A review of the literature. *The 3rd Conference on Standardization and Innovation in Information Technology* (pp. 99-112).

Fidel, R., Davies, R. K., Douglass, M. H., Holder, J. K., Hopkins, C. J., & Kushner, E. J. (1999). A visit to the information mall: Web searching behavior of high school students. *Journal of the American Society for Information Science American Society for Information Science, 50*, 24–37. doi:10.1002/(SICI)1097-4571(1999)50:1<24::AID-ASI5>3.0.CO;2-W

Findings from the A List Apart Survey, 2008. (2009). New York: A List Apart Magazine. Retrieved from http://aneventapart.com/alasurvey2008/

Findings from the Web Design Survey 2007. (2008). New York: A List Apart Magazine. Retrieved from http://www.alistapart.com/d/2007surveyresults/2007surveyresults.pdf

Firer-Blaess, S. (2011). Wikipedia: Example for a future electronic democracy? Decision, discipline and discourse in the collaborative encyclopedia. *Studies in Social and Political Thought, 19*, 131-154. Retrieved April 28, 2012, from http://ssptjournal.files.wordpress.com/2011/08/sspt19b1.pdf

Fitzpatrick, K. (2012). *My view: Are electronic media making us less (or more) literate?* Retrieved from http://schoolsofthought.blogs.cnn.com/2012/02/01/my-view-are-electronic-media-making-us-less-or-more-literate/

Fitzpatrick, K. (2011). *Planned obsolescence: Publishing, technology, and the future of the academy.* New York, NY: NYU Press.

Five pillars. (2010, November 07). *Wikipedia.* Retrieved November 8, 2010 from http://en.wikipedia.org/wiki/Wikipedia:Five_pillars

Flanagin, A. J., & Metzger, M. J. (2000). Perceptions of internet information credibility. *Journalism & Mass Communication Quarterly, 77*(3), 515–540. doi:10.1177/107769900007700304

Flanagin, A. J., & Metzger, M. J. (2007). The role of site features, user attributes, and information verification behaviors on the perceived credibility of web-based information. *Medicine and Society, 9*(2), 319–342.

Flanagin, A. J., & Metzger, M. J. (2008). Digital media and youth: Unparalleled opportunity and unprecedented responsibility. In Metzger, M. J., & Flanagin, A. J. (Eds.), *Digital media, youth, and credibility* (pp. 5–27). Cambridge, MA: MIT Press.

Flanagin, A. J., & Metzger, M. J. (2010). *Kids and credibility: An empirical examination of youth, digital media use, and information credibility.* Cambridge, MA: MIT Press.

Fleckenstein, K. (2007). Who's writing? Aristotelian ethos and the author position in digital poetics. *Kairos: A Journal of Rhetoric, Technology, and Pedagogy, 11.* Retrieved May 19, 2012, from http://kairos.technorhetoric.net/11.3/binder.html?topoi/fleckenstein/index.html

Flemings, H. E. (2006). *The brand YU life: Re-thinking who you are through personal brand management.* Canton, MI: Third Generation Publishing.

Flew, T., Cunningham, S., & Bruns, A. (2008). *Social innovation, user-created content and the future of the ABC and SBS as public service media.* ABC_SBS_Inquiry_Submission. Retrieved from http://eprints.qut.edu.au/16948/

Fogg, B. J. (2002). *Prominence-interpretation theory: Explaining how people assess credibility.* A Research Report by the Stanford Persuasive Technology Lab. Retrieved from http://www.captology.stanford.edu/PIT.html

Fogg, B. J. (2003). Computers as persuasive social actors. In B. Fogg's (Ed.), *Persuasive technology: Using computers to change what we think and do* (pp. 31-60). San Francisco, CA: Morgan Kaufmann.

Fogg, B. J., & Tseng, S. (1999). The elements of computer credibility. *Proceedings of Computer Human Interface SIG Conference,* (pp. 80–87).

Fogg, B. J., Marshall, J., Laraki, O., Osipovich, A., Varma, C., Fang, N., et al. (2001). What makes web sites credible? A report on a large quantitative study. *Proceedings of the SIGCHI Conference on Human Factors in Computing Systems* (pp. 61–68). Seattle, WA: ACM. DOI:10.1145/365024.365037

Fogg, B. J., Soohoo, C., Danielson, D. R., Marable, L., Stanford, J., & Tauber, E. R. (2003). How do users evaluate the credibility of web sites? A study with over 2,500 participants. *Proceedings of the 2003 Conference on Designing for User Experiences* (pp. 1–15). San Francisco, CA: ACM. DOI:10.1145/997078.997097

Fogg, B. (2003). *Prominence-interpretation theory: Explaining how people assess credibility online. CHI '03 Extended Abstracts on Human Factors in Computing Systems* (pp. 722–723). New York, NY: ACM.

Fogg, B. J. (2003). *Persuasive technology. Using computers to change what we think and do*. San Francisco, CA: Morgan Kaufmann Publishers.

Folk, M. (2009). *Then a miracle occurs: Digital composition pedagogy, expertise, and style*. Unpublished doctoral dissertation, Michigan Technological University, Houghton, MI.

Food shortages hit Mat South hardest. (2012, May 16). *The Financial Gazette*. Retrieved from http://allafrica.com/stories/201205200073.html

Forte, A., & Bruckman, A. (2005). *Why do people write for Wikipedia? Incentives to contribute to open-content publishing*. GROUP 05 Workshop: Sustaining Community: The Role and Design of Incentive Mechanisms in Online Systems, Sanibel Island, Florida.

Foster, D. (1997). Community and identity in the electronic village. In Porter, D. (Ed.), *Internet culture* (pp. 23–37). New York, NY: Routledge.

Fottorino, E. (2009). Le Monde: Portrait d'un quotidien. *Le Monde*. Retrieved from http://medias.lemonde.fr/medias/pdf_obj/200912.pdf

Foucault, M. (2001). *Fearless speech*. Los Angeles, CA: Semiotext(e).

Foucault, M. (1997). The masked philosopher. In Rabinow, P. (Ed.), *Ethics: Subjectivity and truth* (pp. 321–328). New York, NY: The New Press.

France, P. (1998, June). The encyclopedia as organism. *European Legacy*, *3*(3), 62–75. doi:10.1080/10848779808579889

Frasca, G. (2003). Simulation versus narrative: Introduction to ludology. In Wolf, M. J. P., & Perron, B. (Eds.), *The video game theory reader*. New York, NY: Routledge.

Freeley, A. (1990). *Argumentation and debate: Critical thinking for reasoned decision making* (7th ed.). Belmont, CA: Wadsworth.

Friedman, S. M. (2011, September/October). Three Mile Island, Chernobyl, and Fukushima: An analysis of traditional and new media coverage of nuclear accidents and radiation. *The Bulletin of the Atomic Scientists*, *67*(5), 55–65. doi:10.1177/0096340211421587

Fritch, J. W. (2003). Heuristics, tools, and systems for evaluating Internet information: Helping users assess a tangled Web. *Online Information Review*, *27*(5), 321–327. doi:10.1108/14684520310502270

Fritch, J. W., & Cromwell, R. L. (2001). Evaluating internet resources: Identity, affiliation, and cognitive authority in a networked world. *Journal of the American Society for Information Science and Technology*, *52*(6), 499–507. doi:10.1002/asi.1081

Frith, S., Goodwin, L., & Grossberg, L. (Eds.). (1993). *Sound and vision: The music video reader*. New York, NY: Routledge.

Frobish, T., & Thomas, W. (2012). Crafting an online political ethos: Resurrecting direct mail tactics on the Web. *Proceedings of the 2012 Hawaii University International Conference on Arts and Humanities, USA*, [CD].

Frobish, T. (2004, April). Sexual profiteering and rhetorical assuagement: Examining ethos and identity at Playboy.com. *Journal of Computer-Mediated Communication*, *9*(3). Retrieved from http://jcmc.indiana.edu/vol9/issue3/frobish.html

Frobish, T. (2006, March). The virtual Vatican: A case study regarding online ethos. *Journal of Communication and Religion*, *29*, 38–69.

Frobish, T. S. (2003). An origin of a theory: A comparison of ethos in the Homeric Iliad with that found in Aristotle's Rhetoric. *Rhetoric Review*, *22*(1), 16–30. doi:10.1207/S15327981RR2201_2

Fukuyama, F. (1995). *Trust: social virtues and the creation of prosperity*. New York, NY: Free Press.

Fulkerson, R. (2005). Composition at the turn of the twenty-first century. *College Composition and Communication*, *56*(4), 654–687.

Fullerton, T. (2008). *Documentary games: Putting the player in the path of history*. Nashville, TN: Vanderbilt University Press.

Fulton, H. (2005). *Narrative and media*. Cambridge, UK: Cambridge University Press. doi:10.1017/CBO9780511811760

Gallego, M. D., Luna, P., & Bueno, S. (2008). User acceptance model of open source software. *Computers in Human Behavior, 24*(5), 2199–2216. doi:10.1016/j.chb.2007.10.006

Gambetta, D., & Hamill, H. (2005). *Streetwise: How taxi drivers establish their customers' trustworthiness.* New York, NY: Russell Sage Foundation.

Gans, H. (2004). *Deciding what's news: A study of CBS Evening News, NBC Nightly News, Newsweek and Time.* Evanston, IL: Northwestern University Press.

Garrett, B. (n.d.). *Dog lovers.* LinkedIn. Retrieved from http://www.LinkedIn.com/groups/Dog-Lovers77136?itemaction=mclk&anetid=77136&impid=&pgkey=anet_search_results&actpref=anetsrch_name&trk=anetsrch_name&goback=.gdr_1327510683790_1

Gasser, U., Cortesi, S., Malik, M., & Lee, A. (2012). *Youth and digital media: From credibility to information quality.* Berkman Center for Internet & Society. Retrieved May 22, 2012, from http://ssrn.com/abstract=2005272

Gass, R. H., & Seiter, J. S. (2003). *Persuasion. Social influence and compliance gaining.* Boston, MA: Allyn and Bacon.

Ghorashi, H., & Boersma, K. (2009). The "Iranian diaspora" and the new media: From political action to humanitarian help. *Development and Change, 40*(4), 667–691. doi:10.1111/j.1467-7660.2009.01567.x

Gigerenzer, G., & Todd, P. M. (1999). *Simple heuristics that make us smart.* New York, NY: Oxford University Press.

Giglio, M. (2011a, Oct 31). The reluctant revolutionary. *Newsweek, 158* (18), 45-45.

Giglio, M. (2011b, Feb. 21). The Facebook freedom fighter. *Newsweek, 157* (18), 14-17.

Giles, J. (2005, December 14). Internet encyclopaedias go head to head. *Nature, 438*, 900-901, Retrieved November 18, 2010 from http://www.nature.com/nature/journal/v438/n7070/full/438900a.html.

Gill, K. E. (2004). How can we measure the influence of the blogosphere? *WWW2004 Proceedings.*

Gillmor, D. (2010). *Mediactive.* Retrieved from http://mediactive.com/wp-content/uploads/2010/12/mediactive_gillmor.pdf

Gillmor, D. (2006). *We the media: Grassroots journalism by the people, for the people.* Sebastopol, CA: O'Reilly.

GIMP Team. (2011). *GIMP – The GNU image manipulation program.* Retrieved November 6, 2011, from http://www.gimp.org

Giroux, H. A. (1992). *Border crossings: Cultural workers and the politics of education.* New York, NY: Routledge.

Gitelman, L. (2008). *Always already new: Media history and the data of culture.* Cambridge, MA: MIT Press.

Glendinning, C. (1990). Notes toward a neo-Luddite manifesto. *The Utne Reader, 38*, 50–53.

Goffman, E. (1974). *Frame analysis.* Cambridge, U.K: Harvard University Press.

Goggin, G. (2011). The intimate turn of mobile news. In Meikle, G., & Redden, G. (Eds.), *News online: Transformations and continuities* (pp. 99–114). UK: Palgrave MacMillan.

Gogoi, P. (2006). Wal-mart's Jim and Laura: The real story. *Bloomberg Businessweek.* Retrieved from http://www.businessweek.com/bwdaily/dnflash/content/oct2006/db20061009_579137.htm

Goldhaber, M. H. (2006). The value of openness in an attention economy. *First Monday, 11*(6). Retrieved from http://firstmonday.org/htbin/cgiwrap/bin/ojs/index.php/fm/article/view/1334/1254

Gong, H., & Yang, X. (2010). Digitized parody: The politics of *egao* in contemporary China. *China Information, 24*(1), 2–26. doi:doi:10.1177/0920203X09350249

Google. (2012). *Search engine optimization (SEO) - Webmaster tools help.* Retrieved February 21, 2012, from http://support.google.com/webmasters/bin/answer.py?hl=en&answer=35291

Govt respects media pluralism: Shamu. (2012, May 3). *Zimbabwe Broadcasting Corporation.* Retrieved from http://eu.zbc.co.zw/news-categories/top-stories/19055-govt-respects-media-pluralism-shamu.html

Govt to curb fertiliser looting. (2012, February 13). *Zimbabwe Broadcasting Corporation.* Retrieved from http://eu.zbc.co.zw/news-categories/top-stories/16461-gvt-to-curb-fertiliser-looting.html

Granka, L., Joachims, T., & Gay, G. (2004). Eye-tracking analysis of user behavior in WWW search. *Proceedings of the 27th Annual International ACM SIGIR Conference on Research and Development in Information Retrieval* (pp. 25–29).

Granka, L. (2010). The politics of search: A decade retrospective. *The Information Society, 26*(5), 364–374. doi:10.1080/01972243.2010.511560

Gratch-Lindauer, B. (2002). Comparing the regional accreditation standards: Outcomes assessment and other trends. *Journal of Academic Librarianship, 28*(1), 14–25. doi:10.1016/S0099-1333(01)00280-4

Graupner, M., Nickoson-Massey, L., & Blair, K. (2009). Remediating knowledge-making spaces in the graduate curriculum: Developing and sustaining multimodal teaching and research. *Computers and Composition, 26*(1), 13–23. doi:10.1016/j.compcom.2008.11.005

Gray, A. B. H. (1960). The ontology of the photographic image. *Film Quarterly, 13*(4), 4–9. doi:10.2307/1210183

Greenfield, P. M. (2004). Developmental considerations for determining appropriate Internet use guidelines for children and adolescents. *Journal of Applied Developmental Psychology, 25,* 751–762. doi:10.1016/j.appdev.2004.09.008

Greimas, J. (1989). *The social sciences. A semiotic view.* Minneapolis, MN: University of Minnesota Press.

Grierson, J. (1933). The documentary producer. *Cinema Quarterly, 2*(1), 7–9.

Griffey, J., & Coombs, K. (2008). *Library blogging.* Columbus, OH: Linworth.

Grodin, D., & Lindlof, T. (Eds.). (1996). *Constructing the self in a mediated world.* Thousand Oaks, CA: Sage.

Grodzinsky, F. S., Miller, K., & Wolf, M. J. (2003). Ethical issues in open source software. *Information. Communication & Ethics in Society, 1*(4), 193–205. doi:10.1108/14779960380000235

Grover, M. (2002). Analyzing the rhetoric of websites. *Kairos: A Journal of Rhetoric, Technology, and Pedagogy, 7*(2). Retrieved from: http://kairos.technorhetoric.net/7.2/binder.html?sectiontwo/grover/

Guan, Z., & Cutrell, E. (2007). An eye tracking study of the effect of target rank on web search. *Proceedings of the SIGCHI Conference on Human Factors in Computing Systems* (pp. 417–420).

Guidance, B. B. C. (n.d.). *Moderation, hosting, escalation and user management, part 1: Hosting, moderation and escalation.* Retrieved from http://www.bbc.co.uk/guidelines/editorialguidelines/page/guidance-moderation-hosting

Gunther, L. (2008, October 2). *Acid3 receptions and misconceptions and do we have a winner?* Retrieved March 28, 2011, from http://www.webstandards.org/2008/10/02/dowehaveawinner/

Gurak, L. J. (1999). The promise and the peril of social action in cyberspace: Ethos, delivery, and the protests over MarketPlace and the Clipper Chip. In Smith, M. A., & Kollock, P. (Eds.), *Communities in cyberspace* (pp. 244–263). London, UK: Routledge.

Gyongyi, Z., & Garcia-Molina, H. (2005). Web spam taxonomy. *First International Workshop on Adversarial Information Retrieval on the Web (AIRWeb 2005)* (pp. 39–47).

Haiman, R. J. (2000). *Best practices for newspaper journalists: A handbook for reporters, editors, photographers and other newspaper professionals on how to be fair to the public.* Arlington, VA: The Freedom Forum. Retrieved December 23, 2011, from http://www.freedomforum.org/publications/diversity/bestpractices/bestpractices.pdf

Hallin, D. C., & Mancini, P. (2004). *Comparing media systems three models of media and politics.* Cambridge, UK: Cambridge University Press. doi:10.1017/CBO9780511790867

Halloran, S. M. (1984). Aristotle's concept of ethos, or if not his somebody else's. *Rhetoric Review, 1*(1), 58–63. doi:10.1080/07350198209359037

Halloran, S. M. (1984). The birth of molecular biology: An essay in the rhetorical criticism of scientific discourse. *Rhetoric Review*, *3*(1), 70–83. doi:10.1080/07350198409359083

Hall, S. (1980). Encoding/decoding. In Hall, S., Hobson, D., Lowe, A., & Willis, P. (Eds.), *Culture, media, language: Working papers in cultural studies, 1972-79* (pp. 128–138). London, UK: Hutchinson.

Hammersley, M., & Atkinson, P. (1995). *Ethnography: Principles in practice*. London, UK: Routledge.

Hargattai, E. (2002). Second-level digital divide: Differences in people's online skills. *First Monday*, *7*(4). Retrieved May 2, 2012, from http://firstmonday.org/htbin/cgiwrap/bin/ojs/index.php/fm/article/view/942/864/

Hargittai, E. (2002). Beyond logs and surveys: In-depth measures of people's web use skills. *Journal of the American Society for Information Science and Technology*, *53*(14), 1239–1244. doi:10.1002/asi.10166

Hargittai, E. (2005). Survey measures of web-oriented digital literacy. *Social Science Computer Review*, *23*(3), 371–379. doi:10.1177/0894439305275911

Hargittai, E. (2007). The social, political, economic, and cultural dimensions of search engines: An introduction. *Journal of Computer-Mediated Communication*, *12*(3), 769–777. doi:10.1111/j.1083-6101.2007.00349.x

Hargittai, E. (2009). An update on survey measures of web-oriented digital literacy. *Social Science Computer Review*, *27*(1), 130–137. doi:10.1177/0894439308318213

Hargittai, E., Fullerton, F., Menchen-Trevino, E., & Thomas, K. (2010). Trust online: Young adults' evaluation of Web content. *International Journal of Communication*, *4*, 468–494.

Hartelius, E. J. (2011). *The rhetoric of expertise*. Lanham, MD: Lexington.

Hart, R. (1998). Introduction: Community by negation—An agenda for rhetorical inquiry. In Hogan, J. M. (Ed.), *Rhetoric and community: Studies in unity and fragmentation* (pp. xxv–xxxviii). Columbia, SC: University of South Carolina Press.

Hartzog, W. (2010). The new price to play: Are passive online media users bound by terms of use. *Communication Law and Policy*, *15*(4), 405–433. doi:10.1080/10811680.2010.512514

Hassanein, K., & Head, M. (2004). *Building online trust through socially rich Web interfaces*. Paper presented at the 2nd Annual Conference on Privacy, Security and Trust, Fredericton, Canada.

Hawhee, D. (2002). Agonism and aretê. *Philosophy & Rhetoric*, *35*(3), 185–207. doi:10.1353/par.2003.0004

Hawisher, G., & Selfe, C. (1997). *Scholarly publishing in rhetoric and writing- The edited collection: A scholarly contribution and more.*

Hawisher, G. E., & Sullivan, P. (Eds.). (1999). *Passions, pedagogies and 21ˢᵗ century technologies* (pp. 268–291). Logan, UT: Utah State University Press.

Hawisher, G., LeBlanc, P., Moran, C., & Selfe, C. (1995). *Computers and the teaching of writing in American higher education, 1979-1994: A history*. Norwood, NJ: Ablex Publishing. doi:10.2307/358464

Head, A. J., & Eisenberg, M. B. (2010). *Truth be told: How college students evaluate and use information in the digital age*. Retrieved from: http://projectinfolit.org/pdfs/PIL_Fall2010_Survey_FullReport1.pdf

Heba, G. (1997). HyperRhetoric: Multimedia, literacy, and the future of composition. *Computers and Composition*, *14*(1), 19–44. doi:10.1016/S8755-4615(97)90036-0

Heim, M. (1988). The technological crisis of rhetoric. *Philosophy and Rhetoric*, *21*, 57–58.

Hendry, D. G. (2008). Public participation in proprietary software development through user roles and discourse. *International Journal of Human-Computer Studies*, *66*(7), 545–557. doi:10.1016/j.ijhcs.2007.12.002

Hernandez, R. (2011, December 20). Robert Hernandez: For journalism's future, the killer app is credibility. *Nieman Reports*. Retrieved from http://www.niemanlab.org/2011/12/robert-hernandez-for-journalisms-future-the-killer-app-is-credibility/

Herring, S., Scheidt, L. A., Bonus, S., & Wright, E. (2004). Bridging the gap: A genre analysis of weblogs. *Proceedings of the 37th Hawaii International Conference on System Sciences*. DOI:0-7695-2056-1/04

Hilligoss, B., & Rieh, S. Y. (2008). Developing a unifying framework of credibility assessment: Construct, heuristics, and interaction in context. *Information Processing & Management, 44*(4), 1467–1484. doi:10.1016/j.ipm.2007.10.001

Höchstötter, N., & Koch, M. (2009). Standard parameters for searching behaviour in search engines and their empirical evaluation. *Journal of Information Science, 35*(1), 45–65. doi:10.1177/0165551508091311

Höchstötter, N., & Lewandowski, D. (2009). What users see – Structures in search engine results pages. *Information Sciences, 179*(12), 1796–1812. doi:10.1016/j.ins.2009.01.028

Höchstötter, N., & Lüderwald, K. (2011). Web monitoring. In Lewandowski, D. (Ed.), *Handbuch Internet-Suchmaschinen 2: Neue Entwicklungen in der Web-Suche* (pp. 289–322). Heidelberg, Germany: Akademische Verlagsanstalt AKA.

Hocks, M. (2003). Understanding visual rhetoric in digital writing environments. *College Composition and Communication, 54*(4), 629–656. doi:10.2307/3594188

Hodapp, E. (2012). *Angry Birds* hits half a billion downloads. *Touch Arcade.* Retrieved from http://toucharcade.com/2011/11/02/angry-birds-hits-half-a-billion-downloads/

Hogle, C. (2000). *Seeking superethos on the web: A guide for technical writers.* Retrieved from http://www.cas.ucf.edu/english/publications/enc4932/connie.html

Hohmann, J. (2011). *10 best practices for social media. Helpful guidelines for news organizations.* ASNE Ethics and Values Committee. Retrieved January, 16, 2012, from http://asne.org/portals/0/publications/public/10_Best_Practices_for_Social_Media.pdf

Holisky, A. (19 Feb 2012). World of Warcraft subscriber numbers dip 100,000 to 10.2 million. *WOW Insider.* Retrieved from http://wow.joystiq.com/2012/02/09/world-of-warcraft-subscriber-numbers/

Homer. (1939). *Iliad* (W. Rouse, Trans.). In W. Rouse (Ed.), *Homer*. New York, NY: T. Nelson and sons.

Hopkins, N. (2011, December 26). Websites targeting Olympics visitors closed down by police. *The Guardian.* Retrieved from http://www.guardian.co.uk

Hornof, A. J., & Halverson, T. (2003). Cognitive strategies and eye movements for searching hierarchical computer displays. *ACM CHI'03 Human Factors in Computing Systems* (pp. 249–256).

Houghton-Jan, S. (2011). Archives. *Librarian in Black.* Retrieved from http://librarianinblack.net/librarianinblack/archives

Houghton-Jan, S. (2012). Miss librarian goes to Washington. *Librarian in Black.* Retrieved from http://librarianinblack.net/librarianinblack/2012/01/sotu2.html

Hovick, S. R., Meyers, R., & Timmerman, C. E. (2003). E-mail communication in workplace romantic relationships. *Communication Studies, 54*, 468–482. doi:10.1080/10510970309363304

Hovland, C. I., Janis, I. L., & Kelley, H. H. (1953). *Communication and persuasion.* New Haven, CT: Yale University Press.

Hovland, C. I., & Weiss, W. (1951). The influence of source credibility on communication effectiveness. *Public Opinion Quarterly, 15*, 635–650. doi:10.1086/266350

Howard, P., Duffy, A., Freelon, D., Hussain, M., Mari, W., & Mazaid, M. (2011). *Opening closed regimes: What was the role of social media during the Arab Spring?* Seattle, WA: Project on Information Technology and Political Islam. Retrieved from http://dl.dropbox.com/u/12947477/publications/2011_Howard-Duffy-Freelon-Hussain-Mari-Mazaid_pITPI.pdf

Howison, J. (2006). *Coordinating and motivating open source contributors.* Retrieved December 13, 2011, from http://floss.syr.edu

Howison, J., Inoue, K., & Crowston, K. (2006). Social dynamics of free and open source team communications. *IFIP 2nd International Conference on Open Source Software.* Lake Como, Italy. Retrieved from http://floss.syr.edu

Hungwe, B. (2009, March 7). Rumours fly after Tsvangirai crash. *BBC News.* Retrieved from http://news.bbc.co.uk/2/hi/africa/7930694.stm

Hunt, K. (1996). Establishing a presence on the World Wide Web: A rhetorical approach. *Technical Communication, 43*, 376–386.

Hyde, M. J. (Ed.). (2004). *The ethos of rhetoric*. Columbia, SC: University of South Carolina Press.

Ishii, K. (2006). Implications of mobility: The uses of personal communication media in everyday life. *The Journal of Communication, 56*, 346–365. doi:10.1111/j.1460-2466.2006.00023.x

Isocrates. (1980). Speeches and letters. In Norlin, G. (Ed.), *Isocrates* (Norlin, G., Trans.). Cambridge, MA: Harvard University Press.

Ito, M., Baumer, S., & Bittanti, M. boyd, d., Cody, R., Herr-Stephenson, B., et al. (2009). *Hanging out, messing around, and geeking out: Kids living and learning with new media*. Cambridge, MA: MIT Press.

Ivanitskaya, L., Brookins-Fisher, J., O'Boyle, I., Vibbert, D., Erofeev, D., & Fulton, L. (2010). Dirt cheap and without prescription: How susceptible are young US consumers to purchasing drugs from rogue internet pharmacies? *Journal of Medical Internet Research, 12*(2), e(11).

Jabr, F. (March 21, 2011). How Josef Oehmen's advice on Fukushima went viral. *New Scientist*. Retrieved from http://www.newscientist.com/article/dn20266-how-josef-oehmens-advice-on-fukushima-went-viral.html

Jackson, G. (2009, March 7). Photographer arrested at crash scene. *SW Radio Africa*. Retrieved from http://www.swradioafrica.com

Jackson, G. (2011, March 7). SW Radio Africa statement on release of CIO names and details. *SW Radio Africa*. Retrieved from http://www.swradioafrica.com/pages/ciostatement010711.htm

Jacob, R. J., & Karn, K. S. (2003). Eye tracking in human-computer interaction and usability research: Ready to deliver the promises. In Hyona, J., Radach, R., & Deubel, H. (Eds.), *The mind's eye: Cognitive and applied aspects of eye movement research* (pp. 573–605). Elsevier Science.

Jacobs, J. E., & Klaczynski, P. A. (Eds.). (2005). *The development of judgment and decision making in children and adolescents*. Mahwah, NJ: Lawrence Erlbaum.

Jansen, B. J., & Spink, A. (2007). The effect on click-through of combining sponsored and non-sponsored search engine results in a single listing. *Proceedings of the 2007 Workshop on Sponsored Search Auctions*. Presented at the WWW Conference.

Jansen, B. J. (2006). Search log analysis: What it is, what's been done, how to do it. *Library & Information Science Research, 28*(3), 407–432. doi:10.1016/j.lisr.2006.06.005

Jansen, B. J., Booth, D., & Spink, A. (2008). Determining the informational, navigational, and transactional intent of Web queries. *Information Processing & Management, 44*(3), 1251–1266. doi:10.1016/j.ipm.2007.07.015

Jansen, B. J., Booth, D., & Spink, A. (2009). Patterns of query reformulation during web searching. *Journal of the American Society for Information Science and Technology, 60*(7), 1358–1371. doi:10.1002/asi.21071

Jansen, B. J., & Spink, A. (2006). How are we searching the World Wide Web? A comparison of nine search engine transaction logs. *Information Processing & Management, 42*(1), 248–263. doi:10.1016/j.ipm.2004.10.007

Jarvis, J. (2007). Networked Journalism. *BuzzMachine*. Retrieved October 11, 2011, from http://www.buzzmachine.com/2006/07/05/networked-journalism

Jasinski, J. (2001). *Sourcebook on rhetoric: Key concepts in contemporary rhetorical studies*. Thousand Oaks, CA: Sage Publications.

Jenkins, H. (2004). Game design as narrative architecture. In Harrigan, P., & Wardrip-Fruin, N. (Eds.), *First person: New media as story, performance and game*. Cambridge, MA: MIT Press.

Jenkins, H. (2006). *Convergence culture where old and new media collide*. New York, NY: New York University Press.

Johnson-Eilola, J., & Selber, S. (1996). After automation: Hypertext and corporate structures. In Sullivan, P., & Dautermann, J. (Eds.), *Electronic literacies in the workplace: Technologies of writing* (pp. 115–141). Urbana, IL: National Council of Teachers of English.

Johnson, K., & Wiedenbeck, S. (2009). Enhancing perceived cedibility of citizen journalism web sites. *Journalism & Mass Communication Quarterly, 86*(2), 342–348. doi:10.1177/107769900908600205

Johnson, T. J., & Kaye, B. K. (2004). Wag the blog: How reliance on traditional media and the internet influence credibility perceptions of weblogs among blog users. *Journalism & Mass Communication Quarterly, 81*(3), 622–642. doi:10.1177/107769900408100310

Johnson, T. J., & Kaye, B. K. (2009). In blog we trust? Deciphering credibility of components of the internet among politically interested internet users. *Computers in Human Behavior*, *25*(1), 175–182. doi:10.1016/j.chb.2008.08.004

Johnson, T., & Kaye, B. (2010). Choosing is believing? How web gratifications and reliance affect internet credibility among politically interested users. *Atlantic Journal of Communication*, *18*, 1–22. doi:10.1080/15456870903340431

Jones, S., & Fox, S. (2009, January 28). *Generations online in 2009*. Pew Internet & American Life Project report. Retrieved from http://pewresearch.org/pubs/1093/generations-online

Jones, S. (1998). *Cybersociety 2.0: Revisiting computer-mediated communication and community*. Thousand Oaks, CA: Sage.

Juul, J. (2001). Games telling stories?—A brief note on games and narratives. *Game Studies, 1*(1). Retrieved from http://www.gamestudies.org/0101/juul-gts/

Juul, J. (2005). *Half-real: Video games between real rules and fictional worlds*. Cambridge, MA: MIT Press.

Kafker, F., & Loveland, J. (Eds.). (2009). *The early Britannica: The growth of an outstanding encyclopedia*. Oxford, UK: The Voltaire Foundation.

Kammerer, Y., & Gerjets, P. (2012). How search engine users evaluate and select Web search results: The impact of the search engine interface on credibility assessments. In Lewandowski, D. (Ed.), *Web search engine research* (pp. 251–279). Bingley, UK: Emerald. doi:10.1108/S1876-0562(2012)002012a012

Kanu, J. M. (2010, November 25). Zimbabwe: Diaspora-untapped growth zone. *The Herald*. Retrieved from http://allafrica.com/stories/201011250083.html

Kaplan, A. M., & Haenlein, M. (2010). Users of the world, unite! The challenges and opportunities of social media. *Business Horizons*, *53*, 59–68. doi:10.1016/j.bushor.2009.09.003

Kapoun, J. (1998, July). Teaching undergraduates Web evaluation. *College & Research Libraries News*, *59*(7), 522–523.

Karahalios, K. (2004). *Social catalysts: Enhancing communication in mediated spaces*. Unpublished doctoral dissertation, Massachusetts Institute of Technology, U.S.

Katz, S. B. (1992). The ethic of expediency: Classical rhetoric, technology, and the Holocaust. *College English*, *54*(3), 255–275. doi:10.2307/378062

Keen, A. (2007). *The cult of the amateur: How today's Internet is killing our culture*. New York, NY: Doubleday.

Ke, F., & Hoadley, C. (2009, August). Evaluating online learning communities. *Educational Technology Research and Development*, *57*, 487–511. doi:10.1007/s11423-009-9120-2

Keller, D. (2007). Thinking rhetorically. In Selfe, C. (Ed.), *Multimodal composition: Resources for teachers* (pp. 49–63). Cresskill, NJ: Hampton Press.

Kelly, D. (2009). Methods for evaluating interactive information retrieval systems with users. *Foundations and Trends in Information Retrieval*, *3*(1-2), 1–224.

Kennedy, G. (1991). *Aristotle on rhetoric: A theory of civic discourse*. New York, NY: Oxford University Press.

Kenney, K. (2004). Borrowing visual communication theory by borrowing from rhetoric. In Handa, C. (Ed.), *Visual rhetoric in a digital world: A critical sourcebook* (pp. 321–343). Boston, MA: Bedford/St. Martin's.

Kenton, S. (1989). Speaker credibility in persuasive business communication: A model which explains gender differences. *Journal of Business Communication*, *26*, 143–157. doi:10.1177/002194368902600204

Kiesow, D. (2011, January 3). Plain Dealer settles comment lawsuit, limits of anonymity untested. *Poynter*. Retrieved from http://www.poynter.org/latest-news/media-lab/social-media/112641/plain-dealer-settles-comment-lawsuit-limits-of-anonymity-untested/

Kietzmann, J. H., Hermkens, K., McCarthy, I. P., & Silvestre, B. S. (2011). Social media? Get serious! Understanding the functional building blocks of social media. *Business Horizons*, *54*, 241–251. doi:10.1016/j.bushor.2011.01.005

Kincaid, J. (2011). *That was fast: Amazon's Kindle ebook sales surpass print (it only took four years).* techcrunch.com. Retrieved from May 12, 2012, from http://techcrunch.com/2011/05/19/that-was-fast-amazons-kindle-ebook-sales-surpass-print-it-only-took-four-years/

King, F. J., Goodson, L., & Rohani, F. (1998). *Higher order thinking skills.* Center for Advancement for Learning and Assessment, Florida State University. Retrieved from http://www.cala.fsu.edu/files/higher_order_thinking_skills.pdf

Kittur, A., Suh, B., Pendleton, B. A., & Chi, E. (2007). He says, she says: Conflict and coordination in Wikipedia. *CHI 2007: Proceedings of the ACM Conference on Human-factors in Computing Systems,* (pp. 453-462). San Jose, CA: ACM Press.

Kjellberg, S. (2009). Scholarly blogging practice as situated genre: An analytical framework based on genre theory. *Information Research: An International Electronic Journal, 14*(3).

Klaczynski, P. A. (2001). The influence of analytic and heuristic processing on adolescent reasoning and decision making. *Child Development, 72,* 844–861. doi:10.1111/1467-8624.00319

Kleinberg, J. (1999). Authoritative sources in a hyper-linked environment. *Journal of the ACM, 46*(5), 604–632. doi:10.1145/324133.324140

Klinkenberg, J.-M. (2000). *Précis de sémiotique générale.* Paris, France: Seuil.

Knight, S. A., & Burn, J. (2005). Developing a framework for assessing information quality on the World Wide Web. *Informing Science Journal, 8,* 159–172.

Kobayashi, S. (2009). *DIY hardware: Reinventing hardware for the digital do-it-yourself revolution.* Paper presented at the ACM SIGGRAPH ASIA 2009 Art Gallery and Emerging Technologies: Adaptation, Yokohama, Japan.

Kogan, H. (1958). *The great EB: The story of the Encyclopaedia Britannica.* Chicago, IL: University of Chicago Press.

Kokis, J., Macpherson, R., Toplak, M., West, R. F., & Stanovich, K. E. (2002). Heuristic and analytic processing: Age trends and associations with cognitive ability and cognitive styles. *Journal of Experimental Child Psychology, 83,* 26–52. doi:10.1016/S0022-0965(02)00121-2

Kolodzy, K. (2006). *Convergence journalism: Writing and reporting across the news media.* New York, NY: Rowman and Littlefield.

Kondo, K. (n.d.). *Bandcamp.* Retrieved from http://bandcamp.com/tag/koji-kondo

Kovach, B., & Rosenstiel, T. (2001). *The elements of journalism: What newspeople should know and the public expect.* New York, NY: Crown Publishers.

Kovach, B., & Rosenstiel, T. (2010). *Blur: How to know what's true in the age of information overload.* New York, NY: Bloomsbury.

Kralisch, A., & Berendt, B. (2004). Linguistic determinants of search behaviour on websites. *Proceedings of the Fourth International Conference on Cultural Attitudes towards Technology and Communication, Karlstad, Sweden* (pp. 599–613).

Kress, G., & van Leeuwen, T. (2006). *Reading images: The grammar of visual design* (2nd ed.). New York, NY: Routledge. doi:10.1016/S8755-4615(01)00042-1

Kuiper, E., & Volman, M. (2008). The Web as a source of information for students in K–12 education. In Coiro, J., Knobel, M., Lankshear, C., & Leu, D. (Eds.), *Handbook of research on new literacies* (pp. 241–266). New York, NY: Lawrence Erlbaum.

La Freebox à l'honneur dans E=M6 (MàJ). (2011, May 17). Freenews. Retrieved February 12, 2012, from http://www.freenews.fr/spip.php?article8300

Lacohée, H., Crane, S., & Phippen, A. (2006). *Trustguide.* Hewlett-Packard Laboratories. Retrieved March 2, 2012, from http://www.trustguide.org.uk/Trustguide%20-%20Final%20Report.pdf

LaGrandeur, K. (2003). Digital images and classical persuasion. In Hocks, M. E., & Kendrick, M. R. (Eds.), *Eloquent images: Word and image in the age of new media* (pp. 117–136). Cambridge, MA: MIT Press.

Landgraf, G. (2012). Code year librarians geek out. *American Libraries.* Retrieved from http://americanlibrariesmagazine.org/inside-scoop/code-year-librarians-geek-out

Landow, G. (1992). *Hypertext: The convergence of contemporary critical theory and technology.* Baltimore, MD: John Hopkins University Press.

Lanier, J. (2010). *You are not a gadget: A manifesto*. New York, NY: Knopf.

Lankes, D. R. (2008). Credibility on the Internet: Shifting from authority to reliability. *The Journal of Documentation, 64*(5), 667–686. doi:10.1108/00220410810899709

Large, A. (2004). Information seeking on the Web by elementary school students. In Chelton, M. K., & Cool, C. (Eds.), *Youth information-seeking behavior: Theories, models, and issues* (pp. 293–320). Lanham, MD: Scarecrow Press.

Larson, C. U. (2004). *Persuasion: Reception and responsibility*. Belmont, CA: Thomson/Wadsworth.

Latour, B., & Woolgar, S. (1986). *Laboratory life: The construction of scientific facts*. Princeton, NJ: Princeton University Press.

Lauer, T. W., & Deng, X. (2007). Building online trust through privacy practices. *International Journal of Information Security, 6*, 323–331. doi:10.1007/s10207-007-0028-8

Lawson, S. (2009, July 13). Clinical reader: From zero to negative sixty with one bogus threat. *See Also: a Weblog by Steve Lawson*. Retrieved from http://stevelawson.name/seealso/archives/2009/07/clinical_reader_from_zero_to_negative_sixty_with_one_bogus_threat.html

Le Monde Forums. (n.d.). Mode d'emploi et règles de conduite. *Le Monde*. Retrieved from http://forums.lemonde.fr/perl/faq_french.pl?Cat=

Le Monde. (2005, March 21). Chats, mode d'emploi. *Le Monde*. Retrieved from http://www.lemonde.fr/services-aux-internautes/article/2003/02/19/chats-mode-d-emploi_309828_3388.html

Le Monde. (2011, May 5). Blogs: Mode d'emploi. *Le Monde*. Retrieved from http://www.lemonde.fr/services-aux-internautes/article/2004/11/10/blogs-mode-d-emploi_386588_3388.html

Le Monde. (2011, May 5). La charte des blogs et les règles de conduite. *Le Monde*. Retrieved from http://www.lemonde.fr/services-aux-internautes/article/2004/12/03/la-charte-des-blogs-et-les-regles-de-conduite-sur-lemonde-fr_389436_3388.html

Leach, J. (2009). Creating ethical bridges from journalism to digital news. *Nieman Reports*. Retrieved from http://www.nieman.harvard.edu/reportsitem.aspx?id=101899

Lee, H., Park, S., Lee, Y., & Cameron, G. T. (2010). Assessment of motion media on believability and credibility: An exploratory study. *Public Relations Review, 36*(3), 310–312. doi:10.1016/j.pubrev.2010.04.003

Leff, M. C. (1999). *The habitation of rhetoric. Contemporary rhetorical theory: A reader* (pp. 52–64). New York, NY: Guilford Press.

Lemley, M. A., & Shafir, Z. (2011). Who chooses open-source software? *The University of Chicago Law Review. University of Chicago. Law School, 78*(1), 139–164.

Lenhart, A., Purcell, K., Smith, A., & Zickuhr, K. (2010). *Social media and mobile Internet use among teens and young adults*. Retrieved February 18, 2010 from http://pewresearch.org/pubs/1484/social-media-mobile-internet-use-teens-millennials-fewer-blog

Lenzini, G., Martinelli, F., Matteucci, I., & Gnesi, S. (2008). *A uniform approach to security and fault-tolerance specification and analysis*. Paper presented at the Workshop on Software Architectures for Dependable Systems, Anchorage, U.S.

Leon, J. (2010). The blog as beat. *Nieman Reports, 64*(4), 9–11.

Lessig, L. (2004). *Free culture: How big media uses technology and the law to lock down culture and control creativity*. New York, NY: Penguin Press.

Lessig, L. (2006). *Code: Version 2.0*. New York, NY: Basic Books.

Levinson, P. (2009). *New new media*. Boston, MA: Allyn & Bacon.

Levi-Strauss, C. (1960). *The savage mind*. Chicago, IL: University of Chicago Press.

Lévy, P. (1994). *L'Intelligence collective: pour une anthropologie du cyberespace*. Paris, France: La Découverte.

Lewandowski, D. (2005). Web searching, search engines and Information Retrieval. *Information Services & Use, 25*, 137–147.

Lewandowski, D. (2008). The retrieval effectiveness of Web search engines: Considering results descriptions. *The Journal of Documentation, 64*(6), 915–937. doi:10.1108/00220410810912451

Lewandowski, D., & Höchstötter, N. (2008). Web searching: A quality measurement perspective. In Spink, A., & Zimmer, M. (Eds.), *Web search: Multidisciplinary perspectives* (pp. 309–340). Berlin, Germany: Springer.

Lewandowski, D., & Spree, U. (2011). Ranking of Wikipedia articles in search engines revisited: Fair ranking for reasonable quality? *Journal of the American Society for Information Science and Technology, 62*(1), 117–132. doi:10.1002/asi.21423

Lim, S. (2009). How and why do college students use Wikipedia? *Journal of the American Society for Information Science and Technology, 60*(11), 2189–2202. doi:10.1002/asi.21142

Lindgaard, G., Dudek, C., Devjani, S., Sumegi, L., & Noonan, P. (2011). An exploration of relations between visual appeal, trustworthiness and perceived usability of homepages. *ACM Transactions on Computer-Human Interaction, 18*(1), 1–30. doi:10.1145/1959022.1959023

LinkedIn. (2011). *About us.* LinkedIn press center. Retrieved May 15, 2012, from http://press.linkedin.com/about

Lin, Y., & Wu, H. (2008). Information privacy concerns, government involvement and corporate policies in the customer relationship management context. *Journal of Global Business and Technology, 4*(1), 79–91.

Lippmann, W. (1920). *Liberty and the news.* New York, NY: Harcourt, Brace and Howe.

Livingstone, S. (2009). *Children and the Internet: Great expectations and challenging realities.* Cambridge, UK: Polity Press.

Lomax, A. (Ed.). (1968). *Folk song style and culture.* Washington, DC: American Association for the Advancement of Science.

Lorigo, L., Haridasan, M., Brynjarsdóttir, H., Xia, L., Joachims, T., & Gay, G. (2008). Eye tracking and online search: Lessons learned and challenges ahead. *Journal of the American Society for Information Science and Technology, 59*(7), 1041–1052. doi:10.1002/asi.20794

Lorigo, L., Pan, B., Hembrooke, H., Joachims, T., Granka, L., & Gay, G. (2006). The influence of task and gender on search and evaluation behavior using Google. *Information Processing & Management, 42*(4), 1123–1131. doi:10.1016/j.ipm.2005.10.001

Luhmann, N. (1979). *Trust and power.* Chichester, UK: Wiley.

Lynch, P. (2010). Aesthetics and trust: Visual decisions about Web pages. *Proceedings of the International Conference on Advanced Visual Interfaces*, New York, NY.

Macdonald, K., & Cousins, M. (1998). *Imagining reality: the Faber book of the documentary.* London, UK: Faber.

MacDougall, R. (2005). Identity, electronic ethos, and blogs: A technologic analysis of symbolic exchange on the new news medium. *The American Behavioral Scientist, 49*(4), 575–599. doi:10.1177/0002764205280922

Machill, M., Neuberger, C., Schweiger, W., & Wirth, W. (2004). Navigating the Internet: A study of German-language search engines. *European Journal of Communication, 19*(3), 321–347. doi:10.1177/0267323104045258

Mackey, R. (2011). 'Gay girl in Damascus' blog a hoax, American says. *New York Times.* Retrieved from http://www.nytimes.com/2011/06/13/world/middleeast/13blogger.html

Mackie, S. (2011, March 15). *Internet Explorer 9 released, but should you care?* Retrieved May 12, 2012, from http://gigaom.com/collaboration/internet-explorer-9-released-but-should-you-care/

Malleus, R. (2011). Whose TV is it anyway? An examination of the shift towards satellite television in Zimbabwe. In Wachanga, N. D. (Ed.), *Cultural and new communication technologies: Political, ethnic and ideological implications* (pp. 128–143). Hershey, PA: IGI Global. doi:10.4018/978-1-60960-591-9.ch007

Mambo, E., & Chitemba, B. (2012, May 11). I'm ready to rule, says Mnangagwa. *Zimbabwe Independent*, May 11. Retrieved from http://www.theindependent.co.zw/political-zimbabwe-stories/2012/05/11/im-ready-to-rule-says-mnangagwa/

Mandell, L. (2011). *Promotion and tenure for digital scholarship.* IDHMC Blog. Retrieved from http://idhmc.tamu.edu/commentpress/promotion-and-tenure/

Mandl, T. (2006). Implementation and evaluation of a quality-based search engine. *Proceedings of the Seventeenth Conference on Hypertext and Hypermedia* (pp. 73–84). New York, NY: ACM.

Mandl, T. (2005). The quest to find the best pages on the Web. *Information Services & Use, 25*(2), 69–76.

Manovich, L. (2002). *The language of new media*. Cambridge, MA: MIT Press.

Manual of style (road junction lists). (2010, November 08). *Wikipedia*. Retrieved December 10, 2010 from http://en.wikipedia.org/wiki/Wikipedia:Manual_of_Style_%28road_junction_lists%29

Manyukwe, C. (2012, May 16). Poll alliance faces hurdles. *The Financial Gazette*. Retrieved from http://www.financialgazette.co.zw

Marsh, S. (1994). *Formalising trust as a computational concept*. Unpublished doctoral dissertation, University of Stirling, Scotland.

Marshall, J. (2002). TPM Editors Blog. *Talking Points Memo*. Retrieved from http://talkingpointsmemo.com/archives/week_2002_12_01.php

Marsh, C. (2006). Aristotelian ethos and the new orality: Implications for media literacy and media ethics. *Journal of Mass Media Ethics, 21*(4), 338–352. doi:10.1207/s15327728jmme2104_8

Marsh, S., & Dibben, M. (2003). The role of trust in information science and technology. *Annual Review of Information Science & Technology, 37*(1), 465–498. doi:10.1002/aris.1440370111

Martin, M. (2007, 25 January). Online journalism ethics: Guidelines from the conference. *Poynter*. Retrieved from http://www.poynter.org/uncategorized/80445/online-journalism-ethics-guidelines-from-the-conference/#ugc

MartinLutherKing.org. (2012, February 14). *Martin Luther King, Jr.: A true historical examination*. Retrieved from http://www.martinlutherking.org

Massachusetts Institute of Technology. (2012). *Oehmen, Josef, LAI research scientist*. Retrieved from http://lean.mit.edu/about/lai-structure/faculty-researchers-and-staff/oehmen-josef

master, n.1 and adj. (2011, March). *Oxford English Dictionary*. New York, NY: Oxford University Press. Retrieved from http://www.oed.com/view/Entry/114751?rskey=DhchNQ&result=1&isAdvanced=false

Matthews, R. (2005, December 23). Wikipedia's search for the truth—The online encyclopedia may soon be as credible as it is popular. *Financial Times*. Retrieved November 18, 2010, from http://www.ebusinessforum.com/index.asp?layout=printer_friendly&doc_id=7931

Matthies, B. (2004). The road to faculty-librarian collaboration. *Academic Exchange Quarterly, 8*(4), 135–141.

Mauer, A. (2010). Using social networks as reporting tools. In Society of Professional Journalists' Digital Media Committee (Ed.), *The SPJ digital media handbook, Part I* (pp. 12-13). Retrieved October 21, 2011, from http://blogs.spjnetwork.org/tech/wp-content/uploads/2010/03/SPJDigitalMediaHandbookV3.pdf

Mavhunga, C. (2009). The glass fortress: Zimbabwe's cyber-guerilla warfare. *Journal of International Affairs, 62*(2), 159–173.

Mazara, G. (2012, May 6). Drumbeat: Can Roki shed his bad-boy image in BAA. *The Standard*. Retrieved from http://www.thestandard.co.zw

McDougall, P. (2012). *Kindle Fire sales hit 6 million in Q4*. Informationweek.com. Retrieved May 10, 2012, from http://www.informationweek.com/news/hardware/handheld/232500684

McGee, M. (2008). Google explains malware warning policy & how to fix your site. *Search Engine Land*. Retrieved February 20, 2012, from http://searchengineland.com/google-malware-warning-policy-15271

McGee, M. (2010). Google news dropping sites, reviewing inclusion standards. *Search Engine Land*. Retrieved February 21, 2012, from http://searchengineland.com/google-news-dropping-sites-reviewing-inclusion-standards-57673

McGonigal, J. (2011). *Reality is broken: Why games make us better and how they can change the world*. New York, NY: The Penguin Press.

McGregor, J. (2009). Associational links with home amoung Zimbabweans in the UK: Reflections on long-distance nationalisms. *Global Networks*, *9*(2), 185–208. doi:10.1111/j.1471-0374.2009.00250.x

McGregor, J., & Pasura, D. (2010). Diasporic repositioning and the politics of re-engagement: Developmentalising Zimbabwe's diaspora? *The Round Table*, *99*(411), 687–703. doi:10.1080/00358533.2010.530413

McKnight, D., Choudhury, V., & Kacmar, C. (2002). Developing and validating trust measures for e-commerce: An integrative typology. *Information Systems Research*, *13*(3), 334–359. doi:10.1287/isre.13.3.334.81

McLeod, K. (2001). *Owning culture: Authorship, ownership, and intellectual property law*. New York, NY: Peter Lang.

McLuhan, M. (1994). *Understanding media: The extensions of man*. Cambridge, MA: MIT Press.

McMillan, G. (2011). Twitter reveals active user number, how many actually say something. *Time*. Retrieved from http://techland.time.com/2011/09/09/twitter-reveals-active-user-number-how-many-actually-say-something/

McNamee, S. (1996). Therapy and identity construction in a postmodern world. In Grodin, D., & Lindlof, T. (Eds.), *Constructing the self in a mediated world* (pp. 141–155). Thousand Oaks, CA: Sage.

McQuail, D. (1992). *Media performance: Mass communication and the public interest*. London, UK: Sage.

Media Monitoring Project Zimbabwe. (2002, July 23). Weekly media update. Retrieved from http://www.mmpz.org.zw

meeblog » ActionScript Developer (aka "ActionScript Assassin"). (n.d.). Retrieved from http://web.archive.org/web/20080502081313/http://www.meebo.com/jobs/actionscript/

meeblog » JavaScript Developer (aka "JavaScript Ninja"). (n.d.). Retrieved from http://web.archive.org/web/20080502110640/http://www.meebo.com/jobs/javascript/

meeblog » Server Side Developer (aka "Server Samurai"). (n.d.). Retrieved from http://web.archive.org/web/20080920042629/http://www.meebo.com/jobs/openings/server/

Mehrabi, D., Hassan, M. A., & Ali, M. S. S. (2009). News media credibility of the internet and television. *European Journal of Soil Science*, *11*(1), 136–148.

Meng, B. (2009). Regulating egao: Futile efforts of re-centralization? In Zhang, X., & Zheng, Y. (Eds.), *China's information and communications technology revolution: Social changes and state responses* (pp. 53–67). New York, NY: Routledge.

Meola, M. (2004). Chucking the checklist: A contextual approach to teaching undergraduates web-site evaluation. *Libraries and the Academy*, *4*(3), 331–342. doi:10.1353/pla.2004.0055

Meriam Library, California State University Chico. (2010). *Evaluating information: Applying the CRAAP test*. Retrieved from http://www.csuchico.edu/lins/handouts/evalsites.html

Metzger, M. J. (2007). Making sense of credibility on the web: Models for evaluating online information and recommendations for future research. *Journal of the American Society for Information Science and Technology*, *58*(13), 2078–2091. doi:10.1002/asi.20672

Metzger, M. J., & Flanagin, A. J. (Eds.). (2008). *Digital media, youth, and credibility*. Cambridge, MA: MIT Press.

Metzger, M. J., Flanagin, A. J., Eyal, K., Lemus, D. R., & McCann, R. (2003). Credibility in the 21st century: Integrating perspectives on source, message, and media credibility in the contemporary media environment. In Kalbfeisch, P. (Ed.), *Communication yearbook 27* (pp. 293–335). Mahwah, NJ: Lawrence Erlbaum. doi:10.1207/s15567419cy2701_10

Metzger, M. J., Flanagin, A. J., & Medders, R. B. (2010). Social and heuristic approaches to credibility evaluation online. *The Journal of Communication*, *60*(3), 413–439. doi:10.1111/j.1460-2466.2010.01488.x

Metzger, M. J., Flanagin, A. J., Pure, R., Medders, R., Markov, A., Hartsell, E., & Choi, E. (2011). *Adults and credibility: An empirical examination of digital media use and information credibility. Research report prepared for the John D. and Catherine T. MacArthur Foundation.* Santa Barbara: University of California.

Meyer, E. (2008, March 27). *Acid redux.* Retrieved March 4, 2011, from http://meyerweb.com/eric/thoughts/2008/03/27/acid-redux/

Meyer, E. (2002). *Eric Meyer on CSS: Mastering the language of web design* (1st ed.). New Riders Press.

Microsoft. (2011). *Internet Explorer - Web browser for Microsoft Windows.* Retrieved November 6, 2011, from http://windows.microsoft.com/en-us/internet-explorer/products/ie/home

Miller, C. (2010). Rhetoric, technology, and the pushmi-pullyu. In Selber, S. A. (Ed.), *Rhetorics and technologies: New direction in writing and communication* (pp. ix–xii). Columbia, SC: University of South Carolina Press.

Miller, C. R. (2001). Writing in a culture of simulation: Ethos online. In Coppock, P. (Ed.), *The semiotics of writing: Transdiciplinary perspectives on the technology of writing* (pp. 253–279).

Miller, K. W. (2005). Web standards: Why so many stray from the narrow path. *Science and Engineering Ethics*, *11*(3), 477–479. doi:10.1007/s11948-005-0017-0

Mills, H. (2000). *Artful persuasion.* New York, NY: AMACOM.

Milstein, S. (2009). Twitter for libraries (and librarians). *Computers in Libraries*, *29*, 17–18.

Miners engage govt on fees. (2012, February 9). Zimbabwe Broadcasting Corporation. Retrieved from http://www.zbc.co.zw/news-categories/business/16363-miners-engage-govt-on-fees.html

Mizzaro, S. (1997). Relevance: The whole history. *Journal of the American Society for Information Science American Society for Information Science*, *48*(9), 810–832. doi:10.1002/(SICI)1097-4571(199709)48:9<810::AID-ASI6>3.0.CO;2-U

Modern Language Association. (2007). Report of the taskforce on tenure and promotion. *Profession*, 9–71.

Moe, H. (2008). Dissemination and dialogue in the public sphere: A case for public service media online. *Media Culture & Society*, *30*(3), 319–336. doi:10.1177/0163443708088790

Möllering, G. (2008). *Inviting or avoiding deception through trust? Conceptual exploration of an ambivalent relationship* (MPIfG Working Paper 08/1). Max Planck Institute for the Study of Societies. Retrieved from http://www.mpifg.de/pu/workpap/wp08-1.pdf

Möllering, G. (2001). The nature of trust: From Georg Simmel to a theory of expectation, interpretation and suspension. *Sociology*, *35*(2), 403–420.

Möllering, G. (2006). Trust, institutions, agency: Towards a neoinstitutional theory of trust. In Bachmann, R., & Zaheer, A. (Eds.), *Handbook of trust research* (pp. 355–376). Cheltenham, UK: Edward Elgar.

Morris, C. W. (1964). *Signification and significance: A study of the relations of signs and values.* Cambridge, MA: MIT Press.

Moyo, H. (2012, May 17). Policy failures render Zim a basket case. *Zimbabwe Independent*. Retrieved from http://www.theindependent.co.zw

Mozilla. (2011). *Firefox Web browser – Free download.* Retrieved November 6, 2011, from http://www.mozilla.com/en-US/firefox

Muhlhauser, P. (2011). Teaching moms and dads to perform the family: Rhetoric and assisted reproductive technology websites. *Computers and Composition Online*. Retrieved May 12, 2012, from http://www.bgsu.edu/cconline/paul-muhlhauser/

Mujuru won't be exhumed. (2012, February 7). *ZimOnline*. Retrieved March 20, 2012, from http://www.zimonline.co.za

Mujuru's death shrouded in suspicion. (2012, February 6). *Bulawayo24 News*. Retrieved from http://bulawayo24.com/index-id-news-sc-national-byo-11859-article-Mujuru%27s+death+shrouded+in+suspicion+.html

Mukundu, R. (2005). *Research findings and conclusions. African Media Development Initiative.* Zimbabwe: BBC World Service Trust.

Murray, J. H. (1997). *Hamlet on the holodeck: The future of narrative in cyberspace*. New York, NY: Free Press.

Napoli, P. M. (2009, April). *Media policy in the era of user-generated and distributed content: Transitioning from access to the media to access to audiences*. Paper presented at the Media in Transition Conference Massachusetts Institute of Technology, Cambridge, MA.

Napoli, P. M. (2010). *Audience evolution: New technologies and the transformation of media audiences*. New York, NY: Columbia University Press.

Napolitano, F. (n.d.). *Pet friendly animal lovers*. LinkedIn. Retrieved from http://www.LinkedIn.com/groups/Pet-Friendly-Animal-Lovers-89012?itemaction=mclk&anetid=89012&impid=&pgkey=anet_search_results&actpref=anetsrch_name&trk=anetsrch_name&goback=.gdr_1327510683792_1

Nardi, B. (2010). *My life as a Night Elf priest*. Ann Arbor, MI: University of Michigan Press.

Nardi, B. A., & O'Day, V. L. (1999). *Information ecologies: Using technology with heart*. Cambridge, MA: MIT Press.

Nayar, P. K. (2010, December 25). Wikileaks, the new information cultures and digital parrhesia. *Economic and Political Weekly*, *45*(52), 27–30.

NBC. (2007, August 14). *Privacy policy*. NBC. Retrieved from http://www.nbc.com/privacy-policy/

NBC. (2007, August 14). *Terms of service*. NBC. Retrieved from http://www.nbc.com/privacy-policy/terms-of-service/

Nemukuyu, D. (2012, February 8). Inquest findings final, say experts. *The Herald Online*. Retrieved from http://allafrica.com/stories/201202080026.html

New constitution's principal drafters must be fired. (2012, February 13). *The Herald Online* Retrieved from http://www.herald.co.zw/index.php?option=com_content&view=article&id=33817

New London Group. (1996). A pedagogy of multiliteracies: Designing social futures. *Harvard Educational Review*, *66*(1), 60–93.

New York Times. (2011, April 18). *Privacy policy*. Retrieved from http://www.nytimes.com/content/help/rights/privacy/highlights/privacy-highlights.html

New York Times. (2011, March 16). *Terms of service*. Retrieved from http://www.nytimes.com/content/help/rights/privacy/highlights/privacy-highlights.html

Newhagen, J. E., & Nass, C. (1989). Differential criteria for evaluating credibility of newspapers and TV news. *The Journalism Quarterly*, *66*(2), 277–284. doi:10.1177/107769908906600202

Newman, R., & Newman, D. (1969). *Evidence*. Boston, MA: Houghton Mifflin.

Nguyen, A. (2011). Marrying the professional to the amateur: strategies and implications of the Ohmy News model. In Meikle, G., & Redden, G. (Eds.), *News online: Transformations and continuities* (pp. 195–209). London, UK: Palgrave MacMillan.

Nichols, B. (1994). *Blurred boundaries: questions of meaning in contemporary culture*. Bloomington, IN: Indiana University Press.

Nichols, B. (2009). The question of evidence, the power of rhetoric and documentary film. In Austin, T., & de Jong, W. (Eds.), *Rethinking documentary: New perspectives, new practices* (pp. 29–37). Milton Keynes, UK: Open University Press.

Nieborg, D. B. (2004, September). *America's army: More than a game*. Paper presented at the meeting of the International Simulation and Gaming Association Conference, Munich, Germany.

Nixon, R. (1998). The feather palace. *Transition*, *77*, 70–85. doi:10.2307/2903201

Nooteboom, B. (2005). *Framing, attribution and scripts in the development of trust*. Discussion Paper, University of Toronto. Retrieved March 20, 2012, from http://www.bartnooteboom.nl/

Nooteboom, B. (2006). Forms, sources, and processes of trust. In Bachmann, R., & Zaheer, A. (Eds.), *Handbook of trust research* (pp. 247–263). Northhampton, UK: Edward Elgar Publishing.

Norlin, G. (1980). *Isocrates*. Cambridge, MA: Harvard University Press.

Norton, R. (1983). *Communicator style: Theory, application, and measures*. Beverly Hills, CA: Sage.

Notability (books). (2012, July 9). *Wikipedia*. Retrieved November 08, 2010, from http://en.wikipedia.org/wiki/Wikipedia:Notability_(books)

Notability (fatal hull loss civil aviation accidents). (2011, March 5). *Wikipedia*. Retrieved November 08, 2010, from http://en.wikipedia.org/wiki/Wikipedia:Notability_(books)

Notability. (2012, November 08). *Wikipedia*. Retrieved November 8, 2010 from http://en.wikipedia.org/wiki/Wikipedia:Notability

Notess, G. (2010). The changing information cycle. In P. McCaffrey (Ed.), *The news and its future* (34-39). New York, NY: H.W. Wilson Co.

November, A. C. (2009). *Empowering students with technology*. Thousand Oaks, CA: Corwin Press.

O'Carroll, L., & Halliday, J. (2012, January 03). Wendi Deng Twitter account is a fake. *The Guardian*. Retrieved from http://www.guardian.co.uk

O'Day, V., & Jeffries, R. (1993). Orienteering in an information landscape: how information seekers get from here to there. *CHI '93: Proceedings of the INTERACT '93 and CHI '93 Conference on Human Factors in Computing Systems* (pp. 438–445). ACM.

O'Keefe, D. J. (2002). *Persuasion: Theory and research*. Thousand Oaks, CA: Sage.

Open Source Initiative. (n.d.). *The open source definition*. Open Source Initiative. Retrieved November 6, 2011, from http://opensource.org/docs/osd

Organisation for Economic Co-operation and Development. (2007). *Participative web: User-created content*. Paris, France: OECD.

Oriella, P. R. Network. (2011). *The state of journalism in 2011*. Retrieved September 27, 2011, from http://www.centroperiodismodigital.org/sitio/sites/default/files/publication.pdf

Owens, J. W., Shaikh, A. D., & Chaparro, B. S. (2011). Patterns of information sharing among inner and outer social circles. *Usability News, 13*(1). Retrieved January 7, 2012, from http://www.surl.org/usabilitynews/131/sharing.asp

Page, L., Brin, S., Motwani, R., & Winograd, T. (1998). *The PageRank citation ranking: Bringing order to the Web*. Retrieved from http://ilpubs.stanford.edu:8090/422/1/1999-66.pdf

Palfrey, J., & Gasser, U. (2008). *Born digital: Understanding the first generation of digital natives*. New York, NY: Perseus Books Group.

Pan, B., Hembrooke, H., Joachims, T., Lorigo, L., Gay, G., & Granka, L. (2007). In Google we trust: Users' decisions on rank, position, and relevance. *Journal of Computer-Mediated Communication, 12*(3), 801–823. doi:10.1111/j.1083-6101.2007.00351.x

Park, R. (1940). News as a form of knowledge: A new chapter in the sociology of knowledge. *American Journal of Sociology, 45*(5), 669–686. doi:10.1086/218445

Park, T. K. (1993). The nature of relevance in information retrieval: An empirical study. *The Library Quarterly, 63*(3), 318–351. doi:10.1086/602592

PatrickBowman. (2010, October 8). *Comment posted to Toward a more thoughtful conversation on stories*. Reuters. Retrieved from http://blogs.reuters.com/fulldisclosure/2010/09/27/toward-a-more-thoughtful-conversation-on-stories/?cp=all#comments

PBS. (2011). *PBS editorial standards and policies*. Retrieved from http://www.pbs.org/about/media/about/cms_page_media/35/PBS%20Editorial%20Standards%20and%20Policies.pdf

PBS. (2011). *PBS privacy policy*. Retrieved from http://www.pbs.org/about/policies/privacy-policy/

PBS. (2011). *PBS terms of use*. Retrieved from http://www.pbs.org/about/policies/terms-of-use/

Pentland, A. (2008). *Honest signals: How they shape our world*. Cambridge, UK: MIT Press.

Pepsi-Cola® Company. (2005). *Over one hundred years of fun and refreshment: The Pepsi-Cola® story.* Retrieved from http://pepsiusa.com/downloads/PepsiLegacy_Book.pdf

Pepukai, O., & Gumbo, T. (2012, May 17). Fresh cholera outbreak worries residents of Zimbabwe's Chiredzi Town. *Voice of America-Studio 7.* Retrieved from http://www.voanews.com/zimbabwe/news/Health-Officials-Confirm-Cholera-Outbreak-In-Chiredzi-Amin-Water-Shortages-151928535.html

Perloff, R. M. (2003). *The dynamics of persuasion: Communication and attitudes in the 21st century.* Mahwah, NJ: Lawrence Erlbaum Associates.

Pew Research Center (PEW). (2011, September 22). *Views of the news media: 1985-2011. Press widely criticized, but trusted more than other information sources.* Retrieved November 12, 2011, from http://www.people-press.org/files/legacy-pdf/9-22-011%20Media%20Attitudes%20Release.pdf

Piao, Y. (2006, May 29). *After all, who is Acosta?* Retrieved July 16, 2012, from http://blog.sina.com.cn/s/blog_49191240010003my.html

Pirolli, P. (2005). Rational analyses of information foraging on the Web. *Cognitive Science, 29*(3), 343–373. doi:10.1207/s15516709cog0000_20

Plantinga, C. R. (1997). *Rhetoric and representation in nonfiction film.* Cambridge, UK: Cambridge University Press.

Plantinga, C. R. (2005). What a documentary is, after all. *The Journal of Aesthetics and Art Criticism, 63*(2), 105–117. doi:10.1111/j.0021-8529.2005.00188.x

Plato. (1925). *Sophist* (H. Fowler, Trans.). *Theaetetus,* Vol. VII. Loeb Classical Library. Cambridge, MA: Harvard University Press.

Pliny the Elder. (1601). *Naturalis historia* (P. Holland, Trans.). Retrieved July 15, 2012, from http://penelope.uchicago.edu/holland/

Ploch, L. (2010). Zimbabwe background. *Congressional Research Report for Congress,* 7-5700.

Poole, A., & Ball, L. J. (2005). Eye tracking in human-computer interaction and usability research: current status and future prospects. In Ghaoui, C. (Ed.), *Encyclopedia of human computer interaction* (pp. 211–219). Hershey, PA: IGI Global. doi:10.4018/978-1-59140-562-7.ch034

Porter, J. E. (1992). *Audience and rhetoric.* NJ: Prentice Hall.

Potter, W. J. (2012). *Media effects.* Los Angeles, CA: Sage.

Preece, J., Nonnecke, B., & Andrews, D. (2004). The top five reasons for lurking: Improving community experiences for everyone. *Computers in Human Behavior, 20,* 201–223. doi:10.1016/j.chb.2003.10.015

Prensky, M. (2001). Digital natives, digital immigrants. *Horizon, 9*(5). doi:10.1108/10748120110424816

Purcell, K. (2011). *Search and email still top the list of most popular online activities.* Retrieved from http://www.pewinternet.org/~/media//Files/Reports/2011/PIP_Search-and-Email.pdf

Purdy, J. P. (2010). Wikipedia is good for you!? In C. Lowe & P. Zemliansky (Eds.), *Writing spaces: Readings on writing,* Vol. 1. Retrieved April 26, 2012 from http://writingspaces.org/essays/wikipedia-is-good-for-you

Purdy, J. P. (2009). When the tenets of composition go public: A study of writing in Wikipedia. *College Composition and Communication, 61*(2), 351–373.

Purdy, J., & Walker, J. (2010). Valuing digital scholarship: Exploring the changing realities of intellectual work. *Profession, 19,* 77–195.

Quintilian. (1958). Institutes of oratory. In Butler, H. (Ed.), *The institutio oratoria of Quintilian* (Butler, H., Trans.). Cambridge, MA: Harvard University Press.

Raessens, J. (2006). Reality play: Documentary computer games beyond fact and fiction. *Popular Communication, 4*(3), 213–224. doi:10.1207/s15405710pc0403_5

Raghu, T. S., Sinha, R., Vinze, A., & Burton, O. (2009). Willingness to pay in an open source software environment. *Information Systems Research, 20*(2), 218–236. doi:10.1287/isre.1080.0176

Rainie, L. (2006, March 23). *Life online: Teens and technology and the world to come*. Keynote address to the annual conference of the Public Library Association, Boston, MA. Retrieved November 7, 2006, from http://www.pewinternet.org/ppt/Teens%20and%20technology.pdf

Ramirez, A. Jr. (2007). The effect of anticipated future interaction and initial impression valence on relational communication in computer-mediated interaction. *Communication Studies, 58*, 53–70. doi:10.1080/10510970601168699

Ramirez, A. Jr, Walther, J. B., Burgoon, J. K., & Sunnafrank, M. (2002). Information seeking strategies, uncertainty, and computer-mediated communication: Towards a conceptual approach. *Human Communication Research, 28*, 213–228.

Ramirez, A. Jr, Zhang, S., McGrew, C., & Lin, S. (2007). Relational communication in computer-mediated interaction revisited: A comparison of participant-observer perspectives. *Communication Monographs, 74*, 492–516. doi:10.1080/03637750701716586

Raymond, E. S. (2000). *The cathedral and the bazaar*. Retrieved November 6, 2011, from http://www.catb.org/~esr/writings/homesteading/cathedral-bazaar/index.html

Read, B. (2006, October 27). Can Wikipedia ever make the grade? *The Chronicle of Higher Education*. Retrieved November 28, 2010, from http://chronicle.com/article/Can-Wikipedia-Ever-Make-the/26960#grading

Reagle, J. (2010/2011). *Good faith collaboration: The culture of Wikipedia*. Cambridge, MA: The MIT Press. Retrieved April 26, 2012, from http://reagle.org/joseph/2010/gfc/

Red Hat. (2011). *Fedora Project homepage*. Retrieved from http://fedoraproject.org/

Reese, S. (2009). Managing the symbolic arena: The media sociology of Herbert Gans. In Becker, L., Holtz-Bacha, C., & Reust, G. (Eds.), *Festschrift for Klaus Schoenbach* (pp. 279–293). Wiesbaden, Germany: VS Verlag für Sozialwissenschaften. doi:10.1007/978-3-531-91756-6_20

Reich, Z. (2011). Source credibility as a journalistic work tool. In Franklin, B., & Carlson, M. (Eds.), *Journalists, sources and credibility: new perspectives* (pp. 19–36). New York, NY: Routledge.

Remley, D., & Erickson, J. (2010). Second Life literacies: Critiquing writing technologies of Second Life. *Computers and Composition Online*. Retrieved May 12, 2012, from http://www.bgsu.edu/cconline/Remley/

Renov, M. (2004). *The subject of documentary*. Minneapolis, MN: University of Minnesota Press.

Renov, M. (2009). First-person films: Some theses of self-inscription. In Austin, T., & de Jong, W. (Eds.), *Rethinking documentary: New perspectives, new practices* (pp. 29–39). Maidenhead, UK: Open University Press.

Reportage bidon sur la Freebox dans E=M6: intervention du CSA. (2011b, October 11). Freenews. Retrieved February 12, 2012, from http://www.freenews.fr/spip.php?article9131

Resig, J. (2011). *Secrets of the JavaScript ninja*. Greenwich, CT: Manning.

Reuters. (2008). *Reuters handbook of journalism*. Retrieved May 27, 2011, from http://handbook.reuters.com/extensions/docs/pdf/handbookofjournalism.pdf

Reuters. (2011). *Privacy policy*. Retrieved from http://www.reuters.com/privacy-policy

Reuters. (2011). *Terms of use*. Retrieved from http://www.reuters.com/terms-of-use

Reynolds, N. (1993). Ethos as location: New sites for understanding discursive authority. *Rhetoric Review, 11*(2), 325–338. doi:10.1080/07350199309389009

Richards, A. R. (2009). Music, transtextuality, and the world wide web. *Technical Communication Quarterly, 18*, 188–209. doi:10.1080/10572250802708337

Richardson, W. (2010). *Blogs, wikis, podcasts, and other powerful web tools for classrooms*. Thousand Oaks, CA: Corwin Press.

Ridolfo, J. (2006). (C)omprehensive (O)nline (D)ocument (E)valuation. *Kairos: A Journal of Rhetoric, Technology, and Pedagogy 10*(2). Retrieved from http://kairos.technorhetoric.net/10.2/binder.html?praxis/ridolfo/index.html

Riegelsberger, J., Sasse, A., & McCarthy, J. (2003). *Shiny happy people building trust? Photos on e-commerce websites and consumer trust*. Paper presented at CHI '03: ACM Conference on Human Factors in Computing, Fort Lauderdale, USA.

Riegelsberger, J., Sasse, M., & McCarthy, J. (2005). The mechanics of trust: A framework for research and design. *International Journal of Human-Computer Studies, 62*(3), 381–422. doi:10.1016/j.ijhcs.2005.01.001

Rieh, S. Y. (2002). Judgment of information quality and cognitive authority in the Web. *Journal of the American Society for Information Science and Technology, 53*(2), 145–161. doi:10.1002/asi.10017

Rieh, S. Y. (2010). Credibility and cognitive authority of information. In Bates, M., & Maack, M. N. (Eds.), *Encyclopedia of library and information sciences* (3rd ed., pp. 1337–1344). New York, NY: Taylor and Francis Group, LLC.

Rieh, S. Y., & Danielson, D. (2007). Credibility: A multidisciplinary framework. In Cronin, B. (Ed.), *Annual review of information science and technology* (*Vol. 41*, pp. 307–364). Medford, NJ: Information Today.

Rieh, S. Y., & Xie, I. (2006). Analysis of multiple query reformulations on the Web: The interactive information retrieval context. *Information Processing & Management, 42*(3), 751–768. doi:10.1016/j.ipm.2005.05.005

Rittberger, M., & Rittberger, W. (1997). Measuring the quality in the production of databases. *Journal of Information Science, 23*(1), 25–37. doi:10.1177/016555159702300103

Robins, D., Holmes, J., & Stansbury, M. (2010). Consumer health information on the web: The relationship of visual design and perceptions of credibility. *Journal of the American Society for Information Science and Technology, 61*(1), 13–29. doi:10.1002/asi.21224

Rockman, I. F. (2003). Integrating information literacy into the learning outcomes of academic disciplines: A critical 21st-century issue. *College & Research Libraries News, 64*(9), 612–615.

Rohrer, C. (n.d.). *Pet industry international*. LinkedIn. Retrieved from http://www.LinkedIn.com/groupsDirectory?itemaction=mclk&anetid=86951&impid=&pgkey=anet_search_results&actpref=anetsrch_name&trk=anetsrch_name&goback=.gdr_1327511692593_1

Rosen, J. (2006, June 30). The people formerly known as the audience. *Huffington Post*. Retrieved from http://www.huffingtonpost.com

Rosenstiel, T. (2010, September 12). A new phase in our digital lives (Commentary). In Pew Research Center (Ed.), *Ideological news sources: Who watches and why. Americans spending more time following the news* (pp. 79-81). Retrieved November 21, 2011, from http://www.people-press.org/files/legacy-pdf/652.pdf

Rosenzweig, R. (2006). Can history be open source? Wikipedia and the future of the past. *The Journal of American History, 93*(1), 117–146. doi:10.2307/4486062

Roszak, T. (1986). *The cult of information: The folklore of computers and the true art of thinking*. New York, NY: Pantheon Books.

Rottenberg, A. T. (2003). *The structure of argument*. New York, NY: Bedford-St Martins.

Rowe, C. J. (1983). The nature of Homeric morality. In Rubino, C., & Shelmerdine, C. (Eds.), *Approaches to Homer* (pp. 248–275). Austin, TX: University of Texas.

Rowlands, I., Nicholas, D., Williams, P., Huntington, P., Fieldhouse, M., & Gunter, B. (2008). The Google generation: The information behaviour of the researcher of the future. *Aslib Proceedings, 60*, 290–310. doi:10.1108/00012530810887953

Rulliat, A. (2010). The wave of suicides among Foxconn workers and the vacuity of Chinese trade unionism. *China Perspectives, 2010*(3), 135-137.

Rusere, P., & Nkomo, N. (2009, March 6). Zimbabwe mourns death of Susan Tsvangirai, PM's wife, in highway crash. *VOA News Studio 7*. Retrieved from http://www.voanews.com

Russo, A., Watkins, J., Kelly, L., & Chan, S. (2008). Participatory communication with social media. *Curator: The Museum Journal, 51*(1), 21–31. doi:10.1111/j.2151-6952.2008.tb00292.x

Sager, R. (2009, August 18). Keep off the astroturf. [Electronic version]. *The New York Times*. Retrieved January 22, 2012, from http://www.nytimes.com/2009/08/19/opinion/19sager.html

Sanchez-Laws, A. L. (2010). Digital storytelling as an emerging documentary form. *Seminar.net, 6*(3). Retrieved from http://seminar.net/index.php/volume-6-issue-3-2010/161-digital-storytelling-as-an-emerging-documentary-form

Saracevic, T. (1975). Relevance: A review of and a framework for the thinking on the notion in information science. *Journal of the American Society for Information Science American Society for Information Science, 26*(6), 321–343. doi:10.1002/asi.4630260604

Saracevic, T. (2006). *Relevance: A review of the literature and a framework for thinking on the notion in information science- Part II. Advances in Librarianship* (*Vol. 30*, pp. 3–71). Elsevier.

Saracevic, T. (2007). Relevance: A review of the literature and a framework for thinking on the notion in information science-Part III: Behavior and effects of relevance. *Journal of the American Society for Information Science and Technology, 58*(13), 2126–2144. doi:10.1002/asi.20681

SBS. (2011). *SBS policies & publications*. Retrieved from http://www.sbs.com.au/aboutus/corporate/view/id/114/h/Terms-and-Conditions

SBS. (2011). *SBS privacy statement*. Retrieved from http://www.sbs.com.au/aboutus/corporate/view/id/113/h/SBS-Privacy-Statement

Scanlon, J. (2007, August 6). Jeffrey Zeldman: King of web standards. *BusinessWeek Online*. Retrieved March 15, 2011, from http://www.businessweek.com/innovate/content/aug2007/id2007086_670396.htm

Schiffrin, A. (2009). Power and pressure; African media and the extractive sector. *Journal of International Affairs, 62*(2), 127–141.

Schlosser, A. E. (2005). Posting versus lurking: Communicating in a multiple audience context. *The Journal of Consumer Research, 32*, 260–265. doi:10.1086/432235

Schneidermann, B. (2000). Designing trust into online experiences. *Communications of the ACM, 43*(12), 57–59. doi:10.1145/355112.355124

Schön, D. (1983). *The reflective practitioner*. New York, NY: Basic Books.

Schumpeter, J. (2010, December 9). The "Internet of things": The Internet of hype. *The Economist*. Retrieved from http://www.economist.com/blogs/schumpeter/2010/12/internet_things

Schweik, C. (2006). Free/open-source software as a framework for establishing commons in science. In Hess, C., & Ostrom, E. (Eds.), *Understanding knowledge as a commons from theory to practice* (pp. 277–309). Cambridge, MA: MIT Press.

Scott, S. G., & Bruce, R. A. (1995). Decision-making style: The development and assessment of a new measure. *Educational and Psychological Measurement, 55*, 818–831. doi:10.1177/0013164495055005017

Segal, J., & Richardson, A. (2003). Introduction. Scientific ethos: Authority, authorship, and trust in the sciences. *Configurations, 11*(2), 137–144. doi:10.1353/con.2004.0023

Selber, S. (2004). *Mulitiliteracies for a digital age*. Carbondale, IL: Southern Illinois University Press.

Selber, S. (2010). Introduction. In Selber, S. (Ed.), *Rhetorics and technologies: New directions in writing and communication* (pp. 1–14). Columbia, SC: University of South Carolina Press.

Selfe, C. (Ed.). (2007). *Multimodal composition: Resources for teachers*. Cresskill, NJ: Hampton Press.

Selfe, C. L. (2004). Toward new media texts: Taking up the challenges of visual literacy. In Wysocki, A., Johnson-Eilola, J., Selfe, C., & Sirc, G. (Eds.), *Writing new media: Theory and applications for expanding the teaching of composition* (pp. 67–110). Logan, UT: Utah State University Press.

Selfe, C. L., & Hawisher, G. (2007). *Gaming lives in the twenty-first century: Literate connections*. New York, NY: Palgrave Macmillan. doi:10.1057/9780230601765

Selfe, C., & Hawisher, G. (2004). *Literate lives in the information age: Narratives of literacy from the United States*. New York, NY: Routledge.

Selfe, C., & Selfe, R. (1994). The politics of the interface: Power and its exercise in electronic contact zones. *College Composition and Communication, 45*(4), 480–504. doi:10.2307/358761

Sha, M. (2011, July 23). *The sweeties my brother brought back*. Retrieved July 16, 2012, from http://blog.sina.com.cn/s/blog_4c497d3a0102dqtj.html

Sha: The Zimbabwe Social Network. (2012). Retrieved from http://www.sha.co.zw

Shamu threatens media. (2012, May 3). *Zimbabwe Independent*. Retrieved from http://www.theindependent.co.zw/political-zimbabwe-stories/2012/05/03/shamu-threatens-media/

Shankar, V., Urban, G., & Sultan, F. (2002). Online trust: a stakeholder perspective, concepts, implications, and future directions. *The Journal of Strategic Information Systems*, *11*(3), 325–344. doi:10.1016/S0963-8687(02)00022-7

Shaver, M. (2008, March 27). *The missed opportunity of Acid 3*. Retrieved March 4, 2011, from http://shaver.off.net/diary/2008/03/27/the-missed-opportunity-of-acid-3/

Shea, D. (n.d.). *CSS Zen Garden: The beauty in CSS design*. Retrieved from http://www.csszengarden.com/

Shea, D., & Holzschlag, M. E. (2005). *The zen of CSS design: Visual enlightenment for the web*. Peachpit Press.

Shen, F., Wang, N., Guo, G., & Guo, L. (2009). Online network size, efficacy, and opinion expression: Assessing the impacts of Internet use in China. *International Journal of Public Opinion Research*, *21*(4), 451–476. doi:10.1093/ijpor/edp046

Sherekete, R. (2012, May 17). Zimbabwe: Official inflation figures 'not a true reflection'. *Zimbabwe Independent*. Retrieved from http://allafrica.com/stories/201205181095.html

Shirky, C. (2008). *Here comes everybody: The power of organizing without organizations*. New York, NY: Penguin Books.

Short, J. A., Williams, E., & Christie, B. (1976). *The social psychology of telecommunications*. London, UK: Wiley.

Sicart, M. (2008). Defining game mechanics. *Game Studies, 8*(2). Retrieved from http://gamestudies.org/0802/articles/sicart

Sicart, M. (2011). Against procedurality. *Game Studies, 11*(3). Retrieved from http://gamestudies.org/1103/articles/sicart_ap

Sillence, E., Briggs, P., Fishwick, L., & Harris, P. (2004). *Trust and mistrust of online health sites*. Paper presented at the SIGCHI Conference on Human Factors in Computing Systems, New York, NY.

Silverstein, C., Marais, H., Henzinger, M., & Moricz, M. (1999). Analysis of a very large web search engine query log. *ACM SIGIR Forum* (Vol. 33, pp. 6–12).

Sina Blog. (2006, June 2). *Acosta: The first grassroots blogger attracting over 10 million hits*. Retrieved July 16, 2012, from http://blog.sina.com.cn/lm/8/2006/0602/1642.html

Singer, J. B. (2009). Barbarians at the gate or liberators in disguise? Journalists, users and a changing media world. In J. Fidalgo, & S. Marinho (Eds.), *Actas do Seminário "Jornalismo: Mudanças na Profissão, Mudanças na Formação"* (pp. 11-32). Universidade do Minho, Braga: Centro de Estudos de Comunicação e Sociedade.

Sisler, V. (2009). Palestine in pixels: The Holy Land, Arab-Israeli conflict, and reality construction in video games. *Middle East Journal of Culture and Communication, 2*(2), 275–292. doi:10.1163/187398509X12476683126509

Six, F., & Nooteboom, B. (2005). *Trust building actions: A relational signalling approach*. Retrieved March 20, 2012, from http://www.bartnooteboom.nl/FrederiqueOS.PDF

Slaton, A., & Abbate, J. (2001). The hidden lives of standards: Technical prescriptions and the transformation of work in America. In Allen, M. T., & Hecht, G. (Eds.), *Technologies of power: Essays in honor of Thomas Parke Hughes and Agatha Chipley Hughes* (pp. 95–144). Cambridge, MA: MIT Press.

Sloane, T. O. (Ed.). (2001). *Encyclopedia of rhetoric*. New York, NY: Oxford University Press.

Smart, A., Tutton, R., Martin, P., Ellison, G. T. H., & Ashcroft, R. (2008). The standardization of race and ethnicity in biomedical science editorials and UK biobanks. *Social Studies of Science, 38*(3), 407–423. doi:10.1177/0306312707083759

Smith, A. (2008). *New numbers for blogging and blog readership*. Retrieved from http://www.pewinternet.org/Commentary/2008/July/New-numbers-for-blogging-and-blog-readership.aspx

Smith, C., & Kanalley, C. (2011, May 25). Fired over Facebook: 13 posts that got people CANNED. *The Huffington Post*. Retrieved May 10, 2012, from http://www.huffingtonpost.com/2010/07/26/fired-over-facebook-posts_n_659170.html#s115707&title=Swiss_Woman_Caught

Smith, A. E., & Blackton, J. S. (Eds.). (1922). *Two wheels and a crank*. London, UK: Faber.

Society of Professional Journalists (SPJ). (1996). *Code of ethics*. Retrieved December 12, 2010, from http://www.spj.org/pdf/ethicscode.pdf

Sokal, A. (1996, May/June). A physicist experiments with cultural studies. *Lingua Franca*. Retrieved June 27, 2012, from http://www.physics.nyu.edu/faculty/sokal/lingua_franca_v4/lingua_franca_v4.html

Solomon, G., & Schrum, L. (2010). *Web 2.0: How-to for educators*. Eugene, OR: International Society for Technology in Education.

Solomon, P. (1993). Children's information retrieval behavior: A case analysis of an OPAC. *Journal of the American Society for Information Science and Technology*, *44*, 245–264. doi:10.1002/(SICI)1097-4571(199306)44:5<245::AID-ASI1>3.0.CO;2-#

Spinuzzi, C. (2003). *Tracing genres through organizations: A sociocultural approach to information design*. Cambridge, MA: MIT Press.

Spinuzzi, C., & Zachry, M. (2000). Genre ecologies: an open system approach to understanding and constructing documentation. *Journal of Computer Documentation*, *24*(3), 169–181. doi:10.1145/344599.344646

Squire, K. (2006). From content to context: Videogames as designed experiences. *Educational Researcher*, *35*(8), 19–29. doi:10.3102/0013189X035008019

St. Amant, K. (2004). International digital studies: A research approach for examining international online interactions. In Buchanan, E. (Ed.), *Readings in virtual research ethics* (pp. 317–337). Hershey, PA: Information Science Publishing. doi:10.4018/978-1-59140-152-0.ch017

Star, S. L., & Ruhleder, K. (1996). Steps toward an ecology of infrastructure: Design and access for large information spaces. *Information Systems Research*, *7*(1), 111–134. doi:10.1287/isre.7.1.111

Stiff, J. B., & Mongeau, P. (2003). *Persuasive communication* (2nd ed.). New York, NY: The Guilford Press.

Stimmerman, B. (2007). *JavaScript ninja : Crisis averted!* Retrieved from http://web.archive.org/web/20081225224424/http://socket7.net/article/javascript-ninja

Stolley, K. (2011) *About me. Dr. Karl Stolley*. Retrieved May 11, 2012 from http://karlstolley.com/about/#more

Stolley, K. (2011). *How to design and write web pages today*. Santa Barbra, CA: Greenwood Press.

Stravinsky, I. (1993). *Poetics of music in the form of six lessons*. Cambridge, MA: Harvard University Press.

Stumpf, S. (1975). *Socrates to Sartre: A history of philosophy*. New York, NY: McGraw-Hill.

Sunnafrank, M., & Ramirez, A. Jr. (2004). At first sight: Persistent relational effects of get-acquainted conversations. *Journal of Social and Personal Relationships*, *21*, 361–379. doi:10.1177/0265407504042837

Sunstein, C. (2005). *Why societies need dissent*. Cambridge, MA: Harvard University Press.

Supreme Court to decide on Zimdollar labour awards. (May 17, 2012). *Bulawayo24 News*. Retrieved from http://www.bulawayo24.com

Svoen, B. (2007). Consumers, participants, and creators: Young people's diverse use of television and new media. *Computers in Entertainment*, *5*(2). doi:10.1145/1279540.1279545

Swaak, M., de Jong, M., & de Vries, P. (2009). *Effects of information usefulness, visual attractiveness, and usability on web visitors" trust and behavioral intentions*, Paper presented at the IEEE International Professional Communication Conference, Waikiki, Hawaii.

Swayne, M. (2012, April 20). Internet hype may blur fiction-fact line. *Futurity*. Retrieved from http://www.futurity.org/society-culture/internet-hype-may-blur-fiction-fact-line/

Swift, M. (2012). *Facebook sees its growth slowing down.* The Tennessean.

SxSW. (2011). *South by Southwest interactive.* Retrieved March 27, 2011, from http://sxsw.com/interactive

Syverson, M. (1999). *The wealth of reality: An ecology of composition.* Carbondale, IL: Southern Illinois University Press.

Tai, Z. (2006). *The Internet in China: Cyberspace and civil society.* New York, NY: Routeldge.

Tai, Z. (2009). The rise of the Chinese blogosphere. In Dumova, T., & Fiordo, R. (Eds.), *Handbook of research on social interaction technologies and collaboration software: Concepts and trends* (pp. 67–79). Hershey, PA: IGI Global. doi:10.4018/978-1-60566-368-5.ch007

Tai, Z. (2010). Casting the ubiquitous net of information control: Internet surveillance in China from Golden Shield to Green Dam. *International Journal of Advanced Pervasive and Ubiquitous Computing, 2*(1), 53–70. doi:10.4018/japuc.2010010104

Tai, Z. (2012). Fame, fantasy, fanfare and fun: The blossoming of the Chinese culture of blogmongering. In Dumova, T., & Fiordo, R. (Eds.), *Blogging in the global society: Cultural, Political and geographical aspects* (pp. 37–54).

Tapscott, D. (1997). *Growing up digital: The rise of the Net generation.* New York, NY: McGraw-Hill.

Teevan, J., Alvarado, C., Ackerman, M. S., & Karger, D. R. (2004). The perfect search engine is not enough: A study of orienteering behavior in directed search. *Proceedings of the SIGCHI Conference on Human Factors in Computing Systems* (pp. 415–422). Vienna, Austria: ACM. DOI:10.1145/985692.985745

Terdiman, D. (2004). Onion taken seriously, film at 11. *Wired.* Retrieved from http://www.wired.com/culture/lifestyle/news/2004/04/63048

Texas A&M University Department of English. (2012). *Department of English tenure and promotion guidelines.* Retrieved from http://dof.tamu.edu/sites/defaults/files/tenure_promotion/Tenure_and_Promotions_Guidlines_English.pdf

The Acid3 Test. (n.d.). Retrieved April 20, 2008, from http://acid3.acidtests.org/

The Guardian. (2009, May 7). *Community standards and participation guidelines.* The Guardian. Retrieved from http://www.guardian.co.uk/community-standards

The Guardian. (2009, May 7). *Frequently asked questions about community on the Guardian website: Everything you've ever wanted to know about community on the Guardian website.* The Guardian. Retrieved from http://www.guardian.co.uk/community-faqs

The Guardian. (2010, March 19). *Terms of service.* The Guardian, Retrieved from http://www.guardian.co.uk/help/terms-of-service

The Guardian. (2011, July 23). *Privacy policy.* The Guardian, Retrieved from http://www.guardian.co.uk/help/privacy-policy

Thorhauge, A. M. (2012). (in press). The rules of the game—The rules of the player? *Games and Culture.*

Tillman, H. (1995/2003). Evaluating quality on the net. *HopeTillman.com.* Retrieved July 5, 2012, from http://www.hopetillman.com/findqual.html

Tillman, H. (2003). *Evaluating quality on the Net.* Retrieved from http://www.hopetillman.com/findqual.html

Timmerman, C. E., & Kruepke, K. A. (2006). Computer-assisted instruction, media richness, and college student performance. *Communication Education, 55,* 73–104.

Timmermans, S., & Berg, M. (2003). *The gold standard: The challenge of evidence-based medicine* (1st ed.). Philadelphia, PA: Temple University Press.

Tong, S. T., Van Der Heide, B., & Langwell, L. (2008). Too much of a good thing? The relationship between number of friends and interpersonal impressions on Facebook. *Journal of Computer-Mediated Communication, 13,* 531–549. doi:10.1111/j.1083-6101.2008.00409.x

Toulmin, S. (1958). *The uses of argument.* Cambridge, UK: Cambridge UP.

Treat, S. R. (2004). *The myth of charismatic leadership and fantasy rhetoric of crypto-charismatic memberships.* Louisiana State University.

Tseng, S., & Fogg, B. J. (1999). Credibility and computing technology. *Communications of the ACM, 42*(5), 39–44. doi:10.1145/301353.301402

Tuchman, G. (1972). Objectivity as strategic ritual: an examination of newsmen's notion of objectivity. *American Journal of Sociology, 77*(4), 660–679. doi:10.1086/225193

Tulley, C., & Blair, K. (2009). Remediating the book review: Toward collaboration and multimodality across the English Curriculum. *Pedagogy, 9*(3), 441–469.

Tummolini, L., & Castelfranchi, C. (2007). Trace signals: The meanings of stigmergy. In D. Weyns, H. Parunak, & F. Michel (Eds.), *E4MAS'06 Proceedings of the 3rd International Conference on Environments for Multi-Agent Systems* (pp. 141-156). Berlin, Germany: Springer.

Turkle, S. (1996). Therapy and identity construction in a postmodern world. In Grodin, D., & Lindlof, T. (Eds.), *Constructing the self in a mediated world* (pp. 141–153). Thousand Oaks, CA: Sage.

Universal McCann. (2009). *Power to the people: Social media tracker – Wave 4*. Retrieved June 4, 2011, from http://universalmccann.bitecp.com/wave4/Wave4.pdf

Universal McCann. (2010). *The business of social: Social media tracker – Wave 6*. Retrieved March 5, 2012, from http://www.umww.com/global/knowledge/view?Id=226

Universal McCann. (2010). *The socialisation of brands: Social media tracker – Wave 5*. Retrieved January 5, 2012, from http://www.umww.com/global/knowledge/view?id=128

Utz, S. (2010). Show me your friends and I will tell you what type of person you are: How one's profile, number of friends, and type of friends influence impression formation on social network sites. *Journal of Computer-Mediated Communication, 15*, 314–335. doi:10.1111/j.1083-6101.2010.01522.x

Valkenburg, P. M., Krcmar, M., Peeters, A. L., & Marseille, N. M. (1999). Developing a scale to assess three styles of television mediation: 'instructive mediation,' 'restrictive mediation,' and 'social co-viewing.'. *Journal of Broadcasting & Electronic Media, 43*(1), 52–66. doi:10.1080/08838159909364474

Van den Hoven, M. (1999). Privacy or informational injustice? In Pourciau, L. (Ed.), *Ethics and electronic information in the 21st century* (pp. 140–150). West Lafayette, IN: Purdue University Press.

Van Der Heide, B., D'Angelo, J. D., & Schumaker, E. M. (2012). The effects of verbal versus photographic self-presentation on impression formation in Facebook. *The Journal of Communication, 62*, 98–116. doi:10.1111/j.1460-2466.2011.01617.x

van der Sluis, F., & van Dijk, E. M. A. G. (2010). A closer look at children's information retrieval usage: Towards child-centered relevance. In *Proceedings of the Workshop on Accessible Search Systems, The 33st Annual International Conference on Research and Development in Information Retrieval* (ACM SIGIR 2010), 23 July, 2010, Geneva, Switzerland, (pp. 3-10).

van Dijk, J. (2005). *The deepening divide: Inequality in the information society*. Thousand Oaks, CA: Sage.

van Dijk, J. (2006). Digital divide research, achievements and shortcomings. *Poetics, 34*, 221–235. doi:10.1016/j.poetic.2006.05.004

Veron, E. (1988). *La sémiosis sociale. Fragments d'une theorie de la discursivité*. París, France: Presses Universitaires de Vincennes.

Vook. (2012). *What's Vook?* Retrieved May 15, 2012 from http://vook.com/whats-vook/

Walthen, C. N., & Burkell, J. (2002). Believe it or not: Factors influencing credibility on the Web. *Journal of the American Society for Information Science and Technology, 53*(2), 134–144. doi:10.1002/asi.10016

Walther, J. B. (1992). Interpersonal effects in computer-mediated interaction: A relational perspective. *Communication Research, 19*, 52–89. doi:10.1177/009365092019001003

Walther, J. B. (1994). Anticipated ongoing interaction versus channel effects on relational communication in computer-mediated interaction. *Human Communication Research, 20*, 473–501. doi:10.1111/j.1468-2958.1994.tb00332.x

Walther, J. B. (1995). Relational aspects of computer-mediated communication: Experimental observations over time. *Organization Science, 6*, 186–203. doi:10.1287/orsc.6.2.186

Walther, J. B. (1996). Computer-mediated communication: Impersonal, interpersonal, and hyperpersonal interaction. *Communication Research, 23*, 3–43. doi:10.1177/009365096023001001

Walther, J. B. (2007). Selective self-presentation in computer-mediated communication: Hyperpersonal dimensions of technology, language, and cognition. *Computers in Human Behavior, 23*, 2538–2557. doi:10.1016/j.chb.2006.05.002

Walther, J. B., & Burgoon, J. K. (1992). Relational communication in computer-mediated interaction. *Human Communication Research, 19*, 50–88. doi:10.1111/j.1468-2958.1992.tb00295.x

Walther, J. B., & Parks, M. R. (2002). Cues filtered out, cues filtered in: Computer-mediated communication and relationships. In Knapp, M. L., & Daly, J. A. (Eds.), *Handbook of interpersonal communication* (3rd ed., pp. 529–563). Thousand Oaks, CA: Sage.

Walther, J. B., Slovacek, C. L., & Tidwell, L. C. (2001). Is a picture worth a thousand words? Photographic images in long-term and short-term computer-mediated communication. *Communication Research, 28*, 105–134. doi:10.1177/009365001028001004

Walther, J. B., Van Der Heide, B., Hamel, L., & Shulman, H. (2009). Self-generated versus other-generated statements and impressions in computer-mediated communication: A test of warranting theory using Facebook. *Communication Research, 36*, 229–253. doi:10.1177/0093650208330251

Walther, J. B., Van Der Heide, B., Kim, S., Westerman, D., & Tong, S. T. (2008). The role of friends' appearance and behavior on evaluations of individuals on Facebook: Are we known by the company we keep? *Human Communication Research, 34*, 28–49. doi:10.1111/j.1468-2958.2007.00312.x

Wang, H., Xie, M., & Goh, T. N. (1999). Service quality of internet search engines. *Journal of Information Science, 25*(6), 499–507. doi:10.1177/016555159902500606

Wang, S. S., Moon, S. I., Kwon, K. H., Evans, C. A., & Stefanone, M. A. (2010). Face off: Implications of visual cues on initiating friendship in Facebook. *Computers in Human Behavior, 26*, 226–234. doi:10.1016/j.chb.2009.10.001

Wardle, C., & Williams, D. (2008). *Understanding its impact upon contributors, non-contributors and BBC News.* Research Report of Cardiff School of Journalism, Media and Cultural Studies. Retrieved from http://www.cardiff.ac.uk/jomec/research/researchgroups/journalismstudies/fundedprojects/usergeneratedcontent.html

Ward, P. (2005). *Documentary: The margins of reality.* London, UK: Wallflower.

Ward, P. (2009). Drama-documentary: The "Flight 93" films. In Austin, T., & de Jong, W. (Eds.), *Rethinking documentary: New perspectives, new practices* (pp. 191–203). Maidenhead, UK: Open University Press.

Warnick, B. (1998). Rhetorical criticism of public discourse on the Internet: Theoretical implications. *Rhetoric Society Quarterly, 28*, 73–84. doi:10.1080/02773949809391131

Warnick, B. (2004). Online ethos: Source credibility in an "authorless" environment. *The American Behavioral Scientist, 48*(2), 256–265. doi:10.1177/0002764204267273

Warnick, B. (2007). *Rhetoric online: Persuasion and politics on the World Wide Web.* New York, NY: Peter Lang.

Waters, L. (2001, April 20). Rescue tenure from the tyranny of the monograph. *The Chronicle of Higher Education*, B7–B9.

Wathen, C. N., & Burkell, J. (2002). Believe it or not: Factors influencing credibility on the web. *Journal of the American Society for Information Science and Technology, 53*(2), 134–144. doi:10.1002/asi.10016

Watson, R. (2009). Constitutive practices and Garfinkel's notion of trust: Revisited. *Journal of Classical Sociology, 9*(4), 475–499. doi:10.1177/1468795X09344453

Weaver, N. (December 18, 2006). What we should learn from Sony's fake blog fiasco: A debate over false marketing practices. *AdAge Agency News,* Retrieved from http://adage.com/article/small-agency-diary/learn-sony-s-fake-blog-fiasco/113945/

Web Standards Project. (2008, March 3). *Acid3: Putting browser makers on notice, again.* Retrieved March 28, 2011, from http://www.webstandards.org/press/releases/2008-03-03/

Web Standards Project. (n.d.). *About - The Web standards project*. Retrieved December 6, 2007, from http://www.webstandards.org/about/

Weinstein, N. D. (1980). Unrealistic optimism about future life events. *Journal of Personality and Social Psychology*, *39*, 806–820. doi:10.1037/0022-3514.39.5.806

Weinstein, N. D. (1982). Unrealistic optimism about susceptibility to health problems. *Journal of Behavioral Medicine*, *5*, 441–460. doi:10.1007/BF00845372

Weinstein, N. D. (1987). Unrealistic optimism about susceptibility to health problems: Conclusions from a community-wide sample. *Journal of Behavioral Medicine*, *10*(5), 481–500. doi:10.1007/BF00846146

Wells, H. G. (1938). *World brain*. London, UK: Ayer.

West, J. (2003, September 29). Meet the new site, same as the old site. *Librarian.net*. Retrieved from http://www.librarian.net/stax/19/meet-the-new-site-same-as-the-old-site/

West, J. (2007, January 1). About. *Librarian.net*. Retrieved from http://www.librarian.net/about/

West, J. (2011a, February 4). More about power than gender. *New York Times*. Retrieved from http://www.nytimes.com/roomfordebate/2011/02/02/where-are-the-women-in-wikipedia/more-about-power-than-gender

West, J. (2011b, November 15). The Kindle lending experience from the patron's perspective "a wolf in book's clothing." *Librarian.net*. Retrieved from http://www.librarian.net/stax/3725/the-kindle-lending-experience-from-a-patrons-perspective-a-wolf-in-books-clothing/

Wheelan, S. A. (2004). *Faculty groups: From frustration to collaboration*. Thousand Oaks, CA: Corwin Press.

Wiggins, A., Howison, J., & Crowston, K. (2008). Social dynamics of FLOSS team communication across channels. *Proceedings of the IFIP 2.13 Working Conference on Open Source Software (OSS)*, Milan, Italy, (pp. 131-142). Retrieved November 6, 2011, from http://floss.syr.edu

Wikipedia. org – Traffic Details from Alexa. (2012, June 29). *Alexa Internet, Inc.* Retrieved June 29, 2012, from http://www.alexa.com/siteinfo/wikipedia.org

Wikipedia. Wikiality and other tripling elephants: Revision history. (2012, April 13). *Wikipedia*. Retrieved June 29, 2012, from http://en.wikipedia.org/w/index.php?title=Wikipedia:Wikiality_and_Other_Tripling_Elephants&action=history WikiProject Logic/Standards for notation. (2009, December 15). *Wikipedia*. Retrieved 10 December, 2010, from http://en.wikipedia.org/wiki/Wikipedia:WikiProject_Logic/Standards_for_notation

Williams, A., Calow, D., & Lee, A. (2011). *Digital media contracts*. New York, NY: Oxford University Press.

Wolf, M. (1985). *Teorie delle comunicazioni di massa*. Milano, Italy: Bompiani.

Woodmansee, M. (1994). On the author effect: Recovering collectivity. In Woodmansee, M., & Jaszi, P. (Eds.), *The construction of authorship: Textual appropriation in law and literature* (pp. 15–28). Durham, NC: Duke University Press.

World Wide Web Consortium. (2007, June 19). *About W3C*. Retrieved December 6, 2007, from http://www.w3.org/Consortium/

WoW Wiki. (2012). *Death knight PvE guide*. Retrieved from http://www.wowwiki.com/Death_knight_PvE_guide

Wright, D. (2010, 27 September). Toward a more thoughtful conversation on stories. *Reuters*. Retrieved from http://blogs.reuters.com/fulldisclosure/2010/09/27/toward-a-more-thoughtful-conversation-on-stories/?cp=1#comments

Wright, A. (2010). A new literacy for the digital age. *Journal of Special Education Technology*, *25*, 62–68.

Wright, B. (2007). *The reality of the virtual (Film lecture by Slavoj Žižek)*. Saint Charles, IL: Olive Films.

Wright, C. R. (1986). *Mass communication: A sociological perspective*. New York, NY: Random House.

Wronko, G. (2008). *Labrador retriever owners*. LinkedIn. Retrieved May 5, 2012 from http://www.LinkedIn.com/groupsDirectory?itemaction=mclk&anetid=1087707&impid=&pgkey=anet_search_results&actpref=anetsrch_name&trk=anetsrch_name&goback=.gdr_1327510683786_1

Wu, S., Hofman, J. M., Mason, M. A., & Watts, D. J. (2011). Who says what to whom on Twitter. *Proceedings of the 20ᵗʰ International Conference on World Wide Web*, New York, USA.

Wu, G., Hu, X., & Wu, Y. (2010, October 29). Effects of perceived interactivity, perceived web assurance and disposition to trust on initial online trust. *Journal of Computer-Mediated Communication, 16*, 1–26. doi:10.1111/j.1083-6101.2010.01528.x

Wysocki, A. (2003). Seriously visible. In Hocks, M. E., & Kendrick, M. (Eds.), *Eloquent images: Word and image in the age of new media* (pp. 37–59). Cambridge, MA: The MIT Press.

Wysocki, A. (2004). Opening new media to writing: Openings and justifications. In Wysocki, A., Johnson-Eilola, J., Selfe, C., & Sirc, G. (Eds.), *Writing new media: Theory and applications for expanding the teaching of composition* (pp. 1–42). Logan, UT: Utah State University Press.

Xie, M., Wang, H., & Goh, T. N. (1998). Quality dimensions of Internet search engines. *Journal of Information Science, 24*(5), 365–372. doi:10.1177/016555159802400509

Xu, J. (2010a, November 20). *State of hibernation during the holidays: More about Niu Niu*. Retrieved July 16, 2012, from http://blog.sina.com.cn/s/blog_46f37fb50100mv5i.html

Xu, J. (2010b, October 29). *New secret to losing weight during the holidays*. Retrieved July 16, 2012, from: http://blog.sina.com.cn/s/blog_46f37fb50100mekc.html

Yamagata, N. (1994). *Homeric morality*. New York, NY: E.J. Brill.

Yang, G. (2009). *The power of Internet in China: Citizen activism online*. New York, NY: Columbia University Press.

Yates, F. (1966). *The art of memory*. London, UK: Pimlico.

Yee, N. (2005). A model of player motivations. *The Daedalus Project*. Retrieved from http://www.nickyee.com/daedalus/archives/001298.php

Yikoniko, S. (2012, May 18). Evicted Teclar shies away from media. *The Sunday Mail*. Retrieved from http://www.sundaymail.co.zw

Young, I. M. (1990). The ideal community and the politics of difference. In Nicholson, L. (Ed.), *Feminism/postmodernism* (pp. 300–323). New York, NY: Routledge.

Young, S., Kelsey, D., & Lancaster, A. (2011). Predicted outcome value of e-mail communication: Factors that foster professional relational development between students and teachers. *Communication Education, 60*(4), 371–388. doi:10.1080/03634523.2011.563388

Yu, F. Y. (2012). Any effects of different levels of online user identity revelation? *Journal of Educational Technology & Society, 15*(1), 64–77.

Zhang, L. F. (2003). Contributions of thinking styles to critical thinking dispositions. *Journal of Psychology (Savannah, Ga.), 137*, 517–544.

Zhang, X., & Chignell, M. (2001). Assessment of the effects of user characteristics on mental models of information retrieval systems. *Journal of the American Society for Information Science and Technology, 52*(6), 445–459. doi:10.1002/1532-2890(2001)9999:9999<::AID-ASI1092>3.0.CO;2-3

Zhang, Y. (2008). Impact of intentional social actions on applications of the Internet technology. [Zhongguo Wangluo Chuanbo Yanjiu]. *China Internet Communication Research, 2*(1), 98–105.

Ziegelmueller, G., Jay, J., & Dause, C. (1990). *Argumentation inquiry and advocacy*. Englewood, NJ: Prentice Hall.

Zimbabwe News Online. (2012). Retrieved from http://www.zimbabwenewsonline.com

Zimbabwe. (2012). *Internet world statistics*. Retrieved from http://www.internetworldstats.com/af/zw.htm

Zittrain, J. (2004). Normative principles for evaluating free and proprietary software. *The University of Chicago Law Review. University of Chicago. Law School, 71*(1), 265–287.

About the Contributors

Moe Folk is an Assistant Professor of Multimodal Composition and Digital Rhetoric at Kutztown University of Pennsylvania, where he teaches a variety of undergraduate and graduate courses in the English Department. His research centers on issues of ethos in digital realms, particularly concerning the relationship that multimodal style has with defining contemporary notions of ethos, and teaching writing with new composing technologies and multiple modes. His previous publications include an article about visual representation in Mediascape and co-authored pieces with Shawn Apostel that appeared in *Computers and Composition Online* and the *Handbook of Research on Computer-Mediated Composition* by IGI Global. His recent work has appeared in *Kairos PraxisWiki* and the Sweetland Digital Rhetoric Collaborative; book chapters on using Google Maps to teach writing, assessing the affordances of multimodal style, and using visual rhetoric to produce and teach graphic novels are forthcoming. His creative work has appeared in *Pank* and *New Letters*.

Shawn Apostel is the Communication Coordinator for the Noel Studio for Academic Creativity at Eastern Kentucky University where he collaborates with communication faculty, Noel Studio research and writing coordinators, technology associates, and student consultants to identify needs and develop instructional/information seminars focusing on visual, oral, and digital communication. A graphic designer by trade, Dr. Apostel has a MA in Professional Communication from Clemson University and a Ph.D. in Rhetoric and Technical Communication from Michigan Technological University. His research interests include visual communication, creativity, digital ethos, e-waste reduction, and instructional use of cloud-based composition programs. His work is published by IGI Global, CCDigital Press, Lexington Books, New Forums Press, and Computers and Composition Online. In Spring of 2013 his co-authored book *Teaching Creative Thinking: Pedagogical Approaches* will be published by New Forums Press.

* * *

François Allard-Huver is a Ph.D. candidate at the CELSA – Graduate School of Communication of the Sorbonne University in Paris and a former visiting fellow of the Annenberg School for Communication, University of Pennsylvania. He is also involved in the RISK project of the ISCC, the French Institute of Communication Sciences from the CNRS. Graduated with a Master in Communication Sciences from the Sorbonne University, he wrote his master's thesis on the question of "Transformation and circulation of the risk notion within the European Commission." He is currently writing his Ph.D. thesis on the question of transparency for risk assessment and risk communication practices in both European and North American scientific institutions.

Kristine L. Blair is Professor and Chair of the English Department at Bowling Green State University, Ohio, where she teaches courses in digital rhetoric and scholarly publication in the Rhetoric and Writing doctoral program. The author of numerous publications on gender and technology, online learning, electronic portfolios, and the politics of technological literacy acquisition, Blair has served as the editor of the journal *Computers and Composition Online* since 2003 and assumed editorship of the print version of *Computers and Composition* in 2011. In 2004 and 2009, she was named the Outstanding Contributor to Graduate Education by the BGSU Graduate Student Senate, in 2007 she received the national Technology Innovator Award from the Conference on College Composition and Communication, and in 2010 she received the national Charles Moran Award for Distinguished Contributions to the Field of Computers and Composition.

Kevin Brock is a Ph.D. candidate in Communication, Rhetoric, and Digital Media at North Carolina State University. His work primarily focuses on the space shared by studies of rhetoric, technical communication, and software development with a special interest in how code (as both practice and text) functions as a rhetorically powerful and significant form of contemporary communication. His dissertation in progress is titled "Engaging the Action-oriented Nature of Computation: Towards a Rhetorical Code Studies."

João Canavilhas is Professor of Journalism in the Universidade da Beira Interior (Covilhã – Portugal) since 2000, where he is also Editor of URBI, the first online university newspaper in Portugal, and deputy director at Labcom – Laboratory of Online Communication. His research work focuses on various aspects of communication and new technologies, particularly in the fields of online journalism, e-politics, social media and journalism for portable devices. He has published widely in national and international scientific journals and is the author of the Tumbled Pyramid model for online journalism.

Aaron W. Dobbs is an Assistant Professor and the Systems and Electronic Resources Librarian at Shippensburg University of Pennsylvania. He earned his M.S. in management from Austin Peay State University and his M.S. in Library Science from the University of Tennessee. His research interests include the intersection of libraries and public policy and practical applications of software to create library services of the future. Aaron recently co-edited *LibGuides: Making Dynamic Web Design and Management Simple for Non Web-designers – A LITA Guide* and he has authored or co-authored publications focusing on the library side of topics including: leadership, *Technologically Indispensable: Leading when you're seen as the tech person*; assessment, *Using Rubrics to Qualitatively Assess Library Services;* and library instruction, *Information Literacy in the First Year: Collaborating, Planning and Assessing at Austin Peay*.

Natasha Dwyer is a Lecturer in the School of Communication and the Arts at Victoria University, Australia, where she coordinates the Creative Industries degree. Her PhD research explored trust in digital environments, for which she was awarded two British Telecom short-term fellowships. Previously she worked as ŒInteractiveDesigner at the Australian Centre for the Moving Image where she worked on video-on-demand systems enabling the public to access ACMI's collection. She is currently working on a project exploring how members of the Samoan community in Victoria perceive trust in the context of health messages. The result will be a series of short-films.

Dmitry Epstein is a post-doctoral fellow with Cornell eRulemaking Initiative at Cornell Law School. His research focuses on online civic engagement, politics of communication platforms, and Internet governance. He received his PhD from Cornell University.

Joe Erickson is an Assistant Professor of English at Angelo State University in San Angelo, Texas, where he teaches courses in web publishing, professional editing, and business communications. He has served as the senior design editor for *Computers and Composition Online* since 2009, and he assumed the book review editor position for the print version of *Computers and Composition* in late 2011. Erickson's scholarship addresses the intersection of literacy studies, disciplinary identity formation, and digital rhetoric, a focus that has led to his current work on the changing landscape of academic publishing as a result of digital media technologies.

Douglas Eyman teaches courses in digital rhetoric, technical and scientific communication, and professional writing at George Mason University. Eyman is the senior editor of *Kairos: A Journal of Rhetoric, Technology, and Pedagogy*, an online journal that has been publishing peer-reviewed scholarship on computers and writing since 1996. His scholarly work has appeared in *Pedagogy, Technical Communication, Computers and Composition*, and the edited collections *Cultural Practices of Literacy, Digital Writing Research, The Handbook of Research on Virtual Workplaces and New Business Practices*, and *Rhetorically Rethinking Usability* (Hampton Press, 2008). His current research interests include investigations of digital literacy acquisition and development, new media scholarship, electronic publication, information design/information architecture, teaching in digital environments, and massive multiplayer online role playing games as sites for digital rhetoric research.

Andrew J. Flanagin (Ph.D., Annenberg School for Communication, University of Southern California) is a Professor in the Department of Communication at the University of California at Santa Barbara, where he also serves as the Director of the Center for Information Technology and Society. His research focuses on the ways in which information and communication technologies structure and extend human interaction, with particular emphasis on the processes of organizing and information evaluation and sharing.

Todd S. Frobish is currently Interim Chair of the Department of Communication at Fayetteville State University in Fayetteville, North Carolina. He received his PhD in Rhetorical Studies from The Pennsylvania State University in 2002, where he completed a dissertation examining ethos and identity issues within computer-mediated communication. He has published widely on the issue of ethos and identity, public address, and scientific and technological discourse. In 2007, he published an edited anthology of the most controversial speeches in American oratory, *Crises in American Oratory: A History of Rhetorical Inadequacy*. He is the 2011-12 University of North Carolina Board of Governors' Excellence in Teaching Award Winner, and the 2007-08 FSU Teacher of the Year. He is a Fulbright Fellow, having participated in the 2006 Summer Seminars Abroad program in Budapest, Hungary. Among other appointments, he has served as the President of the American Communication Association.

Samaa Gamie received her BA in English Language and Literature from the University of Alexandria, Egypt in 1995. She received her MA in Professional Writing from the University of Massachusetts at Dartmouth in 2003 and her Ph.D. in English with concentration in Rhetoric and Composition from the

University of Rhode Island. She is currently entering her third year as an Assistant Professor of English at Lincoln University, PA. She has multiple forthcoming scholarly publications, published poems, and essays.

Nicholas Gilewicz is a Doctoral student at the Annenberg School for Communication at the University of Pennsylvania. He holds a Master's degree in Journalism from Temple University, and a Bachelor's degree in Humanities from the University of Chicago. His research interests include the history of journalism, how journalists construct the social meaning of their work through journalism products, and theorizing new frameworks with which to analyze texts and communities produced by digital journalism.

Abigail Goben is an Assistant Information Services Librarian and Assistant Professor at the University of Illinois-Chicago Health Sciences Library, where she is embedded in the College of Dentistry curriculum. She holds an MLS from St. John's University. Her research focuses on the use of early twenty-first century technologies by librarians in their professional communication and development with special interest in open-access adoption. Ms. Goben is actively involved with the Library Information Technology Association, chairing the Annual Program Planning Committee and participating on the Education Committee, as well as the Library Society of the World. She blogs at *Hedgehog Librarian*.

Ethan Hartsell received his Master's Degree in Communication from University of California, Santa Barbara in 2011. He studies media effects, social media, and credibility, and has published work on children and Internet use, news media bias, and selective exposure.

Nathan R. Johnson is an infrastructural theorist, designer, and builder. His work analyzes the ways that technical systems set the tempo, space, and play of human action. He is currently building an open source digital laboratory to model how scholarly information infrastructure orchestrates research in the discipline of rhetorical studies. This research and design work is rooted in an interest in changing scholarly communication systems to support the needs of modern research practices. He is an advocate of publishing models that support a wider range of media as research, publish then peer review models, and context-specific science. His writing has been published in Archival Science and Poroi. His website is http://nathanjohnson.us. In addition, Johnson is a faculty member in the Rhetoric and Composition program at Purdue University where he teaches courses on research, multimedia writing, and infrastructure.

Misty L Knight is an Assistant Professor of Human Communication Studies at Shippensburg University. She earned her Ph.D. in Speech Communication from The University of Southern Mississippi after defending her dissertation, *Private to Public Life: Rhetoric of First Ladies in the First Year*. Her primary research interests involve political rhetoric, self-defense rhetoric, and humor in communication and her co-authored publications include: *Telic state teaching: Understanding the relationships among classroom conflict styles, humor, and teacher burnout of university faculty; Finger-pointing and federal agencies: Examining the shift in rhetorical riposte between Hurricane Katrina and Hurricane Ike*; and *Humor, organizational identification, and worker trust: An independent groups analysis of humor's identification and differentiation functions*. She teaches courses such as Introduction to Human Communication, Human Communication Theory, Interpersonal Communication, Public Speaking, and Political Rhetoric.

Richard A. Knight began his career in higher education by serving as Director of Forensics and Instructor of Communication at West Texas A&M University. He then earned his Doctorate in Speech Communication with an emphasis in Political Rhetoric at the University of Southern Mississippi, and turned to a full-time career in the classroom. He is currently an Associate Professor teaching in the Department of Human Communication Studies at Shippensburg University, offering courses in rhetoric, nonverbal communication, argumentation, and persuasion. Dr. Knight's published research includes studies in political apologia, post-presidential rhetoric, communication education, and the role of humor in professional organizations.

Laura Kohl is Head of Reference Services at Douglas & Judith Krupp Library at Bryant University in Smithfield, Rhode Island. She received her undergraduate degree in anthropology from Boston University, and received her Master of Library and Information Science degree from the University of Rhode Island. She finds working with the library user to be the most gratifying part of her career and gains great satisfaction from helping the user find that ever elusive piece of information. Laura also often partners with faculty to incorporate information literacy into the undergraduate classroom in an effort to create information savvy graduates.

Dan W. Lawrence is a "fast-track" M.S./Ph.D. student in the rhetoric and technical communication program at Michigan Technological University where he also teaches composition as a graduate teaching instructor. He seeks to explore the lost connection between rhetoric and music, the coupling of which constitutes an ancient art he believes could have more significance than its relegated status as a mere historical curiosity. His current work attempts to apply a Žižekian analysis of ideology to media studies to expand contemporary notions of self-awareness, identity, and universality. Dan W. Lawrence is also an independent musician, dedicated songwriter, cautious multi-instrumentalist, overconfident singer, home-recording tinkerer, quiet thinker, and a native of the Upper Peninsula of Michigan.

Dirk Lewandowski is a Professor of Information Research and Information Retrieval at the Hamburg University of Applied Sciences, Germany. Prior to that, he worked as an independent consultant and as a part-time Lecturer at the Heinrich-Heine-University Düsseldorf. He is author of more than 50 research papers, one monograph and several edited volumes.

Veronica Maidel is a PhD candidate in Information Science and Technology at Syracuse University. Her research focuses on interactive and user-centered information retrieval. She received her MSc in Information Systems Engineering at Ben-Gurion University of the Negev.

Rick Malleus is a Zimbabwean and currently teaches at Seattle University as an assistant professor in the Communication Department. Both his MA and PhD are from the University of Minnesota in Communication Studies with a focus on Intercultural Communication. Malleus has developed and taught a special topics course in *Africa and Communication*, teaches foundation courses in communication theory and media, a communication research seminar and offers several social interaction classes. His current research interests include intercultural re-entry and cross-cultural comparison in technology/media use. Malleus has written several publications for the Zimbabwe Open University.

Alex Markov is a Doctoral student in the Department of Communication at the University of California, Santa Barbara. Alex's research focuses on how people evaluate the credibility of information online, with an emphasis on social and cognitive processes.

Ryan McGrady is a PhD student in the Communication, Rhetoric, and Digital Media program at North Carolina State University where he is working on a cultural history of encyclopedias from ancient Rome to Wikipedia. Research emerging from this has included collective intelligence and the global brain, philosophy of science, history of the book, cultural and media studies, and media history. Before coming to NC, he lived in Boston where he received his MA in Visual and Media Arts from Emerson College and worked on a number of projects at the Berkman Center for Internet and Society at Harvard University, including the Citizen Media Law Project, Center for Citizen Media, Teacher's Guide to Wikipedia, and a study on denial of service attacks on independent media sites.

Ryan Medders (M.S., San Jose State University) is an Instructor at California Lutheran University and a doctoral candidate in the Department of Communication at the University of California at Santa Barbara. His research focuses on the social and psychological effects of the media, with an emphasis on information technologies, political communication, digital credibility, and social identity.

Miriam J. Metzger (Ph.D., Annenberg School for Communication, University of Southern California) is a Professor in the Department of Communication at the University of California at Santa Barbara. Her interests lie at the intersection of media, information technology, and trust, centering on how information technology alters our understandings of credibility, privacy, and the processes of media effects.

Christy Oslund, PhD in Rhetoric and Technical Communication, from Michigan Technological University, has her MFA from Northern Michigan University and her MA from Michigan State. She taught at each of these institutions including: composition, creative writing, technical writing, and incorporated elements of design in most of these classes. Christy also has a background in graphic design; she has worked as an Assistant Editor for *Passages North* literary magazine; she regularly sits on hiring committees. Currently Christy is the Coordinator of Student Disability Services at Michigan Tech. Passionate about social justice issues, Christy volunteers for Amnesty International and with animal rescues. She currently lives with six four legged beings – their only comment on her writing has been to shred her printed drafts or lay on her computer keyboard.

Rebekah Pure (M.A., University of California at Santa Barbara) is a doctoral Candidate in the Department of Communication at the University of California at Santa Barbara, and has completed an external emphasis in Technology and Society. Rebekah's research focuses on how Internet users manage their uncertainty online regarding information credibility, online aggression, and privacy.

Joaquim Paulo Serra has a Graduation in Philosophy, and a Master's Degree and a PhD in Communication Sciences. He is Professor at the Department of Communication Sciences of University of Beira Interior (Portugal), researcher in the Online Communication Lab (Labcom), and Director of the PhD in Communication Sciences. He authored the books *Information as Utopia* (1998), *Information and Sense* (2003) and *Communication Theory Handbook* (2008), co-authored the book *Information and Persuasion on the Web* (2009) and co-edited the books *Online Journalism* (2003), *Online Lifeworld and*

Citizenship (2003), *From the communication of Faith to the faith in Communication* (2005), *Communication Sciences in Congress at Covilhã* (Proceedings, 2005), *Rhetoric and Mediation: From Writing to the Internet* (2008), *Pragmatics: Advertising and Marketing* (2011), and *Philosophies of Communication* (2011). He has also published several book chapters and articles in journals and collective works.

Wendi Sierra is a Doctoral candidate at North Carolina State University. Her research focuses on game design, and in particular the use of game design to create productive learning environments. To that end she has participated as a co-designer on *C's the Day*, a conference game that encourages networking and professionalization and is currently prototyping *Battle Shirts,* a wearable technology gamification of conference spaces. Her work has been published in *Writing and the Digital Generation, Journal of Gaming and Virtual Worlds*, and *Kairos*, and presented at *Computers and Writing* and the *National Communication Association*. In 2012, she served as the co-chair of the Computers and Writing Conference.

Inge Ejbye Sørensen has currently submitted her Ph.D. dissertation about documentary film in a multiplatform context at the department of Film and Media at Copenhagen University, Denmark. Inge is an editor of Audiovisual Thinking, an online journal of academic videos, and has published articles in Danish and international press and academic peer reviewed journals like *Media, Culture & Society* and *Seminar.net*. Inge is also a media practitioner and prior to entering academia, she worked first as a Researcher and later as an award-winning Producer of documentaries and factual programmes for the TV channels Channel 4, BBC Scotland and STV in the UK. Today Inge produces of fiction and her first feature *Timelock* will be released in 2013. Inge is also an accredited Lead Practitioner in Moving Image Education with Scotland's screen agency Creative Scotland.

Ceren Sözeri is a faculty member at the Communications Department of Galatasaray University. She received her Ph.D. from Marmara University with a thesis on "transnational media mergers in Turkey." Her research areas are media economics, media management, competition, diversity and new media business administration. She co-authored the report titled "The political economy of the media in Turkey: A sectoral analysis report" (Turkish Economic and Social Studies Foundation-TESEV, 2011).

Zixue Tai, (Ph.D., University of Minnesota-Twin Cities), is an Associate Professor in the School of Journalism and Telecommunications at the University of Kentucky, USA. He teaches undergraduate and graduate courses in media effects, global media systems, advanced multimedia, and video game studies. His research interests address a multitude of issues in the new media landscape of China. He is the author of *The Internet in China: Cyberspace and Civil Society* (Routledge, 2006), and his numerous publications can be found in premier journals such as *International Communication Gazette, Journalism & Mass Communication Quarterly, New Media & Society, Journal of Communication, Sociology of Health & Illness,* and *Psychology & Marketing.*

Anne Mette Thorhauge is an Assistant Professor at the Department of Media, Cognition and Communication at the University of Copenhagen. She has a Ph.D in media studies with a special focus on video games and communication. She is a consultant at the Danish Media Council for Children and Young People with a special expertise in video games an editor of MedieKultur, a multidisciplinary research journal in the field of media and communication studies. Furthermore, she has been assessor of game proposals in the New Danish Screen game programme.

Dawn Emsellem Wichowski is a Reference and Instruction Librarian at Salve Regina University in Newport, Rhode Island. She holds an undergraduate degree in Political Science from Barnard College in New York City, and a Master's degree in Library and Information Science from University of Illinois, Urbana-Champaign. Her interests within the field of librarianship include educating people to be skeptical consumers of information through the use of information literacy principles and investigating and advocating the role of libraries in promoting civic community. Before entering into librarianship, she held positions in non-profit advocacy and cultural institutions, a web design company, and for a public interest political consultancy at New York's City Hall.

Yonghua Zhang is Professor at the Department of Journalism and Communication, School of Film and TV Art & Technology, and Director of the Center for International Communication Studies and Vice Director of the Film, TV and Media Research Institute, Shanghai University. She is also Vice President of the Chinese Association of Global Communication (CAGC) affiliated with the Chinese Society of Journalism History. She has a M.A. in English Language and Literature (Shanghai International Studies University, 1983) and a Ph.D. in Communication Studies (Fudan University, 2003), Fulbright senior-level research visiting scholar at the Annenberg School for Communication, University of Southern California (1995), Rockefeller resident scholar for one month (August 1999) at the Bellagio Conference Center in Italy, and short-term research visiting scholar (August to November 2010) at the Center for International Communication Studies, Annenberg School for Communication, University of Pennsylvania. Her research areas include: mass communication theories, international and intercultural communication, and Internet communication.

Index